ISIS CODE

Revelations from Brain Research and System Science on the Search for Human Perfection and Happiness

Ariane Page

iUniverse, Inc.
Bloomington

ISIS CODE
REVELATIONS FROM BRAIN RESEARCH AND SYSTEM SCIENCE
ON THE SEARCH FOR HUMAN PERFECTION AND HAPPINESS

Copyright © 2013 Ariane Page.

All rights reserved. No part of this book may be used or reproduced by any means, graphic, electronic, or mechanical, including photocopying, recording, taping or by any information storage retrieval system without the written permission of the publisher except in the case of brief quotations embodied in critical articles and reviews.

The comments made herein are not meant as a source of medical advice. Those seeking medical advice are advised to consult with their medical doctor.

iUniverse books may be ordered through booksellers or by contacting:

iUniverse
1663 Liberty Drive
Bloomington, IN 47403
www.iuniverse.com
1-800-Authors (1-800-288-4677)

Because of the dynamic nature of the Internet, any web addresses or links contained in this book may have changed since publication and may no longer be valid. The views expressed in this work are solely those of the author and do not necessarily reflect the views of the publisher, and the publisher hereby disclaims any responsibility for them.

Any people depicted in stock imagery provided by Thinkstock are models, and such images are being used for illustrative purposes only.

Certain stock imagery © Thinkstock.
Cover and diagrams by Alexander DeBavelaere

ISBN: 978-1-4759-6746-3 (sc)
ISBN: 978-1-4759-6747-0 (hc)
ISBN: 978-1-4759-6748-7 (e)

Library of Congress Control Number: 2012923561

Printed in the United States of America

iUniverse rev. date: 1/8/2013

Table of Contents

Preface xi
Acknowledgments xiii
Introduction xv

Chapter 1 Spring 1
We from Hell, We from Heaven 2
How It All Began: The Quest for the Elusive New Woman 9
The Implicate Order: A Canvas to Every Living Structure and
    Our Inner Compass 19
The Questions: What- Where, When, and How versus Why-
    Which, and Who 24
A Short Story of Science 27
System Science, the Search for a Model behind Energy 36
Consciousness (Individuality) and Awareness (Personality) 40
Coherence and Oscillations 52
Biopsychosocial / Medical Biocybernetics Example: Depression 67

Chapter 2 Why It All Began: The Need to Understand 73
The Brain 77
The Heart 93
And Consciousness 99
In Isis's Footsteps—Isis and Osiris, the Legend 104
Once upon a Time 124

Chapter 3 The Reptilian Brain, the Physical Self,
    and the Instinctual Human 131
Phase of the What and Where Level 131
Jane and John—Birth to Seven Years Old 138
Different Types of Love 151

The "Masculine" Reptilian Brain and the pH Spectrum 162
    Hypothalamus—Amygdala 162
    Genital Spectrum and Homosexuality....................................... 164
The Five Books of Moses.. 181
Chronic Fatigue Syndrome, Cancer, OCD, and Addictions 191
Sleep, Dreams, and Oscillations 199
Men's Brains Age More Rapidly? 216
Locus Coeruleus 217
Women's Masculinization 218

Chapter 4   The Mammalian Brain, the Emotional Self, and the Emotional Human 225

Phase of the *When* Level 225
Relationships 233
Sympathetic Nervous System 245
The Basal Ganglia 248
The Hippocampus 250
The Anterior Cingulate Cortex 254
Face Recognition 256
ADHD and the Reward Circuit 258
Endocrine System 262
Exodus: The Mammalian Brain 265

Chapter 5   The Human Brain of Information Regulation, the Idealistic Self, and the Archetypal Human .. 281

Phase of the *Why* and *Which* Level Developing into the *I Am*—If
    All Phases Develop Harmoniously 281
The Pineal Gland 298
The Master of the Heart.. 306
The Vagus Nerve 323
Effects of Melatonin, Serotonin 329
Subjective and Objective Beauty 349
Fourteen Years Old 351
Lies, Left Dorsolateral Prefrontal, and Archetypes 354
Schizophrenia and Identity 365
Women's Beauty, from Genital and Subjective to Archetypal
    and Objective 371
Symbols: Subconscious Voices of our Ancestors and
    Superconscious Voices of Our Gods 383
On Love 396

Chapter 6   The Analytical Brain, the Rational self,
             and the Personal Human 401
   Phase of the *How Level* 402
   Denial: the Left Hemisphere 403
   The Frontal Lobes 405
   Seth and the Personal Self 407
   Anorexia and the Middle Prefrontal Cortex 413
   The Virtues of Fasting 418
   Levels of Awareness 421
   Science, Versus Religion 423
   Numbers: the Analytical Brain 435

Chapter 7   The Universal brain, the Social Self,
             and the Social Human 447
   Phase of the *Who Level* 448
   The Song of the Nibelungs 451
   Deuteronomy: The Universal brain 454
   The Legend of the Polar Bear 459
   Breathing Exercises 460
   Choir Singing 464
   Ecopsychology 467
   Brain Elements of the Universal Brain 470
   Placebo Effect and the Feminine Polarity 491
   42 Confessions (Papyrus of Ani) 503

Chapter 8   I, Horus, the Androgynous King 519
   The Synarchic Brain, the Conscious Self, and the Complete
     Human (Perfect) 519
   The Perfect Human: I Am 520
   Leonardo da Vinci and the Vitruvian Man 522
   The Sixth Sense of Synergy: *À Mon Seul Désir*
     —The Lady and the Unicorn 527
   Knowledge and the Feminine Polarity 531

Chapter 9   In Isis's Footsteps: Applications 539
   Environment, Urbanism 542
      Paris and the LIFE Biosystem 553
   Sustainable Government: Elected Synarchy or Biosynarchy,
     the Only Possible True Democracy 555
   Prenatal Awareness 564

Epilogue..567
Appendix..601
Notes, References, and Selected Bibliography.................................. 613
Index...639

List of Diagrams and Illustrations

Diagrams

1. Implicate order and fractal structures and functions
2. Egyptian tradition, Tao, and LIFE biosystem on the Tree of Life (Kabbalah)
3. *People of the Book* and LIFE biosystem
4. Central aspect of symbolic level, the rainbow
5. Phases as per LIFE biosystem: age, phases, brain aspect, feminine and masculine polarities with Medical Biocybernetics: cell structures, organs, inhibition and generation functions
6. Different representations of the same auto-regulated system
7. Feminine and masculine polarities in the LIFE biosystem
8. Biopsychosocial model is found in medical biocybernetics
9. Center of self, center of Self
10. Man and woman, receptivity, expressiveness
11. Maturation and evolution of the human brain
12. Healthy and mature LIFE biosystem
13. Phases and their questions. All the phases expressed in harmony = I am
14. Mirroring of the LIFE biosystem of mother and child
15. The different brain aspects of the LIFE biosystem and their physical structures
16. Genital and general spectrum compared to pH
17. Relations: two masculine polarities and two feminine polarities interacting
18. The city of Paris as per medical biocybernetics
19. Genesis and LIFE biosystem, part 1
19.1 Genesis and LIFE biosystem, part 2
20. Taste, main addictions, and stimulation of the LIFE biosystem
21. Masculinization of women
22. Types of love and love phases
23. Ages and phases, midlife depression of women
24. Solomon's Seal and the pentagram
25. Ultimate conventional medicine

26. Master of the Heart, Akasha
27. Master of the Heart as per medical biocybernetics.
28. The spiral of evolution
29. Phases in society when the feminine polarity is deficient

Illustrations:

1. Harpocrates as the child Horus. Silver statue, Ptolemaic dynasty 350-300 BCE. Foundation Calouste Gulbenkian, Lisbon, Portugal
2. Lajja Gauri, Uttâpanad ("she who crouches with legs spread"). Sandstone sculpture, circa 650 CE. Badami Museum, India
3. Venus of Willendorf from 30,000 - 25,000 BCE
4. Sheela Na Gig (Wikipedia)
5. Bharatanatyan, dancer in a trance (Wikipedia)
6. Moses, Michelangelo, circa 1513
7. Vitruvian Man, Leonardo da Vinci, circa 1492. Gallerie dell'Accademia, Venice
8. Lady and the Unicorn, À Mon Seul Désir. The Tapestry Cycle is the title of a series of six Flemish tapestries depicting the senses. They are estimated to have been woven in the late 15th century in the style of millefleurs. (Wikipedia)
9. The Accolade, Edmund Leighton, 1901, oil on canvas (Wikipedia)

Preface

At a time when our civilization seems devoid of direction and faces a global financial and ethical crisis, humankind urgently needs to reassess its priorities. Health is obviously one of them, since in most countries illnesses generate the highest expenditure. Surrounded by all sorts of comforts, we are plagued with more chronic diseases than ever before. Even our mental health deteriorates. This is nonsensical. Could it be that we overlooked something about human evolution? Could illnesses and environmental catastrophes be the only wake-up call nature can voice? And are we hearing this cry now that, thanks to cancer research, we come to realise that we cannot stay healthy and thrive if our environment is not healthy?

Unsurprisingly, and maybe since the brain is our primary tool of communication with this environment, interest in anything relating to it has recently spiked. Unfortunately, this is often the result of a desire to manipulate an object—in this case, the brain—into acquiring what we lust for. One thing is sure though: the brain is holistic and cannot be isolated from the environment. This unit—the environment and the brain it created—should therefore be studied together and get all our attention.

Searching through the prolific and often conflicting data regarding the brain that inundates us, we need to find cohesion. To this end, system science allows a view overarching both the environment and the brain by offering us a system based on traditional Chinese medicine. This one is reliable as it made its proof in the tangible health domain. Decoding it, we now have new elements to investigate and a new perspective seldom born to a sharp and specialized knowledge. Having worked with it for the last twenty-five years, I applied it to my life. It brought me physical health, mental resilience and the understanding that in the modular world we have created, we are now similar to Osiris, a god of ancient Egypt—that is, cut into pieces that can't communicate. We are not whole; we are not complete and not happy. Only Isis, the part in us that strives for wholeness, can put the pieces together. This is her quest, and ours. In her footsteps let us walk; following her we will find ourselves.

I am not a scientist, a psychologist, a priest, or a doctor. I am an autodidact who has a profound love for beauty, nature, life, love, knowledge, and humanity. I believe, however, that this love is a prerequisite to being a scientist, a psychologist, a priest, or a doctor.

I did not wish to write an exhaustive book. Nevertheless, parts of it might appear as such for some who cannot visualize the brain's anatomy. The setting necessary for my demonstration required these references. Therefore, I invite those readers to simply jump over passages which might appeal more to researchers in the field and serve mainly to confirm the accuracy of the LIFE biosystem.

I did not wish this book, either, to follow a linear thinking. My experience has shown me that this approach does not touch the heart; it only dries it up more. This is not how we can comprehend synergy or find Isis. For this, we need to vibrate in the same way a triangle, when struck, will cause similar triangles to vibrate. I therefore chose to paint with broad strokes, merging the conclusions and facts of brain research to the knowledge brought by a universal and perennial system, in the ambiance of a living and evolving natural world. I have lived consciously through this system for the last twenty years of my life. I tried to prove it wrong, to no avail. I hope that the research, facts, stories, illustrations, discoveries, and pain, as well as joys harvested here and there, will act as revealing mirrors, for there are not eight billion Isis goddesses, but only one of whom we all partake.

Acknowledgments

The publication of this book would have been impossible without the combined efforts of all those who revised, questioned, and supported it. Nor without all those who, in the span of twenty-five years, enthusiastically responded to its subject.

The perseverance of those researchers in the field, who sought a path different from the ones already trod upon, inspired me to keep going. In thanks, I dedicate this book to the light of knowledge, for only it can free us from the shadows of ignorance.

To my sons, I hope these pages will be inspiring enough to further your desire for self-knowledge.

To a friend who has continuously encouraged me, may this fruit of my labor fill your soul with joy.

I also thank all of you who took a chance and bought this book. First, because obviously, there is no use in writing if there are no readers. Second, because the proceeds from this book will allow the application of solutions such as prenatal awareness and biosynarchy through the creation of a foundation.

Usually, we follow in the footsteps of someone who did something notable in the world. *Isis Code*, however, is of another nature. It is a quest for the whole, for the inner and the outer reunited.

The image of the grandmother is of one who sews, weaves, and knits us clothes and prepares memorable treats. We feel loved by her. Her attentions warm the heart and feed our hopes. My wish is that through this book, I will accomplish the same for you.

Introduction

Our world lives and breathes. It is a giant being. We are all cells of this great entity. We are not in its head; no, we are more in its heart. Regrettably, this being is very ill. However, sometimes a serious illness can awaken dormant qualities and thus redirect a life, a society, a civilization, even humanity. In some cases, if the message brought by the illness is understood, the illness is no longer necessary.

For humanity, at this point it is not a question of going back to a world that existed only in romantic memory, but to jump ahead. We have played and experimented with matter and have found ways to render our physical reality more comfortable. Unfortunately, what we have gained on one side, we have often lost on the other.[1] What needs lie unfulfilled because of the choices we have made?

Let's face it. At this stage a shrinking minority of individuals controls the choices of an ever-expanding majority. Whatever politicians and some media outlets maintain, I do not see true democracy in action. I do see some lobbying groups, dressed in false moral pretense, limiting the choices in our lives, pushing governments to take actions that are beneficial for a few but detrimental to the majority and to nature.

In a 1936 address at Madison Square Garden in New York City, the then president of the United States, Franklin D. Roosevelt, said, "We had to struggle with the old enemies of peace—business and financial monopoly, speculation, reckless banking, class antagonism, sectionalism, war profiteering.

They had begun to consider the Government of the United States as a mere appendage to their own affairs. We know now that

Government by organized money is just as dangerous as Government by organized mob."[2]

He directly warned us against lobbies profiting from human hurdles. This can be applied to any country and, sadly, is even truer nowadays than it was in 1936.

So what can we individually do to immobilize this ogre we created and prevent it from swallowing more lives? I believe we can transport mountains, one grain of sand at a time. With knowledge and firm resolve, we, the chain of humanity, can work as one heart, in one world, armed with one vision. Individually, though, we first have to become aware of this vision. Although it is mostly unconscious, it is deeply embedded in all of us, in each of our cells. It is the inner compass, notably expressed through the brain. As we will see, this compass, or inner Isis, uses a sense of unacknowledged synergy to communicate with us.

Why Isis? Why not Mary or Venus? Because Egyptians established a civilization that lasted three thousand years.

How did they do it? What was their secret? It sounds very simple: Isis and Osiris, a balanced expression of both feminine and masculine principles, sustained all the spheres of their lives.

The more we learn through archeological digs about this civilization, the more my fascination for it deepens. Of course, civilizations are cyclical. As such, sometimes they will express these two polarities, sometimes they will not, with predictable consequences.

The civilization of ancient Egypt talks directly to the heart, to each one of us. A sculpture of the head of Nefertiti, dating back 3,300 years, is more widely recognized than the face of many presidents or prime ministers dead only forty years.

Isis represents the aspect in us that safeguards the cohesion of our different aspects: physical, emotional, mental, rational, and universal. Without it, there is no organization, no unity, no harmony, no balance, and no health. Isis represents the feminine aspect present in all of us. Her husband Osiris, the legend reveals us, was savagely killed by his own brother, Seth, who cut his body into fourteen pieces and scattered them throughout the universe. Seth, understandably, was eventually associated with hell.

Isis then left on a quest, assisted by her sister and many children, to retrieve the missing pieces of her beloved and reunite them. She eventually found all of them, except for the phallus (the part of Osiris she does not share in her manifestation). As the myth says, Osiris remained, but as the guardian of Hades, kingdom of the dead. One day though, he will live among us, and all, including nature, will rejoice because the masculine and feminine principles will at last be reunited. For us, it will mean we have reached maturity; we will be complete and perfect. Nevertheless, this can happen only after their son, Horus, avenges them. Translated to our human level, this will happen after the human aspect in us is mature and leads. In turn, this cannot happen without our inner Isis taking her rightful place.

Indeed, our inner Isis attempts, during our terrestrial life, to find and reconnect all our parts, which, as we travel the pathways of analysis, accepted values, and material life, have lost their cohesion. The search for the loved one—our complementary aspects—and the pursuit of knowledge will help us find and connect all the pieces. Love is the thread that can guide this quest.

As for now, consciously or not, we are similar to a cut up Osiris, with the heart and its feelings here, the body and its needs over there, the mental living on its own, and the soul deprived of its vessel—lost. We often feel depressed, sad, as if we do not *really* exist. Our spirit is forever circling above, unable to find a proper nest. In turn we cannot manifest our individuality and, unknowingly, we are similar to Sleeping Beauty, forever waiting for her prince.

Not long before he died, during a conversation about the story of the Queen of Sheba, psychoanalyst Carl Gustav Jung said the following: "Somewhere there was once a Flower, a Stone, a Crystal, a Queen, a King, a Palace, a Lover and his Beloved, and this was long ago, on an Island somewhere in the ocean 5,000 years ago. Such is Love, the Mystic Flower of the Soul. This is the Center, the Self." Then he added: "No one understands what I mean. Only a poet can grasp it."

Jung is not the only one who associated love to knowledge of the Self.

For example, in the twelfth century in the south of France, the concurrence of a relative peace and wealth, evangelical pursuits, rejection of the Catholic Church's control, and a desire for refinement led to a new literary genre that celebrated love in its different stages. Many nobles of Occitània (known as Aquitania under Roman rule), in the south of France as well as in Italy, came to realize that love and spirituality could go hand in hand. The fin'amor, a term that designed a code of life, was the convergence of the troubadours' art of love with the Cathar faith and other movements. It enhanced love as a tool for self-improvement, self-knowledge, and self-realization. Chivalry, said to have originated in these areas and more specifically from the region where Charlemagne wielded power, helped spread this ideal of courtly love, and in some case fin'amor (see chapter 7), this, ironically, through crusades against "infidels."

At the time of the troubadours, the lords of the north were unrefined compared to those of the south, who had a desire for beautiful objects, engraved manuscripts, and knowledge. In Catharist Provence, the Kabbalah emerged as a mystical movement. Some Jewish rabbi maintained close relationships with some Parfait Cathars. Then, in 1208, Pope Innocent III ordered a crusade against the Cathars, because their popularity suddenly threatened the teachings of the Catholic Church and jeopardized its expansion. Hence, the people who were, symbolically, the butterflies of the south of France were eliminated. This catastrophe, since also most nobles saw their lands confiscated during the resulting Inquisition, ushered a change of power in favor of the north of France, a decline of the langue d'Oc, and with it, eventually, the loss of an entire cultural heritage. However, the troubadours, through their songs and poems, carried the fin'amor and courtly love across Europe.

The sense of love in its subtler expression, which I name synergy, is truly the manifestation of spirituality in daily life. The mystical flower traveled through time and space right down to us. With this flower in hand we can embark on our personal pilgrimages. For us and for those we care for, this path we dare tread is a testament, a carved stone we bring for the edification of humanity. All of these individual stones will build this new cathedral, this new earth we dreamed of. Gone will be the fascination for the abyss into which we are now absorbed by self-interested powers lying deep in our subconscious and expressed through the inner Seth, otherwise qualified as our hell aspect. We will raise our heads and look at the sky.

At the same time, we will redeem Seth, for without him we can't survive on earth. Through his function of analysis, Seth divides everything. Alas, a flower that has been analyzed is dead and no longer sacred. Seth is not capable of understanding life. Therefore, thinking he saves our material life, we bowed to him in every domain but in the long run he instead leads us to death. His tools are only suitable for superficiality and inert, measurable matter. You could say we began by studying the mouse's tail, believing it would show us how its head functions.

Who am I to speak about love? Only one who is walking its path, one who has craved it since birth, one who tried to isolate and study it. As a woman, I sometimes felt like the Aldonza in Don Quixote[3], —this being the masculine polarity interpretation of what a woman is— because of this need to be loved and to express love. More and more though, I feel like the Lady and the Unicorn[4], expressing *mon seul désir,* —the feminine polarity interpretation of what a woman should be. Through her, I see a clear picture of the woman of the future and can perceive the world in a totally different manner.

The sixth sense of synergy flowers because of this overwhelming need to belong, to love and be loved shared by all humans. Awareness of this sense is attainable only through the experience of our personal heaven and hell characteristics and of our feminine and masculine polarities. It is the center of our personal cross. The one we must bear to "know thyself."

Also, love in all its phases is the *only* door to spirituality. You can stay on the threshold all your life or you can step into the light. When I look back on my life, I see different love types all linked to different needs, all reflections of different stages of maturation, even to different aspects of the brain. I can also see a structure, a system, through which we express ourselves and communicate this love. I believe it could be useful for everyone to become aware of it. Observations of this basic biosystem and my experiences showed me that when the two polarities are healthy, everything else follows harmoniously. When they are not, it spawns all the miseries of the world.

In fact, we are not brothers but sisters. The feminine principle is Isis, which the religious tradition equated to the Divine Mother. This part binds and weaves all human realities through the sense of synergy. The masculine nature is the one that separates through analysis. Useful as it is, it is *not* the bearer of truth or life. The feminine principle is.

So I retraced here Isis's eternal steps. I followed her sacred quest and, like her, I gathered the scattered pieces of my beloved. This pilgrimage is necessary, for if we do not get to know our selves, we will continue to be wandering souls on a barren land.

Truly, there is only one ideal within one heart, resulting in one world. Only this aspect of our brains which analyzes everything consciously divides.

Let us recreate the world together, one grain of sand at a time, in Isis's footsteps.

This book is the unfinished tale of my own pilgrimage. On the winding roads of my anonymous life, I had the grace to witness her presence. My only desire since, has been to live in this presence and that one beautiful spring morning when I look in the mirror, I might see her face, smiling back to me.

Chapter 1

Spring

The clock in the kitchen indicates 5:45 in the morning. Through the window, I see the glistening garden. A translucent cloak, the mist from the river, floats on its borders, softening its perimeter. Soon the day will come. Through the back patio door, I step outside and pause before going down the stairs. Except for the songbirds echoing each other, all is calm. Walking toward a gray metallic garden chair, I presume it is cold and wet with dew. As I pull it back to sit on it, the sound of the pea gravel wakes me up more. I shiver in the early morning spring air. The wind, combing the still naked branches of the majestic maple trees, seems to usher words of love and hope in my direction. I wrap myself tightly in my long blue-gray wool sweater. Attentive to the surroundings, I listen. I gaze around and breathe in deeply. Already, the shadows are escaping, chased by the approaching dawn. Sacredness is tangible. Colors begin to appear, merging pinks and purples. Nature prepares herself for her wedding. The river flows silently southward, carrying away my winter thoughts. My whole being is now focused in its longing for light and warmth.

Suddenly, a bright golden speck flares on the horizon. Silently, powerfully, and vibrantly, the sun rises, little by little, revealing itself in its splendor. It takes my breath away, yet at the same time makes me breathe in profoundly. Motionless, mesmerized, and enveloped in this emerging light, I am now part of it, a drop in an ocean of light. I feel warm, one with the sun, with life, with nature, happy, alive, and free.

I sit in that state for what seems a long time, although only a few minutes passed by. Gradually, alien noises interfere with my serenity.

Isis Code

They become louder and more insistent by the second. Passing cars, newspapermen, as well as my own thoughts of things to do, fragment the calm surface of my mind. Reluctantly, I get up and go back inside, to daily life. I shake my head spontaneously. No, nothing in that daily life can give me the joy I just felt in the pure morning air! Nothing can—except maybe love?

We from Hell, We from Heaven

Ah, love! From a young age we are led to believe that men and women are either the same, with only practical minor differences, or antagonistic, where the male—although it is politically incorrect to say so—dominates the female. This is simplistic and seems more like an excuse for our ignorance and flaws. It sounds like permission to behave badly toward the opposite sex. No, we are not opposites. If we were, we would not so desperately cling to each other! There has to be a part of what we name man in woman, and a part of woman in man. Otherwise, one would automatically annihilate the other as water suffocates fire or as fire evaporates water. We are not the same either. Ask those poor souls who desperately wish for a gender change. Moreover, if we were the same, there would be no dynamism, no motivation between us. This complementary state between men and women helps us get up in the morning and gives perspective to our lives. When we truly love, we are in heaven.

Love reveals us and engages the totality of who we are.

The chosen companion exposes, among other things, our feminine and masculine psychological characteristics. To understand this is primordial, because the perception of feminine and masculine values influences every aspect of human life, whether it is on personal, social, religious, or scientific levels. It even affects choices of government and the health of people. Understanding this gives us control over this mortal aspect of ourselves, this periphery of our being we call the personality or self. This is this "hell" aspect we referred to earlier.

As well, the individuality of the companion we choose reveals our own. To know about the individuality, or core identity, is also primordial, because it gives us a blueprint, the design of our essence for

Spring

this life. At this level, there are no divisions between people; true fusion of hearts can occur. This is the "heaven" aspect.

The personality is modular and mortal while the individuality is immortal and universal. These two elements are in us and are expressed through our feminine and masculine polarities.

We from heaven: there is one being in whom we all merge, from rocks to humans to stars.

We from hell: only we divide.

Since love depends on perception and reception, I consider it a sense. This sixth sense of synergy implies many elements working together. Its holistic quality also indicates that the merging of these elements will manifest a different property, unattainable through the simple addition of its parts.

A sense requires a physiological support: receptors, regulators, the ability to metabolize the information received from the environment and direct it to the brain. It also adds effect to the other senses. Haven't you noticed how things feel, smell, sound, and look different when you are in love? Physiologically, we possess a superstructure associated to the perception of a vast array of subtle energies and of feelings. The cellular matrix, which has been getting more attention lately, is part of this superstructure.

As with any of the other senses, if we have the right conditions, we can improve upon it. We may also be challenged regarding this sense. Research has shown that if you put a man in a situation where none of his senses receive stimuli for a while, his mental state soon deteriorates. The drastic increase in the number of people suffering from mental illnesses seems to indicate that our sense of synergy is deprived of appropriate stimuli. "In 2009, there were an estimated 45.1 million adults aged 18 or older in the United States with any mental illness in the past year. This represents 19.9 percent of all adults in this country."[5]

Our senses give us a feeling of existing in space and time. They show us a *unified* world in which we are. Seeing something while hearing a sound that differs from what is expected shows us how wary a non-unified world can make us feel. Similarly, we have natural expectations of what we should receive through our sense of synergy or love. Deep inside our cells, there is a model of how things should be in a healthy world.

At this time, our life is bathed in a virtual world. Our brains, given the practicality of electronic devices, have been relegated to the back burner. We live in cities, so our senses have atrophied. They are deprived of the stimuli allowing us to feel alive.

We probably all know people who have lost or never had the ability to see, hear, or smell. What about people deprived of a functioning sense of synergy? What about people who have exacerbated this sense? It is a support on which all the other senses merge. It has not been studied before, and we all suffer from it not being acknowledged. Studying it will provide an understanding of our heaven aspect and, by the same token, of our hell aspect. This, in turn, will guide us towards perfection and its subsequent happiness.

To achieve this, since the brain is at center stage of our awareness of a unified world, we should understand its functioning and aim, as well as how it differs depending on gender, ages, and people.

At this time, it appears we are still far away from knowing who we are and why we exist. Why do we get up in the morning? It is difficult to say. Similar to the colossal beasts in biblical times, who were harnessed every morning to a heavy wheel going nowhere, we pace around and around to earn a living. We are told, and believe, that economic "freedom" will bring us happiness. This vision only pits us against one another in a fight for survival that makes a mockery of the ideal we call democracy. Eventually our heart cries, because the hurt—economic or otherwise—we willingly or unknowingly do to others, eventually reverberates back onto ourselves. We then wish for a different world, yet at the same time surmise that the world we live in is certainly the best this planet has ever known. One has to wonder though: Better for whom and at what level? Reassured, we grab a pill and keep marching on.

It seems we do not know how to simply be. We believe that in order to be, we have to do. Indeed, we are often judged on the quantity of doing we can push and shove into our daily lives. We are under the illusion that we are only what we do. Therefore, the more money earned, the more relevant we are and the more envious others might be of us. As such, a teacher does not count for as much as a hockey player (I have nothing against hockey players) or the president of a national company does.

Those who care only about financial gains and appearance tend to scorn those with a functioning sense of synergy. Nevertheless, I wish to believe that each and every one of us, even if for only a split second, dared to dream of a different world where our hearts would vibrate in unison with the whole universe, where our intellect would rise above mere accounting matters, and where our souls, similar to Horus in flight, would embrace the immensity. Unfortunately, present social values indicate to us that this is, rationally speaking, a ludicrous, unrealistic, and unscientific idea.

The world our forebears imagined and created is going through an asthmatic crisis. We are not presented with ideals inspiring enough to motivate us to live, change, or sacrifice our physical lives. We live in a world of endless circulation of information, but with less human communication. Our collective fabric is no more than a tattered rag. Our possible heroes and own heroic nature, who feed on the absolute, can hardly reveal themselves.

Certainly, we live in an exciting world, but these excitements do not fulfill our profound needs. The result is that we are globally more chronically ill, more depressed, and more mentally impressionable.

Whether willingly or forced by life events, those who stop and observe the world go round can feel a malaise. This frenetic race toward promised happiness seems flawed. The world is out of breath, exhausted. In that supposedly rational world, we often feel disconnected from others, from ourselves, and from life.

For the last fifty years, we have believed that our economic and technical advances would lead us to a better and happier world. Instead, similar to a herd, we were driven into a long, dark tunnel. We are at the end of that tunnel. Where is the better world? Where is the promised happiness? There is nothing. I see nothing, except for two tunnels. The first one, going upward, is heaven's tunnel. The second

Isis Code

one, going down, we have already taken many times, through many civilizations. It's hell's tunnel.

If we would collectively dare to take the first one, we would need to accept humility in our lives. Contrary to our interpretation of the Bible and Rousseau's exclamation, man is not master and dominator of nature. He is more its warden, student, and, even more importantly, part of her. Our brains are proof of this. We could take that tunnel upward. Confronted by new horizons, we would finally breathe in freedom. A new order would start its reign naturally, not directed by a shortsighted and consequential economy, but by the desire of allowing future generations to live on this planet, not merely survive on it. To take this tunnel would demand vision, resolve, and determination.

The positive results for everyone, though, would be much greater than the sacrifices we would individually have to accept. It sounds like Utopia, but in my experience, reality sometimes surpasses fiction. It is more a question of being than of doing. Eventually, our *being* will change the way we *do* things. This tunnel would finally bring us home.

By comparison, in the second tunnel we encounter blindness, stubbornness, arrogance, and laissez faire. Outrageous, inhuman speculation goes on and on. "We have only one life to live, let's make the most out of it" is the motto. Alternatively, others say, "Who cares? Forty years from now, we will not be here!" Or, "They'll find solutions, don't worry!" I hear the fat laughs of men satisfied with their own limited, shareholder egos. There is always this international, infernal race to be part of the billion-dollar club. In that race, only a few binge financially at the expense of billions of others and of nature itself. It is like a game of musical chairs when the music stops. To allow those "exciting success stories," how many families live in shadowy apartments in grassless cities? How many pregnant women, reduced to feeding their dreams and shaping their children-to-be on bland and odorless days, with only worries and bills to pay in their futures? How can they joyfully call souls into a world that is increasingly threatened by the destruction of the environment? Burdened by these fears and sadness, what kind of children can they bring into this world?

In this tunnel, we have nothing to change, no efforts to make. This is a parasite's dream. This tunnel is easy. It has a nice downward slope, with hopes of economic gains and recognition for some and the status quo for others. We keep trudging on, eyes closed, ears shut, plundering every part of nature in our wake, akin to those big ships using drag nets. Very efficient, they make a clean sweep of the bottom of the oceans, but in this way destroy nature's ability to replenish itself and to feed the small fishermen, their families, and eventually the world. Choosing this tunnel is akin to replacing those small fishermen and their families with neatly aligned cans of fish on sterilized supermarkets shelves, and eventually, with silent and sterile oceans. It is easy. We only have to repeat: no research has directly proven this.

Because my heart would not survive in the second tunnel, I took the first. Yes, it is steep, and I felt quite lonely at first. Nevertheless, after walking only a few steps I realized that so many had walked there before and that so many are unknowingly ascending it right now!

In that tunnel, I saw Isis's footsteps and followed them. I found there is no savor in life, even no life, without the sense of synergy one finds walking through this tunnel. I have gathered a few twigs from the land on the other side, what I call the land of the living. These twigs, when we analyze them, indicate to us the structure of the world we can build together. As I write, system science is studying these twigs. For those in the first tunnel, these can bring consolation, courage, and knowledge that they are not alone. There is hope.

Those pacing in the second tunnel, if they are not afraid, might take the time to look at the twigs and ponder. They might realize how much they had to allow their brains to be remodeled and frustrated in order to fit in that second tunnel. They can't know what they missed or that they are forgoing heaven on earth for themselves and everybody else. After understanding the message in these twigs, will they prefer to find at the end of the chosen tunnel beauty, health, life, and laughter, and feel they are in part responsible for this, or the destruction, death, and misery they are unknowingly fostering? I am not being an alarmist, only a realist. Indeed, we will feel true joy and inner peace only when we truly know ourselves holistically, and when we accordingly choose our fundamental values.

Isis Code

My eldest son often said: "Question your beliefs." In other words, analyze your synthesis. Too many of us live like monkeys confined to a big cage. Despite loads of bananas regularly thrown at them, even new and improved ones, those monkeys are not happy. Offering them a bigger cage and a wider choice of bananas excites the poor animals for only a short while.

As humans, we have more needs than monkeys do. We must find nourishment not only on the physical level, but also on more subtle ones. These are invisible nourishments, albeit vital ones. True, we do not have all the same needs, qualitatively and quantitatively, but fundamentally they are identical. Those who prefer to eat junk at all levels should not impose their vision on others. In other words, leaf-eating caterpillars should not impose their food on nectar-sipping butterflies. Both caterpillar-type (masculine) and butterfly-type (feminine) people have essential functions in society, and both are parts of us. We need the caterpillar to get to the butterfly, but we need the butterfly in order to have more caterpillars. One is only a more mature form of the other. They should understand each other and work together. Both are essential and complementary.

The economic aspect alone cannot fully satisfy the needs of humans anymore. Their heaven aspect prevents that. That is not to say that the economy is not relevant. Nevertheless, there is a thirst for the absolute in humans which betrays their origin and destination. They also belong to a world different from the solely visible one. This is what defines us as humans. Otherwise, we would only need awareness/consciousness, and not Consciousness. Some individuals are aware of this need while in others it is dormant. That this unquenchable thirst has been embezzled for the gain of our physical selves shows how young, fearful, or decadent our civilization is. Nonetheless, what was okay when we were in the infancy of humanity is not satisfying now that we are more mature. Treat a teenager as if he were an infant and observe the results. He will not develop his potential and will eventually regress. Globally, we are at this stage now. This is a gift from evolution.

In countries affected by wars, this regression was observed. In 1950 for example, someone asked Carl Jung what he thought of the "intellectual precocity" that could be observed from Japan to Central Europe. To this, Jung replied: "I wonder if this is the right

Spring

term ... It is more a type of increased awareness ... observed in all countries touched by wars and revolutions ... It is not a higher degree of consciousness, but on the contrary, it is reduced ... more often, it is a regression towards primitivism."[6] In other terms, Jung says that wars have not created an increase of consciousness in the countries affected, as this would indicate a better expression of the feminine polarity, but at most an increase of physical awareness with a *reduction* of Consciousness, which points to an inflation of the masculine polarity. The intellectual precocity his interlocutor refers to is in fact a lack of control over the Analytical brain, not a sign of mental health (see chapter 6).

Can we choose the first tunnel, which for many seems so steep? Yes. This new order of things will come eventually. It is the march of evolution. Whether it comes because of health catastrophes predicted by the WHO (World Health Organization); by the bankruptcy of a deified economic system; through widespread environmental disasters, terrorist activities, or nuclear war; or due to skyrocketing physical, emotional, and mental anomalies, it is coming. Unavoidable, it will touch each and every one of us.

We live in a world of cycles on every level. Whoever is rich today might become poor tomorrow. As history teaches us, dominant countries eventually lose their hegemony. We cannot escape cycles, even if we use violence and preemptive acts, but we can predict them.

This new order of things, this new heaven, must come via the understanding and willingness of most people to count positively on this earth. Not as parasites, but as human beings. Unfortunately, history shows us that we change by necessity, not by virtue. Nonetheless, we will be forced to change because of the sad state of the earth.

How It All Began: The Quest for the Elusive New Woman

When I turned nineteen, the television station CBC (Radio-Canada) in Montreal, Quebec hired me as an assistant director for a show titled "Femme d'Aujourd'Hui"—"Today's Woman." This wonderful program, which lasted seventeen years, incited women to go out of their comfort zone, out of their houses to find work. The show's host explained that this would enable them to develop all aspects of their personalities. Indeed, research had shown that stay-at-home moms could only regress, were more depressed and felt isolated compared to working women. Women with their own salaries also fared better financially and were less exploited by husbands, who might otherwise control them through indirect or direct monetary blackmailing. The success of this program, broadcasted every weekday at 3:00 p.m., became its downfall. Women did go to work, and eventually there were not enough viewers to sustain the show. I was able to witness the incredible steps women took to free themselves from a role that could restrict them to a subaltern position of servant. I had already decided that I would not go through what my mother accepted. The forever repeating ritual of cleaning the house, preparing the meals, doing the laundry, ironing, cleaning, preparing meals, doing the laundry, ironing ... All for the smile of a man who loves you and the satisfaction of accomplishing your duty. At the time, it seemed like senseless imprisonment to me.

The alternative, the liberated woman, which many of my friends were, did not seem a better solution. Use of the pill or of an IUD meant there were no more cycles or special times for intimacy. We became like men, ever ready. We were bombarded with information regarding how our sexuality should be expressed, how we should live, how we should feel, if we were to call ourselves liberated. I didn't feel inhibited or frustrated. Nevertheless, I viewed relationships as sacred. I could not fathom, and still cannot, that sleeping with a man implied intercourse. I was seeking a soul mate, not a surrogate vibrator.

I also was not comfortable with how we were told we should behave in bed, initiating sex, pursuing men in a masculine manner, and dominating the physical act.

After a night out with my friends, I often saw that their "love" was a speculative affair. Unwittingly, my friends were using sexuality like some men used money and power, hoping to secure an attractive future.

They were looking for an easy life, in which they could be careless, free, and adored. Who could blame them? This was, in fact, an exchange. Males showed off their possessions and social standing, and females checked out the proposed decor. Often those young women thought that after sex, the man would ask for more. These women had the nesting instinct. They were not in love. Rather, they had tasted the ambiance of a possible brighter future, which was highly addictive. They envisioned themselves as couples. Over lunch with me, happy and animated, they would go into details over how they foresaw their future with this new man. They successfully projected their vision of the ideal man onto these men.

It was not the same thing for men. With their predisposition for the most sex with the most women, they moved on after sunrise. Women were the losers, even if after a few times they acted as if it was a normal part of the game, as if they didn't mind the rejection. They were cool girls. The ones who viewed themselves as most liberated were the ones who dared to have two boyfriends simultaneously. Having sex with men, they argued with me, was like brushing your teeth, no more, no less.

Blue Flower and White Goose were the nicknames my comrades at work gave me in those days. They told me I had to get rid of my fairy-tale heart. For a short while, I tried to follow. Result? I did not succeed. I was miserable. I felt I had rolled myself in foul-smelling mud and lost even more contact with my inner self. My needs were obviously different from those of the so-called liberated women. I guess I did not wish to be liberated; I wanted my own Prince Charming and that was it. After two years of trying my friends' diet, complete with the pill and a disastrous love life, I violently rejected the new dogma.

They had grasped the sexual revolution, which heralded the movement toward this new woman, in a way I did not care for. As I understood it, to be liberated meant copying the scattershot sexuality of men, and eventually becoming "above" sex, in the sense that the gender or number of your partners should not be an issue. I remember in Paris, meeting a member of an ancient and well-known French family who was also part of an illustrious secret order. He wanted to educate me on the matter of sexuality, and felt more evolved than me because not only could he exchange the person sharing his bed with no qualms, but having a lover of the same gender was no obstacle to him.

In my view, the pill was the symbol of males' liberation, not mine. With it, boys could pressure girls even more into the sexual act, interpreting their refusal as snobbery, pretension, Puritanism, and, even more personally hurtful, lack of love. In my opinion, this new woman should not have to take the pill and should never be manipulated into using sexual intercourse as something to trade. We are not bonobos!

I came to understand, alas too late, that because of the values adopted by our society, most men—and probably now half of the women—saw things differently. To me, the sexual act was and is sacred. It is the normal consequence of feelings so powerful that the desire for total fusion drives us to physical union. It transcends the physical act, in which we unsuccessfully try to merge with another. Only under exceptional conditions can true fusion be achieved with someone else, but never on the physical level.

Despite love, physical union is deceiving and is in no way synonymous with fusion. It is not the beginning of a relationship but, as many psychologists have observed, often heralds its end. Sometimes personalities are in the way and in any case, our physical bodies are always limiting. At other times our souls, seeking a different or higher love, are in the way. Is the latter such an awful thing? In addition, the feminine polarity motions toward elevation and subtle energies, while the masculine one penetrates into the depths of matter. These two elements create a different goal for men and women in the sexual act, and those different goals are the cradle of all misunderstandings, as I will explain in the further chapter devoted to the Mammalian brain.

When I reached my mid-twenties, I organized my life in order to write a book about relationships, mainly about the similarities and differences between men and women. At the time, however, two things prevented me from embarking on this venture. To begin with, even though I had well researched the psychology and physiology of both sexes, I was still young with little experience of life. I felt as if I was peeking at couples through windows. Then the house I had just moved into burned down. I lost everything and had no more time to write. I stayed in a hotel for a while, and then moved back to my parents' home. After a few months, I had to decide what to do next.

Spring

Nothing except my work was keeping me in Montreal, so I prepared a project to travel to Europe in search of this elusive new woman. My idea was to meet heads of women's groups in nine different countries and interview them. With the blessing of CBC, I left. Unfortunately, as soon as I set foot in England, a strange illness befell me. For three days, I stayed in a London hotel room, exhausted. I had to cancel all the projected interviews, and eventually the project itself. Finally, I gathered enough strength to take a train to Edinburgh for Christmas. There, I felt the happiest. I walked for days, full of stamina. I rejoiced in all I saw: the Christmas carols, the gray-black stones, the sound of Scottish people speaking and laughing while eating baked potatoes with all the trimmings, the mounds on the outskirts of Edinburgh, the castle heavy with history, and the air light with joy. All this had a strange, homey feeling to me. To be outside at Christmas in such warm weather, compared to my frigid Canadian winter, was also divine. Believing myself healed, I boarded a train to Paris. Alas, my ailment started again.

Throughout my life, this strange predicament reappeared, preventing me from doing something or slowing me down in some way. With time, I became conscious of it, and as soon it appeared I would stop and think. If I refrained from following through on some project or changed some part of a plan, the "illness" would vanish.

I decided to quit my job with Radio-Canada and stay in Paris, resurrecting my idea of writing a book on the new woman. I was able to follow my research about this new woman while studying naturopathy and going to Nanterre University in Sciences of Education. I provided reflexology treatments and taught prenatal yoga to mothers to be. As well, I became a founding member of the National Association for Prenatal Education (ANEP) in Paris.

From what I observed, a liberated woman was often not much more than a man adorned with a skirt. The new woman I was in quest of couldn't be a man's copycat. She had to be different. Even in spiritual groups, when a woman was leading she often had to put aside feminine qualities and take on masculine ones. Did we have to let go of beauty, subtlety, and the quasi-magical link we women naturally had with nature to become this new woman? Did we have to become self-oriented, assertive, and ambitious to get anywhere in this world? What function was ours?

To be liberated suggested avoiding feminine values, thus denigrating them and taking on the tools, words, and point of view of men. Did I wish to be part of that world? No. Could I survive without being part of it? I had no choice but to try.

I had suffered from mild anorexia for a few years, and bulimia added itself naturally, chronically. It was not severe. I never embarked on the spiral that brings a person close to death. It seemed more a consequence of my refusal to be and feel like what I saw as *the others*. Something was amiss in my life. I felt ill at ease in my own skin. Through eating, different environments, whether they were physical, emotional, or mental, became a part of me. Something in me refused them. They had been forced on me for too many years, to the point that I just could not find who I was. I couldn't accept that any longer. Each time I felt alien in these different environments, unable to escape, I wanted to get rid of the food I'd just eaten. I was already slim, so it had nothing to do with wanting to become like the skinny models in magazines. It had something to do with who I *truly* was and could not express, even to myself. During those periods, I could observe the wasting away, nothing more.

Women of my age in Paris were often single, unable to find "the right man." Was it men's fault? The women I knew were all intelligent, beautiful, and successful. Was it because we didn't want the same life as our mothers, but still wanted a few things from it? It seemed that when men gave what women asked for, they lost their maleness in some women's eyes. Was our evaluation of what made a man or a woman wrong?

I attended lectures given by medical doctors to pregnant women, thinking I would find there the new woman. Interviewing some of them, I discovered women who mostly seemed lost, who had no control and no real understanding of what had befallen them, victims with their heads full of possible tragedies, of horrific stories they had heard or read about. I often felt their anxiety. No, these were not goddesses walking, conscious and actively partaking of the miracle happening every second within them. Nowhere did I find the expression of *the mother*, a woman connected by all the fibers of her heart to the different kingdoms of nature, breathing in the immensity of the universe, proudly floating in the ocean of stars above at the same time she walked barefoot on this earth.

Spring

No, I didn't find mothers-to-be creative powers, as in "this child will be as strong as ... as beautiful as ... as intelligent as ..." I only found male doctors telling these women what to do, what to think, how to behave "rationally," and warning them not to trust their instincts. In a nutshell, urging them to use their linear, masculine aspects and accept "reality."

Indeed, men saw and analyzed women's sexuality through the lenses of their own. In this way they reduced pregnancy to what they could observe, since they were totally ignorant of the hidden feminine aspect of sexuality and pregnancy. Obviously they were talking about their own reality, the external, observable reality, not the women's.

Conception should be the result of a true fairy tale. Pregnancy should keep or create this magic (a word possibly derived from the prefix *magh* in the Proto-Indo-European language, which means power). Or I should say more appropriately, this sacredness.

In hospitals, many nurses had the masculine qualities I disliked seeing prevail in a woman. Everything was intellectual, linear knowledge, down to earth. There was no poetry, no sacredness, thus no life. When they did try to put some in, *pour faire jolie*—to make pretty—it often felt acted from the outside, empty.

The doctors had fatherly smiles, highlighting that they were the founts of knowledge and security. Women could lean on them, and anyway, they had no choice. A woman had to learn pregnancy either from them or a book. The doctors were the priests of health. Pregnancy was a path full of pitfalls, and only the doctors and their knowledgeable staff could walk a woman through all the possible dangers toward a normal child. Provided, of course, she followed their advice. In their facial expressions, I could see the future mothers felt unprepared for this challenging experience. This left me with a sensation of heaviness. I became fearful of pregnancy. There was no stability or freedom as I had thought there would be. I would not bear a child. I would do anything to avoid falling into that!

My research for the book on prenatal education resulted in an invitation to attend a medical congress of neonatology in Paris. J. P. Relier, who was responsible for the invite, was a specialist of sudden death syndrome in newborns. The renowned Dr. Minkowski and others discussed new discoveries in obstetrics. Although I found the talk intellectually stimulating, I could not help but notice one of the doctors sitting on the panel. She was pregnant, well into her seventh month, and was chain-smoking. In that moment, the gap between intellectual knowledge, what we call reality, and its application to real life became obvious to me.

I also observed that every group is pyramidal. That is, whether a classroom, a gang, a group of researchers, a church, a sect, a political party, a music group, or even a country, the one at the top is the alpha male. The other men bow to him while many secretly hope for his dismissal or death, so they can swiftly replace him at the top. Women are naturally attracted to the top of their favorite pyramid. It is instinctual and often not conscious. Men have therefore a natural tendency to hate pyramids they cannot conquer.

If the alpha male seeks a woman, she becomes the most beautiful and desirable woman. Without any effort, she is promoted to the top of that pyramid. Moreover, she often emotionally dominates that man. This not-so-new woman, through her sexual liberation, can now seek the alpha male even more relentlessly.

In the philosophical group I frequented on Sundays, I saw some men's self-serving maneuvers. They would play the mystic card, the *I am a highly evolved being* card. The one who portrayed himself as such could easily find young women eager to receive his "pure" manly emanations. The most successful were of course those seen in the vicinity of a guru. Those men would sincerely argue they were merely tools used by superior energies to contact us women, who were of the imperfect gender, and in this way improve the earth. They instinctively understood how young women craved the absolute.

Through my research, I found the lectures of a French philosopher of Bulgarian origin: Omraam Mikhaël Aïvanhov. Reading those he had given since 1937 when he arrived in Paris, I realized that many of these lectures regarded the qualitative aspect of feminine and masculine principles. In one of them, I found his idea of men and women sharing different polarities on physical, emotional, and intellectual levels. I was helping Pierre C. Renard, a doctor in naturopathy, to write a book on prenatal education. He used Aïvanhov's thesis without expounding upon it. I started observing men and women through this viewpoint, interested to see where it would lead. I found it quite enlightening. Soon afterward I discovered the works of Carl Gustav Jung, the Swiss psychiatrist, founder of analytical psychology. Often considered the first modern psychologist to do so, Jung stated that the human psyche is "by nature religious,"[7] and observed humans through the lenses of two polarities, feminine and masculine, which he labeled as anima and animus. Eventually, I realized that all traditions were more or less built on this observation.

In France in the 1980s, many female psychoanalysts and psychologists pronounced how the new woman ought to behave, think, and feel, and why. To be credible and taken seriously, this new woman had to use accepted tools and viewpoints of how things ought to be. In a nutshell, for the new woman to be accepted, she had to be a man.

I often felt uncomfortable and frustrated when reading their essays. Many felt vindictive. This did not resonate with me. Nevertheless, they were a rich source of anthropologic lore, and I reveled in the sometimes contradictory interpretations the different authors gave different facts. The main ideas put forward were that sexuality was mostly a cultural bias and that the patriarchal societies and their religions used it as a tool to repress women. Parents transmitted this bias to their children, who would eventually do the same to theirs.

This new finding sprang from the observation of hermaphrodite twins. Some were brought up as boys while others were raised as girls. At the time, most thought that this would not incur any problems since it was believed that biology determines everything and we emerged from matter. Now we know it had dramatic consequences for some. We can't generalize. These are individual situations, and involve too many elements to allow for a general, one-size-fits-all statement. Even though I did accept the patriarchal point of view as the source of conflict, from my personal experience I rejected the idea of external factors as the sole source of my femininity.

Sure, my mother had wished she hadn't conceived for a fifth time, and when the inevitable happened, she did wish for a boy. Truly, I was more attracted to my brother's toys than to my own. I vividly recall playing with his green garbage truck, his electric train, his

Isis Code

Meccano toys and brick sets, and his telescope. This, I surmise, was not due to a hidden desire on my mother's part. She always dressed me in cute little dresses and desperately tried to tame my wild mane. Still, except for one occurrence, I hardly remember looking at my sisters' dolls. I do remember, though, my joyful discussions with my cat, the wind, and bees. My Christmas requests were mostly for musical instruments, drawing and painting tools, and the occasional toy I'd seen on TV. When I was older, I was smitten with William Tell, Batman, Spock of *Star Trek*, and Captain Troy. When I played with my neighbor, I was the hero, who obviously was masculine. As I remember well, I did not prefer that character because he was male but because he was heroic. Women were simply not considered heroes.

I also loved to pretend I was an animal and could run as fast as a horse, or lie in the sun and be in a trance like a cat, or curl on the monkey bars as fluidly as a snake, or swim underwater as beautifully as a dolphin. In all this, I *was* the cat, the snake, the dolphin. When I tried to jump from the top of a ladder to show my father I could fly, however, he wisely prevented me. I did not feel a division between animals and myself, and even less of one between masculine and feminine. This came only at puberty, for obvious physiological and social reasons. Neither my mother nor my father ever seemed to have the type of connection I had to nature, so they did not influence me. This was personal.

In the mid-1980s, I met with Dr. Alain DeBavelaere, a well-respected Parisian medical doctor, teacher, researcher, and lecturer on medical biocybernetics. The first professional I met who really had an understanding of masculine and feminine polarities, he used, following Jung, the terms animus and anima. At the time, I was scheduled, as he was, to give a lecture on the subject in Paris. My approach was solely through what I had experienced and my research for the book on prenatal education. He opened my eyes to a different take on Jung's approach, as seen through some elements of traditional Chinese medicine. Merging what he showed me with my little diagram on polarities, a structure emerged. More amazingly, after one week under his care, my anorexia and bulimia stopped, despite having lasted for ten years, and they have never returned. I was impressed. For years, he taught me his personal approach of medical biocybernetics (MBC). I used it in my understanding of Man and Woman.

The Implicate Order[8]: A Canvas to Every Living Structure and Our Inner Compass

The search for a general theory of everything, dubbed ToE, has existed for as long as humans. It springs from our profound need to hold mentally all that there is in a comprehensive form. This vision imparts coherence and harmony to our lives. It is necessary to our well-being.

General relativity considers that space itself expands. The common image is of a balloon with pieces of paper stuck on it. The pieces of paper move apart as the balloon is inflated. Scientists theorize that the whole universe was not even the size of an atom before it expanded in a bang, instantaneously to the size of a grapefruit. The microwave radiation, accidentally discovered in 1965 because of communication satellites, is the best proof of the Big Bang theory. The goal now is to find a set of equations, a theory of everything (ToE) that would explain all.

The first scientists did not express their theories in broad light. The notions that formed them were considered sacred and were therefore hidden. It was thought that those who were blind to sacredness might misuse the theories and empty them of their most salient feature: their essence. Therefore, masters would transmit these verbally and in complete secrecy to a few chosen apprentices. Echoes of these theories still exist in some sacred books. Those perceptive men should not be mocked, just as we should not mock modern scientists who, by rigorous reasoning and the extraordinary tools they have at their disposal, try to achieve the same thing: understanding. After all, if the human brain is what it is now, it is in part thanks to such people who have sacrificed their lives reflecting on and living these subjects. One of those seekers said we must relentlessly pursue knowledge, even if this pursuit led us all the way to China. (At the time, in the seventh century, China was considered the end of the world.)

For all I know, the universe we live in could be relatively similar to one cell of my body and located in a giant being, surrounded by millions of other similar beings in the image of humans. These could be a cell inside another being, ad infinitum. What matters to me, though, is the coherence and harmony this world we live in must have. Without coherence and harmony, it would not have generated creatures that require it to stay healthy. Whatever this universe is, part of multiverses or alone, its simplicity and complexity are mindboggling and beautiful.

So what do our scientists, the intuitive ones of the past and the analytical ones of modern times, have to say about this universe?

There are many cosmogonies, all in affinity with the people who propound them.

The Brahmanda suggests that the universe expanded from a concentrated point, a Bindu. The universe, it proposes, as with any living being, is bound to a perpetual sequence of birth, death, and rebirth, cycling between expansion and collapse. This we can read in the Hindu Rigveda dated 1500–1200 BC. It is similar to the Friedmann-Einstein theory, which considers time as endless and without beginning, and the universe forever oscillating between expansion and contraction. We are faced here with two different approaches, each using a different tool of understanding but nonetheless resulting in the same conclusion.

We can also read the following:

> When the universe first came into being, at the Big Bang, 13 billion years ago [now it is estimated at 15.8 billion[9]], it was enormously hot. At first, all four forces acted in exactly the same way (so there was really **one** force). But as the universe cooled, its symmetry broke, and the forces split into individuals with distinguishable characteristics. Our present universe, with its four forces, is an imperfect shadow of that elegant original—the result of three broken symmetries.[10]

> We find echoes of this beginning in the Bible, Gen. 2:10 (King James): "And a river went out of Eden to

Spring

water the garden; and from thence it was parted, and became into four heads."

Scientists have calculated the probability of the universe being as it is with a viable earth as a statistical possibility of 10^{99}.

In the cosmology offered to us by superstring theory, the expansion is said to have begun with nearly infinite energy within a small volume of space. Another approach arising from quantum theory depicts a flat universe in which material forms are bathed and made of a plenum (more exact than the notion of a vacuum), and in which information travels twenty thousand times faster than the speed of light.[11]

Thus, at the beginning, there was this quiescent plenum (vacuum), this sea of undifferentiated quanta, and now there is still this plenum. This is similar to the field of cosmic consciousness; the emptiness/fullness one can experience as the supreme principle of existence.[12] I believe this is what atheists call *nothing*. In a sense, they are right.

From hitherto, energy rose, similar to a first note produced by a singer. Was there an explosion? Did the intertwined forces at work, because of the restrictions inherent to time and space, battle against each other and eventually cause that explosion? What we know is that the speed at which the galaxies drive apart is accelerating. The Big Bang theory cannot account for this. The gravitational pull of matter should diminish the expansion, but it does not. We now say that 4.6% of the universe is known matter, about 70% is dark energy, and the remainder is dark matter.

An interesting theory by Jose Beltran Jimenez and Antonio L. Maroto[13] shows that the presence of the electromagnetic field on a super–Hubble scale generates a constant that could account for the observed acceleration. Also, coherence of cosmic ratios as well as the fact that matter and antimatter are not in equal proportion is troublesome. Nonetheless, scientists understand that the universe is totally coherent and astonishing in the fine tuning of its constants.

One thing is now certain. William Heisenberg, the German theoretical physicist who made foundational contributions to quantum mechanics and won the Nobel Prize in Physics in 1932 for the creation of quantum mechanics, declared: "The world is built as a mathematical, and not as a material structure."

System science is, therefore, the best tool for truly understanding this universe as it emanates from the highest point of theoretical science. The system used throughout this book echoes this view. One force divides into two polarities, each one existing as a fractal within the other (2 × 2). Any of those two polarities can further subdivide into the same. The world is thus considered to be infinitely divisible into two polarities, represented in TCM (Traditional Chinese Medicine) by the Taijitu symbol (diagram 1). The whole world, including humans, consists of those fractal structures and function. In geometry, a well-known figure called the Sierpinski triangle can be used as an example. This fractal shape looks exactly the same no matter its size. A fractal object shows self-similarity. The object need not exhibit exactly the same structure at all scales, but the same type of structures, as well as functions for organisms, must be potential on all levels.

This is what the implicate order refers to and what we name the LIFE biosystem. Einstein's protégé, the American physicist David Bohm, gave this name—implicate order—to the deeper reality suggested by quantum theory and intuitively grasped by all the great traditions. To him, this is an undivided holistic realm, beyond the notions of space, time, matter, or energy. Like a seed, it holds the folded fractal potentialities of the universe. The hologram, a metaphor to explain this implicate order doubled with its explicate aspect, served to illustrate his concept.

"Princeton physicist David Bohm (1980) became a believer in holographic systems during his study of plasma systems. He found that when a gas became a plasma, the individual electrons began behaving as a unified whole. The electrons became engaged in a process of self-organization. Bohm became disillusioned with quantum theory because it attempted to isolate cause-and-effect relationships from the universe as a whole. He maintained that only a holistic view would explain the electron co-ordination in high energy plasma systems."[14]

Similarly, Genesis states: "So God created mankind in his own image, in the image of God he created them; male and female he created them." In other words, with a few rules informing and modeling a limited set of elements, unfolding in time, we can attain infinite complexity and retain order. Since I am not knowledgeable in algorithms, I took the liberty of comparing the structure of the Tao and its inherent biocybernetic system with the structure of the Five Books of Moses, which are, after all, the basic tenet of Western tradition.

Established correlations already exist between Hindu tradition and Kabbalah,[15] the discipline and school of thought concerned with the esoteric aspects of Rabbinic Judaism. Kabbalah was systematized in the eleventh to thirteenth centuries in Hachmei Provence (Southern France) and Spain. It was reshaped in the sixteenth century in Ottoman Palestine following the expulsion of the Jews from Spain in 1492. Hassidic Judaism, its popularized form, flourished in the eighteenth century.

Are they all fractals of the same system? Do they all share the same structure and function (diagram 2)? It appears so.

As I will demonstrate in this book, the structure of the biocybernetic system links the traditions of the "People of the Book" (Jewish, Christian, Muslim) to the Asian tradition. I therefore concluded that main religions are all aspects of the same one, seen under different phases (diagram 3). Consequently, disagreements between faiths have no link with religion per se. In reality, they are only a consequence of an immature Human brain. (See chapter 5 on the Human brain.)

Regarding the Tao, Carl G. Jung had this to say:

"Taoism formulates, as you know, psychological principles which are universal. These principles are so vast that they can be applied to the entire humanity. What the Occidental man needs is to experiment *facts* which cannot be replaced by words. What are important to me are principally the methods which will allow the Occidental to access the facts upon which the Tao notions are based. Everywhere, it is the one and same truth, but I must say that Taoism is one of the most perfect formulations of truth I was ever given to see."[16]

The Questions: *What, Where, When,* and *How* versus *Why, Which,* and *Who*

"Who am I?" and "Why am I?" are legitimate questions. Unfortunately, they cannot be answered by observing materiality and using linear deduction alone, since we are not solely four-dimensional material beings. Even with a quark, if we consider it simply as an object moving along in one dimension, we ignore its internal structure. We have no tools to measure or analyze that aspect. For us, its inner does not exist. But this is not true. We can measure its implied mass, its life expectancy, its charge and spin.

Rationality is a linear tool influenced by the limited awareness we have of the phenomenological world. It can only analyze what it can separate from the whole; what is submitted to time and space and is in the field of its awareness. It is unaware of the reality of inner ordering. Influenced by past events, it is oblivious to the big picture and to the future. It has the modular function of a particle, not a wave function. Ultimately, these questions of "who" and "why" are not best answered by sitting at a desk at school. Although mathematics can help us with the four dimensions of space and time, it appears a fifth dimension associated to information is missing.

> We normally perceive things as existing in the four dimensions of space-time. Holographic theory, however, presumes that there is at least a fifth dimension that represents a more fundamental aspect of reality. Normally, we do not possess the sensory skills to perceive this dimension, and it remains hidden from our awareness. The holographic model of reality stresses the role of beat frequencies in our construction of reality. Suppose the fifth dimension consists of extremely high frequency energy far outside our range of normal perception. When two or more wave fronts interact, a third frequency is created that consists of the difference in frequencies between the two waves. Since the beat frequency is all we can perceive, we construct reality based on these illusory waves without any awareness of their true source.[17]

Spring

The simple description for this arrangement is that one-dimensional space can be thought of as a line, two dimensions as a flat surface, and three dimensions as a cube that has length, width, and height. To imagine four-dimensional space, we might imagine ourselves moving in time through three-dimensional space, though as stated above, this would not be an exact picture of four-dimensional space which joins "space-time" into one larger whole. Within four-dimensional space, we are at a particular location in physical space at a particular time.

I wondered if there are theories explaining why sensory and motor tracts decussate?

It appears to have been proven mathematically that in organisms, without decussating, a central relay (the brain) cannot sense and reproduce a three-dimensional environment accurately if the networks are composed of more than one thousand or so neurons. This would explain why smaller tracts in humans, such as the olfactory and reticulospinal ones, are not decussated. How about sensing a fourth and a fifth dimension? All dimensions touch the physical one in some way, although they are not all dimensions of the space itself. Creatures are the vector on the physical level. The more evolved the creature, the more subtle the dimensions it can consciously access and express. The first and second dimensions have no interior; they are line and plane. These are accessed by the Reptilian and Mammalian brains, and they are active and reactive. The third dimension, such as a cube or a sphere, has an interior. I thus associate it to the rationale/nutritive, as this one is at the crossroad between outer and inner. The fourth dimension (time) is a succession of points to eternity; and the fifth, to the mature Human brain, is the world of thoughts or information, which has an access to the physical, although is not of the physical realm.

Accessing the fourth dimension, we enter a world in which information and time reign over space, in which function reigns over structure. The Kabbalah appears to mention ten dimensions. For now, humans can hardly be conscious of three, let alone five. Everything in this reality goes through our fractal physical body. If we wish to be aware of more dimensions, we need to refine and strengthen our bodies, thus improving the quality of our genes. For those of us who are already born, it implies better hygiene on all levels. For the world in general, it implies improving the quality of the natural, emotional,

Isis Code

mental, and symbolic environments, and of conception, pregnancy, and the life of young children.

What, *where*, and *how* am I are getting more answers through amazing technological advances. We now know that the whole universe is a gigantic collection of organized photons acting in a real wave and particle dual fashion. The highest energy form of this electromagnetic spectrum is the gamma ray. Since energy can never be destroyed (the first law of thermodynamics), the energy created at the Big Bang still exists and will continue to exist forever. At the Big Bang, energy was converted into subatomic particles. This explains why during fission of an atom, for example, gamma rays are emitted. Therefore we find it as matter ($E = mc^2$), which ultimately is similar to confined light. The basic structure of what we call matter involves charged particles (electromagnetic) bound together in many different ways. We can thus say that bodies, in essence, are made of condensed or slowed down and organized photons, or trapped light.

A long time before we understood that we are mainly organized energy, the intuitive notion that the microcosm must somehow correspond in its structure and function to the macrocosm became the cradle of all the greatest civilizations. It opened the door to vertical hierarchical religions and a path to horizontal linear science with its edification of general laws. We are a product of these religions and sciences. This is the subconscious mental cross we have to bear, collectively as well as individually.

Today, we marvel that Babylonians, some four thousand years ago, knew how to solve a quadratic equation. We remain astonished by the almost industrial milling machines used by Chinese of the fourth century, or their precise clock of the eleventh century. We are bewildered by the Great Pyramid of Giza. We wonder how Euclid, possibly a student at Plato's Academy in Athens, was so right in his mathematical postulates. Most remarkably, his fifth postulate,[18] considered a mistake for so long, was finally proven right in the nineteenth century!

It took all that time to make sense of it.

Evolution in science, as in nature, doesn't follow a strict linear progression, but a spiral, a wave pattern.

Spring

Likewise, physical structures evolved to better apprehend and control tangible reality. Our actual brains and the cities we built vouch for this. But this evolution also drove us away from the intuitive knowledge we had of nature. This happened because for the analytical mind to work, it had to pretend it was *outside* of nature. One cannot observe oneself objectively. We had to nail ourselves to time and space, so to speak. Therefore, connection with the dynamic aspect of reality diminished as awareness of the condensed aspect of reality increased. This is in harmony with the progress of evolution, and it will eventually lead us to become what our ancestors would have called "gods." The connection between the intuitive and the physical will be reinstated and added to our awareness of this world, once we reintegrate into the whole. This too is part of evolution. It takes us from the collective unconscious to the collective, universal consciousness. It will then be the end of this actual world, namely the end of time in this world of separation, and the birth of a new heaven (mind) and a new earth (body). In this, Isis is our silent guide.

A Short Story of Science

The modern story of science goes like this: a very bright physicist or mathematician unveils a truth. For years, sometimes hundreds of years, that truth remains hidden in a corner, unused, attracting little attention other than scorn. Suddenly it will be "discovered," and will become a cornerstone or an indispensable detail in someone else's theory. Thus, there is a constant flux between experimentation and theory. Scientists only go on resolving one level of disagreement to the other. Despite the lofty ideals of some who label themselves as "scientific," by which they mean they are only open to certainties and objectivity, uncertainty is the fabric of science.

The example of the black raven, as put forward by Carl Hempel, shows us why science has such difficulty freeing itself from its perception of matter and from old accepted principles. A member of the Vienna Circle, he proposed that if we were to establish a law that all ravens are black, every time someone found a black raven, that confirmed the theory. Fine. It then becomes so engrained in everyone's mind that if a raven happened not to be black, either nobody would see it because no one could make the connection that this was a raven, or anyone saying they saw a white or brown raven will be ridiculed.

In other words, to find something new requires openness to the unique and to something new. *What we do not believe in, even if it is the truth, is therefore out of reach to us.* It is wiped from the slate of possible awareness. Unsurprisingly, novelty is the realm of the right hemisphere of the brain, which bears feminine characteristics.

From the fifteenth to the twentieth century, we believed that to know the universe we had to disassemble it down to its smallest component until we found life or the soul in it. A little like me as a young child, marveling at a transistor radio and pleading for permission to take it apart in order to find the origin of the sound. This is the realm of the left hemisphere of the brain, which bears masculine characteristics.

Galileo helped with this trend as he showed that the planets could be viewed as inanimate objects moved by mechanical forces. Impressed, other scientists started viewing all nature, even people, as objects subject to the laws of physics, and they began explaining everything mechanistically. This replaced the four-thousand-year-old vitalistic approach. Biology became nothing more than a conglomerate of clever mechanisms. Nature was simply a dissectible object. Sacredness was lost.

This was a necessary step. To objectively analyze a part of our selves requires the ability to "have a look at it." This was a proof that we were entering, on a global scale, a more mature, rational phase.

Sadly, science based itself on what it accepted as law in the physical world and dismissed the knowledge patiently accumulated concerning the inner self. Science just negated its existence. Arguing that the inner is not the sort of thing physics deals with was, in a sense, dealing with it. This was the message received by the general population. Therefore, the outer self-developed while the "irrational" (read nonlinear) inner self, viewed with great suspicion, lay ignored. It became taboo, even laughable, to believe in anything linked to it. The world of the external saw unprecedented evolution, but our internal beings shrank until they were left, largely underdeveloped, in the hands of religions.

With the theory of atoms, a positivist philosophy (logical positivism) of science arose. Positivists argued that science should deal with what is visible and tangible, with what can be measured and observed. It made sense since what was not visible and tangible was for them of the religious domain. Hypothetical reasoning was undesirable. In the science of the nineteenth and early twentieth centuries, uncertainty was unacceptable since science's aim was to understand matter. Thus science would predict everything and, more importantly, control everything, since everything in the eyes of science is material. This explains why science became the favorite of great monetary powers, which then slowly but surely abandoned religion.

General laws had to be generally accepted. "These laws are qualitative descriptions of phenomena that have been observed to occur unfailingly in nature. We do not know why they hold—the laws are simply postulated and have always been confirmed when tested, and so we are confident using them."[19]

Eventually, theory contained elements that physicists were almost certain existed in reality, but which they could not obtain through their experiments. A gap opened between what theoretical sciences said should be and what the experiments could reveal in material, observable reality: science's ability to predict became approximate.

Indeed, the American physical chemist Gilbert Lewis had theorized that light consisted of photons. Nobody ever saw a photon. The encounter between one of these photons—or quanta—with an electron became a quantum event. Max Born, a German-British physicist who won the 1954 Nobel Prize in Physics with Walther Bothe, demonstrated that this encounter cannot yield only one possible outcome. On the contrary, it yielded many. According to Wolfgang Ernst Pauli—an Austrian theoretical physicist and one of the pioneers of quantum physics, who won the Nobel Prize in Physics in 1945—this meant that one needs to choose to measure either the momentum (energy) or the position (space) of a quantum element. Science had to choose one or the other. What Bohr added is that any measurement disturbs, thus influences, the system measured. Heisenberg would prove that measure of one aspect of the system closes the door on what else could be found.

Isis Code

A scientist cannot start from scratch and expect to discover all there is to find in one short life. Despite attracting highly individualistic personalities, science commanded a fraternity between researchers in order for any progress to be made. Each new theory has to manage the susceptibilities and sensibilities of well-established scientists. A scientist's new theory should not throw overboard what the dominant group has previously accepted. (The same applies to religions.) Quantum physics, however, overthrew the rules with the Heisenberg uncertainty principle; that is, that the observer, by the mere fact of observing, changes the observed. It was a paradigm shift. With relativity and quantum physics, a new era for science and religion had begun, closing the gap between matter and energy. Once it surfaces in the general awareness, it will have profound effects on humanity.

As the experimental biophysicist Mae-Wan Ho confirms the following:

> Quantum physics is the culmination of a long series of attempts to fragment reality into the smallest particle; only when physicists got down to the infinitesimal, indivisible quantum, they find that the whole exercise was futile: it cannot be done at all! It turns out that in order to have a consistent representation or theory, it must be supposed that <u>observer and observed are one indivisible system</u>, and that the very act of observation transforms reality from an indefiniteness of multiple superimposed states of being to a state of definiteness, which however, cannot be predicted in advance. Moreover, the same act of observation can simultaneously determine the state of a system which is widely separated from the one observed, as though reality were indeed, **an organic, universal whole**. This has prompted David Bohm and his colleagues to reformulate quantum theory on the basis of universal wholeness: every particle or being is embedded **in a field**, or quantum potential **consisting of the influences from every other being in the universe**. From this perspective, **wholeness and interconnectedness are actual and primary, just**

as fragmentation and separation are illusory.[20]
(Selected text made bold and underlined by me)

This field, as we will later see, is similar to the field represented by the Master of the Heart (diagrams 26 and 27) as mentioned in the Chinese traditional texts and medical biocybernetics, and understood through the LIFE biosystem.

The same can be said of the mental tools we use to comprehend reality. We can use one tool, which is synthesis, and see the big picture; or use the other tool, which is analysis, and see the particular. The particular is not the whole truth. The human brain functions using both views. In western society, since the fourth century BC (the time of the ancient Greek philosopher Democritus), we have favored analysis, endowing it with rationality. Not surprisingly, this is the worldview of the left hemisphere of the brain, the one with more masculine properties. Nevertheless, if we used both views alternately, the observations we made would acquire more depth.

Consider how the two eyes function. The separate images received by each one combine into a single one. The slight angle difference of the images received by the two eyes gives depth to the image. Moreover, people think and learn best in three dimensions. When scanning a text, the brain will absorb approximatively 100 bits per second, while a three-dimensional image will yield 1 billion bits per second. To see life through analysis *and* synthesis gives it dimension. This fact now orients science toward system thinking.

This is the direction I have taken for the last twenty odd years, for I believe it is the only way to have a more truthful and organic understanding of the world around and inside us. Truly, nature uses those two points of view in all processes. It is the essence of homeostasis.

Science wished to control human destiny by knowing all the secrets of the visible. Religion wished to control human destiny by knowing all the secrets of the invisible. Because of the natural evolution of humans, both went through a phase of masculine polarity governance. Both have a desire to understand all so they can predict and control all. Both have created along the way theories that observation cannot entirely corroborate. This is normal, since they stem from the same biological apparatus that has not seen yet the summit of its evolution.

Before, religion and science were one. The observation of the natural world, some archeologists say, brought people to religion. I believe more accurately that it brought them to science. Science has created its separate and incomplete domain. By the same token, science is incomplete. The sacred texts rose as much from the cradle of science as they did from religion. The difference is that knowledge (religion + science) was intuitive, unconscious, sprouting from the phase humans were in, associated to the feminine polarity. Human evolution requires that we add consciousness to this knowledge and this could not be done without first going through a transitional state in which the masculine polarity prevailed. Now, since this phase is almost completed, we need to acquire a wholistic point of view which is possible only through the reintegration of the feminine polarity into our lives.

Both science and religion view causality as an underlying immovable principle. When the causality is only statistical, as in the case of the latest physical theories concerning the world of the infinitesimal, we do not suddenly call this science "religion" because we cannot see what we are talking about. The divergence between science and religion is in the *where* of this causality and the definition, the *what*, of reality. On the theoretical level, they meet.

Modern science determined the cause to be at a lower level of laws. With the advent of theoretical sciences, this is changing. We now understand that there is a theoretical world that embraces many laws and many more phenomena. This is truly a phenomenal revolution; a true paradigm shift. Science and religion need to merge in order to be truthful to their roots. They both refer to a cause, which is unity of the multiplicity. The only real difference lies in the tools they use, the point of observation and reference, and the consequences their interpretations have on humans' physical and psychological lives. Both science and religion bear the same human frailties and misconceptions.

Man created a God and a Science in his image. And still does. With the discovery of America, we could have expressed a new system of thought. Instead, we reproduced the same old patterns. Science and religion need to work hand in hand if we are to get closer to a certain truth. The relatively new world of system sciences elevates the debate and the cause. There, science of the outer and science of the inner can meet and enrich each other. Wolfgang Pauli, who is known for his work on spin theory and exclusion principle (underpinning the structure of matter and chemistry), vividly demonstrated this. His opinions were vital to the scientific discoveries attributed to William Heisenberg and Niels Bohr. He was influenced, as were so many others, by Carl Gustav Jung.

This is what Pauli has to say about the nucleus: "The radioactive nucleus is an excellent symbol for the source of energy of the collective unconscious. It indicates that consciousness [awareness] does not grow out of any activity that is inherent to it; rather, it is constantly being produced by an energy that comes from the depths of the unconscious and thus has been depicted in the forms of rays from time immemorial."[21]

Religion (and when I say religion I refer to the science of the inner since too often religion is a manmade potpourri of theories, politics, history, and emotions) uses the tools of insights and connection to the whole to grasp the invisible. It uses the internal to interpret the external. Linked to human mortal nature, however, it cannot bring us to the truth if we stay with its superficial, time-related words. The human condition of the medium and receiver limits the accuracy of the religious message. Only the substrata, the matrix of these writings, the *system behind the writings*, can enlighten us on human reality. As for belief in God, I share Pauli's views, as illustrated by Werner Heisenberg's recollection of a friendly conversation among participants of the 1927 Solvay Conference. Paul Dirac was protesting against prominent scientists who still believed in God. To which Pauli answered: "Well, I'd say that also our friend Dirac has got a religion and the first commandment of this religion is 'God does not exist and Paul Dirac is his prophet.'"[22]

And also this thought of Carl G. Jung:

> If we speak of belief, it is that we have lost knowledge. Belief and non-belief in God are only substitutes. In his naivety, the primitive *does not believe*, he *knows*, because he gives, with reason, as much value to

> inner experience as to outer experience. He does not have theology and did not let his spirit be obscured by stupidly astute concepts. He orients his life –through necessity- following exterior facts as well as interior ones which, contrary to us, he does not feel and live as separated. He lives in a whole world, and we live, us, in half a world and only believe, or not, in the other half. By what we named "intellectual evolution" we have masked it [this other half]. In other words, we live at the light of electricity which we have fabricated ourselves and—summit of irony—we either believe or not in the sun...[23]

Science, on the other hand, uses the analytical dissection of the outer world, relying on the 4.6% of known matter. It uses the external solely to draw conclusions regarding the internal. Where one of these approaches becomes powerless, the other becomes almighty. They are the two sides of the same coin. Both are necessary to attain knowledge if we ever want to understand the whole of human reality. A productive approach requires true experience, which must include both external and internal experiences.

The chemical-mechanistic theory has provided a better understanding of chemical reactions and mechanistic problems, guiding us toward an extraordinary technological revolution. For example, we have wonderful new tools for surgery and diagnosis. Unfortunately, in this solely analytical view of the smallest components of molecular biology, too many physicians have lost sight of life and of the fact that individuals are unique. Humans have become mere machines in need of fixing. Dignity and sanctity of life nowadays are only romantic notions for dreamers and poets and useful simply to fund research. Research as well is too often an indirect subvention to large, profit-oriented pharmaceutical/chemical companies. In their turn, these companies, unfortunately, seem primarily interested in pleasing their shareholders. As stated on this web posting:

> Despite imminent generic erosion of blockbuster brands, the schizophrenia market will retain its value, supported by the uptake of premium priced depot formulations and new pipeline therapies to offset the loss in sales revenue. Datamonitor forecasts the

> schizophrenia market to grow at a compound annual growth rate of 1.0% (2009–2019). Historic (2009) and future (2010–2019) sales dynamics of schizophrenia brands, generics, and pipeline agents ... several major unmet needs remain, including improved patient compliance, solutions for partial responders or treatment refractory patients ... the opportunity remains for developers to address the lucrative patient population suffering from negative and cognitive impairment.[24]

In this day and age, we need to find a model that will solve the problems that arose from science's way of thinking. Obviously, one cannot solve a problem using the kind of thinking that generated the problem. We must first admit that scientists have always relied heavily on intuition, visions, and hunches as much as on analysis in the design of their theories, even though they seldom admit that.

For example, Niels Bohr, who gave us the atomic model, posed the concept that electrons, negatively charged, gravitated around a positive nucleus in order to maintain harmony in the atom. These electrons, he also proposed, can transit from one orbit to another. William Heisenberg, reporting a conversation he had with Bohr, said that the complex atomic models did not come to Bohr through rigorous classical mechanics but had come to him intuitively, "as pictures." Of course, such intuition doesn't happen to an unprepared and undeserving mind.

Another example is that of August Kekulé, the German chemist, who devised the ring structure of the benzene molecule. He admitted the inspiration came to him while he was half-asleep in his fireside armchair, dreaming about snakes catching their tails. We normally call this armchair science. Elsewhere, in a conversation with Einstein, who disliked the matrix mechanics, Heisenberg protested that the strange developments relating to it had been forced on him since he had to devise a theory avoiding unknowable internal dynamics. Wasn't that the way Einstein had proceeded to get to his special relativity? Heisenberg asked. "Possibly I did use that kind of reasoning," Einstein grumbled. "But it is nonsense the same."

Despite this, Einstein himself famously stated: "Imagination is more important than knowledge." Such examples abound in the scientific literature.

As I stated before, many of the physicists who entered the world of special relativity and quantum physics became familiar with the ideas of psychoanalyst Carl Gustav Jung. Of these, we can name Albert Einstein, Wolfgang Pauli—the architect behind wave-particle complementarity—Pascual Jordan, Niels Bohr, and others.

The problem with modern science dogma is that its models build on the possibility of making general statements. They are thus dependent on testability (reproduction of data) by repetition. This, of course, excludes from the beginning any reference to something unique. Consequently, something crucial is missing in this rational point of view, since *one of nature's most important particularities is the production of organisms that are all unique.*

If we simply look at ourselves, we know that this is true. No two humans have the same fingerprints, or the same voice. No two humans are exactly alike.

In addition, as scientists entered the world of the intangible, of molecular invisibility, they had to rely on seemingly more subjective theories. Laplace's ideal of perfect predictability would never be fulfilled. In 1902, Rutherford had already complicated the debate with his transmutation theory. The supposedly indivisible parts of the elements, the atoms, were not permanent! Pauli's goal became to find a worldview that was true enough to incorporate this qualitative uniqueness. System science was, and is, the only possible answer.

System Science, the Search for a Model behind Energy

While I was working on the book about prenatal education in Paris, I became increasingly conscious of how everything in the manifested world is composed of and sustained by energy. Even if there are many types of energy, we can ultimately visualize them as similar to the same primordial electromagnetic spectrum. We only have to consider the action of the sun on a flower to understand this.

Spring

All is linked because, ultimately, it is of the same nature. This is fascinating.

The sciences of biology, chemistry, and physics would gain—and we would as well—if they worked toward fitting in the same common universal system. They developed in different directions but should ultimately show us a unified world. After all, aren't they dealing with the same universe? We specialized ourselves so much that every scientific field became like a different country with its own language, history, culture, belief system, corporations, heroes, and foes. There is little chance, when one resides outside of these countries, not to be perceived as a stranger and a possible threat. Luckily, globalization, through multidisciplinary contributions and research, is slowly changing this.

In fact, even with the Big Bang theory, we have to accept that everything in the physical world has a dual existence, sometime behaving as a particle, sometime as a wave. Our world is part of a gigantic electromagnetic field.

We should therefore be totally in tune with this energy that makes up our bodies. True, we are made of stardust, but this stardust is energy. Therefore, it is not surprising that geneticists use direct-current pulses to jumpstart mitosis and eventually produce clones. Recently, alternating electrical fields were used in the fight against brain cancer, by interfering with the division of malignant tumor cells. (In the brain, normal cells do not divide, but cancer cells do.) It is also not surprising that light transforms into protein in plants, or that electromagnetic impulse is the "language" of our brains and hearts. Photons are the units of light. They have no mass, but great consequences.

Of course, there are different types of energy—the ones linked to mass and the ones linked to radiations and the ones linked to information. This last type is gaining interest. David Bohm, the great physicist, called it "in-formation," implying a process, seen as a field, which forms the recipient. For us, it is the implicate order carried out by the general Master of the Heart. This field is the latest addition to the list started by Michael Faraday, which now comprises the zero point-field, the universal EM-field (electromagnetic), universal G-field (attraction), and the universal Higgs field (mass). Will some of these fields eventually merge and be understood as one?

This universe is changing. The whole of the universe evolves, not just us, the inhabitants of planet Earth. At its birth, it seems Earth was awash in a sea of cosmic neutrinos. Well, about 10%. These sub-atomic particles zip around and through us at nearly the speed of light.

Microwave light seen by WMAP (Wilkinson Microwave Anisotropy Probe) from when the universe was "only" 379,000 years old, shows that, at that time, atoms made up 12% of the universe, dark matter 63%, photons 15%, and dark energy was negligible. In contrast, estimates now show the current universe consists of 4.6% atoms, 23% dark matter, 72% dark energy, and less than 1% neutrinos. What does it say about the aim of evolution? It seems to point toward a general transformation into dark energy. Evolution's direction is then not toward more mass but more energy. There are tremendous implications to this.

Consulting a lecture given by Michael Faraday, subsequently published in 1849, I was impressed by how he could communicate his love for nature through his observation of different phenomena. He presented himself as a natural philosopher and distinguished between vital powers and elementary physical ones. Why this distinction? Because at the time, the physical body was thought to be separate from its environment, as if in a bubble, and the energies it contained had nothing to do with the energies of the lowly physical world. Man was a special creature of God and had nothing to do with the soulless lower realms. Scientists were specializing in areas that became increasingly disconnected from each other in the terms they used, and even by their interpretations of the physical world.

In this lecture, Faraday marvels at the fact that the powers governing all phenomena in nature are so few. He summarizes these forces as such: all bodies possess an attraction, every one toward another. This he names gravity. He goes on to talk about the chemical attraction of particles, which are in affinity, about the attraction of different type of particles, and about the power of heat (energy) to expand or contract bodies. He summarizes the attractions as gravity, chemical affinity, adhesion, and electricity. He then goes on to talk about magnetism.

Magnets have the attraction of gravitation, the attraction of cohesion, and a certain chemical attraction. It is a separate and dual attraction, which is as continuous as the gravity attraction. He then shows a magnet and its two ends, the one containing a power of attraction and the other a power of repulsion. These powers are throughout the mass of the object but express themselves only at the ends. He showed that the powers can be changed from one into the other. Thus these two powers, repulsion and attraction, are one, used in two different expressions.

He ends by saying:

> What study other than of the laws by which this universe is governed is there more fitted to the mind of man than that of physical sciences? And what is there more capable of giving him [the man] an insight into the actions of those laws, a knowledge of which gives interest to the most trifling phenomenon of nature, and make the observing student find tongues in trees, books in the running brooks, sermons in stones, and good in everything.

He took the last sentence from Shakespeare. Viewed in this manner, this science, from the phenomenon, elevated Man toward the sacred. That knowledge, I deeply love.

My mother used to repeat "L'homme n'est grand qu'à genoux"—"Man is great only on his knees." Popularized by Napoleon Bonaparte, this is a quote from Pythagoras, albeit transformed. Pythagoras meant that Man is at his greatest when he protects and takes care of the weakest and most innocent of society, while Napoleon meant that man's openness to the superior and sacred made man himself superior. In a sense, both are saying the same thing, as one causes the other. As we will later see, this openness to the sacred is one of the key elements that distinguish us from other mammals.

From time immemorial, we have tried to understand ourselves, our purpose, where we come from, where we fit in the scheme of things,

Isis Code

and how to best live these short lives of ours. One of the first things we learn is that we are profoundly social creatures and that it is easier to survive in a community than alone. Many animals have this socializing instinct. Nevertheless, we humans seem to have made human exchanges necessary to our mental health.[25] Men are more affected by social exclusion, and women need positive emotional exchanges with community members.

Those who reflect upon the human condition have used analysis—dividing something into different parts—to convey a sense of understanding, through the study of physical laws, to a greater group of people. This was a step forward compared to the reactive nature of the emotional level. It allowed a better comprehension and a universalization of the concept of what a human is. Humans became not only those living in the same tribal community, but also included strangers.

At this phase of human evolution, however, we have to ask: Can we truly understand the sum through the exclusive study of the visible parts, as we did until now to define humans? If we consider ourselves as inanimate, yes. Unfortunately, it is not that easy. Life imparts a holistic condition to matter, resulting in the sum being more than the compilation of the different parts. Laws now have to be integrated in the system which gave them birth.

Consciousness (Individuality) and Awareness (Personality)

Science is now indirectly showing us that because of the inherent order of this world, intelligence, — or Consciousness—is an unavoidable force in the operation of all phenomena. In a nutshell, energy—or information, order, intelligence, consciousness, whatever name we want to give it—gave rise to life and awareness through evolution. The essence of this Consciousness, as associated with a possible fifth dimension, can be seen as a more fundamental aspect of reality, while the tangible world would be only its reflected aspect. This is in agreement with the latest scientific discoveries concerning the brain and quantum physics. Consciousness would be this fundamental aspect of reality, this reference frequency of a holographic universe.

In agreement with this, the brain appears as a complex frequency analyzer, an oscillatory system merged with the environment. Stanford neuroscientist Karl Pribram observed that when brain-injured patients had large portions of their brains removed, they did not suffer a loss of any specific memories.[26] Karl Lashley (1950) trained rats in maze running and subsequently tried to surgically remove memories related to the mazes, but to no avail. The rats retained their maze-running knowledge. In 1977, Pribram concluded that memories are distributed throughout the whole brain.

Personally, I see memories as informing the cellular matrix throughout the whole body. The brain has elements associated with certain memory-retrieval features, but the whole body memorizes. There is the memory you can voice and analyze (expressed by the brain) and the one that is on the subconscious and cellular level (associated in part to the cellular matrix). Like Bohm, Pribram used the holographic paradigm to explain the functioning of the brain. In particular, the fact that information is distributed throughout the hologram, so that each piece of it contains information about the entire image, showed him how the brain could encode memories. In Isis and Osiris, when Isis finds the different part of Osiris's body, she can recreate the whole likeness of Osiris from every piece. (See the section in chapter 2 entitled, Isis and Osiris, the legend.)

Paul Pietsch of Indiana University (1981) did not agree with Pibram's theory, so he set out to disprove it. Using salamanders, he did everything imaginable to their brains. In a series of 700 operations he sliced, flipped, shuffled, subtracted, and even minced the brains, and still their behavior normalized a short while after the brains were returned to the salamanders' skulls.[27] After all these surgeries, he concluded that the mind perceives and stores information by encoding and decoding complex interference patterns. All objects within perceived reality are thus endowed with a reflection of Consciousness. The difference is in its manifestation and expression. Therefore, all objects are interrelated and interdependent. Matter should be viewed as presenting evolved properties of that one primal substance. Consciousness is an essential property of the universe, being present at the beginning as well as developed later as an epiphenomenon. The study of brain oscillations confirms this as György Buzsaki, a Board of Governors Professor of Neuroscience at Rutgers University pointed out:

> It is the brain's interactions with the body and the physical-social environment that provide stability and meaning to a sub-set of all possible spontaneous states of the brain. The brain gradually acquires its self-awareness by learning to predict the neuronal performance of other brains. In other words, acquisition of self-consciousness requires feedback from other brains.[28]

This has profound implications.The Egyptian tradition divided the human into nine different parts, bearer of nine different functions or attributes of this underlying Consciousness. The scholars who study Egyptian philosophy do not agree over the exact meaning of each part, so I will not attempt it either. What is of interest to us here is that the belief system of ancient Egypt is the closest to the idea that everything is energy, more or less condensed, with the body and its visible matter being only one-ninth of a total human. It is merely the tail of the cat, so to speak.

Egyptian high priests deduced that energy (also known by the Hindu as prana, and chi by the Chinese) comes and inhabits matter in a condensed form. Without a constant feeding of these energies, the matter loses its organization, dissolves, and eventually reintegrates into the environment, while this parcel of Consciousness, similar to a ray of light, has stopped shining on it. This concept triggered the Egyptians' many funeral rituals of embalmment, to hold onto this vehicle in the advent this parcel of Consciousness shone back to earth. For all the embalmed mummies, they certainly noticed that none came back to life, so the knowledge behind embalmment is probably more profound than we are ready to admit and is possibly associated to a concept of reincarnation.

To translate the Egyptian theory into Westerner and Christian language, we would say that matter is condensed spirit and that there is nothing outside this spirit. Everything is a crystallization of it, and is thus sacred. As such, physical bodies are simply structured crystallizations. This is in harmony with the fact that we cannot precisely define what energy is, except as potential of information in space and time. Physics describes it as working potential. We only know how to label its various manifestations. In the same way, we cannot see a photon or electricity or magnetism. We can only see their

Spring

manifestation. One remark must then be made: it is therefore not the gigantic which opens the door to infinity, but the infinitesimal. Infinity and infinitesimal are, after all, of the same Latin origin: infinitus. The fifth dimension does not shine from without but from within.

Scientific and religious men tried to find a cause behind the different phenomenon.

Religion chose God as a cause, which it then proceeded to define and thus automatically limit and reduce. This went against religion's own principles, as God is not an object. He subsequently became one in man's image. I thus call this "god" an idol. Science, which considers the universe a closed system, chose natural laws, which it also proceeded to define. By doing so, science limited man to matter. I thus call "man" a thing. We erred both ways.

Many people are unwittingly using themselves or their situations as an ultimate cause. Some say, "It is because I acted this way that this happened and I was punished [or rewarded]." Or, "It is so and so's fault that this happened." This gives them an explanation which becomes controllable. They then tell themselves, "I monitor my behavior, and in this way I control my destiny. If I act well, nothing bad will happen to me." Whatever problem we have or whatever we want, we tell ourselves that God or science will handle it. This is like a child who still does not know right from wrong and need his parents to show him the way. With his developing frontal lobe, the child often tries to manipulate his parents into doing what he wants.

Since the strict control of behavior does not necessarily equal peace and happiness, the reward for good behavior became a place and was displaced to the future. It had to be better than dreams, so that all the pain endured in this life would be worth it. This is the necessary "paradise." You suffer: you get rewarded after death. You don't suffer? Then you can hardly be a saint.

Reincarnation uses a similar stratagem, displacing in time not only reward but also punishment. Hell is here and paradise is the future; or it is the reverse, depending on how we act or acted. In any case, too often we consider that the world is ours to use and abuse and God will help us. We created this god, inevitably, in the image of

Isis Code

human imperfections. This is an egocentric, immature point of view, leading to an egocentric, immature religion.

On the scientific side, many scientists still use the material world as the ultimate cause. What if the imaginary boundary we have put on the universe to allow computations was, in fact, imaginary? What if the ten dimensions of the superstring theories influenced one another? Some believe that chemistry, hormones, the brain are the cause instead of the medium of feelings, thoughts, inspiration. They see thoughts as emerging from the body. Still, science now admits that "the neural mechanisms of generating novel constructs remain an enigma."[29] In the same vain, they add: "Exactly how knowledge is preserved in the brain despite the constant changes is a fascinatin and profound question, one that neuroscience has not even begun to tackle directly."[30] Of course we express thoughts using a physical body, but they are not sourced uniquely in it. Otherwise we should believe a radio is the source of the sound it emits.

Some also believe that we had discovered everything and nothing is left to find. They naively consider they control everything, since they have in their hands a few formulas that they trust summarize the universe.

This leads to a world subjugated to matter and upon which we have no real influence. New phenomena will require new actions, new funding. This forces us to live in a somewhat chaotic mental world, since many phenomena have not yet been explained. They are piling up in a corner of the human mental egregore. Some believe thoughts and feelings are only objects, figments of the imagination that emerge from physical conditions. These people live in fear of the unknown and wish to feel the firm ground under their feet. Whatever problems arise, they believe science will solve it. The world is ours to use and abuse, and science will help us. This science, created by us, is inevitably in the image of human imperfections. This is also an egocentric and immature point of view leading to an egocentric science.

Both science and religion have codes of belief. Both proceed from different aspects of the brain to appraise life. Science, unknowingly, uses tools from the emotional level and from the logical level with an emphasis on the logical. Religion as we know it uses tools from the emotional level and tools from the logical level with an emphasis on the emotional. Depending on people, on their maturity, motivation and experiences, some will lean more on one side than the other, and never be complete. However, a new trend surfaces. In the same manner electricity becomes magnetism, science is becoming metaphysical. In Pauli's words: "to find a new language that could make the hidden dimension in nature accessible to the intellect ... neutral with respect to the distinction between psyche and matter."[31]

Attracted by the Kabbalah, I observed how closely related it is to the Egyptian tradition. Is it really surprising when we know Moses grew up surrounded by the Egyptian and then the Babylonian cultures? The universal laws expressed in the Kabbalah, of course, predate Moses and even the Kabbalah itself. It seems that in most traditions, including the scientific one, we can find a similar universal skeleton on which all is built.

Religion and science have the same root, the same DNA. Therefore, we can find it. This skeleton, which the Egyptian tradition named Ma'at, is very close to the Da'at of the Kabbalah. Da'at is the expressed aspect (Thot in the Egyptian tradition, the masculine aspect of Ma'at), the conscious knowledge of Kether. It is the sefirot (house) of unification. Ma'at in the Egyptian tradition is the law, the truth, "an extended grid against which all could be measured and balanced."[32] In the LIFE system, I name it the Master of the Heart. In the Greek tradition, Pherecydes of Syros in his writings represents it as a meshed veil sprinkled with the elements of creation Zeus presented to, and wrapped around, Chtonie (the physical world)[33]. It sets the order of the universe from chaos, when creation or manifestation occurs, and is the personification of the fundamental order of the universe. It is the sieve through which the unmanifested receives organization. It is similar to the dharma and the Tao, a vast field of consciousness and potentialities that constitute the primary reality of the universe. It is unified and nonlocal. It is the implicate order.

This system expresses itself first in the vacuum, which, as science has demonstrated, is a plenum from which the quanta will be retrieved. In the Hindu cosmogony, this plenum is similar to Aditi, the highest Akasha, womb of space, synthesis of all things, and divine wisdom. In Kabbalah, it is Binah, the higher mother and cognition.

(Hearing is associated with Binah.) In the Egyptian tradition, it is Tefnut.

Now the divisions I will put forward could pose problems for those who have no experience with and therefore have no belief in a subtler world. Some people who deem themselves scientific cringe (which is an emotional reaction) at the idea of something remotely linked to "irrationality." Wholeness is thus dismissed. I will do my utmost to use in part what our materialistic science has discovered. This is the reason why I refer to the brain, vector, and fractal element of this implicate order. The terms become limiting when we use divisions in the world of imponderable matter. Bear with me and take it as a cultural discovery of ancient traditions.

I will attempt to use accepted notions as much as possible to divide the human being into parts. I will use the popular term of energy in the sense of "a potential for work or information." Instead of speaking about dissimilar bodies, I will refer to different functions and structures in a closed system—which scientists say the universe is—and in the open system—which a human is. Let's keep in mind that I propose these ideas as a frame, as done in any science. They are schematic, not exhaustive. This book is written with the desire to share some findings and hopefully orient us toward more discoveries and a better understanding of ourselves.

As the French chemist Antoine Lavoisier demonstrated in the late 1700s, nothing is lost, nothing is created, all is transformed. A long time before Lavoisier, Socrates said the same in other words: everything comes out of everything. Thus, every action eventually becomes a cause, with repercussions on the whole. Scientists therefore consider the universe a closed system, although we now know that new space is regularly created. The remaining question is that since this "closed" universe is made of energy and this energy is, at the source, electromagnetic, where in this universe do we place thoughts and feelings, which obviously have the potential to create, through the appropriate medium, a reaction as strong as would any physical agent?

One intriguing aspect of the superstring theory is the concept of extra dimensionality. Extra dimensions consist of dimensions of both space and time. As such, I consider them part of the "material" world. Could there be other dimensions not affected by time and space? Potential dimensions? According to the latest in M-theory, an extension of string theory, the higher dimensions are not affected by time. Time is a dimension in itself, so it only applies in the lower dimensions. Thoughts could therefore be of a "higher" dimension, not affected by time or space.

Newton's third law states that forces come in pairs of equal magnitude and opposite direction. Said differently, for every action, there is an equal and opposite reaction. We also know that nothing affects the speed of light in a vacuum. This light is an electromagnetic radiation ranging in wavelength from the shortest—the gamma rays—to the longest—the radio waves, which have the lowest energy photons. The earth bathes in these energies.

The Maxwell-Boltzmann distribution of molecular speed for molecules of different mass taken at different temperature shows us that lighter molecules have higher average speeds than heavier ones. They contain a higher energetic level. In comparison, gamma rays have a higher energetic content than visible light. Translated by analogy to a human being, the subtler aspects will consist of higher energies, endowed of a natural propensity toward denser ones. The subtler aspects consist of higher energetic properties with stronger informative potential and stronger effects although these will take longer to manifest in the denser world. As with a gas condensing to become a liquid, thoughts might therefore eventually "condense" into actions. An egregore is a mass of these thoughts or feelings, similar to Jung's collective unconscious. To offer an example: at the death of Princess Diana, the more people were made aware of other people's feelings, the more the event seemed to take a life of its own. Emotions have a high energy content and are somewhat closer to the physical world than ideas. Hence, they motivate us toward action more rapidly than ideas can.

We also know that the molecules of a solid are not free to rotate or to move through space, but they can oscillate, *storing energy in that manner.*[34] We may thus divide this world of ours into two broad categories of matter: one that transfers its energy and one that receives it.

Isis Code

These become charged elements, opposed in time and complementary in space (since the receiver can eventually become the giver). Indeed, Charles-August de Coulomb, in the late eighteenth century, observed that charge manifests itself in two opposing forms. This discovery led to the well-known axiom that like-charged objects repel and opposite-charged objects attract.

Of course, the aspect or element that gives its energy can eventually become a receiver of energy. To simplify my future discourse, I will therefore use a "+" sign for the energy, matter, or structure that *gives* energy and a "–" sign for the energy, matter, or structure that *receives* energy. These are always, we should remember, *in comparison* of one aspect to the other in time, but *not* an absolute. As an example, 5°C is cold if the temperature is normally 25°C but feels warm to us if the temperature is normally -15°C. Since energy and mass are ultimately manifestations of the same thing, as Einstein said, then everything is in comparison on a scale from the most energetic (gamma rays, which are the most +) to the heaviest mass (black holes, which are the most –). This is also the realm of the physical world (known matter), which does not mean that one day we will not find something more energetic than gamma rays. *All scientific theories are built upon 4.6% of matter. This determines our present point of view of the world, our conceived and accepted ideas. Those ideas are not an absolute.*

To look at the example of temperature again. The coldest temperature possible as established by science is -273.15°C or absolute zero. At that temperature energy has been totally extracted from the atom. We do not know the warmest temperature, which must have been during the Big Bang. Contrary to dense matter, temperature, as we remember, does not affect light in a vacuum. Light is above matter, energetically speaking.

I will have to use correspondences for lack of a nonsubjective science of the invisible. To quote Peter Atkins: "All our actions, from digestion to artistic creation, are at heart captured by the essence of the operation of a steam engine."

This is a system. In the following chapters I will describe this basic system linked to Ma'at. It appears to me as *the highest science since it takes into account the amazing uniqueness of nature.* This is

Spring

what Pauli, Jung, Bohm and many others were searching for. Pauli, in his correspondence to Carl Gustav Jung had in mind to find "a description of nature integrating both physis[35] and psyche."[36] Nature is self-correcting, self-regulating, and self-modifying. Thus, we can change as the environment changes. We humans have the greatest effect and control on this environment. We can therefore control our own evolution, for better or for worse. It is a serious responsibility, and implies that the cautionary principle should be used in all actions directed toward the environment. In a sense, God has nothing to do with it. He already did His work. We now live the unfolding. Roles are predetermined. Who fulfills them and how well is not. The system is unchanging in its structures and functions. The objects of its application, however, can evolve.

In this system, we divide manifestation into five horizontal levels with imaginary boundaries: physical, emotional, rational/logical, universal, and idealistic/symbolic. They are not distinct, but rather flow from one to the other, as would the different colors of a rainbow. To continue with the rainbow analogy, purple would be analogous to the symbolic aspect, and red, more materialized, to the physical one. If we join two rainbows together to complete a circle, we can observe that the purple is central and the red peripheral (diagram 4). The symbolic aspect—some would say spiritual—is consequently the center of our being, the Self. We cannot express this Self in the material, polarized world without a physical body that contains the peripheral (red) physical mass. This most condensed aspect of ourselves, and thus the most receptive to higher levels of energy, is the medium we use in this world. It is not the origin but the consequence. It is the grail of our symbolic level.

If the body is a crystallization of those subtler aspects, then its features should be linked to, and the expression of, these subtler aspects. Many ancient traditions developed their medicine around this concept. My own experience is with the theory of the law of the five elements pertaining to TCM. The system I use in this book, called the LIFE biosystem (Law Inherent to the Five Elements), derives from TCM and other traditions (Sumerian, Jewish, Egyptian, Hindu, medieval—both Christian and Muslim), with added elements from my continued work with Dr. DeBavelaere in medical biocybernetics, as well as results from the most recent brain research and my own observations.

Isis Code

This understanding of the five elements seems to have been with humanity since its beginning. Even if we take only the Western branch, the pre-Socratic Greek elements, we can follow these throughout the Middles Ages and the Renaissance, as they persist and deeply influence European thought and culture.

This is, in fact, a holistic knowledge system by which we live, similar to that of our ancestors when science and religion were still united. It is a science and a religion combined, as they should be; a way of life. The research and reflexion pertaining to it provide a better understanding of our personal and individual natures on different levels. I did not create a melting pot, but sought the skeleton that every tradition dresses up, depending on time and political and social developments. This is the sieve I was referring to previously (diagram 5). This skeleton projects in nature itself. Of course, this system does not belong to a country, a race, an association, a school, or a person. It simply is, in the same way the elements in nature are, in the same way physical laws exist. It is universal. It can be expanded on to the infinite, which is what many cultures did, but the essential aspects remain the same. As I explained in the introduction, in this book I only express the DNA—or should I say the RNA?—of this system as applied to the human.

Someone said that all the great religious traditions have many points in common not because they are right or refer to the same God, but because they emerged from the same cells within the brain. It is an astute observation and has some obvious truth in it. (Similarly, radios all emit sounds when currents goes through them.) It could similarly apply to science to explain both its findings and limitations. Nevertheless, I would not dismiss greatness on the account that it has a limited expressor; i.e., humans. Quite the opposite. We all express ourselves through the same natural laws, the same type of brain, the same biosystem, so all our creations more or less have these same imprints and limitations attached to them. This also allows us to appreciate (or not) other people's creations. They "talk" to us. The wonderful thing about this is that genius can be understood or at least felt by all of us. It can awaken dormant knowledge and help accelerate our evolution.

If two people are in affinity with the same model, they are also in affinity with each other. This is what religion tries to achieve. This also explains the necessity for a society to look toward a same model to succeed in strenghtening its social fabric. It is the zeroth law of thermodynamics[37] applied to our psychology.

We can observe that we have developed elements that do not make sense in a uniquely emerging and Newtonian-type evolution. Take this notion of sacrificing for something that does not positively affect our lives, genetic pool, or the lives of our close ones. Sacrificing for an idea or the future, for example. This becomes more apparent with the development of the prefrontal lobe, although we can find the structure of sacrifice in lower orders of the animal world, such as in the world of insects. The bee will sacrifice itself for its community, and the ant has a second digestive pouch for the benefit of hungry nest mates. In the same way, a vampire bat that had a successful night of feeding will regurgitate blood for its less fortunate companions. In the animal world, we find numerous stunning displays of sacrifice to protect and provide the best care for offspring. Some risk their lives without a moment's hesitation, as if they were protecting themselves. Where does sacrifice fit in? As the acorn of the oak tree contains the blueprint of the mature oak tree, for every kingdom potentialities are all in the gene (genetic) or in the Master of the Heart (epigenetic). These potentialities include, for every kingdom, the feminine polarity.

We appear to be more than a blob of cells evolving because of wants/needs/genes and confrontations with a changing environment, although we are also that. Are we only shaped by opportunity, or is there a direction to evolution? In accordance with the evolution of the universe, it seems that the most evolved is also the most energetically efficient, needs little but the most subtle fuels. The fact science found that those humans who eat less live longer points in this direction. It appears that, unlike dinosaurs, the most evolved can process adequately different types of energies or information as well as vast quantities of them. Nature also seems to favor multiplicity and uniqueness. We can see this through the example of the hummingbird. Hummingbirds are the second most diverse bird family on earth, they can fly backward and are the only birds to do so, they are brightly colored, their wings are hollow and fragile, and it is a mystery how they survive. These little birds can fly nonstop 800 km (500 miles)!

Isis Code

Does every organism have, imprinted in its cells, the ultimate collective and optimal blueprint of its species? I believe so. We know that only the mutations that are best adapted have remained. The question is, best adapted for what? The LIFE biosystem points to the answer.

Nature is both periodic and self regulated. Every living being expresses itself through an autoregulated system, a mirror to the implicate order. This system we qualify as biocybernetic, cybernetics being the "science of information and its regulation among machines and living beings"[38] and bio meaning life. LIFE biosystem sprouted from medical biocybernetics as taught by Dr. DeBavelaere. This elementary and general biocybernetic system is applicable and common to all living beings. This is observable even at the molecular level. It can be applied as well to evaluate structures, organizations, and products created by humans. This book is not a manual of MBC, but I use some elements of it in the LIFE biosystem approach. The central point is that it allows us to understand that behind every manifestation of life, there is a subtle structure that follows the general plan of the implicate order. This is a nonlocal, invisible structure. Thanks to this unseen matrix, as organisms become more complex, they are more apt to process larger amounts of information/energy. The human brain is the most complex, with emerging information processing abilities never before seen in life realms on earth. I believe that the goal of evolution is general Consciousness, unfolding in a dimension of time and space.

David Bohm agrees, stipulating that the state of a quantum is not random. It is channeled by an underlying system. He named this system Q. Q emerges from an unobservable domain, guiding the observed behavior of particles. Bohm calls this the implicate order and he names the manifested world the explicate order.[39] The biocybernetic system is the implicate order as expressed through Chinese philosophy. We have here a perfect example of undivided science and religion.

Coherence and Oscillations

Perennial religions follow an implicate order that they tend to localize "up there." Studying them, I could see they share the same

Spring

system. This is an implicate order as named by Bohm, similar to the one described in the Tao. An implicate order is nonlocal. It simply is. Religion is the exoteric aspect (explicate order, exoterism, the personality aspect) of a knowledge base (implicate order, esoterism, the indivisble aspect). Unfortunately, without the esoteric aspect, religion is a somewhat empty shell, devoid of life and susceptible to the viruses of imperfect physical, emotional, mental, and social realities.

This implicate order, unhindered by our physical, emotional, mental, and social imperfections, gives us coherence. That is, an optimal expression of all these levels of human expression: health. In heart research, this physiological coherence can be seen through a wavelike pattern in the rhythms belonging to the heart, as well as an increased parasympathetic activity, a better heart-brain synchronization, and harmony within the body's systems. Unfortunately, since a human is an open system to the environment as well as to collective and personal subconscious information, this state of physiological coherence is rarely maintained. We must always strive to attain it. (In religion they call this the right path.) As an image of God, we carry this potential of undivided yin and yang, of feminine and masculine. We are a reflection, a hologram, if you wish, of this implicate order.[40]

In physics, *coherence* is used to describe two or more waves synchronized to manifest a waveform. The laser is such a construct, in which multiple light waves phase-lock to emit a unified, coherent energy wave. In physiology, coherence describes a state in which two or more of the body's oscillatory systems, such as respiration, heart rhythm patterns, and brain cortical oscillations, become synchronous and operate at the same frequency. We call this entrainment.[41] The natural advantage of this is that this dependence of systems develops synchrony and a powerful team effort to realize a common goal. Think of a hockey team.

With the LIFE biosystem, we have five such waves working together, consistent with a nonlinear model of the control system[42]: the physical, emotional, mental, collective (environmental) and symbolic (archetypal) oscillations.

> A number of psychological phenomena argue in favor of the idea that these cognitive events require hierarchical processing....Each oscillatory cycle is a temporal processing window, signaling the beginning and termination of the encoded or transferred messages, analogous to the beginning and end signals of the genetic code. In other words, the brain does not operate continuously but discontiguously, using temporal packages or quanta.[43]

With Einstein's discoveries, it becomes essential—and a prerequisite to further evolution—to consider matter as condensed energy or energy as matter returned to its original form depending on our seat of observation.

As with all natural phenomena, we can also observe that living matter submits to cycles born from movements of energy from their manifestation to their resorption. Looking at the seasons, modeled on this implicate order, we observe in winter that the energy seems condensed to a maximum. Spring comes and the energy rises, expanding and manifesting itself. Then it spreads to the surface of the earth, revealing itself through vegetation. Summer is here. Then, it seems, time is suspended as the two momentums cohabitate for a brief moment, until the energy drains toward the inside of the earth. This is what TCM teaches us. Our own breathing cycle seems modeled on these movements, as is the wave of the electromagnetic spectrum. This cyclical movement of energies is inherent to all living processes. It is the oscillation of life. The brain is central to receive these oscillations. It is thus essential to a true understanding of life and health. Unfortunately, research dismisses this element since laboratory analysis, until now, allowed only the study of phenomena outside of these cycles. It studies space (matter) outside of its frame of time. It nails matter in order to observe it "objectively," thus block its oscillations on the different levels. Things are changing, though, as interdisciplinary and intercultural studies become more prevalent. Oscillations are now studied in brain science.[44] We need to reintegrate the time frame in our observations (diagram 6).

In system science, we observe that in a basic system of three elements, in which A triggers B, B triggers C, and C triggers A, A becomes both cause and effect. Cybernetician Heinz von Foerster

Spring

called this circular causality. It is also known as a recursive circuit. The steam ball regulator illustrates this.[45]

As we look back at evolution, or down through a microscope at minuscule organisms, we can see that life on earth is allowed through energies manifesting the information they carry in certain structures, thus allowing particular functions.[46] In this we can see that reproduction, defense, and nutrition are basic functions shared by all living organisms. If we develop a recursive circuit with A being reproduction, B being defense, and C being nutrition, this is not complete enough to allow autoregulation, because living systems are open on many levels. If it was a closed circuit, it could work. A living system reproduces, fights, and eats. These three elements compose the basis of humans' expressive (+) action/structure. The eat/analyze aspect belongs to the intermediate.

Nonetheless, no organism can survive without a minimum of exchanges with the environment and without being able to process a minimum of information, both internally and externally. These two elements, exchange and process of information, can be added to the basic mechanical circular causality, giving us a picture of an autoregulated living system of five phases or elements. These two last elements, or functions, added to a part of the intermediate element (the "eat" aspect) compose in humans the basis of their receptive (-) nature, that is, perception (diagram 5).

The receptive nature is judged feminine, and the expressive one is deemed masculine. I will thus speak of masculine and feminine polarities.

As day gives way to night, in every living process masculine (+) protects and then melts into feminine, and feminine (-) feeds and then merges into the masculine (diagram 6).This reminds us of the electromagnetic wave. The feminine polarity is receptive comparatively to the expressive masculine one and vice versa. We could compare it to the process of breathing. To inhale requires the use of a different function than to exhale. Together they make a cycle. If we add the electromagnetic wave with its phases of expansion-acceleration, retraction-deceleration to the main elements of this elementary structure, we find a general biocybernetic system. We can see here the oscillatory property of living matter. Many, if not all, essential cellular and metabolic processes are compartmentalized in time. A fundamental theoretical conclusion concerning oscillations is that it requires a negative feedback loop (an odd number) with at least three elements (see recursive circuit). These elements have to be connected by mutual inhibition to obtain the required oscillation.

This is what the biosystem expresses. The nutrition "house" is the central value of the oscillation, and we have five elements (diagram 6).

These oscillations occur in nature at all time scales. To name but a few: cardiac rhythm, neuronal oscillations, hormonal oscillations (cortisol is highest in morning, melatonin is highest at night), circadian rhythms (metabolites oscillate), predator/prey population cycles, communication in animal and cell populations (fireflies can synchronize their flashing, bacteria can synchronize in a population, girls' menses can become synchronized), blood pressure (lowest just after midnight), breathing alternating from one nostril to the other (1 to 5 hour cycles), cognitive performance (best in midafternoon), and so on.

As demonstrated in 1998 by Gu and Spitzer,[47] oscillation frequency controls the activation of distinct sets of transcription factors and the expression of different genes. In this way, oscillations direct cells along specific developmental pathways. This is important to remember. The organs that have a function of oscillation, such as heart and lungs, are therefore primordial to evolution's direction. *Their oscillations are strongly influenced by our psyche.* Frequency-dependent gene expression is certainly widespread.

In neurons, oscillations in cells can be blocked for days on end; when unblocked, firing will resume to the rhythm previously followed. The circadian-rhythm firing is an inherent property of individual neurons. Coupling between many biological- or chemical-phase oscillations is an essential dynamic feature of observed processes, and represents organisms' adaptive response to a fundamentally cyclical, oscillatory environment. We also can observe entrainment of frequency between these, which is different from a simple pattern due to stimulus response. These can be phrased in mathematical language through ordinary differential equations, as seen in the Kuramoto model.

What does quantum physics has to say about biocybernetic?

> Instead of consciousness collapsing a quantum superposition in a succession of quantum jumps, we have consciousness offering a quantum plenum of superposed possibilities to the match with the more restricted possibilities of sensory input. Instead of a salutatory world line in the Heisenberg succession of objective tendencies and actual events, there is a continuous unfolding of worlds from a holoworld. In the U/Y [Umezawa, 1993] formulation of quantum field theory and Yasue's extension of quantum field theory to quantum neurophysics model, there is no consciousness with a random core. Instead *consciousness is cybernetic*.[48] (My emphasis)

The cell being a microcosmic template of the macrocosm, we should find structures and functions manifesting this implicate order in the cell. We therefore have

1. the nucleus and DNA supporting the genetic energy, a.k.a. reproduction;

2. the mitochondrias and lysosomes as the defensive energy;

3. the RNA, which reads the DNA in order to synthesise proteins (enzymes) and allows the expression or repression of different genes acting as an information regulator (epigenetics);

4. the endoplasmic reticulum and cytoplasm supporting the nutritive energy and general regulator of the system, a.k.a. nutrition; and

5. the membrane filtering the exchanges and environmental energies.

It is not mere coincidence that science has accepted DNA as the cause of many expressions in the living organism. It is more stable than RNA, which can explain why it was ultimately chosen by evolution as a grail. At the DNA level, and in the biocybernetic system as well, energies are condensed to a maximum. As we have learned, nevertheless, energies don't initiate from genes. A corpse still has all its genes but most probably no more oscillations. We now know RNA might have preexisted DNA, as the biocybernetic model also implies (diagram 5). We find RNA elements in DNA. This observation leads us to consider epigenetics to be as influential as genetics, if not more so. RNA appears to be a phase prior to DNA. I believe this is because of the Master of the Heart, this field of the implicate order that bathes and gives rise to every structure.

TCM (Traditional Chinese Medicine), similar to the Egyptian tradition, has qualified the environmental phase (5 in diagram 5) as the *mother of energies*. Both traditions therefore use evening as the *beginning* of a new day. This is also the case for the Jewish tradition and many others. It also can be observed that some genes are expressed only if certain environmental conditions arise. It appears that in evolution, RNA came first, being able to replicate. Then perhaps to secure the information, DNA took over some functions of RNA, but kept mutation as a tool for diversity, (although I believe epigenetic has a better chance of allowing positive diversification). DNA is like the library where RNA goes to pick and choose information. Interestingly, in our biocybernetic model RNA is connected to information regulation, whereas DNA is associated with reproduction. As diagram 5 shows us, information regulation (RNA) and exchanges with the environment work as a pair and pertain to the receptive, thus feminine, nature of our biocybernetic system.

Researchers from Purdue University and the University of Texas at Austin have used the crystal structure of the molecule of a primitive fungus to understand how life evolved from the simple to the complex. This shows how RNA evolved. "It's thought that RNA, or a molecule like it, may have been among the first molecules of life, both carrying genetic code that can be transmitted from generation to generation and folding into structures so these molecules could work inside cells," says Purdue structural biologist Barbara Golden. "At some point, RNA evolved and became capable of making proteins. Then, proteins started taking over roles that RNA played previously—acting as catalysts and building structures in cells."[49]

This is in harmony with our observation of the LIFE biosystem, and with the idea that matter is condensed energy. How interesting that the RNA in our system shares a phase with the heart aspect, will be discussed later. The RNA importance is also supported by a recent discovery. A study published in the July 2009 issue of the journal *Human Mutation* shows the results of research by scientists at McGill University in Montreal. Morris Schweitzer of McGill's medicine department, who led the study, confirmed that there are major genetic differences between blood and tissue cells, shredding the long-standing assumption that blood DNA mirrors tissue DNA. Genetic research based only on blood samples might be flawed. In our LIFE biosystem, this discovery makes sense since RNA is in the regulation of information domain, as is the heart (blood). The blood shouldn't necessarily have the same DNA. Nevertheless, blood DNA points to the future of tissue DNA.

Another example is the molecule that allows plants to grow flowers at the right time and the right place. It could be similar to the RNA messenger Flowering Locus T. Virtually all eukaryotes (organisms with nuclei) have innate circadian rhythms. These are usually in harmony with the species and the natural environment.[50] In plants RNA messenger seems to be in charge of this photoperiodism, this inner clock, this oscillator.

> Intense activity in the field of circadian rhythms has led in recent years to a basic understanding of how an endogenous clock is generated. Oscillating products of the period (per) and timeless (tim) genes, which feedback to regulate their own synthesis, and transcription factors, which activate these genes, combine to generate a molecular loop that apparently drives behavioral and physiological rhythms. The best-characterized component of this system is the "per gene," with considerable effort directed towards identifying the mechanisms that regulate cyclic expression of RNA and protein. Since the cycling of PER protein is controlled largely by post-transcriptional mechanisms, the relative importance of RNA versus protein cycling has been addressed in several studies. However, it now is clear that regulation of per cannot be dissociated from that of tim, since they are co-dependent components. The overt behavioural phenotype likely depends upon the effect that any perturbation has on both components, rather than on either alone. Major features of the

feedback loop appear to be conserved, from fruit flies to mammals. One difference between the two systems is the manner in which the "molecular clock" responds to light. In flies, levels of TIM protein are reduced in response to light, while in mammals, per RNA is induced. The pathway that conducts light to the clock is poorly understood but there is increasing evidence in support of a dedicated pathway for circadian photoreception, as opposed to the sole use of the visual transduction system.[51]

We have shown elsewhere[52] that the pineal gland, which is part of the Master of the Heart system (a dedicated pathway for circadian as well as subtle energies reception) and feminine polarity, receives input even in blindness situation (A. L. Rogers, M. Menaker, and R. Van Horn). This will be discussed in chapter 5.

Genes are a manifestation, a coding, not a cause. In the same way, written letters are not the ideas themselves, but only a support of the ideas we wish to express. The same idea can be expressed in many languages. The written letter is therefore the receptive aspect, a symbol or matrix, in harmony with the idea. The idea is the expressive, informative agent. The idea precedes the letter. To assume the letter is both the expressive agent and the source is wrong. The letters are the DNA. The reader, or the RNA, will choose to read the text or not.

We now know that the genetic code is dynamic, not static. Twins do not share the same genetic code, which points to the importance of epigenetics. We inherit our genetic code from our parents. We also inherit the epigenetic aspect expressed by the environment we are born into and expressed by our mothers' LIFE biosystem. At all levels we are influenced by it, and the different constitutions as described by homeopathy are the signature of this interaction between genetic and epigenetic.

After the publication of the book on prenatal education, I returned to Quebec to promote the book. When I returned to Paris, the naturopath and writer Pierre C. Renard, to whom I was assistant, asked me to attend a lecture by Dr. DeBavelaere. Renard had met him and was intrigued by his ideas concerning the animus and anima. He had

Spring

a feeling important discoveries might be revealed during the lecture. After the lecture, Dr. DeBavelaere invited me to join his group of friends at an upscale Parisian brasserie called Terminus Nord. In this boisterous atmosphere of frenetic servers clad in black vests and white aprons, surrounded by the late crowd and its happy laughter, I asked him how he'd reached his observation of the feminine polarity. I understood that his interpretation of TCM was unique in that sense, differing from the teaching of Chinese scholars and his teacher, Dr. Jean Claude Darras. He nevertheless respected and admired the latter for the valuable work he did. Indeed, during the 1980s, Dr. Darras and Pierre De Vernejoul repeated Dr. Hans's experiment using radioactive tracers on human beings. They proved again the existence of meridians, channels in which vital energy travels.[53]

In the experiment, they injected and then twirled radioactive technetium into the acupuncture points of various vounteers. Using nuclear scanning equipment, they followed its flow. They also injected nonacupuncture points. In the latter, the radioactive tracer diffused outward from the injection site in circular patterns. When the acupuncture points were injected, the radioactive technetium followed the exact meridian pathways. This revealed a trajectory neither linked to blood vessels nor to the nervous or lymphatic systems. How humans managed to describe these same channels intuitively and build a science around it is truly mindboggling, and it goes to show that we have abilities we do not develop because we don't believe or recognize them.

They also found that when acupuncture needles were inserted into distant points along the same tracer-labeled meridians, the rate of flow of the technetium through the meridians changed. This observation supported the claim that stimulation of certain points with metallic acupuncture needles affected the flow of vital energy (chi) through the body's meridians.

Dr. Darras was also one of the first to have the source Chinese text translated. While Dr. DeBavelaere was studying TCM, a particular sentence that was regularly repeated throughout the source text intrigued him. He asked Dr. Darras about it, but the text was dismissed as filler of no medical value. This was an unsatisfactory answer. The text that interested him was: "The young Yellow Emperor was receiving a teaching concerning Tao. When the text was crucial,

Isis Code

he would say: 'This is of the utmost importance, I will go and purify myself.'"

While reading biographical notes on the Muslim philosopher-scientist Avicenna, I found an echo to this behavior. While he was studying philosophy, whenever Avicenna found notions too abstract or alien to him, he would put down his books and perform ablutions and prayers. Understanding eventually would come to him, and he could then continue his studies. Even in his dreams he would pursue the problems he was wrestling with, and their solutions inevitably would come to him.

This notion that purification and dreams are useful, if not necessary, tools for the proper understanding of some abstract notions went against the mechanistic vision imposed on medical students. It intrigued Dr. DeBavelaere and eventually led him to another approach to TCM and directly to the understanding of the feminine polarity role. He related to me that TCM is very different from how knowledge is taught in Western medical schools. It appears very difficult, sometimes even impossible for an analytically educated mind to switch to a synthesist approach. In his five years of teaching medical acupuncturist doctors, he noted that some never succeeded in this transition and thus couldn't grasp the depth of the different notions involved. That explains why acupuncture and homeopathy have difficulty entering the mainstream arena of sciences, although they are scientific. They are wrongly considered placebos. Simply put, their action is one of synthesis, not analysis. It is not a question of efficiency.

After assembling organic structures under two polarities—feminine and masculine—Dr. DeBavelaere observed that modern illnesses generally pertain either to disrupted feminine energies or overwhelming masculine ones. These thrust the system so much out of balance that symptoms have to be expressed as an output to safeguard life and regain homeostasis. He regrouped the organs of the Chinese system under these polarities. Summarized, the feminine polarity (-) contains lung, heart, Master of the Heart, pancreas (-), and associated elements; while the masculine polarity(+) comprises kidney, liver, traditional digestive system (triple warmer), pancreas (+), and associated elements (diagram 7). After this revelation, his teaching to medical doctors changed. As he related to me, however, each time he addressed the feminine aspect of the system during his lectures, or explained how the trivialization and disdain toward the feminine polarity created all sorts of physical and emotional problems, Dr. Darras would get up and leave the audience. To Dr. DeBavelaere's chagrin, Dr. Darras never tried to discuss the subject, to argue or to prove him wrong.

As with any other established institution, with time TCM became rigid. Eventually it too considered the feminine aspect of the biocybernetic system as of secondary value. To explain, I will elaborate what constitutes the biocybernetic system as taught by Dr. DeBavelaere. Below is an except from a lecture he gave to the International Society for System Sciences in July 2006.

> "I have experienced in medicine that if we use a procedure similar to Einstein's, which is to start our reflection from a theoretical level, this brings some obvious advantage. Not only are we able then to give account of most phenomena, that is to say symptoms in medicine, but we also have a holistic view of physiology and physiopathology. More, we can give account of some unexplained phenomena, for instance, numerous functional symptoms. We finally have the ability to define health in terms other than "absence of disease."

Clinical observation is not the level where the questions arise, but the level at which our theory is confirmed.

As with steam machines, certain elements inhibit or control other elements in order to maintain the cohesion of this autoregulated system.

The cybernetic principle discovered by engineers is a general one: if all the variables are tightly coupled, by manipulating one of them we can indirectly control the system. This principle plays on the universal nature of systems. Latil writes, "The regulator is unconcerned with causes; it will detect the deviation and correct it. The error may even arise from a factor whose influence has never been properly determined hitherto, or even from a factor whose very existence is unsuspected." In our biocybernetic system, the regulator

Isis Code

is the nutrition function, also named the pancreas, as its phase sits between the phases of introversion and extroversion. If the regulator cannot control the system, symptoms will occur, as research has shown with cancer. This reveals an evident link between cancer and the pancreas phase of the biocybernetic system.[54]

These biocybernetic structures appear to be universal since they can be adapted to molecules, to the elements of the cell, to the organs, as well as to any viable complex structure built by Man (cities, society.

> Interestingly, this idea resembles the dynamic model of biological systems based on Chaos theory [Gohara, 1996]. In this model, input signals are processed by interactions among the partial systems, and when the parameters of the partial systems are changed, the output signal changes from steady state to period two, four, or chaos. In the TCM model, the exogenous factors are processed by interactions between the five phases, resulting in various symptoms. If we assume the five phases are the partial systems in the mathematical models of biological systems based on chaos theory, the TCM model is quite similar to the biological model. This implies that TCM attaches greater importance to function than to structure (i.e., anatomy) and regards the human body as a dynamic system.[55]

Here, TCM diverts and often considers only three main forces or energies as essential (issued from the one primordial one), which will sustain, protect, and regulate the body: the nutritive energy (pancreas; ying qi), the defensive energy (liver; wei qi), and the ancestral energy (reproduction; kidney; yuan qi). They do not consider information (heart) and exchanges (lung) as important. Unfortunately, these two form the feminine polarity, as per Dr. DeBavelaere's teaching. As previously stated, it would not be normal for a system of thought to go through an era of masculine dominance—communism as it was practiced being a fine example—without any detrimental effect on the feminine aspect.

TCM has a long and rich history. This form of medicine appeared in parallel with Egyptian, Babylonian, and Indian (Ayurveda) medicine.

Legend implies that it dates from the time of the Yellow Emperor (2698–2596 BCE), although scholars have adopted the view that it is most certainly at least two thousand years old. All these traditions were part of an oral tradition, so exact dating is difficult to achieve. I personally consider that, as with many great discoveries, they all rose from the same period and sprang from the application of same biocybernetic system.

It looks like acupuncture was first publicized to the American population in the 1970s following reports in the press relating the experience of a reporter from the *New York Times*. After he became ill with appendicitis while visiting China, a Chinese acupuncturist performed an appendectomy on him without anesthesia. Of course, the Western medical institution replied that the technique was successful because of the placebo effect. I would like to see them try it. As with homeopathy, this has been refuted since animals (who can't respond to suggestion) also succumb to the analgesic properties of acupuncture and do clearly respond to properly prescribed homeopathy. Further, I could argue that countless drugs proffered by Western medicine have the placebo effect since a powerful group is behind them and they are recognized. This has a more potent placebo effect than that of therapies that are vaguely acknowledged and mostly regarded as quackery. Also, even with a patient open to homeopathy, if the accurate remedy is not prescribed, the symptoms will be not affected. The other point often put forward is that we still don't know how homeopathy works. Just to give one example, we do not fully understand how Dexedrine, Ritalin, Adderall, or other stimulants used to treat ADHD clearly work, but the fact that they appear to strengthen links between the frontal lobe and other regions of the brain[56] is enough effect to us. Or maybe it is only the placebo effect?

The WHO (World Health Organization) enumerates over one hundred different ailments for which acupuncture treatment has been shown effective. These range from chronic pain to addictions, allergies, asthma, colds, environmentally-induced illnesses, flu, gastrointestinal disorders, Meniere's syndrome, migraines, osteoarthritis, pesticide poisoning, sinusitis, ulcer, stroke, sciatica, and many more. Also, in the 1970s, lowered resistance values for over 50% of acupuncture points (acupoints) along the large intestine meridian were identified (Dr. Robert Becker et al.). Dr. Becker suggested that the points acted as amplifiers of a semiconducting direct current traveling along the perineural cells. These cells wrap around every nerve of the body. This DC system became more negative as it traveled to the ends of fingers and toes and more positive as it returned to the trunk and head, as expressed in the TCM theory. He also found acupuncture points are more positive than the surrounding skin. The insertion of a needle short-circuits this battery effect and generates a current of injury lasting for a few days. Further electrical activity occurs because the twirling of the needle provokes low frequency pulses of electricity and an ionic reaction between the cellular matrix and the metal needle. This generated electrical energy flows along this system to the brain and is analogous to the chi described by classical TCM.

The structure beneath ancient ways of life is that there are universal elements—earth, fire, sky (the ether, chi, and subtle energy), water, and air—and that if one of these building blocks is missing, then the manifested world as we know it ceases to exist. Therefore they represent a system. In those traditions, man is *made of* these five elements, not something beside them. There are analogies between the traditions, as they mainly all divide their theory into five fundamental elements or stages. The Egyptian model reveals it through the story of the five main gods born outside of time, but who nevertheless saw five days dedicated to their memory. This is reflected through drawings on the walls of many of their pyramids. In the Ayurvedic system, alchemy, and TCM medicine, we observe it in their five main elements, phases, and their correspondences.

Within complex and ancient systems such as TCM and Ayurveda, one can observe confrontations between different schools, as they all pretend that their interpretation is the best and the only relevant one. I use Dr. DeBavelaere's interpretation because I find it the simplest and most complete, therefore the easiest to understand. Also, as I discovered while preparing this book, it is in harmony with the most recent scientific discoveries pertaining to the brain. Further, as I have witnessed its application over the last twenty-five years, I have seen impressive results.

My understanding relies on a set of elements. Simply put: five phases, in affinity with five elements, five aspects of the brain, five functions, five levels of awareness, five periods of life, understood through two polarities in the oneness of manifestation. This is what I name the LIFE biosystem. Medical biocybernetics comprises the medical aspect, while LIFE biosystem, not being a medical therapy, emphasizes socio and psychocybernetics.

This fractal interpretation of the human body could play a pivotal role in diagnosis and treatment. It is difficult to accept such an idea since the different parts seem to be unrelated anatomically. However, considering the human as a fractal structure opens new possibilities in prevention, treatment, and understanding of imbalances. These five phases are similar to many school levels. For example, before accessing grade five we have to complete the previous grades. So I see the five phases as: from conception to seven years old deals with physical development, the onset of adult teeth expressing the end of that period. The motor aspect is mature. The period of seven to fourteen years I associate with work on the emotions; the hormonal system is forming. Fourteen to twenty-one years is with the elaboration of personal identity, while the phase from twenty-one to twenty-eight years old is devoted to the elaboration of intellectual concepts. The frontal lobe maturation closes this period. From twenty-eight years old to thirty-five, the adult has an idea of who he is and what he wants. The consciousness of "who am I?"—what some would call the spiritual aspect—has landed. This period closes with the maturation of the corpus callosum. It is the social phase where the individual exchanges with his environment in a meaningful and conscious manner. None of the phases can be jumped over. The entire education system would gain in adopting such a system based on natural cycles. Further, these periods exist in the system of medical biocybernetics and are linked to the energetic expression of physiological organs and their associated energy structures (diagram 7).

Biopsychosocial / Medical Biocybernetics Example: Depression

In 1977, psychiatrist George L. Engel at the University of Rochester posited in an article in *Science* "the need for a new medical model." The term biopsychosocial (BPS) was born. [57]

To my knowledge, though, no definitive, irreducible biopsychosocial model has been published. I hereby suggest for this to utilize the LIFE biosystem model.

The advantage of the LIFE biosystem is that it takes into account the different aspects of man and does not contradict elements of psychiatry (dissociative states) or the BPS observations. On the contrary, it encloses them and gives them new significance (diagram 8).

Let us take depression as an example. Depression is expected by health authorities worldwide to become the second highest cause of incapacity in the world.[58] In the Wikipedia entry on BPS: "An example is that depression by itself may not cause liver problems, but a depressed person may be more likely to have alcohol problems, and therefore liver damage. Perhaps it is this increased risk-taking that leads to an increased likelihood of disease."

In medical biocybernetics (MBC), the model not only helps us understand the cause, but also how to treat it most effectively. There are three types of depression:

1. Reactive, which is conditioned by an external event

2. Neurotic, inner and outer triggers that originate from a defective psychological structure, and

3. Psychotic, which is the least environmentally dependent. The medical doctor who uses medical biocybernetics does not treat depression per se, but treats a patient who expresses the symptoms associated with depression.

The society we live in has an imbalance in many aspects. The general imbalance is of the feminine polarity confronted by an out of control masculine one. This negatively influences the environment, the phenotypic aspect linked to lung, mother of energies of our system. Since lung feeds the kidney aspect, we all begin life at a disadvantage. The kidney will not have sufficient energy to control the heart aspect. The first general treatment is to enhance in society the characteristics pertaining to the feminine polarity. Research has shown that depression occurs when limbic elements are overactive. This is explained by our system through the fact that an energetically depleted lung aspect will lose its ability to control the Mammalian brain.

On a personal level, the energy on information regulation, or Psy, linked to the heart, is responsible for the transport and treatment of information. It sends orders of acceleration and inhibition of oscillations throughout the system depending on the programming of the kidney (brain).

If this programming (the window for this programming shuts at seven years old) is one of troubled feminine or masculine polarities, *all the corresponding functions will be affected*. This will eventually put a lot of pressure on the aspect that regulates the system (pancreas). Its hyperactivity can be seen in our modern society: obsession with food, obsession with economic matters, obsession with objects, general obsession, and general analytical obsession. Instinctively, our society promotes elements that stimulate the pancreas aspect, leading to its demise. The taste for sweets and the resulting obesity pandemic is a direct consequence of this.

By the year 2020, depression is projected to reach second place of the ranking of disability-adjusted life year (DALY) worldwide, calculated for all ages, both sexes. Today, depression is already the second cause of DALYs in the age category 15–44 years for both sexes combined.[59] Worldwide, India already has the highest rate of depression at 36%.[60] This is catastrophic. Once the pancreas has no more energy to regulate the system, there is a general burnout, expressed differently by each patient.

Depression, as the word implies, is a general lowering of the dynamic potential, as chronic fatigue syndrome and burnout are, but where the nervous and psychological aspect are more apparent. Every depression comes with asthenia and a general perturbation of the MH system (see chapter 5). Since the lung energy is blocked, the depressed person suffers symptoms of sadness. An understandable attraction toward alcohol, which transitionally and artificially stimulates the liver, only adds to the problem. Alcohol and cigarettes go well together as one will stimulate the Lung function while the other the Liver. Both bring an immediate relief but eventually cause an aggravation (diagram 20).

Second, the homeopathic[61] diathesis, expressing a predisposition to certain illnesses, will influence the three different constitutional types (genotype + phenotype), and highlight the main underlying aspect of the depression. The psoric type is the most resistant to depressive states. The tuberculinic, with its constitutional lung energetic depletion, is vulnerable to the environment. The sycotic has a strong obsessive tendency, while the luetic is subject to nervous imbalances. A total reassessment of life conditions and hygiene is necessary to reactivate the Master of the Heart (MH). This can be achieved through appraisal of personal values and goals, physical environmental, and an analysis of habits and patterns at all levels. A general promotion of the feminine polarity in all sectors of life could effect a primordial change and is a necessity. Homeopathy, with its action on the MH system and on the different "houses," would be invaluable. Acupuncture, as well as nutrition, osmotic contact with nature, and breathing techniques would all be effective in regaining balance. These treatments, added to the conventional treatment if need be and adapted to the uniqueness of each case, will determine success. However, the best nutrition and exercises can do little in the case of a dysfunctional Master of the Heart.

Patients with the same disease might require different homeopathic medicines and treatments. Success in homeopathy is difficult to reproduce patient to patient, as it is a medicine of information (dynamic and synergetic), individualistic, and dependent on the medical and psychological expertise of the therapist to find the right remedies. Some symptoms can be so subtle that the patient is not always aware of them. Therefore, the doctor must spend time getting to know his patient and needs to be inquisitive. A medical homeopath regards symptoms as positive evidence of the body's inner intelligence. He thus works with the system, not against it. On the other hand, conventional drug prescriptions mostly work against the system, by shutting down the alarm system and leaving the thieves in, so to speak. The side effects of these drugs are well reported, because if we mug one aspect of the system, the other aspects will react to circumvent the assault.

For these reasons, in several countries homeopathy is covered through national insurance to different extents, while in others, it is fully integrated into the national health care system (as it should be). In many countries, the laws governing the regulation and testing of conventional drugs do not apply to homeopathic remedies (and they shouldn't). Conventional drugs are used as one compound versus one symptom, while in homeopathy the treatment (informative agent) pertains to the whole system.

With the holistic trend, medical doctors and therapists can be portrayed as a man holding many balloons. Each balloon is a specialization, or a therapy, or an aspect of the patient. These are all different tools we can use to help the patient regain health. As Dr. DeBavelaere pointed out to me, the man holding the balloons is trying to do a synthesis of different analyses. Two points should be observed here:

1. The multiplicity of elements (balloons) is difficult, indeed almost impossible, to manage. It is the equivalent of a government confronted by a problem of alcoholism in the community, and finds that the only solution is to create comities to fight alcoholism.

2. The sum of therapies is supposed to create in the patient the sense of unity necessary to regain health. To have different specialists for different "parts" of the patient only furthers the schism, instead of bringing back the lacking cohesion and coherence. Health relies on harmony between all the oscillatory systems of the organism.

I do not believe in "just in the head" illnesses. Their manifestations are physical; therefore they *are* also physical. Their cause and the treatment might not be only physical. However, even if the physical damage does not clearly show, the existence of energy meridians indicates that illnesses are at first energetic and then are expressed in a denser form. Illnesses are never only physical either. Similar to everything in the universe, imbalances and their attached frailties expressed through illness touches many levels. For example, an allergic reaction or sensitivity to dairy products will generate physical reactions, but it can also create symptoms of anxiety. Human beings are complex and by ruling out any dissociation we are promoting general health at all levels. Medical biocybernetics should be practiced by medical doctors who have furthered their knowledge in alternative medicines, psychology, and nutrition, as seen through the medical biocybernetics model. This approach is priceless because it establishes a space-time continuum within the human being, from

Isis Code

the molecular biology to the neurology, and from the physiology to the psyche.

This model could be used in every domain, from education to architecture, to insure our creations are life enhancing for all the inhabitants of earth and in harmony with natural balance. We can create this paradise. We have the means.

A doctor of medical biocybernetics is similar to a man riding in a hot air balloon. He did an analysis of a synthesis (system science). This provides him with flexibility and adaptability. The therapeutic elements are elements of one therapy, which is the synergy of many therapies. The result is a truly holistic approach.

I would add that medical biocybernetics, similar to nature, is holistic. Nature is not trees + animals + humans + sky + water, etc. It is One. All interacts with all. All is within and without all.

In the same way, MBC is not TCM + homeopathy + naturopathy + traditional medicine, etc. It is One. Every element fits in the general system, and through their synergy the system acquires correspondences and brings new understanding to a plurality of symptoms, which previously did not seem to have any significance or connections. The integration of this new understanding concerning the brain and synergy, instead of limiting us to a type of neurophrenology, brings a theoretical clarity and cohesion. Research can add it to its arsenal of tools of understanding.

Also, the success Dr. DeBavelaere has in helping his patients, as well as the ease with which he can explain why one person has particular symptoms and how these are rational, relevant, and meaningful is quite remarkable and reassuring. I found the harmony, intelligence, and structure of his approach both beautiful and liberating.

Chapter 2
Why It All Began: The Need to Understand

My parents were religious, strict Catholics. Good people with strong principles who were rudely tried by life. Our family conditions were not happy ones, although we could consider ourselves middle-class on the financial scale. My father was a dedicated, hard-working man who lived true to his beliefs. He was an inspector with the provincial police. Nothing could bring emotion into his voice as much as talking about "his boys," as he called them. These were police officers, young men, who had decided to put their lives on the line for the good of all. I saw a picture of these men in uniform, kneeling for prayer. It is truly moving.

I believe my father sincerely loved his wife, my mother, and was certainly faithful to her. He was a beautiful man, tall, blue eyes, and black hair. He had a Rock Hudson/George Clooney type of beauty, with a demeanor à la Cary Grant. Women surely found him attractive, and in his position he could have received their generosity without having to give anything in exchange. That is how I remember him.

My mother was pretty, petite, lively, determined, disciplined, courageous, and devoted. She could have come across as harsh or dry, but her belief in the Virgin Mary gave her a soft side that brought out her aristocratic German beauty and nobility type of beauty. As she got older, she became even more beautiful. A professional nurse, she had tasted freedom before getting married at thirty years old.

My father dreamed of a life on a farm. Knowing what sort of life farming entailed, having grown up on one, she advised him to forget about her. They got married and she carried five children, of which

Isis Code

I am the youngest. I remember them walking side by side, hand in hand, or reciting prayers when they went to bed. Difficulties did not break them apart; through their faith, it cemented them more. They bought a house in Montreal when I was six months old. The house no longer exists—a church has replaced it—but my memory of it is still vivid.

I had three sisters and a brother, who was the eldest of the family. As I grew up, I realized he walked with crutches. When I was in my late teens, my parents had no choice but to send him to a specialized institution, where his days passed, long and dull. There, he felt so lonely that almost every day he would phone and ask, "When is someone coming to visit me?" He had dreams of becoming a plane pilot, would preferably be called John, even though he was French speaking; loved the khaki and beige color worn by the military; and deeply admired the USA. He also believed that if someone would take him to Fatima, he would heal. He never went.

I remember him recounting to my mother how other kids teased him at school. Painfully aware of how conscious he was of his limitations, I found it all very cruel. One might ask: How could a benevolent God allow this? Curiously, I never questioned the intelligence of life. What was, was.

One day, when I got on a bus, I saw him sitting at the back. He called my name in his typical slurred fashion. People turned to him and then to me, staring. I ignored him as if he were a stranger and sat at the front. I still regret it.

For whole weeks his bedroom door used to remain shut because, I was told, he was ill. I can still remember my father's funeral; my brother in a wheelchair, hardly noticing anything because of the drugs he'd been given. The pain strangled me. However, I also can recall his laughter when he was younger, and how smart I thought he was when he showed me the first radio he built by himself. The few times I went to his room, he would intensely stare at me as he had me listen to the shortwave. Meanwhile, my parents probably thought this attraction was unhealthy, because one day and without any explanation, I was forever forbidden to come close to him. For many years I wondered why. The question painfully lingered in my heart.

The hypothesis to explain his struggles was that his troubles arose from concomitant factors of illnesses and possibly over immunizations. Although specialists affirmed his condition would not worsen, they were wrong. Despite medical visit after medical visit, no one could do anything but witness the worsening of his condition.

With time, his intellectual faculties dwindled, and medications transformed him into an easy-to-care-for patient. With the knowledge I later acquired, I did wonder about the repeated immunizations, as my mother was particularly keen on us receiving all of them, as well as the booster shots, but that is another story. When she saw the results we have with homeopathy administered with a deep knowledge of the medical biocybernetic system, my mother felt guilty for not consulting with a homeopath about my brother. A medical doctor who practiced homeopathy was available at the hospital, but others frowned on this unconventional, somewhat bizarre type of medicine. (Not much has changed, and really, vaccination is similar in its approach to homeopathy.) Now she will never know.

He was the eldest and a beautiful boy with a great sense of humor. The fact he was the eldest and in this condition weighed heavily on my parents' shoulders. They seemed to feel responsible for the pain he had to endure, although they were not. On the contrary, I strongly believe they were the reason he stayed alive and could still laugh.

When I was young, real communication with my two older sisters was scarce. Because of my age, it is as if there were a canyon between me and the rest of the family. For me, that canyon had a name: Monique. My first memory of her is an upsetting one. Probably about five years old and really wanting someone to play with, I went to her as she stood alone in our backyard. Two years older than I, my sister was of my height, with beautiful blue eyes, slightly curled red hair, hands as white as porcelain, and little feet like a Chinese doll. She refused to run, saying she could not run well. I explained to her that since she had been a horse before, she only had to pretend she still was and then would run easily. She shook her head, red curls from her ponytail bouncing from side to side. Being so young, I couldn't understand why she didn't want to run with me. My father came up to me and spoke in a very firm tone: "You can't run with her! Leave her alone!"

Isis Code

I did not understand. I felt hurt, as if I had wanted something bad. I was mortified.

Later I understood Monique also had problems, the source of which was unclear. Nevertheless, she would never be my playmate. I did not ask questions. I spent my childhood feeling mostly ignored. Since I was the sort of child who needed cuddles to feel alive, my life was lonesome. When I went to my mother one day for a hug, she said, "Oh, you. You have an unnatural need for affection." Did this mean I had problems too? I wondered, and stayed away. When I was in my thirties, she told me she had lost her mother, my grandmother, when she was only four years old. She had no memory of her other than of a woman lying down on a bed in a dark room. Additionally, she had lost most of her nine brothers and sisters to the dreaded Spanish flu. Then I understood; her emotional evaluation of normalcy had been biased by her life experiences.

My parents took good care of me physically, and through their example kept the door to faith open in me. Given their situation, they did more than one might have expected. I am very grateful for all they did. One day, when I reacted badly because my brother had what I considered "my share" of something, my father said, "It's normal, he has a difficult life." I felt egoistic and hated myself.

It seemed to me my two older sisters were always together, thus that they "had" each other. I was down the line, so far away! Sometimes I would forget that and would burst in on them with an idea, an observation, or an emotion, and they would either gently laugh at me or simply ignore me. I was not really conscious of this at the time, only taking it in. I became shy and reclusive until I was fourteen.

For both my brother and sister, the frustration they would express vocally came from the desires and potentialities they could not translate into actions. Their bodies were like broken pianos. The symphony existed, I could sometimes see it in their eyes, but they could not play it. Regularly, in my dreams, I identified with my sister. When I would wake up, my relief would give way to a feeling that somehow, in some strange way, I was like her. Anxiety was my daily companion. As a teenager, I ran every evening and felt a prisoner of a senseless world. This was when the mild type of anorexia settled in. Not eating, I then felt lighter, happier, more in control, and therefore beautiful.

Why It All Began: The Need to Understand

Now I know this anorexia was a defensive mechanism, an attempt to reset the pancreas structure (see chapter 6) and a normal aspect of my then generally frugal personality.

If our consciousness is only an emerging property of the brain, as some suggest, why then did my brother and sister suffer? Their physical bodies did not hurt. They would not know. How could my brother feel, and thus long for, the realization of potentialities he never expressed? When the sidewalk was slippery in winter, he would tell *me* to be careful, because for him, walking with crutches made it even more dangerous. We tend to project onto others who we are. Their physical bodies were the conditions that prevented the expression of their psychological DNA. We all have nightmares in which we want to run but can't; we want to scream, but nothing comes out. This was my brother's and sister's daily lives. The psychological pain they endured was linked to the level of awareness they had of their limits. Only they truly knew the level of that suffering.

The brain to me is the main vector and expresser of who we are in this finished dimension. As our heart, it is placed at the crossroad between our inner and outer lives, between manifestation and potentiality. It defines not who we are but what we can manifest or not. Because of my family background, I had to question myself early on about the body.

Our brains are a projection of our bodies in the same way our bodies are a projection of our brains. The five main aspects of the LIFE biosystem are therefore projected in it, as it is a complex adaptive system modeled on the biocybernetic system.

The Brain

The brain. Is there three pounds more shrouded in mystery? For the Celts and the Templars, the brain was the door to the soul. For ancient Egyptians, it was scooped out and discarded in their ritual of embalmment, while the heart was respectfully preserved. For secret doctrines, every part of the body is epitomized in it ; and conversely, every aspect of the brain is in the heart, considered the first and last link to life and often as the seat of the soul. Surprisingly, our medical

Isis Code

science does acknowledge important reciprocities between heart function and the brain.[62]

For Carl G. Jung, it was most probably a "decoder which would have as function to transform the tension or the relative infinite intensity of the psyche [archetypal world] in us unto perceptible frequencies."[63]

Our brain is divided into two hemispheres. We are told that the global hemisphere, the right one, corresponds to the left side of our body and is activated each time we treat information globally, each time we use colors, images, symbols, rhythms. It is also used each time we pay attention to emotions in the language of someone we are interacting with.

We could say that this hemisphere is more feminine in its approach, since it is akin to synthesis. When we are born, this is the hemisphere we preferentially use until we are around three years old. After that time, we are able to consciously recall our memories.

The elaboration of this hemisphere happens from conception to seven years old. Its maturation is mostly expressed after 35 years of age and controlled by the quality of the 14-21 phase (Human brain). I associate it to the feminine polarity.

The logical hemisphere is located on the left and corresponds to the right side of the body. It is used each time we treat information in a sequential manner, in details, each time we talk, write, count, read, and analyze. This hemisphere is elaborated between conception and fourteen years old, and from twenty-one to twenty-eight years old. I associate it to the masculine polarity.

In the face, the left eye sees details, the right one the whole picture. The left eye indicates the masculine polarity and the right, the feminine one[64].

From time immemorial, people have acknowledged that we are more than mere puppets subjected to the unyielding forces of matter. We have wondered if the soul exists, and if it does, what is it and where is it. These questions arise from the masculine polarity. Alas, the soul, if we hypothesize that it exists and follow the universal, traditional description of it, could not be limited by space (what and where) nor

Why It All Began: The Need to Understand

time (when). It is therefore nonlocal and eternal. A sacred vessel, it acts as a medium between spirit (individuality, Self) and mortal self (personality, self). As the body is a skin to our soul, the soul is the skin to our spirit.

For the ancient Egyptians, the soul is the immaterial heart (Ab), the interface, matrix, field between spirit and matter; the Master of the Heart of our LIFE biosystem. It is the psychic (as defined by Carl Gustav Jung) center of every human being. The physical heart system, the lower heart (Hati), receives impulses from it. The ancient Egyptians attributed emotions to the quality of this lower, mortal heart. Science dismissed emotions as having a possible link to the heart, but is now somewhat recanting. The heart seems to process and regulate the information contained in emotions.

The pineal gland was also in the running for seat of the soul, as many texts in different traditions seem to indicate.

Where in the brain does the soul, if it exists, exert its influence, and how? Today, the debate has shifted to a certain extent and transferred the concept of a soul to Consciousness. Perusing scientific literature concerning the brain, I have often read that those who suggest an "I" exists after the body dies are delusional. The main problem, I believe, is one of definition. When I refer to Consciousness, it is not the same as the consciousness of our daily reality. That awareness is obviously the result of our brains interacting with our bodies. No body, no interactions. The conscious experience itself, which brings a thought or a sensation into focus, is the result of an excitation of the concerned network with a sufficient intensity, duration, and spatial extent. This is the manifestation of a level of consciousness, not ultimate Consciousness. Nevertheless, there can't be consciousness, even awareness, without a primordial, universal Consciousness. But Consciousness is, even without a consciousness to manifest it. The interesting point is why, in the sea of data we are bombarded with every second, do we choose to focus on some things and not on others. This depends on *who* (Universal brain) and *what* (Reptilian brain) we are.

The goal of human experience becomes the expression and total integration into this Consciousness, as snow crystals revert to water once the sun shines. A large amount of information received on many levels never reaches consciousness. We would otherwise go mad from the onslaught of data. We filter which information become conscious based on who we are, by affinity. In the same way, we can hear our names spoken even in the mumble of a crowd of people talking together. The more Conscious and sensitive and healthy we are, the more information we can receive, process, and express.

If I were to make an analogy, to me, the body and brain are only the piano. The pianist plays to express the music of his choice. The pianist is not the piano and is not the music, although he needs both to be a pianist. When he sits and plays, he uses the piano but he is not the piano. The piano can be studied in terms of *what, where, why,* and *how.* In our LIFE biosystem these are respectively associated to the Reptilian, Mammalian, Human, and Analytical brains. The pianist himself, in contrast, is understood under the why and who. These pertain to the Human and the Universal brains. In some way, both the pianist and the piano are an extension of the music. The essence behind the music, its potential, in some sense has created the pianist and the piano. It will still be, potentiality, even if pianist, piano, and sound cease to be.

Note these three quotes about Mozart, the first two from Karl Barth and the third from Franz Niemetschek. "Mozart creates music from a mysterious center, and so knows the limits to the right and the left, above and below. He maintains moderation." "Mozart's music always sounds unburdened, effortless, and light. This is why it unburdens, releases, and liberates us." "His imagination held before him the whole work clear and lively once it was conceived. One seldom finds in his scores improved or erased passages." It appears that not only objective Beauty, as seen by Plato, can be found in this implicate order, but also objective Music (see chapter 5).

If you have no interest in or respect for the piano, you can't expect to play a great symphony. Of course, if the goal is only to strike a few keys and make noise, even a broken piano may suffice.

As of yet, our science has not developed the tools to understand either the pianist or the music. This is because our grasp of the universe is essentially anthropic. In this sense, we believe the music is created by the piano in the same way we thought the sun turned around the earth. We therefore close our imaginative window to anything else. If I am to believe we have free will, then somehow "I" must defy the laws of nature as defined by our present scientific dogma. Surely this would not be accepted by a rational—that is, linear—point of view. But then, I believe, the property of quantum offers the same challenge. So why not postulate we are also a pianist, a soul, or an independant Consciousness as a beautiful, joyous, and inspiring theory? This is a necessity anyway if we want to ask the right questions regarding the piano. Otherwise we will only find what we already know, reassuring us in our fractioned view. In any case, where is the harm in fostering such a speculation?

As illustration, here is a little personal anecdote. It happened to me at the fragile age of sixteen. My mother, full of good intentions, dragged me to see an orthodontist. With my two front teeth bigger than the others because they were positioned somewhat in front of them, I looked very British, as we used to say then in Quebec. I thought my teeth looked rather nice, but my parents were thinking about my future and that this could impact it adversely. So here I was with my mother, entering the doctor's office. After the exam we went back to his office. With me standing there, my mother asked, "So, doctor, how long will she have to wear the braces?" He answered, "At least four years." After pondering, he slowly added, "Maybe five."

I don't know what happened at that moment, but I found myself staring at the doctor's back from above—although at the time I did not realize it was from above—as he bent over my still body, lying on the floor. Then I saw him run to an adjoining room, open the door to a small cabinet above a sink, grab a vial, and swiftly return to my motionless body. He bent over me again. As a nurse, my mother was very collected. The next thing I remember, we were leaving the office and walking down the hall toward the elevator. I was confused, but at the same time I had felt how concerned he was for me. I had a feeling he was a good man. I told my mother how touched I had been to see him running so hastily to the other room and getting the bottle. She stopped walking, turned to me, and said with irritation, "What are you saying? You couldn't have seen him do that! You were unconscious!" Only then did it hit me that what I had experienced just didn't make any sense. I had seen what I couldn't have seen. I had not yet realized that I was "unconscious." Even though I had seen my body lying flat on the floor, I had been emotionally detached and unconcerned about that, so I hadn't questioned it. But all the details

Isis Code

were there, vivid. It had *not* been my imagination. Unconscious? Really?

Science, confronted by a plethora of such experiences by healthy individuals all over the world, tells us that we can recreate something that seems like out-of-body experiences (OBE) by stimulating the Reptilian brain with magnets or by stimulating the right parietotemporal lobe. Individuals with lesions in the right temporal or parietal lobe can experience events that seem like OBEs. In the LIFE biosystem theory, these are part of the feminine polarity. This does not by any mean prove that all OBEs are the results of neurological damage of the feminine polarity brain aspects. As well, it can't explain my spontaneous devoid-of-feeling experience, nor that I saw him go in a room I could not have known existed. As with many other inexplicable phenomenon that happened to me over the years, I opened a drawer in my mind and shoved it in there.

Through all the new discoveries in brain research, an exciting image of the brain as an energetic manifestation and holographic info-energetic system is emerging.

> The pattern or organization of any biological system is established by a complex electrodynamic field, which is in part determined by its atomic physiochemical components and which in part determines the behavior and orientation of those components. This field is electrical in the physical sense, and by its properties it relates the entities of the biological system in a characteristic pattern and is itself in part a result of the existence of those entities. It determines and is determined by the components. More than establishing pattern it must maintain pattern in the midst of physiochemical flux; therefore it must regulate and control living things. It must be the mechanism, the outcome of whose activity is wholeness, organization and continuity. The electrodynamic field then, is comparable to the entelechy of Driesh, the embryonic field of Spehmann, and the biological field of Weiss.[65]

Could this electrodynamic field be the implicate order we describe expressed through the Master of the Heart's structure? I believe so.

The holonomic theory of Karl Pribram uses the mathematical wavelet function, based on evidence that the dendritic receptive fields of the sensory cortex can be described as such by Gabor functions. It seems that the added inverse Fourier transformation, which gets us back in time-space, can be used. This is accomplished most likely by movement. Bohm and Pribram have worked together to the edification of a vision of quantum minds. Even if these are debated by some who believe in multiverses, we will eventually come to an understanding that there are either more levels to ourselves, or more universes where aspects of us live, or both. One thing is sure. We cannot see ourselves as limited as before. As Pribram says, "What the data suggest is that there exists in the cortex, a multidimensional holographic-like process serving as an attractor or set point toward which muscular contractions operate to achieve a specified environmental result. The specification has to be based on prior experience (of the species or the individual) and stored in holographic-like form."[66]

Whether the ten houses of Kabbalah (diagram 2) are the different universes, strings, or the different layers of our personal holograms being processed by the mind remains to be seen. It appears to me, nonetheless, that there is information stored within a universal holographic system, in a type of field [67] that has been understood by most of humanity's religious traditions and hinted by science itself. Hindu religious tradition tells us that this field, called Akasha, is available for perusal to everyone, provided their personal decoder is developed enough. I will add that the maturation, development, and health of one's feminine polarity determines if someone can develop the ability of accessing and properly interpreting and processing the data hidden in this field.

The brain is highly plastic, as shown by the fact that the maps of normal body parts change every few weeks following our life experiences. It evolves phylogenetically (evolutionary development; genotype) and is modified ontogenetically (phenotype) by experience.

For example, when we learn a bad habit or give in to an addiction, it takes over a brain map. Each time we repeat the habit, we reinforce it, as if saying "This is really who I am and who I want to be." The habit claims more control over that map and prevents the use of it for other more constructive habits. Unlearning is harder than learning, especially if the reward system of the brain has been activated (see chapter 4). Actions, feelings, and thoughts define our physical reality. As with a garden, when it comes to allocating brain processing power, brain maps are governed by competition for precious resources, and recycling the unused is an important tool. What we do, feel, think, and believe *can turn some genes on or off* as certain parts of the brain are affected.

We first believed that neurons do not regenerate. Scientists believed that we stopped producing new neurons when we were two years old. Where would we put them otherwise? A nice mechanistic point of view. Now we know this was false. In the 90s, research showed that even in old age, with basic conditions provided, the brain keeps on producing new neurons. Using terminal cancer as volunteers, a research team injected a tincture designed to color only new neurons (Eriksson et al., 1998). Autopsies later performed revealed many new neurons. It was also demonstrated that these new neurons were not the result of mental exercices, but were the result of the body cells being engaged, such as during aerobic exercises. Sitting in front of a computer or a TV will not help the production of new neurons, but dancing, moving, breathing, walking, loving will. Neurogenesis is intimately associated with the workings of our Universal, social brain (see chapter 7). Several neuropsychiatric conditions have been associated with altered rates of neurogenesis in animal models, including Alzheimer's disease, temporal lobe epilepsy, ischemia, and depression, to name a few. All are also associated with a deficient feminine polarity.

In the October 1999 issue of *Science* magazine, Elizabeth Gould published an article[68] asserting she had found new neurons were produced by three areas of the macaque brain. These three associative areas of the neocortex[69] were the prefrontal cortex, the inferior temporal, and the posterior parietal cortex. This was hard for many in the scientific field to swallow, as these are *newly* evolved areas implicated in superior cognitive functions, such as decision making, short-term memory, facial recognition, positioning of objects

Why It All Began: The Need to Understand

in space, etc. The plasticity of the brain has been extensively demonstrated, and we now know that some circuits previously believed hardwired can be modified in some cases. Somehow, throughout our lifetimes, the psychological patterns we accept have a definitive effect on our brains, confirming some aspects and changing others. Depending on the individual and his belief system, the plasticity of the brain contributes to human flexibility or rigidity. This sacred vessel reproduces the world around us, filtered and tainted by our unconscious and conscious selves.

Until the mid-twentieth century the scientific intelligentsia viewed the brain as an amazing but rigid machine. The first approach of the brain was localizationist (one point, one function); and then neurotransmitter specialist (maps change under the effect of neurotransmitters); and then behavioralist (we act in certain ways because of our brain structures and past experiences that transform the brain); and then atheist (we emerge from our brain cells); and now, finally, systemic, with an understanding that all of the above have their value and that the brain can't be substracted from its environment. In a quantum-field theory of the world, there is no doubt that in this view, our comprehension of the brain will profoundly change in the present century.

Two important observations need to be emphasized: we do not know how knowledge is conserved in a brain undergoing constant changes, and we cannot explain why some people with damaged brains do not exhibit symptoms normally associated with those deficiencies.

The most extraordinary example of brain plasticity is hydrocephaly. In this condition, the pockets of cerebrospinal fluid called ventricles expand so much that the entire brain is reduced to a small layer close to the skull. Without treatment, hydrocephaly may lead to brain damage and death. Dr. John Lorber, a British neurologist, has documented over 600 scans of people suffering from this condition. He divides them into four categories. In the most acute form, 95% of the cranial cavity is invaded by the cerebrospinal fluid. In this research, half of those presenting this most critical form were profoundly retarded. The other half, unexpectedly, had IQ levels higher than the average population; that is, greater than 100. One

Isis Code

of them even had an IQ of 126 and had received first-class honors in mathematics.[70]

In another case, Dr. Goldberg, clinical professor of neurology at the New York University School of Medicine, relates the case of Sister Mary, one of the School Sisters of Notre Dame in Mankato, Minnesota. Until her death at 101 years old, Sister Mary performed well on all cognitive tests. Surprisingly, the postmortem analysis of her brain revealed all the hallmarks of Alzheimer's disease. How can an Alzheimer's brain generate an intact mind? Most participants (86%) of this study, who at their last physical exams before death met the criteria for the "excellent" category of healthy aging, had neither brain infarcts nor met criteria of neuropathology symptoms. However, similar to Sister Mary, 12% of the analyzed brains belonged to sisters who were in the "excellent" category, but after death, autopsies showed that their brains had the physical signs of brain infarcts or Alzheimer's disease.[71] How could this be, if the brain causes our behaviors and intelligence?

The poignant story of Catalan poet and scholar Pedro Bach-y-Rita is another amazing one. The victim of a stroke when he was sixty-five years old, he was left paralyzed on half of his body, unable to care for himself and unable to speak. Doctors believed he would not recover, and his son was advised to find an institution to take care of him. Refusing to follow this advice, the son moved his father back to Mexico, where he started all over. With the help of his gardener, he taught his paralyzed father how to crawl, and then went through never-ending exercises to help him regain the use of his body. It worked so well that at the end of the year, the father was able not only to speak properly but resumed teaching full-time. Six years later, on a visit to friends in Bogota, Columbia, he went hiking in the mountains at an altitude of nine thousand feet. That proved fatal to his heart. He was seventy–two. His other son, Paul, who was a medical doctor and researcher in rehabilitation medicine, asked Dr. Mary Jane Aguilar to perform an autopsy on his father's brain. She invited him to come and view the results. His father's brain was lying on a table, sliced open. Overcoming his revulsion at seeing his father's brain thus unceremoniously exposed, what he saw was unbelievable. His father's brain had *never healed*. The lesions were mainly in the brain stem. Ninety-seven percent of the nerves from the cerebral cortex to the spine were destroyed, as well as the major brain centers that controlled movement. Nevertheless, he had recovered all their functions.[72]

Dr. Lionel Feuillet and colleagues at the Université de la Méditerranée (Mediterranean University) in Marseille, France, wrote a letter to the *The Lancet* medical journal[73] about a man who was a father of two and worked as a civil servant. The brain of this forty-four-year-old man was little more than an outer shell. What was normally packed with neural tissue was filled with cerebrospinal fluid.

Another extraordinary finding concerns experiments that have shown time and again that somehow the brain can unshuffle signals from crossed nerves, and can use a different area to support a function that has lost its structure, due to a stroke for example.[74] Such a finding, appearing in the *Stroke* journal, showed imaging scans of an adult brain affected by a stroke. The language function, linked to a stroke-damaged area of the brain, had shifted to the opposite, undamaged area. These are exceptions but indicate a wider function to the brain.

It, especially the right hemisphere, knows when something is wrong and tries to overcome it. The humonculus directing everything in our lives from inside our brains has died. Now it seems more like there is an energy field that organizes matter around a preordained plan and uses matter as a plant would use soil. Again, could it be expressed by the meridians system used in acupuncture and the Master of the Heart field?

Brain and body are one. The "I" is a conglomerate, using the body as a tool as long as the tool is usable. It is as personal as our fingerprints. Nevertheless, there are some general aspects that as humans we all share. Another thing we have learned: developing brains are far more vulnerable to the outside world than we are ready to admit.

Here is one example. The brain is comprised of at least 60% fat and is the fattiest system in your body.[75] Pesticides, herbicides, and other chemicals deposit themselves in fatty tissue such as the nervous system.

Isis Code

"Toxicity to the brain is not routinely included in testing pesticides," Philippe Grandjean of the Harvard School of Public Health and the University of Southern Denmark told Reuters. "Because many of them are by design toxic to the brain of insects, it is very likely that they are also toxic to human brains," he said of a review of almost 200 scientific reports worldwide about the brain and pesticides.

The study indicated that ethylenebisdithiocarbamates, organophosphates, carbamates, pyrethroids and chlorophenoxy herbicides, as well as pesticide chemicals, have a cumulative effect and therefore could be damaging to the brain.[76] It is unreasonable to shut ourselves to these facts.

The brain can be studied in many ways. We have at our disposal a technological arsenal to help us in our analytical endeavors: positron emission tomography (PET), functional magnetic resonance imaging (fMRI), electroencephalograph (EEG), and magneto encephalograph (MEG). There are two main problems with the conclusions of the different researches, however. The first problem is that familiarity with a task implies a drop in signal strength. Therefore, an easy or habitual task might not generate a detectable signal. Only the cognitive, motor, or sensitive tasks that require effort will generate a reaction. This is only a minor aspect of our "mental" life. The second problem arises in part from the first. We sometimes get conflicting or downright contradictory results when we compare some research data. We find ourselves in danger of becoming like the apocryphal scientist who affirmed that frogs hear through their legs. (The doctor trained a frog to jump on an audio signal. When he cut the frog's hind legs off, it no longer jumped on cue. He therefore deducted that frogs hear with their hind legs....)

After a year of reading about all these contradictory research results, I concluded that since the brain is so plastic, as unique as our fingerprints, and changes so much in the span of a few years, even weeks, research has to be more detail specific about the subjects undergoing the tests. It is important to know their ages, their physical and psychological gender (sexual spectrum, see chapter 3 on the Reptilian brain), their health status, their predominant side. It also needs to be clear which area is tested—front (anterior), back (posterior), ventral, dorsal, rostral, caudal, ipslateral, contralateral, left, right, gray matter, white matter, etc.—as these have different functions. Often, dorsal and posterior are linked to perception, ventral and anterior to action, gray matter to computation and action, white matter to communication and perception, and left and right depend on what and who is being tested. Also, a precise definition and the use of the same word for the same brain area would help. The sheer number of results would eventually give rise to a better understanding. Also, participants in the research cannot pretend they do not know they are being tested. This fact already has affected many results. Lastly, the brain is so individualistic and plastic, it is almost impossible to extrapolate general results without those appearing to be controversial. Meanwhile, my division and distribution of brain sectors to the different phases of the biocybernetic system remains incomplete in its minute details, but corresponds to the organizing energy field applied to the brain structures and to the different levels of human manifestation. I use it as indicator. It is not set in stone as we all evolve.

It would be puerile to assert a fixed mapping of function into structures without considering the dynamic changes experienced by the brain throughout life and evolution. Indeed, a structure used for one function nowadays might be used for a different one further along in evolution or maturation, as well as used differently depending on the gender of the individual. The most ancient structures of the brain are modular, while later-appearing ones are more heteromodal since their function is to integrate information streaming from many modalities.[77] Elkhonon Goldberg developed a theory of cognitive gradient to try to organize the potpourri of information concerning the brain. In the brain, it seems, massive continuous interactions take place with modest preordained function of the parts. Although probably right, Goldberg's model (back to front, bottom to top) could benefit from the dynamism of the biocybernetic model. Phrenology tried to analyze the human psyche through cranial bumps, and modern science through lesions, which results in a dissociative model. The next step points to the study of associations.

Brains, unlike computers, do not have a central processing unit. Selective damage will rarely result in a complete loss of function. It appears then that functions as well as inhibition and control are distributed across the brain. The similarity with the phases of inhibition and control of the LIFE biosystem would be difficult here to ignore.

Isis Code

Traditional Chinese Medicine shows us functions as dispersed to five "houses" and maintained by comprehensive functional interactions. "Therefore, in contrast to modern western medicine, brain diseases are regarded as systematic diseases in TCM, and their treatments are aimed to normalize not only the activity of the organs, but also the balance of functional interaction."[78]

To treat the brain, we therefore have to treat the whole individual, and our efforts should first go to prevention.

In the LIFE biosystem approach, we divide the brain into corresponding zones and phases associated to our different levels of expression. Each level acts through a different structure and function, inhibiting in the materialized world a different type and level of manifestation. As seen in evolution, the function can change which structure it uses, as examplified by the basal ganglia. It was in charge of executive control for millions of years, but eventually relinquished that position to the neocortex. Brain evolution is gradiental in the sense that it forms an evolving and continuous spiral (schemas 11–28). Each phase can also be associated with life stages, from conception to brain maturity. These cycles of seven years have been used by many civilizations. I refer to them here because they correspond closely to the maturation of physical realities. Of course all the elements of every stage interact simultaneously, and the newborn comes with all these structures already in place. Nurturing and preacquired elements will determine the maturation and evolution of these structures. Phases, or ages, promote the development of one stage over others. We can't jump stages or use the energy allocated for one on a different one without ensuing detrimental effects. That would eventually lead to a halt in evolution and sometimes regression or illnesses. We will also observe how women and men favor different parts of their brains for the same action. The genders are different, all the way down to the cell.

Echoed in brain research we find a dichotomy, underlined here by Walter Freeman:

"My latest preoccupation had been Arthur Wigan's conviction that two personalities live inside one head, one in the left cortex and a different one in the right. Wigan (1944) anticipated by more than a century very similar thoughts from Joseph Bogen, Peter Vogel,

Why It All Began: The Need to Understand

Roger Sperry, and Michael Gazzaniga—four near-geniuses and not one of them mad!

Their gist was:

1. Your two minds often get along with but sometimes contradict each other,

2. Your right is not heard but inferred when you manage math, geometry, music, and the visual arts,

3. Your left arranges language for you and generates new combinations of words. Your left also explains to you, and helps you explain to others, what you already accomplished. *It doesn't lead your conduct but follows along behind by 300 milliseconds.* Similarity and influence achieve adaptive outcomes *but injure either partner and the remaining one becomes noticeably erratic even in its specialties.* (Kandel, Schwartz, and Jessell, 1991; Gazzaniga 1998)

Patterns of brain activity are simply unrepeatable; every perception is influenced by all that has gone before. As Walter Freeman goes on to show, recent developments in nonlinear mathematics can contribute to some understanding of these *nonrepeatable brain activities.*"[79] (Emphasis added)

I consider this a capital observation: in the brain, when one of the partners is hurt, the other one can't function properly. Therefore, the recent quasi-general attribution of negativity to the right hemisphere is wrong, as it is always observed in ailing brains. Since the right hemisphere is the one that seeks synergy, it is normal that in case of separation or of lesion on the left hemisphere, negative emotions would be expressed by the general system. Not because the right hemisphere is "negative," but because it is incomplete and is therefore suffering. It is an important aspect of the feminine polarity, and is even expressed in the Bible, Gen. 3:16: "your desire shall be for your husband, and he shall rule over you."

"Husband" here means the masculine polarity, which we all have, and not men. God is not a male chauvinist. As observed by neurolo-

Isis Code

gists, the left hemisphere follows but controls the right one while the right one alerts the left one to discrepancies.

One last point about some results in brain research. From my reading of texts from different sources concerning damaged brains and gifted people, it appears that much of the research ends up with the same linear and false conclusion. It is also a rumor I could read in many blogs dedicated to brain and intelligence. I cite here one blogger who summarizes the thought: "There's probably a good reason why most modern brains aren't as optimal as they could theoretically be. There's got to be a cost to genius: some get lucky and don't have to pay the piper; others may end up in institutions or dead ... or maybe something more prosaic, such as failing to procreate."[80]

I have seen this way of thinking in my office. A renowned artist from New York came to consult with me about nutrition. After we talked, I surmised that his problems were partly self-inflicted. He lived, it appeared, in a psychic structure of turmoil that constrained him to focus and enhance certain elements, with the result that his Mammalian brain was overstimulated. "Why?" I asked. Eventually, the truth came out: "I have to be mentally ill. Otherwise I can't be a genius!"

This to me is a "gift" from the romantic era, in which only tortured minds and damaged emotional selves could be deemed creative, since many creative (and recognized) artists were so. This is now reinforced by the idea that the right hemisphere is linked to doom and gloom. Combine that with the notion that women and creative minds have to be irrational, and you have in your head a formula that spells disaster.

In this new theory of the self, which LIFE biosystem is, instead of labeling each element as if it were a new pagan god, I named general groups as unities or phases associated to a theoretical level. At this point, the neural mechanisms allowing an explanation or confirming emergence of Consciousness have not been scientifically discovered. I doubt it will ever be.... My advantage in this research is that I am unencumbered by any particular scholastic preconceptions. I base my opinions on the LIFE biosystem, observations, common sense, brain science, and life experiences.

In a complex system, regulation is provided by several independent factors. This way, when one controlling factor goes awry, others can compensate for it. The image of the beehive is eloquent. Different functions will be maintained by the bees. During their lifespans, the same bees will perform diverse functions.

By dividing the brain into corresponding zones associated to our different levels of expression, we can observe how women and men manifest themselves differently at each stage. This knowledge will eventually allow a better understanding of, and be remedial to, our gender-based communication difficulties.

The Heart

Over half of heart disease cases are not explained by the standard risk factors, such as high cholesterol, smoking, or sedentary lifestyle.[81] What causes heart diseases then?

Embryogenesis teaches us that the cardiac precursor cells migrate toward the midline of the neural tube and fuse into a single heart. The heart tube undergoes a right inward looping to change from an anterior/posterior polarity to a left/right one. The future brain bulges between fifteen to twenty-one days after conception, whereas the heart starts beating at twenty-one days after conception. It will expand and contract one hundred thousand times a day until our last breath. Between twenty-nine and thirty-five days after conception, the brain will divide into five vesicles. The fetus is then ¼ of an inch. The heart emerges in the buccopharyngeal membrane area where the crescentic mass of the ectoderm and endoderm come into direct contact. In front, where the mesoderm fuses in the middle line, is the pericardium, energetically associated to the Master of the Heart. The heart is intimately in affinity with the five vesicles, the three embryogenic layers and all of the body structures. It is situated more on the left of the chest and is thus in affinity with the right hemisphere of the brain, the feminine polarity. This has been proven in many ways.

The evoked potentials recorded from the right hemisphere during various cardiac events significantly differ, while they don't with the left hemisphere. Also, perception of cardiovascular activity seems to

Isis Code

be processed more effectively in the right hemisphere.[82] I therefore believe that the state of health of the feminine polarity has a direct impact on heart function.

It has been shown that the heart can affect the electrical activity of the brain.[83] The normal fluctuations in cardiac activities coincide with changes in the setting in which the two cerebral hemispheres receive information. AEPs (average evoked potentials) recorded from the right hemisphere are larger at diastolic pressure, and AEPs recorded from the left hemisphere are larger at systolic pressure.[84]

As an example of this interaction between brain and heart, we tend to believe that depression is uniquely a problem with neurotransmitters in the central nervous system. Disturbance in the brain is, we are told, the main culprit. Now research indicates that poor left ventricular heart function and depression are related in patients who suffered a myocardial infarction. Some depression might not be caused by problems in the brain but by a defective heart.

Hormonal, chemical, rate, and pressure are different types of information translated into neurological stimuli by the heart's nervous system, and are sent to the brain through several afferent pathways. These are the same paths used by pain signals and other feelings. These bits of information are then accessed by the brain through the medulla, which is situated in the brain stem of the Reptilian brain.

We also now know that it is possible to die of a broken heart.[85] I have witnessed it.

What did ancient scientists have to say about the heart? In the fourth century BC, Aristotle established the heart as the most important organ, since according to his observations of chicken embryos it was the first to form. To him it was the seat of motion, sensation, and intelligence. His conclusions are similar to what we find in TCM, in that he said the heart is a hot, dry organ.

Galen of Pergamon, a Roman physician, surgeon, and philosopher, proposed that the veins connected the operations of the liver to the heart, thereby circulating vital spirits throughout the body via the arteries. In TCM, liver structures energetically feed heart structures. Here again there is concordance between East and West.

Avicenna (Ibn Sīnā) in his Canon of Medicine (AD 1012) expanded on Aristotle's ideas and wrote that the heart allowed the faculties of nutrition, life, apprehension, and movement (that is, pancreas, lung, kidney, and liver) through many other members, and was the source of all faculties. In harmony with TCM, he believed that the heart controlled the lung (a truth, energetically). If you have experienced arrhythmia, you probably observed that it could incapacitate your breathing function. For Avicenna, breath was the vital power, the "innate heat," which had nothing to do with the air we breathe. He believed that the heart controlled and directed all other organs. He was not totally wrong, as the heart is the structure that is the most energetically expressive.

Nowadays, relationship between the brain and heart is well documented in medical literature.

Patients with subarachnoid hemorrhage (the subarachnoid is the area between the arachnoid membrane and the pia mater surrounding the brain), for example, may develop impressive electrocardiographic changes and present with new left ventricular dysfunction and biochemical evidence of myocardial injury.[86] In TCM, the brain, associated to kidney function, inhibits the heart.

Human embryogenesis teaches us that the three germ layers give rise to different cells and structures. *The ectoderm* sees the skin cells transform into neurons and give birth to the nervous system, including the brain. Without BMP-4, a growth factor known as cytokine, the ectoderm cells would automatically develop into skin cells.[87] These transformed cells are, in a way, an inner skin, carrying information about internal and external events. The heart carries forty thousand neurons that sense, learn, and recollect.

In our LIFE biosystem, as well as in TCM, the phase lung has a direct effect on skin. Not surprisingly, therefore, breathing exercises have a tremendous effect both on skin health as well as on stress regulation.

The mesoderm gives rise to the extracellular matrix (connected to the lymphatic system and to the cerebrospinal fluid): the heart, liver, kidney, and pancreas stem cells. Finally *the endoderm* is responsible for the lung, stomach, and pancreas cells.

As we have seen, between fifteen and twenty-one days, when the human embryo is about one eighth of an inch, the future spinal cord forms, the future brain bulges, and the primitive heart's tube appears. They are then a unit. Around the twenty-second day after conception, the heart starts beating in a regular cadence. This pulse is imprinted on every newborn cell, every structure that will develop afterward, never stopping until death do us part. The 50 to 70 trillion cells that form our bodies are all bathed in the rhythmic rocking of the heart. First the rhythm will be the same as the mother's, since the child's LIFE biosystem is a clone of his mother's. It will then accelerate gradually, plateauing around the eighth week after conception and then decreasing to about double the mother's rhythm at term.

TCM considers structurally shallow organs as male and solid organs as female. As such, the brain is female (which is why TCM links it to kidney, the most receptive organ), and the heart male. The heart, in fact, is the most expressive organ of our biocybernetic system. Not surprisingly, the brain has more white matter than gray matter, confirming its receptive capacities. In the LIFE biosystem, the heart is the masculine aspect of the feminine polarity, and kidney is the feminine aspect of the masculine polarity.

The heart's rhythm acts as a powerful force to bring the rest of the body, including the brain, into a similar rhythm. If the pattern of your heart rate is one cycle every ten seconds, then it pulls the brain into a state of synchrony. That is, your brain harmonizes with your heart and both operate at higher levels of efficiency and effectiveness. This is what researchers at the Institute of HeartMath found. They also conducted a study in which they showed participants a series of images while monitoring their heart and brain activity. The images were either peaceful, pleasant photos of pets, children, nature, etc., or unpleasant photos of violence, trauma, and threats. They found that participants' hearts registered an emotional reaction to the disturbing photos *five to seven seconds before* the image appeared on the screen and *three seconds before the brain reacted.* [88] How can our conventional science explain this? It can't.

"The heart is a highly complex, self-organized information-processing center with its own functional "brain" that communicates with and influences the cranial brain via the nervous system, hormonal system, and other pathways. These influences profoundly affect

Why It All Began: The Need to Understand

brain function and most of the body's major organs, and ultimately determine the quality of a person's life."[89]

John and Beatrice Lacey found that the heart seems to have its own peculiar logic that frequently diverges from the direction of the autonomic nervous system. The heart appears to send messages to the brain. In their view, not only are those messages understood, they are heeded by the brain, affecting a person's behavior. Similarly, the right hemisphere, associated with the heart, perceives information before the left hemisphere does.

The autonomic nervous system is divided in three: the sympathetic system, the enteric, and the parasympathetic system. The sympathetic nervous system is always active at a basal level since it maintains homeostasis.

Shortly after the Laceys' observation, neurophysiologists discovered a neural pathway and a mechanism whereby input from the heart to the brain could either inhibit or facilitate the brain's electrical activity. Then in 1974, the French researchers Gahery and Vigier, working with cats, tweaked the vagus nerve (which carries many of the signals from the heart to the brain) and found that this action reduced the brain's electrical response to about half its normal rate. In our system, the vagus nerve is associated to the feminine polarity and thus to the Master of the Heart. The vagus nerve supplies motor parasympathetic fibers to *all* the organs—save for the suprarenal (adrenal) glands—from the neck, where there are acupuncture points dedicated to the regulation of information (heart), down to the second segment of the transverse colon. It is part of the manifested aspect of the Master of the Heart, which will be discussed in chapter 5. To summarize, evidence suggests that the heart and nervous system are not simply following the brain's directions, as previously assumed. Moreover, the heart has its own intrinsic nervous system that processes information independently of the brain or nervous system. This is in accordance with biomedical cybernetics, which asserts that the heart is a type of information processor. "This is what allows a heart transplant to work. Normally the heart communicates with the brain via nerve fibers running through the vagus nerve and the spinal column. In a heart transplant, these nerve connections are severed and do not reconnect for an extended period of time, if at all."[90] However, the transplanted heart is functional through the capacity of its intact, intrinsic nervous system. Since nature does not create structures aimlessly, the connections between the heart and the nervous system must have a function. The brain and the nervous system must then receive information from the heart in a similar manner. Since receivers of a heart transplant can function well, there must be energetic paths overarching the structures, feeding them and still expressed, despite the missing physical structure. The meridian system points to the answer. The heart is a key element of our sixth sense, which I will discuss later. It is part of the Master of the Heart and therefore of the sense of synergy (see chapter 5).

In 1981 a hormone produced and released by the heart called atrial natriuretic factor (ANF) or ANP (atrial natriuretic peptide) was isolated by a research team in Kingston, Ontario, Canada. As a result, the heart was reclassified as an endocrine or hormonal gland. The hormone is produced, stored, and released by cardiac myocytes (heart cells that have an internal rhythm) in the upper chamber of the heart. It is secreted following certain signals received by the heart, such as high blood pressure, hypervolemia, strong emotions, exercise, caloric restriction, and immersion of the body in water. A powerful vasodilator, it is also responsible for the homeostatic control of body water, sodium, potassium, and fat (adipose cells). This control of the circulatory system decreases blood pressure, pulmonary capillary wedge pressure. It also inhibits the effects of catecholamines (used in fight or flight response). Its effects are wide spectrum: on the blood vessels, on the immune system,[91] on the kidneys and the adrenal glands, and on a large number of regulatory regions in the brain, including the thalamus, the hypothalamus, the pituitary gland, the pineal gland, and the olfactory bulb.[92] ANF also has an inhibitory action on the sympathetic nervous system (sympathoinhibitory action).[93]

Dr. J. Andrew Armour, pioneering neurocardiology researcher of the University of Montreal who first introduced the concept of a functional heart brain in 1991,[94] found with his students that the heart contains a cell type known as intrinsic cardiac adrenergic (ICA). It is classified as adrenergic because it synthesizes and releases catecholamine (norepinephrine and dopamine), neurotransmitters once thought to be produced only by neurons in the brain (norepinephrine mainly by the right hemisphere and dopamine mainly by the left) and by ganglia outside the heart.[95] More recently, researchers discovered that the heart also secretes oxytocin, commonly referred to as the love or bonding hormone. This is in harmony with our LIFE biosystem, which associates the medial prefrontal cortex, where we find oxytocin receptors, with the heart aspect. Indeed, it was found that the mPFC (medial prefrontal cortex) also releases noradrenaline (norepinephrine), especially during REM sleep.[96]

In the LIFE biosystem, we make a distinction between the lower heart, which is the physiological and energetic heart system and stage, and the higher heart which is the sum of all the stages when well developed, the polarities working harmoniously, and the entire system acting in synergy as a biosynarchy,[97] in harmony with the implicate order. In the Chinese traditional texts, this would be referred as Dantian functionality. It therefore becomes the center of the Self, while the pancreas aspect remains the center of the self (diagram 9). The lady with the unicorn—*À mon seul désir*—is an expression of this reality (see chapter 7, illustration 8).

Many living functions, including the collective behavior of ant colonies (Goodwin, 1994[98]) are nonlinear. Before, scientists considered them unpredictable and erratic. The term coined for this is deterministic chaos. The unrepeatable electrical activities of the brain, as well as frequencies patterns, are of this class. Nonetheless, during an epileptic crisis, the spectrum is seriously impoverished (Kandel, Schwartz, and Jessell, 1991). The heart is also much more irregular in healthy people than in cardiac or elderly patients who present a loss of complex variability in their heart rate (Goldberger, 1991). I believe that the seemingly chaotic patterns are because functions are global, associated to many levels, and can self-regulate their rhythms to accommodate general homeostasis as well as effects of the psyche.

And Consciousness

A few etymological sources will help us understand the original meaning of the words we use to express some ideas, such as science, consciousness, and soul-spirit.

Science: borrowed from classical latin, "scientia," meaning knowledge

Isis Code

A knowledge, by definition, has to be more than a virtual copy and paste event. It has to become cellular, so to speak. If you have never admire an orange fruit in all its phases, from bud to fruit or if you have never smelled, seen, and eaten an orange, do you truly know what an orange is?

Consciousness: feminine name borrowed from the latin "conscientia," meaning knowledge in common[99]

This consciousness has a wide range of meanings, depending on the phase we apply it to, from awareness of sensations of the physical body—which I prefer to call awareness, passing by Locke's definition of consciousness as "the perception of what passes in a man's own mind"—all the way to Consciousness, which is global, a source of inspiration, and thus in affinity with the feminine polarity and its collective aspect. In my view, it is similar to the soul and to the Master of the Heart. This Consciousness, when received by a proper vehicle and joined to the personality (masculine polarity) is the Consciousness we should long for and aspire to. This is only possible when our social selves and all of our different levels have harmoniously developed and work together toward one universal goal. When this is done, we become such as the light of a candle, and we may enlighten those who yearn for intimate knowledge of this Consciousness.

Spirit: "mid-13c., 'animating or vital principle in man and animals,' from O.Fr. *espirit,* from L. *spiritus* 'soul, courage, vigor, breath,' related to *spirare* 'to breathe ...'" (From the Online Etymology Dictionary)

The spirit to me is a masculine aspect within Consciousness, which uses the vehicle of the feminine polarity (heart/courage and lung/breath) to manifest itself through the conjunction of the two polarities expressed at the pancreas phase. This is why the spiral or ladder is the symbol of the pancreas phase.

Can the whole be devoid of what the parts contain? No. Since humans who emerge from nature can attain Consciousness, I postulate that nature has Consciousness.

In order to manifest itself, Consciousness needs an appropriate vehicle. The soul, as a fractal element of the implicate order, organizes

Why It All Began: The Need to Understand

its denser aspect—the physical body—in all its dimensions, eventually allowing the spirit (Consciencio) to vitalize and inhabit the crudest levels of materialization. The soul's nature is of the feminine polarity. The soul tends and opens to her beloved, the spirit. She is the mother of all that is living as she supplies the medium for the expression of the spirit in this dimension, the Master of the Heart. He tends toward the soul to vivify it. Spirit and Soul are one outside of this dimension, but polarized, like a beam of light, in this manifested dimension.

If we consult the Bible, we find the *Spirit* of God hovering over the waters. In French, the word *breath* is used instead of spirit. This shows that in the beginning, the two polarities act as one. The verse from Genesis, "And the evening and the morning were the second day," again indicates the precedence of the lung aspect and feminine polarity, which in the Tao pertains to the evening and to breath. It is also the "mother of all energies." Already, this shows that the two polarities are *potentially* divided, and from our earthly point of view are potentially undivided.

The masculine polarity with its feminine aspect became the universe as we know it. The feminine polarity with its masculine element at this point of creation is not manifested, but ready to divide for the subsequent manifestation. God, therefore, is the first to self-divide. So "God" actually is also our universe and the mother of all universes, infinite and timeless. God's verb is Consciousness. He is information, as well as the subject, verb, complement, and, through manifestation of the universe, object. I therefore avoid naming God as a separate entity. Otherwise I reduce him to his aspect of object.

We are all finite parcels of God, similar to drops in an ocean. The Bible shows how Consciousness created the implicate order in five parts (diagram 19, 19.1). The feminine polarity became manifested as God saw that it was not good for the masculine polarity to be manifested alone. "Not good" meaning only that it was not in harmony with the implicate order, the image of Consciousness. "Goodness" refers to the *heart aspect*, which is the masculine element of the feminine polarity.

What cannot bear Consciousness is condemned to disappear. As soon as the feminine polarity was expressed, when Eve was created,

Isis Code

the center where masculine and feminine connect also became manifested. In our system that corresponds to the Analytical brain and the pancreas aspect. So we can now say that, in reality, Eve allowed manifestation and allowed the realization and expression of Consciousness in our manifested, polarized world. Then the center, which was the heart aspect (tree of life), is put aside as manifestation in space and time necessitates a division (the devil), allowed by the tree of knowledge of good and bad. Nevertheless, we can reconnect through this center where masculine and feminine join, climb back to Eden so to speak, although as long as we are physically manifested, we are in the world of separation. Dante would appropriately say that we live in Hell.

Some research scientists who have devoted themselves to the study of the mind-brain believe Consciousness is merely an epiphenomenon of the electrochemical events occurring within the brain. I believe they refer to awareness and consciousness; not Consciousness.

In harmony with this and our system, neuroimaging studies reveal that many of the brain areas involved in the generation of emotion existed much earlier than the neural circuitry that allows for the awareness of, and control over, these processes (Damasio, 2000). Involution precedes evolution as cause precedes consequences. Information precedes the formation of the structure that will manifest and allow the expression of this information in a manifested world, as well as the eventual evolution of this structure through the expression of the matching information. Reception precedes expression. Consciousness is the fundamental energy.

Astronaut Edgar Mitchell, returning to Earth from his Apollo 14 moonwalk, had this reflection:

> In a peak experience, the presence of divinity became almost palpable and I knew that life in the universe was not just an accident based on random processes. The knowledge came to me directly-noetically. It was not a matter of discursive reasoning or logical abstraction. It was an experiential cognition. It was knowledge gained through private subjective awareness, but it

> was—and still is—every bit as real as the objective data upon which, say, the navigational program or the communications systems were based.[100,101,102,103]

In his book discussing Consciousness, Dr. Roger Blomquist, head of radiology and nuclear medicine at the Sevier Valley Hospital in Richfield, Utah, proposed[104],

> This results in a hyperawareness state that is both introspective and extrospective. (One has a greater awareness of information that relates to "self" as well as information that relates to "nonself.") It would also appear that the greater the ratio of true information stored to false information stored, the more closely any new higher-order concept will approximate reality. The most probable explanation would seem to be that there is information stored within a universal holographic system, that it is potentially available to anyone, and that the more global one's level of consciousness, the more information that becomes available to consciousness for processing.

If we get back to our tunnels, as more people take the lower tunnel, the one linked to death, the less rewarding it will be to do so, since the environment will be so polluted for everyone, that no amount of money will allow anyone to buy a place to flee and hide. Besides, as quantum physics has demonstrated, we are all *really* connected. We cannot escape feeling what other beings feel, even if we receive the information only on a subconscious level. Our physical selves, composed of quanta, experience this link directly.

We can never hide in a castle and expect not to feel anything, even if we are a male crocodile. Even this insensitive beast, figuratively, would suffocate in its own polluted waters.

As more people take the lower tunnel, we will not be happy, and will be forever rationalizing to find the culprit. We will point at everything in our physical, emotional, mental, and social environments, when the problem resides somewhere else. Yes, our inner Isis is there, silent, keeping vigil.

Isis Code

Collectively we need the same vision, for on earth, quantity counts. This vision is already there, within us. Isis in us waits to be recognized and awakened from her long sleep. Sure, the pendulum would eventually bring us to the first tunnel, but wouldn't it be better to skip some of the suffering? Yes, the more we know about the state of the environment, the more we understand that we all now live in hell. However, we are all from heaven. This is where evolution leads us. The synergy of those two aspects, working hand in hand, could manifest miracles.

In Isis's Footsteps—Isis and Osiris, the Legend

Ancient civilizations used simple images to express abstract notions relating to the creation of the universe. They employed common knowledge so all their people could relate to the same story. These images used in the respective cosmologies have similar characteristics across the globe. With time, these stories took on a political flavor and came to express the power of a group thought to be descended from or to have lived near the greatest gods. Through these stories, people could feel unique, great, and powerful, as well as guided by a godly leader. More importantly, these stories allowed individuals to position themselves in time and space, in history, in a community, and in the universe. This was a stable base upon which peoples' identities could be built. These mythologies became part of the human collective unconscious and are now an integral aspect of our collective psyche. We could also presume that these stories were similar because we emerged from a primordial collective psyche, a Consciousness.

In the ancient world, cosmogony often involved geographical areas that individuals could see or visit during their lifetime (as it still exists in many faiths). In Greece, for example, this was a significant factor in the healing process of the ailing. In fact, the external world was an effective tool with which to touch, transform, and harmonize all aspects of man. Intuitively, man was considered a fractal aspect of the universe. Through the world he lived in, he could thus access the divine, of which he was but a shadowy image.

The most powerful people were those who were said to be descended from the greatest gods. Gods, then later on saints, whether they physically existed or not, left powerful imprints on the physical and psychological worlds. We can still summon their virtues because we can still vibrate to them through our feminine polarity. Their history and reality were never contested in ancient times, as involution from gods to humans was an accepted fact, in the same way evolution from primates to humans is accepted by our civilization. One does not contradict the other. These are two movements, happening always and simultaneously in the world. In fact, gods are archetypes and, therefore, part of the foundation of the human psyche. They are elements of the Self which we project on the external world. We cannot put them "aside" without generating detrimental effects on our psyche.

Yes, we can still see gods acting in the world today if our eyes are open. They act through nature, including humans. Pilgrims of old could visit the places where the gods had evolved, attuning themselves to their manna, awakening their own inner godly aspects. They would then vibrate these qualities through the external world, in turn evolving toward becoming gods themselves. They were called Initiates. When the earth became Seth's kingdom, when humans were definitely nailed to the cross of time and space, we could still hear a whispered "Recite," as this was the only thing left to them. Seth —as we will see— always tried to go against the reality of inner gods so he could take for himself the power of humanity. Nevertheless, if Seth reintegrates the natural order, he will regain his name of *divine ladder*, as he can either initiate or block the road to evolution. As he is now, he killed Osiris, so he can't see the truth. He ignores Isis, so he has no contact with life and, consequently, he destroys the environment. He abuses Nephthys, his wife, so he is sterile. He fights Horus, thinking through his deception prone mind and jealousy that he, Seth the Clever, should be king. He is as good as a dead slave, but he naively believes he is truly the only leader possible, and a free and powerful one with that. What he is in fact is truly delusional.

In this story of Isis and Osiris, Isis will be our guide. We will not be pilgrims of the outer world, for unless we act differently in the future, the outer world is not sacred anymore. Nature, which was the seat of sacredness, has been defiled everywhere by the Seth aspect of humans. At this time, and as long as we do not resacralize this earth, we can no longer reach the deepest parts of our selves

Isis Code

through nature. Moreover, our analytical, rational, and divisive nature (Seth) is ruling so much over our minds that it censors everything associated with the marvelous and the sacred. It does this so well that we cannot see it anymore. Also, because Seth is a part of ourselves, he blinds us to our true identities, and we are confined to the dungeon of a frigid matter. What a miserable world we have created! Our masculine polarity is not useful in our actual quest; quite the contrary.

We therefore have to become patient pilgrims of inner realms. For this, we now need to become receptive; therefore we need a functional feminine polarity.

As Isis told me the story, I will recount it to you. Open your heart and you will breathe freely. Breathe and you will read with understanding. Understand and you will hear and feel your truth.

Here, because of an eternal and universal point of view, two secular traditions are allowed to merge naturally: the Egyptian with its Jewish/Christian/Muslim development, and the Taoist one. The Tao, in the spirit of Lao-tzu, says that the followers must recognize manhood and protect womanhood in order to be "the river-bed of the world."

"*Ex aliis numquam unum facies, quod quaries, nisi prius ex te ipso fiat unum*"

"*You will never find the One you seek out of the other, unless you become One yourself!*" Gerhard Dorn

"*In some sense man is a microcosm of the universe; therefore what man is, is a clue to the universe. We are enfolded in the universe.*" David Bohm

Isis and Osiris, the Legend

A long time ago, before crocodiles roamed the marshes, before marshes were, before the earth itself existed, Ptah was. Ptah, Ain Sof in the Kabbalistic tradition, bearer of all the names but of no name, bearer of the inherent qualities of existence, void and presence together, total darkness and primeval infinite light, projected like a breath, Atum, in which all is. To signify Atum, the ancient Egyptians used the symbol of a crown, in the same way that the ancients of the Kabbalistic tradition used it to signify Kether. In the Taoist tradition, the Taijitu, the symbol of the eternal Tao, had the same function. Like a cell transformed by mitosis, Atum brought into existence the twins: Shu, he who rises up, and Tefnut, she of moisture. In the Taoist tradition, they are named Yang of dry expansion and Yin of humid retraction.

It is said that the eternal Tao, Ptah, Ain Sof, begets One (Atum, Kether, Tao), and One begets two (Tefnut and Shu, Yin and Yang). In Kabbalah, Tefnut is named Binah, the womb; and Shu is Hokmah, the point, the beginning. These are the two main principles. We can call these masculine for Hokmah/Yang/Shu and feminine for Binah/Yin/Tefnut (diagrams 2, 6).

These brought into existence, like nesting Russian dolls, Geb and Nut, father and mother of the main gods, also known in the Kabbalah as Gevurah—which we translated as rigor—and Hesed, known as grace. The names of gods given to archetypes are an easy way to visualize a cosmogony. It is a way to explain where we come from, which is the eternal and universal question inherent to human nature. It is also a tool to link seemingly insignificant beings to gigantic energies, thus giving them a place and usefulness in the great scheme of the universe.

In Tao, as in ancient Egyptian or Indian traditions, personification is made of archetypes, principles, or levels of informative involution. In this way, as with the Kabbalah, we can extract abstract and systemic notions, which otherwise

Isis Code

would not convey understandable information. Now, we can use simpler images without a multiplication of principles, or gods, at each level of incarnation.

Tao, bearer of yin and yang, is also bearer of the ten celestial trunks (the different sephirots in Kabbalah) and of the twelve terrestrial branches (the astrological influences). In the Egyptian tradition, we have Ptah, inside of which is Atum, which contains all, visible and invisible. The twelve terrestrial branches are the twelve astrological signs depicted in Ramses IX funeral chamber. These are influences that act directly on the mother and influence the child to be. What we read in the weekend newspapers has nothing to do with that astrology. Unfortunately, most of the knowledge associated with it has been lost, although it can be regained.

Agrippa expressed this general structure by a pentagram inside two circles. There is no outside or inside, unless through a mistaken analytical viewpoint. All that is, is in Atum. Our conception of a god somewhere out there, far away, is illusory because we are used to considering our bodies as "one" and all that is outside our physical space as "other." It is like someone considering his big toe as not being part of him because his protruding stomach hides it from his view. I also can say "toe," separating it from me in some way, but the word is not the reality (diagram 6).

As such, the ten celestial trunks in Taoism correspond to a structure, while the twelve terrestrial branches correspond to functions. Kabbalah therefore has ten sephirots or houses. By extension, goddesses are often called "lady of the house." The lowest five correspond to the five main Egyptian gods. They are inherent to the structure of the universe.

Why It All Began: The Need to Understand

If we carry on with our Egyptian/Kabbalah story, it is said that Geb and Nut begot five children. The firstborn was Osiris/Hod, then a place was set apart for Horus/Tipheret. Seth/Malkout was violently born, followed by Isis/Netsah and Nephthys, who is assimilated to Yesod. We can further this by attribution of the five organs (phases), planets, or elements of the Taoist five elements, which would give us: Osiris/Liver/Mars, Horus/Heart/Sun, Seth/Pancreas/Earth, Isis/Lung/Venus, and Nephthys/Kidney/Moon.

Interestingly, painted on the wall of the funeral chamber of Ramses IX, we can see the future being as a god (Amon) with a hidden double. Above this godly figure is an oval containing five red shapes, red being the color of incarnation. The central one is a small vertical oval shape, similar to a heart. The four others bear the hieroglyphic sign that means "piece of flesh." This image is surrounded by a solar disk. The Egyptian tradition considers the sun (not the physical one, but its essence) as the origin of the created gods—the archetypes—and it is true that in the Kabbalah, the sun (or Tipheret) has direct access to all sephirots, safe for Malkout, which to me is dedicated to/Seth/pancreas/Earth. From the earth, access to Tipheret is available only through the other sephirots, these being sacred.

So, this is the story of the beloved children of Geb and Nut, as recalled throughout the ages. Geb and Nut are twins, children of twin Shu and Tefnut. Both were born from Atum. Nut is usually depicted as carrying a vase of water on her head.

***

Thoth, god of wisdom, had to find a solution. Gods are given to the manifested world so they can bring what is outside of time and space into the dimension of time. They can also organize the physical world to reflect the eternal intelligence. How would Nut conceive gods if she was not allowed to give birth during the days of a year? This was a difficult problem to resolve; one Thot had never encountered before. How could this newborn universe be prevented from reverting to chaos without gods being born into it? Impossible. There was no time to waste. Thoth devised a game he would play with Khonsu, god of time. The prize of the game? Light days. Khonsu loved games! So they played and played and played. And Khonsu lost every round!

Eventually irritated, Khonsu decided he did not want to play games anymore. Thoth quickly gathered up the precious light he had won and made it into five extra days. These were added to the normal year of 360 days. This is why in ancient Egypt those five days were deemed most holy.

On those light-days outside of time, Nut was allowed to give birth.

On the first of those days, her eldest son was born. Like thunder, a vibration echoed over the plenum, saying: "The lord of the entire universe is born." She named him Osiris, a combination of the words "holy" and "sacred." Nut prophesized: "A just and fair god, he will bring culture to the people of Earth. No other god could have been born without him being born first. Because of this, he will be king and leader of the gods in the universe. He is the throne receiving the sun. On Earth, his emblem will be a grain sprouting through the gift of light and warmth. I give him earthly spring and all the energy of expression that goes with it. He will manifest himself through what will become people's protective energy. To him belong motivation, action, and defense. He will show them how to cultivate the ground, how to take care of the land. He will show them how to transform the products from the earth to fulfill their needs. Eventually he will teach them an invisible law in order for them to live in harmony one with another and with the whole. Osiris, my son of the many eyes, is the one who observes. Because of his origin, men will depict him as being dark. In the sacred processions honoring him, a vase full of water will always open the sacred march. This is so men remember that Osiris would not exist without Tefnut, his grandmother and my mother, as the day would not exist without the night first receiving it. So be it."

Nut was pleased. The visible and the invisible rejoiced.

The second day was set aside. Fair god Horus, son of Osiris and Isis, will be born on that day. His name signifies "he who is above." Nut prophesized: "He will be represented on Earth by the living pharaoh, a manifested god. Those in humanity who choose to be his disciples will enlighten the darkest corners of Earth, bringing forth a source from their hearts. They will be kings of love and wisdom."

Horus is the sun on the throne. Although he is already born above, he can manifest hereafter only through Isis being united to Osiris. Tipheret of the Kabbalah is his house, as will be Da'ath, which will manifest one day in humanity. Horus will eventually have to avenge his father if Consciousness is to sacralize the earth through humanity. He is the god who feels as he is an image of the One. Although his time is there and was always, his time will manifest in the dimension of Earth, when all that is written has unfolded. Joined with Isis and the nutritive aspects of his brother Seth, he will be expressed through the feminine polarity of humanity. This aspect of man will wish to express beauty, benevolence, and wholeness, as does the greatest god. But he will have to claim humanity, even if that means sacrificing himself for a time. Only then will he replace Seth in the hearts of men. His time will come. For now he is given the heart aspect of humanity. Horus, Master of the Heart, will be. As he is his mother's favorite, to him summer is given. He is both son and grandson.

On the third day, Nut's second son was born, Seth, by tearing her side as he came out. Because of his lack of sensitivity, Nut gave him dominion over what divides and separates. He is the trickster, the schemer, the plotter, the liar of the group, and thinks highly of himself. Nevertheless, he is also the center of the material energies of evolution and involution of the world. Nut knows that despite all the pain he will inflict, one day he shall be redeemed, for he is a son of hers. Since nutrition necessitates unbinding elements, he is also linked to it. He will love giving extravagant feasts and will shine with his earthly mind and power. To him was ascribed the pancreas aspect of man with its analytical gift. Since he belongs to the two polarities of humanity, Seth will submit both males and females to his genital nature. Separated from the unborn Horus, he has no respect for the heart and its world. Without Osiris to control him, he might destroy and corrupt everything! Nut couldn't give him a season, so she gave him harvesting time, hoping he will learn that greediness and pretension might destroy nature and his own reason of being. Without generation and love, there cannot be harvest.

On the fourth day her precious daughter Isis first saw the light. Nut could breathe again. Isis the sacred saw Osiris and recognized herself in him. Seeing his sister for the first time, he joyously exclaimed: "She is truly the flesh of my flesh!" She is a hidden part of Osiris, since her name is contained in his.

Why It All Began: The Need to Understand

Because of this, pharaohs and kings will call their wives "sister." Before their births, Isis and Osiris knew they would one day conceive Horus. They were destined to each other even before their material birth. Her function is to give immortality by burning impurities through her breath. As much as Seth is a divider, Isis is a gatherer. Her hieroglyph is also a throne. The season of fall is given to her, as she is mistress of introspection. It was declared that how she is treated by humanity will define the world, as she becomes the mother of all, perfect vessel of Tefnut as Osiris is the perfect model of Shu.

Nut's second daughter, Nephthys, was born on the fifth and last day. Nephthys was destined to Seth. Poor Nephthys! As retracted as death and as unconscious as the moon, she is nonetheless the door to life, to manifestation. She goes where life takes her. Left to herself, she would become the great prostitute and unleash the subconscious shadows on earth. As her name indicates, with his name contained in hers, Seth is a hidden part of her. The earthly season of winter is dedicated to her for its closeness to nonmanifestation. She would never conceive with Seth despite his rapes. She will have a child from Osiris. He will be named Anubis and become regent of the underworld. To her Nut gave the virtue of regeneration on earth. She is the shadow from which everything is manifested. The kidneys of men, the house of their reproductive energy, will manifest her. She precedes Osiris but is still very much in the invisible world. With Osiris and Seth she forms the masculine polarity manifested through humanity.

Nut could now rest.

The curse of Re was therefore both achieved and beaten. The days on which her children were born belonged to no year. Nut and Geb rejoiced.

Heaven proclaimed to all that Osiris, the good and fair king, was born to bring happiness. The people of Earth would feel this joy through the health of their bodies. When they came of age, Osiris and Isis were duly married so they could reign together. They represent respectively the active and the receptive aspects of humanity. Because her name is included in his, Isis manifests on Earth the expression of Osiris's qualities and has an impact, through their love, on his inner self.

Nephthys was given to Seth. Poor Nephthys longed for Osiris. She wanted him so much! But the name of Seth can be found in her name. Thus, Seth is in Nephthys and he controls her destiny. He manifests on earth the expression of her qualities and has an impact on her manifestation, just as Isis is in Osiris and controls his destiny.

Right from birth, Seth despised Isis and was jealous of Osiris. He did not care for Nephthys either, love being an alien concept to him. He thinks: *What is love, what is the use of it, except as a tool to abuse others into thinking you care about them. A tool you use to manipulate them at your will, the fools!* He laughs, and his laugh can be heard all over the earth, sending shivers of sadness through all humans. He would have sex with Nephthys to show her he is above her. He will have sex with anything so he can dominate, but he is sterile. Who cares? He needs no one! Similarly, Seth will not know Horus. Only when it is time for Horus to avenge his father will Seth finally see him.

He sincerely thinks the kingdom should be entrusted to him! After all, did he not save Re from being swallowed by Apep, the great snake of chaos? Is he not standing at the prow of the solar vessel and defeating the enemies of Re? By this, is he not upholding Ma'at, the divine order? Is he not the most apt with calculations and measures? Is he not the most

Why It All Began: The Need to Understand

rational and unemotional? If he were finally chosen as sole leader, he would make those men walk his line without them even realizing it!

When Osiris had taught humanity how to live peacefully and how to grow their sustenance in a sustainable way, he entrusted the kingdom to Isis so he could wander to teach the rest of the world. Thanks to the efforts of Isis and Osiris, men had learned poetry and music and they were not the brutish animals they used to be. Leaving Isis with the people, he knew they were safe.

With time, Seth envied Osiris and despised Isis even more. The more Osiris was good, the more he was praised and loved, the happier mankind became and the stronger grew Seth's desire to kill Osiris and rule in his place. He wanted to be recognized as the One.

When Osiris left Egypt, Seth did not attempt to seize the throne while Isis was watching over the land. He just pretended she did not exist. He ignored her. To hurt her feelings even more, when Osiris returned from his travels, Seth was obsequious and charming to him. He knelt in front of the good god his brother, bidding him good life and prosperity. It was only to prevent suspicion. Isis knew better, but she couldn't bear conflict so she kept silent.

Alas, already Seth was plotting the demise of his elder brother. He had plans. Aided by seventy-two acolytes and a queen jealous of Isis's popularity, he would utterly destroy Osiris! He plotted with the seventy-two connivers, like seventy-two heartbeats in a minute, like the seventy-two paths of the Kabbalah. Secretly, Seth obtained the exact measurements of Osiris's body and ordered the construction of a most beautiful chest, made with everything that is sacred on earth. *Osiris would be captivated by it!* he thought. And only Osiris could perfectly fit in it. The chest would become his coffin. It was too easy! He became agitated with anticipation. He would spare no expense, so the chest was fashioned from the rarest and most sacred woods: cedar brought from Lebanon and ebony from Punt at the south end of the Red Sea. He had it adorned with precious stones and paintings of the most beautiful scenes on earth. The names of the greatest gods, Osiris's ancestors, were engraved on all sides. Seth knew, this chest will be irresistible to his stupid brother!

Then Seth gave one of the great banquets he was famous for, in honor of his beloved brother Osiris. It was the greatest feast ever given. When the heart of Osiris was made glad with feasting, songs, and dance, the chest was brought in, and all were dazzled by its beautiful details. Osiris marveled at the sacred cedar inlaid with ebony and ivory, gold and silver, and the paintings with figures of his family, birds, and animals, of all that he loved. He desired it greatly.

"I will give this chest to whoever fits it most exactly!" Seth exclaimed.

At once the conspirators took turns to see if they could win it. However, they were too fat, or too thin, or too short, or too tall. None fit it.

"Let me see if I can fit into this exquisite, marvelous piece of artwork," said Osiris, and he lay in the chest while all gathered around it.

"I fit exactly. It is therefore mine?" Osiris asked joyously.

"Yes!" Seth shouted.

Before Osiris could rise, the accomplices slammed the lid down, nailed it, poured lead to seal the top and, in the same way, his destiny. Seth and his companions took the chest and cast it into the river. Never would he be seen again, walking the land of the living! Never!

As lead is the heaviest metal and water is of the physical world, Osiris was trapped in the physical world but with no means to communicate with it anymore.

Hapi, the river god, carried Osiris out into the Great Green Sea, until it came to the shore of Phoenicia, near the city of Byblos. Finally, to escape, Osiris passed away and his spirit went to Duat, the Place of Testing. He could not yet pass beyond it to Amenti, where the pure souls live forever.

In the meantime, Isis was in great anguish. She had always known that Seth was filled with wickedness and jealousy, but Osiris would not believe in his brother's brutality. Isis knew Osiris was gone, although no one had told her.

She also had been warned that her brother was like a madman. He should never be killed no matter what damage he was doing, as all was already written. He was the divisive element necessary for humanity to one day reach perfection. But she was deeply suffering from being separated from her twin soul.

Meanwhile, men accepted the unacceptable by closing their eyes, by becoming insensitive to the plight of their brothers and sisters. They ignored their own souls, their own origins, and they anointed Seth as their king and master, believing that since he wanted that so much, he would care about their freedom and happiness. They were mistaken.

Seth, nicknamed the red hippopotamus because of his uncontrollable appetite to own and to control, wants to swallow the earth. He has either weakened or killed the other gods and has ignored Isis, thereby depleting her of her vital energies. Similar to a cancer, he spreads, multiplying his divisions in the head of Man, taking all life into him. Geb and Nut see all this, but they also know he is sterile. Isis received an allotted amount of energy from her mother, she passed it on to Nephthys, who served herself and passed it on to Osiris, who passed it on to Horus. Only then would it pass to Seth, and then back to Isis for another cycle of life. So Geb and Nut know very well Seth will eventually have no more energy to steal, and the physical earth will engulf him with all his vain schemes and illusions. By killing Osiris, the circle of life has been broken, but Seth is ignorant of life. In his image, humans destroy their house—the earth—and the housing of their spirits, their bodies, reflects this.

Isis fled deep into the marshes with baby Horus safely hidden on her heart, far away from the gaze of her destructive brother Seth. She found shelter on a little island where the protective goddess Buto lived and entrusted her beloved child to her. To further safeguard her son against Seth, Isis loosened the island from its foundations, letting it drift, preventing in this way anybody finding this sacred island. She did not fear for her person. Seth could do nothing against her essence, as they didn't live in the same world.

She started her search for the coffin containing her beloved twin soul. Far from there, the waves had cast the chest into a tree that was growing on the shore. The tree shot out branches and grew leaves and flowers. It became a mighty tree as the power of Osiris still existed, though he lay dead. The branches had grown together and hidden the chest in the trunk itself. Soon, all wondered where its beauty and fragrance came from. As the tree became famous throughout the land, people came from afar to see it, but none knew it hid the body of a god.

Meanwhile, Isis wandered, crying, all over the land. Despite asking everyone, she found no one who had seen the chest. In that world of the finite, her powers were of no use.

She questioned children who were playing by the riverside. Indeed, they said, they saw such a chest floating past them on its way toward the Great Green Sea. Isis blessed them and decreed that children shall speak words of wisdom and sometimes tell of things to come.

In the meantime, an illustrious king heard of the tree and desired for it to become one of the five pillars of his palace. There Isis came and sat down by the seashore. The maidens who attended their queen had come down to bathe. When they emerged from the water they saw Isis and asked her to teach them how to braid their hair. They had never seen hair like hers before. When they returned to the castle, a wonderful scent floated in the air. The queen asked how they had got that fragrance and how their hair became so beautiful. The servants told her of their encounter. At once the queen bid them to bring to her this lady of the seashore. When Isis arrived, the queen entrusted her two sons to her, and she graciously accepted. The youngest was ill but soon he became strong and well, although she did no more than carry him around in her arms.

When this was accomplished, Isis reverted to her own form. The queen fainted when she saw the shining goddess and suddenly understood who she was. The king and queen offered her all the treasures of their kingdom, but Isis asked only for the great pillar that held up the roof as she felt it

Isis Code

contained the remains of her beloved. When it was finally given to her and opened, she saw her beloved devoid of life and wept for her loss, powerless. Fleeing and hiding the chest in the marshes of the delta, she hurried back to the floating island where Horus was waiting. Horus was growing in beauty and dignity.

Widowed and having lost her reason to live, Isis implored Thoth, guardian of secrets, to help her. She asked him how to bring Osiris back to his body. After all, he had brought his own self into being by speaking his name. Thoth, lord of knowledge, searched through his powers. He knew that Osiris's spirit had departed his body and was therefore lost. To restore Osiris, Thoth had to help his spirit recognize his body and rejoin it. Thoth and Isis together created the Ritual of Life, which allows us to live forever when we shed our mortal selves. But before they could act, Seth discovered them.

Seth had been hunting wild boars with his dogs, and he was hunting by night since darkness served him well. By the light of the moon he recognized the chest. At the sight of it, hatred and anger surrounded him in a fiery cloud, and he stormed like a dragon. He stole the body of Osiris and tore it into fourteen pieces, scattering them all over Earth. Now he was sure Osiris would never see the light of day again! Standing atop a mountain in the desert he shouted: "No one ever destroyed the image of a god. Yet I, Seth the Great, did! I have destroyed Osiris!" His laughter echoed through the earth, and all who heard it trembled, hid, and wept uncontrollably. Isis sighed.

Isis pleaded with her sister Nephthys for help, for she was mistress of the world Seth controlled and she watched over the door to the underworld.

Nephthys left her wicked husband and joined her sister in her quest. Anubis, son of Osiris and Nephthys and king of the underworld, assisted them. Isis searched among the many streams of the world, traveling in a boat made of papyrus. Slowly, piece by piece, she recovered the fragments of her beloved. The crocodiles, in their reverence for her, touched neither the pieces of Osiris nor Isis herself.

Wherever she found a piece, she formed by her love the likeness of his whole body and inspired the priests to build a shrine and perform funeral rites in his honor. And so there were thirteen places that claimed to be the burial place of Osiris. In that way she made it impossible for Seth to further destroy the body of her husband.

Nonetheless, there was one piece she could not recover, for it had been eaten by certain fish. Isis, however, did not entomb any of the pieces where the shrines dedicated to Osiris stood. She gathered the pieces, rejoined them through her virtues, and made a likeness of the missing member so that Osiris was complete again. However, nothing that has shed its mortal self, not even a god, may use his mortal envelope once the link to manifestation has been severed. So Osiris went to Duat, the abode of the unborn. Anubis yielded the throne to him. There he stands in judgment over the souls of those who lived on earth. After the last terrifying battle, when Horus will reign over Seth in the hearts of men, Osiris will return to Earth once more.

The child Horus was growing in beauty and power. Many times the spirit of his father came to teach him how a noble heart should behave.

One day Osiris asked his son, "Tell me the noblest thing a man can do."

Horus replied: "To avenge his parents for the wrong done to them."

This greatly moved Osiris. He felt his son was ready to fulfill his mission. He kept on: "And what animal would be best suited to help you in this task?"

"A stallion" replied Horus, standing proud.

Osiris nodded but then added, "Wouldn't a lion be better?"

"A lion suits one who needs help, but a stallion suits one in pursuit of a vanquished enemy."

Osiris knew then that the time had come for Horus to defeat Seth. Re himself, shining father of the gods would come to his side in his own celestial boat that sailed across the heavens and through the dangers of the underworld.

The war raged, and is still raging. The last and greatest remembered battle was at Edfu, where the temple of Horus to this day stands in memory of it. The forces of Seth and Horus drew near to one another. Among islands and rapids, the forces of Seth, in the form of a red hippopotamus of gargantuan size, sprang up on an island and uttered a terrible curse against Horus and Isis. He yelled, "Let the waters of the Earth flood and destroy my enemies!" And his voice was like the thunder rolling across the infinite. At once a storm broke over the boats of Horus and his army, the floods of an uncontrollable reptilian world roared, and the water rose into tidal waves. Horus was unmoved, keeping the direction, his ship shining through the night, reflecting the benevolent light of Re.

Why It All Began: The Need to Understand

Enraged, Seth then turned and stood at bay, blocking the whole stream, so enormous a beast he was. But Horus took the form of a beautiful young man, twelve feet in height. He held a harpoon thirty feet long. The blade was six feet wide. Seth opened his mighty jaws to destroy Horus and his followers, but Horus threw his harpoon. It struck deep into the head of the earthly hippopotamus, deep into its brain. That one blow destroyed Seth the wicked, enemy of Osiris and of the gods. The storm passed away, and once more the sky was as clear and blue as Horus's eyes. The people came out and welcomed Horus, singing sincere songs of praise repeated by the priests ever afterward.

Short was the reign of Horus the brave, as men quickly tire of peace. When Horus passed from earth for he reigned no more, he appeared before the assembly of the gods. Seth came also in spirit, and both contended in words for the rule of the world. And they still do.

Egyptians believe that the last battle is still to come. Seth was vanquished on the physical level, but now he needs to be defeated on the subtle ones. They believe that one day Horus will permanently defeat Seth. Osiris will then cross the threshold of material life back to Earth, bringing with him all those who were faithful followers of Ma'at the implicate order. Because of this belief, Egyptians embalmed their dead and laid their inert bodies beneath impressive pyramids of stone. The tomb chambers, analogous to their passage from chaos to manifestation, are as many doors onto the underworld. There, their physical aspect awaits the pure souls returning from Amenti, ready to receive them again. This is how it will be: with the good god Osiris, Isis his queen, led by Horus and now protected by Seth and Nephtys, they will live forever.

Isis Code

Once upon a Time ...

Jane, twenty, and John, twenty-two, fell in love. Whatever role they occupy in society, they are still works in progress. Their brains will reach maturity as they approach thirty years old, with their frontal lobes expressing this. Meanwhile, almighty hormones have been pushing them one toward the other since their teens. Before he turned fourteen, you could read on John's bedroom door No Girls Allowed; after that: Girls' Night Every Night. Please, please please! Twenty-eight is normally when men start being less in the thrall of their reproductive organs and their powerful hormones. By then, many are married, have children, and suddenly realize this is not exactly what they really wanted or needed.

For now, John and Jane are oblivious to this. They don't know exactly what "being in love" means. But ever since they embarked on this magic carpet, everything feels so extraordinary! Their hormones will allow them to fly for free for six to eight months. Nature knows how to get us interested in reproduction. Depending on the health of their polarities and which phase of the system had its say in their mutual attraction, their story is already written.

Jane bears her ovules and has since her own birth. Now that her body is apt to conceive, an ovule is released every month. John produces spermatozoids at will. These are tremendously influenced by his daily environment. Already there is a difference. In her physiology, Jane is linked to the past and to the future and is cyclical, while John is concentrated in the present and ever ready to fertilize. This has a profound effect on their psychology.

What happens during conception?

The spermatozoa know where to go in order to find and fertilize the ovule. A mindboggling 200 to 500 million spermatozoa will race toward it. A few barriers along the way are erected to temper this passion, and only the most resilient will survive. Fertilization happens in one of the fallopian tubes, in the ampulla of the uterine tube. This ampulla has three layers. The most external is called serosa. We find a serosa mainly in three other areas: the external membrane of the heart (pericardium); surrounding the lungs, where it has an active function in breathing; and in the abdominal cavity covering most

Why It All Began: The Need to Understand

organs and in contact with the extracellular matrix. Interestingly, in our system as well as in the TCM, these are all attached to the Master of the Heart, and therefore represent the feminine polarity. These are doors used by the subtle level to influence our physical aspect.

What triggers the gametes to choose one fallopian tube over the other? It has been accepted that the phenomenon is chemotaxis. This means that chemical components motivate the motility of the spermatozoid. I would not be surprised if in a few years we learn that it is also a difference in potential between the spermatozoid and the ovule that is responsible for this attraction. The pH of the uterus is alkaline while the vaginal fluids are acidic. Semen has a pH of 7.2 to 8, so it is slightly alkaline. An acid has a positive charge while an alkali has a negative one. With a sea urchin, for example, we can observe at conception a change of the electric charge through the surface of the ovule. This is a primordial step to avoid polyspermy. This charge is present at the beginning. As soon as a spermatozoon enters the ovule, the polarity changes, preventing entry of other genetic material.

Contrary to urban legend, the spermatozoon does not pierce the ovule and the first to arrive is not necessarily the winner. As I learned in 1982 at a medical convention, if the spermatozoon is in chemical affinity with the ovule, the ovule seems to liquefy its membrane, *allowing* the spermatozoon entrance into the inner sanctum after it has left its shoes (its mitochondries/energy reservoir) at the door. Adenosine triphosphate production (ATP) from mitochodries has a predominent role in biological electrochemical energy transfers within the pellucid zone. Two Russian scientists, Vera Schroeder and M. Koltzoff, were the first to identify a ephemeral glowing hoop at the moment of contact between spermatozoon and ovule, like an instantaneous firework.[105] Could this be similar to the direct current used by geneticists to jumpstart mitosis? Also in 1995, research by Dr. Kenneth Glander of Duke University, Durham, found that in Muriquis monkeys, vaginal electric potential of mothers to be will modify in order to promote fertilization of the ovule by male or female spermatozoa according to the need to preserve the male/female equilibrium of their group. He noted that female howlers consumed certain plants before and after copulation that they did not eat at any other time.[106]

Scientists believe that the binding of a sperm to an ovule involves a host of other molecules. The exact mechanisms are not yet adequatly understood. We know nevertheless that in human, ovule metabolism undergoes variations in sodium and calcium concentration, therefore influencing pH and potentiality.

So the spermatozoon has entered the inner sanctum, forgoing elements that would prevent the fusion. They are now one. All the cells originate from this new being. All are daughters of this unique corona radiata ovule. All will be in affinity with that first goddesslike cell. All will be in affinity one with the other. We should always keep this oneness in mind.

In 1933, Schroeder and Koltzoff published a research in *Nature*[107] magazine, in which they recorded their use of electrophoresis.They concluded that depending on whether they carry a chromosome X or Y, spermatozoa have opposite polarization. The X spermatozoa—the feminine chromosome—has a negative charge, and the Y spermatozoa—bearer of the masculine chromosome—has a positive charge. This confirmed many other studies that revealed that a weak electrical current passing through a solution containing spermatozoa provoked the X chromosome to be attracted to the anode (+) and the Y chromosome to be attracted to the cathode (-). Following this, in June 1992, the department of biology at the Scientific University of Tokyo made public the results of a study carried out by five Japanese scientists (Ishijima, M. Okuno, H. Odagiri, T. Mohri, H. Mohri) titled "Separation of X and Y chromosome-bearing murine sperm by electrophoresis."[108] Their research agreed with the results of the two Russian scientists. We have to agree the idea is interesting.

On the physical level, John and Jane are affected by their biology. Each of their cells reminds them that biologically, John is male and Jane is female. Complex hormones will reiterate this throughout their lives. The shapes of their bodies alone tell them a simple truth: the male has an organ used to penetrate, insuring perpetuity of life. The essence of this penetration thus gives it its nature—it is expressive. The female has an organ to receive the semen and may in this way bear a child, ensuring also perpetuity. She is receptive on the physical level. The essence of this reception is thus receiving in nature. She will eventually become a giver at the birth of her child, who will be received by nature and society. Those two functions, receptive and expressive, are at play at each level of the individual. They are associated with masculine for the expressive function and feminine for the receptive function.

Through the different aspects of our beings, we are oriented similarly to an electromagnetic wave. If a person is expressive on his or her physical level, the emotional level will be receptive, the rational/analytical level will be expressive, the collective level will be receptive, and the symbolic level will be expressive (schemas 10–16). Expressive and receptive are at play on all levels, simultaneously. This is what we name sexuality. It does not concern only the physical level.

This leads us to a first observation: homosexuality does not truly exist. The sexual automatically involves two polarities. There are only physically barren couples. We have qualified as "woman" the human who has female genitalia, and "man" the one who has masculine genitalia. We judged humans only through this crudest aspect. It is a partial and limiting point of view, extremely reductionistic. Since our point of view of the world is highly materialized through the predominance of masculine values, this was normal but far from the truth. There is more to us than our physical skin. Since we bear the two polarities on different levels, I believe we can all become homosexuals in the literal definition of the term, given some environmental factors. I will further discuss this in the chapter relating to the Reptilian brain.

So John and Jane will be parents. For nine months Jane will do her best to build a temple to this godlike being they called into their physical world. Whatever society tells her, the more her pregnancy progresses, the more Jane can feel a connection to this soul. There is undeniably something sacred and transformative about this experience. The child is not a copy of herself, her heart tells her that. The fetus already has an independant life inside her and influences her behavior. Their hearts are synchonized, and everything she feels, he experiences. The reverse might as well be true. Applying a novel concept from physics and nonlinear dynamics to their data, Van Leeuwen et al. uncovered a hitherto unknown phase synchronization between the individual heartbeats of mother and fetus—a marker of coupling between their autonomous cardiac systems despite continuous noisy fluctuations in the beat-to-beat intervals.[109] They also demonstrated that maternal–fetal cardiac coupling are probably not mediated by maternal breathing. One thing is sure. The child to be is bathed in the magnetic field of her heart, constantly.[110]

In the biocybernetic system, the phase they are both going through is lung—in which the environment is almighty—toward kidney, when the baby will be born from the waters of her womb with his first breath into this world. He will cry, not laugh, coming into this reality. What is he leaving behind? I say that when babies are born, they draw instinctive smiles and joy from us because they come from the higher heart. They still contain the light from the world they left behind and we somehow recognize this. For the Egyptian tradition, everything starts in this period before birth: the lung. Egyptian priests, in the same way as Kabbalists, druids, and Chinese sages did, considered that the day started the previous evening, which, in accord with TCM, corresponds to the lung. The following phase, Kidney, is linked to the night and to the crossroad between manifestation and potentiality, what we labeled life and death. This is echoed in our religious tradition: "And there was evening, and there was morning." The soul gave up a former body in this phase and a new one will be given as a medium in this phase also.[111]

The kidney, or house of reproductive energy, is really at the crossroad of physical life and death. This might explain why the genital center in males' brains, the amygdala, is also central to aggression.

Paul D. MacLean, the American neuroscientist and physician who devised the evolutionary triune brain theory, observed that the brain could be divided in three stages of evolution. This theory, although not accepted by all comparative evolutionary neuroanatomy researchers, is very popular among psychiatrists as a valuable tool. Also, recent development in brain mapping and embryology tend to give credence to Dr. MacLean's theory.

Taking the fractal aspect into account, I have divided the brain considering the LIFE biosystem approach. Evolution follows a spiral, which is similar to the triurne brain theory with some details added, and in harmony with the most recent discoveries of brain processes. We thus find the cerebellum and Reptilian brain; the limbic system and occipital region of the Mammalian brain; the Human brain with its parietal lobe and medial prefrontal cortex; the Analytical brain of the rational self and its frontal lobe, with emphasis on the left hemisphere; and finally the social and Universal brain, with its emphasis on the right hemisphere and temporal lobes (diagram 15). When I mention the name of an organ, I imply also its attached energetical structures as described in TCM as well as in MBC (Medical biocybernetics). Let us never forget that these divisions are only general, since in humans *all levels of the brain and of the rest of the body are constantly and simultaneously communicating.*

We will now embark on a pilgrimage through the different phases of our personal cybernetic system.

Chapter 3.
The Reptilian Brain, the Physical Self, and the Instinctual Human

"*If we could raise one generation of children with unconditional love, there would be no Hitlers.*" **Elisabeth Kubler-Ross**

"*In any way we use our might against weakness, we are promoting, in no matter how seemingly insignificant a fashion, the spirit of war.*" **Leo Tolstoy**

Phase of the What and Where Level

The Kidney Aspect

- structuration phase: from conception (or previous phase) to seven years old (teeth and speech)
- main sense: hearing (from previous stage completion) toward vision
- building blocks: from conception to birth
- nourishment: nutrition of physical elements and subtle elements (love, affection)
- aspect prepared: physical, frame of emotional
- cybernetic phase and function: from genital to defense
- brain aspect: Reptilian brain—brain stem, spinal cord, right amygdala, right hippocampus, cerebellum, anterior hypothalamus/pituitary gland, entorhinal cortex (from lung to kidney), ventral tegmental area, coreleus neurons, and more
- cybernetic associated main organs: kidney and bones, attached structures, feeding liver

Isis Code

- level inhibition and control of human self: parietal lobes, fronto medial
- level controlled by nutrition, later by psy energies (energy from the psyche) and environment
- cybernetical gender structure: part of the masculine polarity (the receptive aspect of the masculine polarity). This phase is priviledged by men and is part of the masculine polarity. Women receptive, men expressive.
- cell structure: DNA
- energetical function: reproduction
- stimulant: emission of low vibration and sounds (drums, heavy death-invoking sounds, exhaust motor sound) physical violence, money, pornography, masturbation, demonstration of physical power (sports, military), strenuous exercises, the salty taste in nutrition. (Overstimulation brings a depletion of the corresponding energies of those structures and a subsequent enslaving to that phase.)
- gift of that phase: the moving animal (freedom from space)
- motivation of that phase: need (survival of the biosystem)
- type of attraction/relation: dominance/genital/libido
- main relational attitude: binary
- psychological: seeking
- emotional: freeze/fear, fight/flight
- Level of awareness: personal unconscious added to collective unconscious
- house analogy: foundation following the blueprint
- Night
- Bible equivalent: Genesis
- Egyptian god: Nephthys
- Blue indigo/black
- figuratively: pH
- element: water
- Love type: instinctual, physical, egocentric, genital

The dictum "structure defines function" fails in the face of evolution as the evolved brain uses ancient structures differently and for other functions. Furthermore, for the same function, men and women will use different brain structures.

This period of the Reptilian brain concerns the question *what*. As an unconscious successor of the *I am* on the spiral of life, the

The Reptilian Brain, the Physical Self, and the Instinctual Human

Reptilian brain lives in the instant. It is the phase of all ends and of all beginnings, the door to materiality. It is the first question we ask when faced by novelty. It is the question we ask when we wish to deepen our knowledge. But the *what* limits itself to the observable and immediate.

In our biocybernetic system, the kidney aspect concerns the most retracted, concentrated, and receptive phase. We therefore assign to it our most condensed aspect: the physical body.

Under the control of the Reptilian brain, I also call it the subconscious motor brain (diagram 12). All other brain aspects will be built on this one, as we know that motor skills allow for the development of higher brain function. It is the foundation of our physical, emotional, rational, and collective home. Figuratively, it is the basement of our house.

Through it, information from the body, including from the heart (neurologically, hormonally, electromagnetically, and biophysically) constantly flows.

The Reptilian brain serves our will to power and makes our physical bodies the focal point. All our other levels will tend to concentrate on it for the length of a lifetime.

Well adapted to territorial dominance and genitality, this system expresses itself through automatic, subconscious reflexes. It is said that we all have in us the tail of a reptile. To cut the tail is to cut any possible manifestation. As we know, brain stem trauma can lead to coma and death.

This most primitive aspect, which Dr. MacLean also named the R-complex, is composed of the mesencephalon, which connects the hindbrain to the forebrain, and the rhombencephalon (hindbrain). Mainly, we will talk here of the brain stem and spinal cord. The tectal-tegmental interface of the midbrain is, without contest, the most ancient precursor of reactional expression.

The mesencephalon directs hearing, pupil dilation, body and eye movement, and controls response to sight. The hindbrain relates to attention, autonomic functions, complex muscle movement, maintenance of muscle tone, conduction pathway for nerve tracts, reflex movement, simple learning, arousal, balance, cardiac reflexes, blood circulation, and sleep. In it, the pons relays sensory information between cerebrum and cerebellum. The myelencephalon—the part of the brain stem that contains the medulla and some nerves, including the vagus nerve, which will be discussed later—controls the autonomic functions, breathing, digestion, heart rate, swallowing, vomiting, and sneezing.

It reacts to direct stimuli.

This brain complex is not interested in emotions per se, other than fear, rage, and arousal. Those, in fact are binary reactions, not what I would call emotions. Emotions are modulations, points of convergence between the subconscious and awareness levels, and are of a more elaborate nature. The Reptilian brain's job is to help us survive in the immediate and this material world. It has no interest in niceties. It is as binary and as eloquent as a spasm.

Emotions, on the other hand, carry time, and thus personal memory. The brain aspect adapted to emotions controls the means to discharge these emotions from the system: the voice for the Mammalian brain and action for the Reptilian brain. Women will favor the former, men the latter.

The particularity of the Reptilian brain is it's instantaneousness. Time for the Reptilian brain is punctual and finite. It doesn't exist. Don't expect joy, happiness, or remorse at this level. There is no tomorrow. The most you can get is survival impulse or the relief felt after the release of a tension. Nevertheless, we cannot impede the development of this aspect without endangering our future.

When we observe the behavior of reptiles, fishes, and snakes, we see that they share a predominance for the left/right, sideways body movement. Backward is not their forte, and immobility is preferred in the image of the Kidney energy. The Mammalian brain, with the predominance of hormones and neurotransmitters and with the awareness of cycles, will prefer down/up/down movements associated to the forward ones. The constantly upright body position will be preferred by the Human brain. With each of these phases, different abilities emerge that allow the entity to rise above the limitations inherent to its awareness of this world, whether those relate to space (right/left) and action (forward and out), emotion (down/up and backward),[112] thought (up/down), awareness (in/out/in), or time (past/present/future) (diagram13). Of course, in humans all of these are present all the time from a young age.

The "primitive" man still in the primordial lung phase did not dissociate psyche and external world. He lived in a type of space-time continuum. On this subject, Carl Gustav Jung says,

> With the primitives, the frontier between the unconscious and the external world disappear.... We can little speak of a contact between an "I" and the environment since there is not yet in him an "I" in the sense we give to this word. His consciousness is a state of immersion in a flux of events in which the surrounding world and the inner world are not, or only very vaguely, distinct.[113] (My translation)

The world of the Reptilian brain is thus the world of the territorial in the instant, a world of separation. The here and now. For this aspect, time and space are still merged but in *one point outside*, toward which all the brain's attention converges and focuses. Space will develop when time is added with the Mammalian brain. The action of time within the implicate order eventually brings awareness of space (matter). This is the basis of functioning of a brain; continuously but discontiguously, "using temporal packages or quanta."[114] In the same way embryology teaches us that we are born with synesthesia, merged senses that gradually become modular.[115] This is also true for all the other levels. In the Reptilian brain senses are merged, but they are unconscious and therefore not expressed. This signifies that for a baby, for instance, sensory deprivation *is* emotional deprivation. Perception takes place on an unconscious and profound level and is reacted upon on an all or nothing fashion. Although the Reptilian brain is totally dominated by the past, the subconscious, and the environment, it is not aware of them. It has no memory other than the one being recorded by the cells. It is a world focused on a point in space and time.

In the LIFE system, the Reptilian brain is related to the phase spanning from birth to defense, part of the masculine polarity. This corresponds to the period from birth until seven years old. In the

Isis Code

traditional law of the five elements, it is linked to the kidney aspects toward the liver aspects and its innate immunity.[116] These are, in this instance, not only organs but energy systems endowed of all the different levels (physical, emotional, deductive, collective, symbolic). Although we usually consider only what we can flatten between two glass plates under a microscope, there are more dimensions to everything, but they cannot be accessed by solely using analysis.

Our different senses are tools for translating stimuli (sounds, smells, light, touch, taste, love) into electrical signals—the language of the brain. In the same way, the different aspects of the LIFE biosystem translate external and internal stimuli into electrical signals that the different brain aspects and the body will decode and react upon. In TCM, the kidney is considered the producer of marrow and the brain is called "the sea of marrow." Therefore, the brain is energetically a tributary of the kidney in the sense that it is qualitatively and quantitatively dependent on the genetic energy inherited from the parents.

Can we find a link between kidney function and brain health? Interestingly, a meta-analysis of thirty-three prospective studies comprising 284,672 subjects showed that a low glomerular filtration rate at baseline was independently related to incident stroke.[117]

In another research, a total of 375 subjects—335 with chronic kidney disease and 40 with essential hypertension—were included. Among causes of chronic kidney disease, hypertensive nephrosclerosis had a robust association with silent brain infarction while recognized risk factors didn't.[118]

In evolution, subcortical structures developed before the cortex. For millions of years they guided the complex behavior of many organisms. Two sets of structures could be identified: the thalamus, in charge mostly of receiving and processing information; and the basal ganglia, in charge of motor behavior and action. In humans, I associate both with the Mammalian brain, having evolved from the Reptilian one.

The amygdala, which I place chiefly within the Reptilian brain because of its instantaneity, regulates the interaction of the organism with the outside world (right amygdala), in its quest for survival of the organism and the species. Like the Reptilian brain, it is binary in its approach to the environment.

The cerebellum copies the anatomy of the whole brain. It is necessary for coordinating fine movements and sensory data, and plays a part in regulating thought and human language. The cerebellum and the basal ganglia work hand in hand with various areas of the cortex and neocortex. The cerebellum also carries about half of the number of brain neurons (about 50 billion). It has been shown to have a close bond to the frontal lobe, a definite role in complex planning, and to coevolve with the dorsolateral prefrontal cortex (DLPC). This is confirmed by the LIFE biosystem theory, which endows the Analytical brain, in which we find the prefrontal cortex—and particularly the left hemisphere and DLPC—with the ability to control and inhibit the Reptilian brain. This is also in harmony with both TCM and BCM, in which the pancreas aspect controls the kidney aspect. Furthermore, it explains the astounding number of fibers these structures receive in humans compared to nonhuman primates, as well as the lack of drive witnessed in dorsolateral frontal disease patients.

Subconsciousness should not be underestimated. I consider that our level of awareness is similar to the known matter of the universe, 4.6%. This explains Libet's brain experiment: when you touch the aspect "hand" in the somatosensory area of the brain, the reaction is less direct than physically touching the hand. The brain starts by planning and executing a movement *unconsciously*. This is the world of predilection of the Reptilian brain.

Try to remember the earliest memory etched on your consciousness. What is it? There is a distance between when the memory is coded and when one retrieves it. To remember is to become more aware of something. It could be there, in the subconscious before it ascends as awareness. Then you remember. This happens because from the period of conception to around seven years old, many of our lives' events are stored directly in the subconscious. They do not need to be traumatic events. During this time, it is as if the senses record everything, like a sponge. The conscious being, relatively narrow at this time, uses the subconscious as a box. Consider a tourist who packs away all the pictures he takes while on vacation. When he gets home and looks at the pictures, he might be surprised to see that he had forgotten a few sites, and that these are attached to events that he would have forgotten if he hadn't taken the picture. The Master of the Heart is responsible for the coding of memories. Our brain is responsible for retrieving them.

What happens in this first period of life is decisive for the quality of the individual's feminine polarity. Traumatic events from this time, including during pregnancy, may influence the whole life and, notably, will color the ambiance of one's daily life. We can compare it to a painting; the artist starts by covering the whole canvas in white paint. This will be the backdrop for the scene the artist wishes to portray. The unconscious memories of the first seven years and nine months are like this backdrop. Or it could be compared to the atmosphere music for a movie. You don't actually pay attention to it, but it is there all the time and grants every image an emotional hue.

The brain stem, the most phylogenetically ancient of our brain structures, is the home base for the brain chemicals known as neuromodulators. These include mainly dopamine, but also serotonin and norepinephrine (noradrenaline). Researchers have determined that the ventral tegmental area (VTA), a tiny nucleus sitting in the ventral midbrain, is the origin of dopaminergic pathways. It is probably an amplifier and conveyer of signals computed elsewhere in the brain. This aspect of the brain is suspected to be the root of many psychiatric disorders, such as addiction, schizophrenia, and Parkinson's disease, and possibly cancer, as we will later see.

Jane and John—Birth to Seven Years Old

Jane and John have just become the proud parents of a little boy. Was it sadness, fear, or rage Jane heard in his first cry? She can't say. After the infant was cleaned and all the measures were taken, the nurse finally handed the child to the new, beaming mother. How should I hold him? Jane asks anxiously. He is so small!

She wants to hold him against her heart and never let him go. Having the infant physically removed from her even for only a few minutes made her feel empty, nervous, and anxious, even though John was right there beside her, smiling. He looks happy and relieved, but also so pale and tired!

Her breasts feel like they will burst, as does her heart, overwhelmed by tender feelings for her baby. He is so beautiful! She can't take her eyes off him. John feels a little neglected. As she breastfeeds him, the nurse tells her she will need to rest after. Everything is so fuzzy. Then she hears the nurse say she will bring the baby back in one hour and a half, promise. Jane doesn't want to, but she has to let the infant go. Too soon, the nurse leaves with her baby and John in tow. Her first motherly pain, a needle pricking her heart, surprises her by its brutality. Alone, she feels as if she is floating away, in a half dream, in search of her child. She sees her baby, feels him as in the last months of her pregnancy. She compares what she feels from him now to eminent men of the past. Somehow she knows him. And now she knows the name he should bear. For the next seven years, John and Jane will shield this child against the immorality of the world by surrounding him with their love.

"A particularly critical period lasts from approximatively ten to twelve months to sixteen or eighteen months, during which a key area of the right frontal lobe is developing and shaping the brain circuits that will allow the infants both to maintain human attachments and to regulate their emotions. This maturing area, the part of the brain behind our right eye, is called the *right orbitofrontal system*."[119]

Allan Schore, a psychiatrist from California and a leading researcher in the field of neuropsychology, has suggested that the quality of early mother-infant relations influences the quality of the frontal lobes later in life.[120] Likewise, various studies report that stress in infancy might eventually impair frontal-lobe function. The mother/child bond that began in utero will create in the future adult the aptitude to control his Reptilian brain. Also and most importantly, this aspect and phase of the system is the most receptive of all.

The right hemisphere of the brain, related to the feminine polarity, remains dominant in the child until the age of three and a half years, as if responding to the previous phase of the lung, in which the right hemisphere dominates. Of course, when we are born, we already have all the elements necessary to live. They are not fully developed, though, and will continue to mature until we turn thirty. For women, this period extends also into pregnancies. The dominance of the right hemisphere explains why four-months-old infants, when presented with female faces, show enhanced gamma electrical activity in their right hemispheres.[121] It is as if they are seeing themselves in a mirror.

The period of childhood, from conception to fourteen years old, is a long one. During a conversation, a doctor told me that, subjectively, twelve years old is our half-life. I could not find the research concerning this as it was in France a while ago, but research tells us now that: "The subjective duration of an interval of real time varies inversely with the square root of the total real time (age)."[122] This is to say that for a one-hundred-year-old individual, time flies by. How can this be? When the brain slows down, time subjectively speeds up. When the brain speeds up because more excitatory neurotransmitters are at work, time subjectively slows down. Hence the slow motion and enhanced memory of a new love experience. Not only are we young then, but many neurotransmitters are at work when we see our beloved! Consequently, for a ten year old, one hour of waiting is exceedingly long indeed. It is excruciatingly long for a baby.

When we force our children to sit down without physical activity or playtime for hours on end, this is detrimental to their Reptilian and Mammalian brains, and consequently to all their different brain structures. I am not surprised that ADHD affects more boys than girls, since boys express their masculine polarity on the physical level.

We are changing the brains of these kids by forcing detrimental behavior on them. They should be free, in contact with nature, surrounded by love, at least until they are seven years old. Before then, learning is okay, but only through play and not *instead* of the necessary expression of their biological needs. After seven, formal teaching can start, but with respect for the ongoing developments of the child. Finland is leading the world on this. They have one of the best educational systems in the world. One key feature of this system is that they start primary school at seven, not before. Anu Kara, a public-school teacher there for the last fifteen years, commented: "I teach less and they learn better because they are interested." They might start later than in other countries, but they learn faster and better.[123] I wish all kids could have this chance.

Love, given by biological parents or surrogates, is a must. In the LIFE system, we learn that the heart aspect of the mother controls the

The Reptilian Brain, the Physical Self, and the Instinctual Human

lung aspects of the child. Further, her lung aspects feed his kidney aspects, and enhance or slow down the distribution of energies throughout his body in all its dimensions (diagram 14). The child's LIFE system is a fractal, a clone, or a mirror of hers. The more she is in agreement with the general LIFE biosystem, the better for the child. She is thus the first and foremost epigenetic element influencing the child. Although the umbilical cord has been cut, this energetically mirrored connection prevails until the child is about three years old, greatly influencing the child until his personality (the ego) is built. It lasts until the mother passes away. Leonardo da Vinci said that mother and her unborn child share the same soul. He was right in that they share the same space/time in the Master of the Heart field. Quanta born to the child will always be connected to those of the mother.

Time for a baby is subjectively the longest. So I have to wonder, what animal in nature separates the newborn from its mother at birth? This is a crucial period. What animal refuses to breastfeed its baby?

There are good reasons why the breasts of a human mother are positioned so the infant can hear the rhythms of her heart, and not placed lower as those of many other mammals. Feeding becomes the natural phase, echoing the oscillations that bathed the child before birth. With it, the child feels he is "back home." It allows him to continue his development extra uteri as a natural prolongation of the inner. It helps him steer clear of the division of inner and outer. In normal development, touch (heart aspect), smell (lung aspect), taste (pancreas aspect), sight (liver aspect), and movement (kidney aspect) are integrated during breastfeeding and encoded in the developing brain. The three first elements express the feminine polarity and integrate these to the masculine one. It has been found that even short-term separation from his mother leads to elevated cortisol levels in a baby, indicative of negative stress.[124]

"In fact, after one full day of separation, infant rats already show altered brain organization of chemical receptors. A similar rat study revealed that one day without mother actually doubled the number of normal brain cell deaths."[125]

I do not see why human infants would be less sensitive than infant rats. They are more sensitive. In fact, newborns separated from their mother at birth will soon build a resistance to touch and nurturing, and the ability of the brain to assimilate the meaning of touch will be impaired. This fits with the Master of the Heart, which will be discussed in the fifth chapter.

Bonding and breastfeeding are two natural laws that exist for good reasons. That we took the liberty to think they are not of prime importance only goes to show the dominance of our masculine polarity. Biologically and socially, we are damaged as a consequence of welcoming our children into this world in this way. For nine months, the child was never alone. Suddenly, he is abandoned in an alien, aggressive world, separated from a part of himself—his mother. All of his receptive aspects are similar to a sponge until he is seven, and most importantly for the first three years. The information he receives, or is deprived of, during that time will model his cells. We will never know how we might have developed given a proper rearing. Nonetheless, it is not too late for the generations to come, although it takes three generations to erase the effects of deficient nurturing.

René Spitz (1946/1965) documented that even with medical and physical care, infants that are isolated in cribs and who receive little or no physical tenderness can and do die from an emotional wasting away, a depression that he called marasmus.[126] These infants are, of course, bottle-fed. Tryptophan, an essential amino acid that converts into brain serotonin, is richly present in colostrum and breast milk but absent from formula milk. This lack could lead to a depressive state. Medically, however, the health problems of these babies are linked to the fact that they cannot assimilate nutrients. That seems to be a consequence, not the cause.

Dr. Cicely D. Williams, a Jamaican pediatrician, introduced the name kwashiorkor through her 1935 article in *Lancet*.[127] The illness is similar to marasmus, except that it manifests itself in babies older than eighteen months. Marasmus is observed in younger infants, and involves an edema seen in the typical distended abdomen popularized by food aid agencies. Kwashiorkor can be translated as "the sickness the baby gets when the new baby comes."[128] It is an acute form of marasmus. Before, 90% of the victims of this illness would not survive. Apparently, despite foreign aid, the numbers have not improved that much nowadays. These children present different symptoms due to an enlarged liver with fatty infiltrates and lack of lung energies, such as anorexia, dermatosis, and irritability. Peripheral edema and skin lesions predominate. As Dr. Latham, director of the International Nutrition Program at Cornell University pointed out: "When flaky-paint dermatosis is seen in a malnourished child with edema, it is pathognomonic of kwashiorkor."[129] The small victims also fail to produce antibodies following vaccination. Often their systems for protein assimilation and synthesis, itself made of proteins, become deficient, leading to death. The introduction of food is therefore, alas, too often an ineffective treatment.

Several studies concerning foundling homes have noted that many of the children sent there died. For instance, from 1775 to 1796, 10,272 infants were admitted to the Dublin Foundling home. Only forty-five survived. The foundling wheel, which meant babies could be abandoned at the home anonymously, was dismantled in 1826. Popular during the Middle Ages to prevent infanticide, baby hatches were reintroduced in 1952, and since 2000 they have been used in many countries. In May 2010 in Vancouver, Canada, one was opened. It is called Angel's Cradle.[130]

Deprivation negatively impacts maturation of the Reptilian and Mammalian brains. If it starts soon after birth and lasts six months or longer, it results in the most profound disturbances in a child's social and emotional development. If deprived later in life, children show symptoms of septal nuclei destruction. The septal nuclei composed of structures that lie below the rostrum of the corpus callosum, plays a role in reward and reinforcement. Those deprived children will crave social contact, but will behave in socially inappropriate manners, exhibiting either shyness or bullying.

> "The medial amygdala and later the cingulate and septal nuclei are the most vulnerable during the first three years of life. If denied sufficient stimulation these nuclei may atrophy, develop seizure-like activity or maintain or form abnormal synaptic interconnections, resulting in social withdrawal, pathological shyness, explosive and inappropriate emotionality, and an inability to form normal emotional attachments."[131]

Isis Code

Unmothered girls who become mothers might express symptoms similar to those seen in bilateral amygdala destruction. Their amygdalae become unresponsive. H. Harlow, who ran the primate lab at the University of Wisconsin in the 1950s and '60s, in 1965 found in primate research that: "After the birth of her baby, the first of these unmothered mothers ignored the infant and sat relatively motionless, hour after hour. Other motherless monkeys were indifferent to their babies or brutalized them ... and nearly killing them until caretakers intervened. Despite the consistent punishment, the babies persisted in their attempts to make maternal contact."[132]

René Spitz, who worked as a consultant in a foundling home during World War II, witnessed the sequel to human-contact deprivation and named the syndrome anaclitic depression. Beginning in 1950, he was a mental health consultant to the World Health Organization. In 1965 he described the four different emotional phases an infant younger than age two would go through within this syndrome.[133] The first phase lasts days and sometimes weeks, and involves the infant crying and screaming. If the separation continues, in the second stage the child will not cry anymore and will lose interest in its environment. With the third stage, the infant does not respond to affection anymore. He often sits and stays motionless. If continued, the separation would bring further deterioration, illness, and possibly death.

Montague Francis Ashley (1905–1999), otherwise known as Montagu,[134] was born in England under the name of Israel Ehrenberg and was an anthropologist and humanist. He presented a petition in the World Court to outlaw male genital cutting and female genital cutting of children. A prolific writer and recipient of the humanist of the year award in 1995,[135] he exposed American sources opposed to mother/child bonding. These sources, some of them noted here, have established wrongful child rearing practices for the past century. These were exported worldwide. Unfortunately, their influence is still felt and their code sometimes applied internationally. When my children were born more than twenty years ago, I was told to ignore my babies' cries. Of course I simply couldn't. In his book *Psychological Care of Infant and Child*, published in 1928, John Broadus Watson stated: "a sensible way of treating children ... Never hug or kiss them, never let them sit on your lap. If you must, kiss them once on the forehead when they say good night." Luther Emmett Holt, the leading pediatric authority of his day, stated in his 1894 textbook: "To induce sleep, rocking and all other habits of this sort are useless and may be harmful." In 1916 he advised that the crib should not rock in order that "the unnecessary and vicious practice may not be carried on." He could not have been in greater error, as we now know that gentle rocking of the infant/child is essential for normal brain-behavioral development and bonding. When that truth is shown to us, what previously seemed truthful but was not, although it was "scientific," suddenly appears as it truly is: inhuman.

Similarly, Dr. Ferber, a well-known American pediatrician, noted in 1985: "If your child is like this, you may be comforted to know that head banging, body rocking, and head rolling are very common in early childhood and, at least at this age, are usually normal. If your child exhibits any of these behaviors there is little need for concern about emotional difficulties or neurological illness."[136]

Can someone be more wrong?[137] But he can be forgiven; he was only speaking from what was known then. It only goes to show how deprived the children were in 1985, to believe that these behaviors were "normal." We now know for sure that body rocking and other stereotypical behaviors are expressions and consequences of sensory and emotional deprivation (Somato Sensory Affectional Deprivation, SSAD). With the discovery of the sense of synergy we understand why lack of touch and movement has been shown to stunt nerve-cell growth. We can also imagine that lack of love would have the same effect. The brain literally craves movement, sensory stimulation, and love for normal development. Why is it so difficult to accept? Because we can't measure what is immaterial: love. The masculine polarity, hence our science, cannot deal with it.

Also, movement stimulates the vestibular sensory system, the auditory system of the inner ear, which in turn connects to the cerebellum. At all ages, and more so during brain development, movement promotes corporal coordination, integration of binocular vision, hearing, circulation of CSF liquid, and activation of the frontal lobes. All bodily functions and all autonomic functions are thus affected by movement.

At night, I was petrified and used to rock myself to sleep. I rolled my head from side to side. I suffered from insomnia and continually felt emotional pain and sadness. When I would finally fall asleep, exhausted, I had recurrent nightmares. I was not the only one in my family who exhibited this type of behavior.

In our system, the pancreas is associated to rhythmic movements and is part of a security system intended to safeguard the homeostasis of the organism at all levels. Rocking is a natural movement for self-soothing and is carried on from before birth. The repetitive movement triggered by obsessive compulsive disorders are in fact indicative of ineffective efforts by the system to regain balance.

One has only to witness the happy faces of children when they are on a swing to realize that movement is crucial to their well-being. This need is modeled on the original motion a fetus feels from its mother walking and breathing, with her heartbeat punctuating her movements. These movements, felt all the time by the growing fetus, literally bond infant and mother before birth. The infant who experiences movement knows it is alive and that it is connected to the mother, creating a nurturing feeling of safety. So intense is this need to bond that young animals raised in isolation have been known to form attachments to inanimate objects such as television sets, even to animals that might maul and kill them. Babies will frantically seek connection, even with mothers who physically reject and abuse them.

The rhythmic movement is an appeal to synchronize systems that might have lost their synchrony. Who has not, in their pain, rocked themselves to comfort?

Joseph Chilton Pearce in *Magical Child* (1977) underlines the importance of bonding: "Bonding is the issue, regardless of age. Bonding is a psychological-biological state, a vital physical link that coordinates and unifies the entire biological system. Bonding seals a primary knowing that is the basis for rational thought."

In our LIFE biosystem, bonding relates to the heart aspect and to the feminine polarity. Lack of feminine polarity generates lack of ability to bond in adults. Drs. William Mason and Gershon Berkson (1970) conducted experiments with monkeys. They compared monkeys raised alone in cages with fixed "mother" cloths and bottles, with monkeys reared by swinging "mother" cloth and bottle surrogates.

The monkeys in the cage with the still, lifeless surrogate developed some of the symptoms manifested by isolated baby monkeys, including depression, withdrawal, aversion to touch, rocking, chronic toe and penis sucking, self-mutilation, and pathological violence as juveniles and adults. If a new monkey was introduced into the cage with a monkey brought up with the static surrogate, the resident money would attack the new monkey if it tried to be friendly. In other words, instead of accepting physical touch, the monkey reared with a still mother would violently push the other monkey away and defend itself at all costs.[138]

Monkeys reared by the moving surrogates developed normally with only minor comfort-seeking behaviors, such as thumb sucking. If another monkey was introduced in the cage, the first monkey reacted calmly and gently, letting the new monkey smell, touch, and play. Baby humans need even more relational interaction than monkeys do.

A 1967 study by Dr. Mary Neal showed positive effects for premature infants who were rocked in swinging bassinets. These premature babies showed accelerated neuromaturation, gained weight faster, lifted their heads sooner, had stronger grasps, and were dismissed from the hospital sooner than those who had not been rocked.[139] Despite these impressive results, this need has been overlooked and newborns are still placed on fixed mattresses. Older people would also benefit from rocking movements, to enhance blood circulation and comfort. I have seen my ninety-six-year-old mother develop bedsores in the hospital because of the lack of movement and blood circulation, not to mention the emotional ill effects she has had to endure because of forced immobility.

Pioneering studies at McGill University in the 1950s and '60s documented that rearing puppies in social isolation resulted in abnormal brain maturation and functioning, as well as deficient emotional and social behaviors.[140] In humans as in monkeys, after compiling studies on the subject, I could observe that lack of movement (kidney aspect) coupled with lack of touching (heart aspect) could lead to dysfunctional behaviors in humans such as: alcohol/drug abuse and addictions; anger/rage; aversion or hypersensitivity to touch; chronic stimulus-seeking (obsessive-compulsive) behaviors such as rocking, thumb sucking, and self-mutilation; dependence;

depression; hyperactivity and hyper-reactivity violence; impaired pain and pleasure perceptions; impaired sexual (pornography seeking) and bonding sense (sense of synergy); and social alienation with antisocial behaviors that include bullying, violence, suicide, and homicide.

James W. Prescott, PhD, perhaps motivated by his own early life experiences in an orphanage, began his career researching the effects of maternal deprivation. He examined the neurobiological mechanisms involved in the loss or lack of mother (or surrogate) love, and its effects on the structural and functional development of the brain. These studies documented both structural and functional abnormalities of brain-cell development.[141] *The Origins of Love and Violence* is the gathering of his results.

The three sensory modalities in the mother-infant relationship are as follows:

1. Movement—vestibular-cerebellar system, Reptilian brain, feminine element of the masculine polarity

2. Touch—somesthetic system, Human brain, heart aspects, masculine aspect of the feminine polarity

3. Smell—olfactory system, Universal brain, lung aspects, feminine polarity

The synergistic aspect of the mother's feminine polarity, expressed through the functioning of her sixth sense of synergy, will convey to the child—or not—the necessary emotion and subtle energies to develop basic trust, bonding, and belonging. When separations occur regularly, they can lead to early and chronic releases of high levels of stress hormones, not to mention low expression of favorable ones, as previously discussed. All these practices have been promoted as a preventative measure: don't spoil the child. Delayed bottle feedings, physical separation during the day and night, and detachment of parents when confronted by their babies' stress were, we thought, expressions of our good intentions toward the future well-being of the child. Unfortunately, negative stresses in infancy potentially have the most profound, persistent, and negative effect on adult resilience, hormone regulation, and behavior.[142] On the other hand, early physical contact and maternal responsiveness can lessen genetic predispositions and enhance the child's resilience.

Megan Gunnar, a researcher in child development, and her colleagues did infant studies that confirmed animal research findings.[143] They found a lower production of cortisol in infants who received affectionate care. Eighteen-month-old toddlers who were deemed insecurely attached because they had received low levels of attachment expression revealed elevated levels of stress hormone. This continued at age two, and they started expressing more inhibitions and fearfulness. Dr. Gunnar concludes that the level of stress experienced in infancy permanently models the brain's stress responses. These in turn will affect memory, attention, and emotion. Chronically elevated cortisol levels in infants lead to heightened responses to stress throughout life. High blood pressure and heart rate—which we explain also through the fact that the Reptilian brain will not be able to control the heart aspect—are two examples of this pattern, as well as of the Reptilian brain feeding the following phase of the Mammalian brain.

Occasional spikes of cortisol throughout the day are natural, but endlessly elevated stress-hormone levels in infancy are associated with permanent detrimental effects on brain development. Cortisol elevates blood pressure and heart rate, increases blood sugar, and interrupts digestive and kidney functions. It is also linked to overstimulation of the sympathetic system through a lack of control from the feminine polarity. The many negative symptoms and illnesses linked to a constantly elevated level of cortisol have been listed through research: high risk of heart disease and adult-onset diabetes, memory and spatial learning deficits, lower scores on mental and motor skill tests, premature onset of puberty (which increases a person's risk of developing cancer), more depression, more behavior problems, lower scores on intelligence tests, anxiety disorders, anorexia nervosa, obesity, Alzheimer's disease, accelerated aging symptoms, decreased immune system functioning, sleep problems, and worsening of skin conditions such as eczema. All of these can be attributed to a deficient or defective feminine polarity, which forces the defense reaction of the Reptilian brain to remain turned on and the Mammalian brain to be overexcited. As an adult, such a person may demonstrate type-A personality behaviors and subsequently fall into depression. A high level of cortisol does not cause depression per se. An ensemble of chain reactions does.

Our brain cells unconsciously and automatically encode the physical, emotional, deductive, collective, and symbolic environment we encounter. We see and experience life through the patterns encoded in these cells. They are tinted glasses that censor what we see, altering our perception of the world. Whatever happens during that coding will positively or negatively affect our masculine and feminine polarities, and will be responsible for the beliefs and patterns that shape our lives.

The HPA (hypothalamic–pituitary–adrenocortical) axis is the chief regulator of stress reactions. Breastfeeding mothers produce significantly less stress hormones than those who bottle-feed their infants. Without regular closeness to a comfort giver, the infant not only suffers from elevated stress hormones, but also receives less benefit from oxytocin and other positive biochemical influences. As previously stated, a brain that develops in a stressful environment overreacts to stressful events and poorly controls stress hormones throughout life. The effects, however, reach way beyond a person's blood pressure and ability to deal with stress. In other words, when parents heed the desire to be close to their infants and respond quickly to their needs and desires, their nurturing gives nature the chance to develop sensitive, conscious adults. Withholding attention from an infant weakens the whole system and wounds the feminine polarity. A message of abandonment is sent to the organism. Vital chemical messengers quickly diminish, and the organism prepares itself for death. This is worse than what a message of hunger or cold could do.

Some people seem to think that babies are calculating, that they can manipulate us into giving them more care. However, in babies, the part of the brain linked to deception is not mature yet. The first aspect to shut down from lack of affectionate care is the feminine polarity.

The human genome consists of some thirty thousand protein-producing structural genes, enough for developmental potentialities, but certainly not enough to replace nurturing. Nurture is how the central nervous system develops. Genetic determinism is far from absolute.[144] Recently, researchers seem to be modulating their discourse concerning genes, saying that the whole organism is in charge of its genes, instead of the other way around.

The Reptilian Brain, the Physical Self, and the Instinctual Human

One point I never saw discussed in studies is the problem of overbearing mothers. As with love withdrawal, this situation arises from inadequacies of the sense of synergy. If we consider that science can discover the existence of something only through its measurable manifestations, then as I said, *love* cannot be properly described and analyzed by masculine-oriented science. As an example, there is this enlightening text by Allan N. Schore, who tries to scientifically describe love:

> During the bodily based affective communications of mutual gaze, the attuned mother synchronizes the spatiotemporal patterning of her exogenous sensory stimulation with the infant's spontaneous expressions of endogenous organismic rhythms. Via the contingent responsivity, the mother appraises the nonverbal expressions of her infant's internal arousal and affective states, regulates them and communicates them to the infant.[145]

Different Types of Love

This is the problem of humanity. The masculine polarity has to go to great lengths in order to analyze and often ineffectively describe subtle energies, and can only do so through descriptions of material elements, which is truly inadequate and obviously reductionist, to say the least. To circumvent this problem, the LIFE biosystem, because it uses theoretical notions that exist before the division into masculine and feminine polarities, provides different tools. It brings an understanding that *love* expressed through different phases will stand for different things, although one will verbalize it the same way: "I love you." For instance, using this system we understand that the same energy travels throughout the system, expressing itself differently in accordance with the predominant phase. The *love* from the Reptilian brain is a possessive reaction because it is self-centered and territorial. From the Mammalian brain, it is instinctual, sexual, and linked to hormones. It is attraction *and* repulsion between sexes. For the Human brain, it is an emotion based on self-image and exists for the promotion of a person's good image of himself. For the Analytical brain, it is mathematical calculation for a person's own good. For the Universal brain, it is loss of the frontier self/nonself in a desire for fusion. If the heart aspect—the Human brain—evolves properly, all these will work together as a biosynarchy beyond space and time; and will generate true love and thus promote what is best in both partners. All these phases are constantly interacting and generate all levels of love. This implies that when we say *love*, it can mean a thousand different things and is highly dependent on how our LIFE wheel turns or not. Therefore, when someone says regarding love, "whatever love means"—as Prince Charles, in his honesty, did—he either demonstrates a certain reflection on the subject and, rightly so, an inability to find a one-size-fits-all definition of love; or he sees love from a rational point of view. It does not exist for him. When we look at a couple from the outside, it is our rational, Analytical brain that sees. This is not a good judge of other people's feeling. As we will see in chapter 6, the Analytical brain has no empathy. The French phrase *Honi soit qui mal y pense*—Shamed be he who thinks evil of it—is aimed toward the rational brain.

As with any living entity, couples are born, grow, and whither once the seed is manifested. Couples follow nature's life cycle. When two people join, they become a third entity, for every person who enters your heart changes your personality by awakening subconscious aspects of your psyche. This can be observed with children at home and then in society. Parents believe their child is the same when at school, and are often surprised to discover, when talking with the child's teachers, aspects of their little one they didn't even know existed. It is the same when we are with different people. Loved ones awaken dormant aspects in us. Two people entering a relationship are similar to two galaxies colliding. There is far less damage than one would expect, and many new stars are created in the process.

As with people, all couples are unique. On the physical level, we should limit ourselves to one partner at a time, the one who shares our physical space. On the other levels, you can enter more relationships. The ultimate felicity would be to find someone with whom you vibrate on all levels, but this is an exceptional occurrence. Also one has to think of the result: immobility. So couples have their own inherent cycles. First there is an attraction, which could be on one or many different levels. This attraction is a meeting in time and space of two harmonious recognitions. The lung aspect is the soul aspect, the feeling you have when you meet someone that you somehow *know* this person; or that you were expecting this person, even though he or she is not necessarily "your type." This link will last for as long as these people have something to settle (good or bad). It could be for a few days or a lifetime. These are strong links, and avoiding them can lead to illness.

After the lung phase, the couple goes through a seeking period when the partners need to see each other. It is similar to an addiction. The couple enters the kidney phase. The pressure to mate is very strong at this point. The couple can have sex right away and often this will be the end of the relationship, or they can keep this energy and put it in their energy bank. The liver phase that follows involves doing things together and for the other. Perhaps she cooks and makes herself beautiful, and he wants to give her the moon. They then become inseparable, which is the heart phase. Their personal identity rests on the other. Eventually, they have children, a house, a dog, and a lawnmower. This, not surprisingly, is the pancreas aspect. Because he needs the social world to express who he is (identity through status) and the children are growing, they have more contact with the community. They have come full circle and are back to the kidney phase. The kids are grown and gone; they begin a period when they have to decide if they start another cycle together or if they go their separate ways. They can choose not to live together, which does not mean they cease loving each other. The couple will always exist in their children, and they must respect this. They will not have to break the subtle links if they eventually share their lives with others. The heart is vast. They have to accept they are no longer the people they were before they met. They are who they were, plus the qualities the person they love—or loved—awakened in them. They are two galaxies that have collided and created new stars.

For those who have accessed the evolved heart phase and stayed together, the positive link goes on to another level, another spire, despite the disassembly of the physical structure that we inappropriately call "end of life." Higher, mature love is eternal.

To summarize, the developing brain is encoded with values inherent to the backdrop painting realized since conception and during the first years of existence. These become learned behaviors rooted in biology (Montagu, 1971) and the health of both polarities depends on them. For over a generation now in America, the third leading cause of death in the 15–24 year age group has been suicide.[146] Is it that

Isis Code

surprising when we know that the Reptilian brain has control over the Human brain (diagram 12)? As well, the suicide of children aged seven to fourteen years old has doubled in the last generation. What a sad statistic! Is it surprising when we know that the Universal brain, mother of energies, controls the Mammalian one? Without enough energy in the feminine polarity (lung) to control the liver aspect, there are outbursts in the Mammalian brain that have been associated with depression (diagram 12).

Many perversions and depressions can be explained in terms of the plasticity of the brain allowing it to model itself on unresolved childhood conflicts. Shouldn't we start to focus on the pre- and perinatal periods of life, and provide couples with support for the benefit of the whole society, following the practice of many evolved civilizations? This would be the wisest humanitarian and economic decision we ever made. The consequences would be widespread: fewer people in prisons and in hospitals, less drag on our universal health system, less psychiatric care, and a happier, stronger society. In a phrase, less misery and a far stronger economy!

Differences between male and female brains do not appear only at puberty, in the emotional phase, when gender-related hormones increase significantly. We can observe a difference in children as young as three years old. Girls will prefer dolls and static toys while boys will prefer moving ones. Gender-specific differences during tasks have been demonstrated through different brain activation of the thalamus.[147] On average, women have 55–58% of the upper body strength of men and are on average 80% as strong for the same body weight. At age three, males already can throw a ball farther and with more accuracy.[148]

Infants are all ears. They are attentive to sounds and focus on every word they hear. Sounds are known to have an effect on the heart. An external auditory stimulus, such as a clicking sound used in dog training, will varied the heart rate. When it hears a loud noise, a baby reacts by large movements of limbs and torso, a startle response, even while in the womb. Until eighteen months, infants will also react strongly to distress sounds from other infants. They are not separated beings yet and this will continue as long as the right hemisphere clearly dominates, which is up to around three years old.

Using brain scans, researchers at University College London witnessed the process of noise effects on adult test subjects. The volunteers were presented with unpleasant sounds combined with images. Even inaudible, fearful stimulus had the capacity, at the unconscious level, to trigger activity in the attention center of the cerebral cortex. The fear response was then channeled to other parts of the brain, preparing a cascade of reactions through the sympathetic system. How could it?

Lead researcher Jorge Armony said, "It makes perfect sense—you can't stop and think about certain things, you have to react."[149]

Studies in Frankfurt, Germany, have shown that the closer children lived to noisy airports, the lower their intelligence was.[150] Also, in public housing high-rises above the Dan Ryan Expressway in Chicago, the lower the floor the children lived on—that is, the closer they were to the highway—the lower their intelligence. It makes sense. Each time we hear a sound pulse, every neuron of the auditory cortex is excited.[151] Michael M. Merzenich, professor emeritus neuroscientist at the University of California, San Francisco, confirms that all these neuron firings result in massive hormone releases. His research has shown that in such conditions, baby rats were predisposed to epilepsy. After they matured, exposing them even to normal speech caused them to have epileptic fits. The animals had undifferentiated brain maps and indiscriminate neurons that got turned on by *any* frequency. Merzenich now has an animal model for autism. His findings have been confirmed by recent brain scans of autistic children. They do process sound in an abnormal way. Merzenich attributes the difficulty in learning and focusing to undifferentiated neurons. When pushed, these children might experience confusion, hearing a buzzing that forces them to withdraw into a protective shell. He thus warns women against pregnancy in a noisy environment.[152]

When we encounter a real or perceived threat, for instance a dog on the loose, suddenly barking at us during our morning walk, our threatened feminine polarity receives the information through the senses. The Master of the Heart stimulates our masculine polarity for defense by contacting the hypothalamus, a tiny region of the Reptilian brain and part of the masculine polarity, which sets off an alarm system in our entire body. Through the sympathetic system released by the controlling centers, this structure will prompt the

Isis Code

adrenal glands, set atop the kidneys, to release a surge of specific hormones. The order of the emotional and physical reactions has been well described by Jeffrey A. Gray and is in harmony with our cybernetic system. In fine-tuning with Walter Cannon's popular 1929 "fight or flight" formulation, a more thorough sequence is *freeze* (overwhelmed Reptilian), *flight* (overwhelmed Mammalian), *fight* (Mammalian if flight is impossible or if defending others or property), and *fright* (Reptilian/Universal). The latter is an instinctual release, a "play dead" response when the fight outcome appears not to be favorable to the victim.[153] Because it is an automatic response from the Reptilian brain, this could appropriately be called a survival tactic.

The first stage, freeze, is one of hyper-vigilance, the Reptilian brain concentrating on one point—stop, look, and listen. Also, immobility during threat is an instinctual reflex which allows the prey to avoid being detected, given that mammalian carnivores tend to detect moving objects. The kidney element is the most receptive of the whole system, thus retraction is its defense tool.

The next step is an attempt to flee. If that is judged impossible, then the victim will try to defend himself. The fright response can be observed in all mammalians, including humans. This immobility, the "play dead," is also an ancient reaction. If you are presumed dead by the predator, then it will let down its guard and you might have then the chance to flee. Again, it is a normal reaction of the Reptilian brain. It is *not* because you agree to the act. Victims of sexual assault who exhibit extreme passivity during the aggression are not "saying yes," as alleged by some rapists.

Starting in the eighth month of pregnancy, there is a normal increase in the production of cortisol. This initiates production of fetal lung surfactant, promoting maturation of the lungs. Cortisol also controls and inhibits functions that would be nonessential or detrimental in a defensive situation. In a threatening situation, cortisol will alter immune system responses and suppress the digestive system and the reproductive system, as well as growth processes.[154] This complex alarm system also communicates with the regions of the brain in control of mood, motivation, and fear, and is balanced and controlled by the right hemisphere (feminine polarity). It appears thus that a functional feminine polarity balances the level of cortisol and noradrenaline (norepinephrine) in the organism, keeping them at optimal levels. Accordingly, ADHD has been associated with diminished growth hormone and prenatal excess cortisol,[155] indicating a problem with the feminine polarity.

Almost all body processes are disturbed by chronic activation of the stress-response system, which implies a subsequent overexposure to cortisol and other stress hormones. Stress management strategies include the stimulation and protection of the feminine polarity aspects of the system, as well as:

- being reared in a physically and emotionally healthy environment;
- eating a fresh, healthy, pesticide-free diet (feminine aspect of the pancreas, feminine polarity), getting regular exercise, contact with nature, and plenty of sleep (purification of kidney and liver aspects);
- practicing relaxation and breathing techniques (lung, feminine polarity);
- fostering healthy friendships (lung, feminine polarity);
- having a sense of humor (lung-heart, feminine polarity);
- and seeking a holistic belief system and goal for one's life (know thySelf; heart, feminine polarity).

People exhibiting psychopathy have a dampened response to emotional stimuli. Since the development of the amygdala (Reptilian brain) leads to the development of the ventromedial prefrontal cortex (Bechora et al., 2003)—that is, the Human brain—early damage to the amygdala might result in a failure to activate conditioned responses or memory of an emotional event, as well as an inability to accurately judge the facial expressions of others. The brain will reorganize itself to compensate for this abnormality. Psychopathy is associated with hypo-functioning of temporo-limbic circuits (universal-mammalian axis), which are normally associated with emotion processing. Perhaps frontal brain regions (Analytical brain) are then engaged in some compensatory manner (Hare 1988). A person might enhance this aspect as a countermeasure to his deficit. This might also trigger a need to abuse some substances (Blair, 2003). In this sense, it will be difficult to understand the functional neuroanatomy of psychopathy, since the different structures might have developed or been used differently by the psychopath.

Consequently, psychopaths do not appear to present psychotic symptoms or disabilities in intellectual function —read analytical function. Kiehl et al. (1999) found that compared to criminal nonpsychopaths, psychopaths have difficulty processing abstract information (emotional or other). They also seem to have an appetitive response to negative stimuli; a nonexistent startle-response point; an impaired function of the frontotemporal area, the anterior temporal lobe, and the amygdala; and an inadequacy in error monitoring [associated to an inefficient feminine polarity] (Bates, Liddle, and Kiehl, 2003).[156] Studies have also shown that the volume of the right anterior cingulate is positively correlated with harm avoidance[157] (Pujol, Lopez, Deus, Cardoner et al., 2002, Hare,1991).[158]

The amygdala is a key center for coordinating behavioral, immunological, and neuroendocrine reactions to environmental threats. If it has learned that chaos is the customary atmosphere, then disharmony will become its baseline. The person so affected will feel at home in incoherence. The frontal cortex will then mediate decision according to this model.

The imitation of facial expression requires the combined work of the insula—a part of the feminine polarity and Universal brain, which connects action representation with limbic elements—and the amygdala. As we will later see, this is shown in our system by the Universal brain controlling the Mammalian one. It appears that in psychopathy, there is a break out of the white matter between the frontal lobes and the amygdala, as well as less gray matter in the frontal lobes. Since these should control the amygdala, we can see a problem emerging. The result is the development of a neurodissociative brain in which the phases are not in synchrony.

I consider that the amygdala, often regarded as a part of the basal ganglia, has a different function than the basal ganglia. As the amygdala regulates binary instinctive reactions (reptilian) and has an energizing behavior in response to anticipated rewards (mammalian), I associate it with those two phases. The pleasure of the reward associated with the amygdala differs from the enjoyment of the reward (Arana et al., 2003; Gotfried, O'Doherty, and Dolan, 2003), which makes the amygdala closer to the kidney phase than to the liver one. It is a small almondlike structure located in the anterior temporal lobe. Thus in our diagram 15, it is in part controlled by the lung aspect for the left amygdala and by the pancreas for the right amygdala. It is rich in dopamine and alerts the rest of the system via the hypothalamus.

Author John Robbins writes in his Pulitzer-Prize nominated *Diet for a New America* (1987):

"The way we treat animals is indicative of the way we treat our fellow humans. One Soviet study, published in Ogonyok, found that over 87% of a group of violent criminals has, as children, burned, hanged, or stabbed domestic animals. In our own country, a major study by Dr. Stephen Kellert of Yale University found that children who abuse animals have a much higher likelihood of becoming violent criminals."

A 1997 study by the Massachusetts Society for the Prevention of Cruelty to Animals (MSPCA)[159] reported that children convicted of animal abuse are five times more likely to commit violence against other humans than their peers, and four times more likely to be involved in acts of vandalism.

Here are a few examples that allow us to wander in the mind of people who basically have a nonfunctioning feminine polarity. It appears that in the mammalian phase, these boys already displayed symptoms of cruelty. Living creatures were mere objects to them, so that later in life, humans would also be viewed as "things." We will observe also that most of time, this cruelty was of a genital type, and therefore the Reptilian brain was involved and dissociated from the other phases. I offer many examples, but I believe these few are clear enough to understand that the masculine polarity without a functioning feminine polarity is dangerous and not in touch with reality, even if the rational aspect appears functional. It does not make people "more masculine" if their feminine polarity is not functional. It only makes them remorseless and dangerous. Some violent acts performed by the likes of these people were too gruesome for me to describe here. Many websites quote the same examples, so the source is unclear. These have been gathered across the web.

- Russell Weston Jr. tortured and killed twelve cats. He burned and cut off their tails, paws, and ears; poured toxic chemicals in their eyes to blind them; forced them to ingest poison; and hung them from trees, with the noose loose enough to

- create a slow and painful death. Later killed two officers at the United States Capitol Building in Washington, DC.
- Jeffrey Dahmer staked cats to trees and decapitated dogs. Later he dissected boys and kept their body parts in the refrigerator. Murdered seventeen men.
- Kip Kinkel shot twenty-four classmates, killing two, in Springfield, Oregon. He killed his father and mother. Said he blew up a cow once. Set a live cat on fire and dragged it through the main street of town. Classmates rated him as Most Likely to Start World War Three.
- As a boy, Albert DeSalvo, the Boston Strangler, placed a dog and cat in a crate with a partition between them. After starving the animals for days, he removed the partition to watch them kill each other. He raped and killed thirteen women. He often posed bodies in a shocking manner after their murders.
- Eleven-year-old Andrew Golden and thirteen-year-old Mitchell Johnson tortured and killed dogs. A friend of Golden's stated that "he shoots dogs all the time with a .22." On March 24, 1998, in Jonesboro, Arkansas, Golden and Johnson shot and killed four students and one teacher during a fire drill at their school.
- Theodore (Ted) Robert Bundy was forced to witness animal cruelty by his grandfather. He later killed at least thirty-three women.
- Michael Ross, Connecticut strangler of eight women, was forced by his father when he was five years old boy to strangle the sick chickens on their poultry farm.
- Henry Lee Lucas killed numerous animals and had sex with their corpses. He killed his mother, common law wife, and an unknown number of people
- Richard Speck threw a bird into a ventilator fan. Killed eight women.
- Peter Kurten, dubbed the Vampire of Dusseldorf, tortured dogs, and practiced bestiality while killing the animal. Murdered or attempted to murder over fifty men, women, and children.
- The Kobe Killer, an as yet unnamed fifteen-year-old boy in Japan, beheaded a cat and strangled several pigeons. He decapitated eleven-year-old Jun Hase, battered to death a ten-year-old girl with a hammer, and assaulted three other children in separate attacks.

- Richard William Leonard's grandmother forced him to kill and mutilate cats and kittens when he was a child. He later killed Stephen Dempsey with a bow and arrow. He also killed Ezzedine Bahmad by slashing his throat.
- When nine years old, Eric Smith strangled a neighbor's cat. At thirteen, he bludgeoned four-year-old Derrick Robie to death. Smith lured the little boy into the woods, choked him, sodomized him with a stick, and then beat him to death with a rock.
- David Berkowitz, "Son of Sam," poisoned his mother's parakeet out of jealousy. He later shot thirteen young men and women. Six people died and at least two suffered permanent disabilities.
- Michael Owen Perry decapitated a neighbor's dog. He later killed his parents, his infant nephew, and two neighbors.[160]

In 1993, it was estimated that for every 1% of aggressive violent acts perpetrated in America, the cost to the country was $1.5 billion (Resis and Roth, 1993), and it is certainly greater today. This represents money that could be better spent on providing healthy beginnings for humans.

In all those abominations, sexual arousal is present, but not fear. The arousal may come from rage or nervous excitement, as suggested by study of rapists. A study of sexual assaults on elderly women might clearly demonstrate what rape is all about. We would most probably finally understand that rape and animal abuse are more about power and the Reptilian brain than sex and the Mammalian one. Child sexual abuse or sexual abuse of young and senior women—is there really much difference? The most weak and innocent, the ones most associated with the feminine polarity appear to be a rapist's most vulnerable and thus most empowering victims.

Anthony Walsh, PhD, in his book *The Science of Love* also offers evidence that deprivation of emotional care affects the functioning of emotional centers in the brain, leading to disruptive behaviors. "*In fact, the physiological line of thought reasons that socialization and the development of conscience (the internalized control of behavior) are largely a function of autonomic conditioning in childhood.*"[161] A.P. Weil (1985) also notes in his research "*that experiences induce neurophysiological structuring is increasingly recognized.*"[162] (See

Isis Code

also Rourke, Bakker, Fiske, and Strang, 1983.) Sociopathic and psychopathic criminals appear to share some genetic, neurological, as well as psychological abnormalities. The real question we now have to ask ourselves is: Do these precede or proceed the behavioral?

The "Masculine" Reptilian Brain and the pH Spectrum

Is there a sexual dimorphism between the Reptilian brain of man and woman to account for my labeling this part of the brain *masculine*? I believe so.

His sexuality is of a genital type in the sense that it is one of conquest (kidney) and spatial dissemination (kidney/liver). It is also automatic, reflexive, and powerful, even somewhat obsessive, especially before maturation and control by the frontal lobes. Hers is protective (liver), selective (heart), and temporal (liver/lung). The amygdala and hypothalamus have a larger volume in men. Activity preponderant in the right amygdala further suggests that heightened amygdala reactions to angry faces may be uniquely involved in the expression of reactive aggression in men, not in women.[163]

The practice of Hindu chanting is connected to the left amygdala (liver phase) and associated with bliss, ecstasy, and other positive emotions.

Hypothalamus—Amygdala

We are the only animals provided with a brain that triples in size during the first two years.

The hypothalamus is the integration center of the autonomic nervous system. It regulates body temperature and the endocrine function. The anterior hypothalamus is linked to the parasympathetic activity while the posterior hypothalamus is associated to sympathetic activity, such as the stress response. I thus associate the anterior hypothalamus, which is 2.5 times larger in men than in women, to the feminine aspect of the masculine polarity (kidney aspect). It is normal that it would be larger in men than in women since it is part of the masculine polarity, the Reptilian brain, and the physical self.

Adjusted for total brain size, the amygdala is significantly larger in men than in women. It modulates the storage of memory from emotional events, and does so through interactions with endogenous stress hormones released during those events. In women, studies have indicated a preferential involvement of the left amygdala in memory for emotional material (generally visual images). This is in harmony with the fact that women use their masculine polarity on the emotional level. On the other hand, in men there is a preferential involvement of the right amygdala in memory. So it seems the amygdala has laterality—women left and men right. Even though I do not dichotomize brain function, I can observe that structurally, functionally, and anatomically, those functions mostly follow the lines of masculine and feminine polarity. For example, when it is time to encode emotions, women will use their hippocampi and men their amygdalae. In accordance with our system, in men the stronger memories are often linked to genitality and threats. The male amygdala bears many testosterone receptors, which accounts for males' short fuse, often observed after the onset of puberty. It heightens a man's response to what he considers irritating, and anger is never far from the surface, especially when the liver is overworked with inadequate lung energies. With time, the frontal lobes will take over and the amygdala will become more controlled, as our diagram and observations from researchers indicate (diagram 15). Men avoid emotion and women avoid confrontation and anger. In a threatening situation, women will be more vocal (Mammalian brain and liver aspects) and men more physical (Reptilian brain and kidney).

Furthermore, men and women perceive pain differently. In studies, women require more morphine than men to achieve the same level of pain diminution. I will analyze this later. Women are also more likely to ask for treatment, analyze their pain, and even vocalize these pains. Again, I explain this through the fact that generally, women are masculine (expressive) on the emotional level, housed in the Mammalian brain, while men are feminine (receptive) at that level. The area of the brain that is activated during pain is the amygdala, and again, women left, men right. The right amygdala has more connections with areas of the brain that control external functions, which suits the Reptilian brain of the physical self. The left amygdala has more connections with internal functions of the emotional self. This difference probably partly explains why women perceive pain

Isis Code

more intensely than do men and suffer anxiety four times more. They will manifest general imbalances, preferably through anxiety, psychosomatic pains, verbalization, etc., while males will tend to act, to express themselves physically.

The receptors morphine binds to are mostly of mu, kappa, and lambda types. The brain sites for these receptors are numerous, but they are mainly found in areas of the masculine polarity (Reptilian and Mammalian brain), as well as in the digestive tract and spinal cord. The Reptilian brain and dopamine are associated with the masculine polarity. This therefore appears to explain why morphine is not as effective in women.

Regarding specificity of femininity/masculinity, one would expect that gender-based differences would be associated with inherently feminine or masculine characteristics. However with our LIFE biosystem, rating masculine and feminine characteristics is a quantitative and qualitative method for assessing a spectrum. It concerns many levels and phases, in contrast to the dichotomy of "male versus female." Data from research indicate correlations between morphology (masculine polarity), social perceptiveness (feminine polarity), and the degree (quantity) of femininity, not only in females but also in males. Associations with femininity are therefore specific and not simply a result of physical gender.[164]

Genital Spectrum and Homosexuality

What are the implications of the potentials, plus and minus, of masculine and feminine? First, we must keep in mind that these are used comparatively. One is plus when compared to the other, which is minus. As such there is a gradient between minus and plus that covers a wide spectrum. The pH measure can be used to illustrate this because in living matter, all exchanges are affected by potentiality. Since water is the largest component of the body, pH is the best tool to express physical balance relating to our Reptilian brains. The maximum of alkalinity, which is caustic, is 14. On the pH scale, the maximum of acidity, corrosive, is 0. Although a stronger acid substance may exist, a superacid, but there is no more aqueous solution. It therefore is measured through the Hammett acidity function (HO). At pH 7, a solution is neutral.

In an organism, acid/base homeostasis is crucial. Outside of the pH level that is compatible with life, our bodies cannot sustain themselves. The electrolytic balance, the result of the pH balance, is important for all higher lifeforms. It keeps the equilibrium between the intracellular and extracellular milieu, and it is critical for nerve and muscle function. The main organs regulating electrolyte and pH balance are, not surprisingly, the kidneys. They receive blood directly from the heart and filter it. They have a crucial function in homeostasis and pH regulation, and relate with other organs through the endocrine system. Accordingly, in the LIFE biosystem, the physical level is paired to the kidney aspect of our system.

All of our biological aspects could be looked upon through this balance between minus and plus, receptive and expressive. This duality, taken at the physical level, implies a gradient in genitality. On the left side we could draw a large minus sign, thus receptive and more basic, at pH 14, going all the way to the right, where it becomes a big plus sign, ph 0, more acidic and emissive. This is because female is the default setting in humans (diagram 16). This minus-to-plus would express our genital type from mainly receptive (feminine) to mainly expressive (masculine). I say genital instead of sexual because sexuality in humans is more complex than genitality, as it concerns other levels as well, and not solely the physical one. As a figure to express this, I could say that at pH 7 you are physically neither male nor female, or expressing both as a DSD syndrome (disorder of sex development). Your other levels will determine, later on in life, on which side of the spectrum you are on, through which gender you manifest yourself physically. This shows that the more subtle levels have a higher potential of information. The physical is merely expressing these.

We have heard of the sometimes catastrophic results of believing that our psyche emerges from the body, instead of understanding that our psyche uses the body as a tool of expression. Some bearers of DSD have been castrated in their infancy and raised as girls. The rationale, purely mechanistic, was that since the body determines everything—or so we thought—the child would develop normally, aided by science's knowledge of hormones. A few suicides later, science realized it had erred. Obviously, biology can limit an expression, but humans are mostly top-down beings. When researching the subject, I came across the thoughts of Anne Fausto-Sterling,

PhD, Professor of biology and gender studies at Brown University in Rhode Island. She asserts that sex is a continuum,[165] therefore similar to our genital spectrum, rather than a collection of discretely defined phenotypes. The occurrence of hermaphrodites and pseudo-hermaphrodites, called sexually mosaic conditions, in her estimate is about 1.7% of the population.

In our spectrum, at pH 3 you are physically a male and at pH 10 you are a female. I use this image so we can understand that we are not either all male or all female, but that there is a gradation. As magnetism can be transformed into electricity, in the embryo stage, female can transform into male. Women living through stressful situations have more chance of giving birth to girls than to boys. Similarly, in our physiological cybernetic system (system of regulation, inhibition of energies) drawn from the Taoist law of five elements, the lung is considered the mother of all energies, an important part of the feminine polarity, as well as the beginning of our incarnation. Is it a mere coincidence?

Embryology corroborates this by teaching us, as previously stated, that the default gender is female. If the spermatozoon contributes a Y chromosome, then the zygote will develop as a male. Unlike the X chromosome, the Y contains little genetic information. It does contain a gene, SRY, that will switch to androgen production, but at a later stage. So let us say that, as with pH gradation, the gradation of our pH spectrum goes from 14 female to 0 male, 7 being neutral. All is a question of comparison. Physically, to a female at level 12 a human at level 8 would seem manly and, although a woman, attractive.

Nevertheless, as Leonardo Da Vinci pointed out, nature doesn't favor extremes.

If I am a female at 14, it doesn't necessarily signify that I am more feminine than someone at 8. As previously stated, feminity and masculinity are determined by all the different layers of our beings working together, from the physical to the synarchic, and added to the culture in which we evolve. To prove that point, a Swedish study noticed that homosexual men have brains similar to that of straight women, and homosexual women have brains closer to heterosexual males, in brain behavior unrelated to sexual activity.[166]

For me, homosexuality is a mistaken term. It is not a third gender, as some think. Attraction exists *only* between two polarities. The polarity spectrum explains this (diagram 16–17). In the manifested (space+time) world, polarity is unavoidable. The Reptilian brain is the first expression of this. In harmony with this, it has been reported that "homosexuality" is defined between the ages of three and five, and is not reversible after seven. Hence, the subject is viewed only through the physical structure. However, no research has shown that homosexuality is a solely biological matter. Physical and hormonal differences can change your physical position on the spectrum. Nevertheless, defining someone solely through his or her reptilian period requires foregoing all the other levels, thus our inherent freedom as human beings. This renders us prisoners of genitality, which is the crudest aspect of humanity.

We can't define ourselves only through biology. Emotions, thoughts, ideals, dreams, aspirations, socialization exist, so we are not only primal animals. Humans have an inclination to define everything, even living beings, solely through their physical aspects in a bid to stay "objective." This is reductionism, a serious error resulting from the dominance of our Reptilian brains and instinctual selves, as well as of our frontal lobes and analytical minds, all of which are associated with the masculine polarity. This error is reiterated over and over again in our conception of the world and of ourselves. The environment we live in vouches for this destructive habit we have.

Take the example of a male at pH 1 and another at pH 6. To the male at level 1, the male at level 6 will seem attractive. In those couples, the one who has the male function is obvious. There are different types of sexual attraction, and the question is altogether complex and simple. This is only at the physical level, the *what* level. Adding the emotional, intellectual, collective, and symbolic levels complicates the matter. The image of the spine figuring our different expressive layers is a good one to use. As with our system, there is a driving force in the spine to find balance, to compensate for any imbalance. This compensation will produce symptoms at different levels, results of the need of balance. But let us carry on for now with the physical level.

The top-down process from psyche—and for me this includes symbolic, collective, rational, and emotional aspects—to physical widens the genital score of humans. In other words, on our sexual spectrum, the physical aspect is dominated by the brain, which can have a different sexual gender projected from the emotional, rational, collective, and symbolic functions. Therefore, though the physical determines absolutely the gender in animals, in humans, the plasticity of the brain supports the polarized information expressed by the four upper levels of consciousness.

Of all the possible twosome relationships, one type is seldom mentioned. It is the one relating to the male genital act performed by a male at level 0, thus strong acid. Males at level 0 can only assert dominance, and therefore often rape their victims. This is genitality, not homosexuality. In any cases, the victim will undergo profound psychological changes. The perpetrator, on the other hand, reinforces the pattern in himself. This also brings up the problem of child adoption by homosexuals on the right side of the spectrum. (See Frank Lombard, the Duke University administrator accused of molesting and offering his five-year-old adopted son for sex, via the Internet.) Male rape is far more common than we might presume, since men have a natural tendency not to report it. Despite this underreporting, about 3% of American men—approximately 2.78 million men—have declared to have been victims of rape at some point in their lifetimes (Tjaden and Thoennes, 2006).[167] For women, the rate climbs to 5%. This raises a burning question. By suggesting that people can find true love in homosexuality, and by following that with a general promotion of homosexuality, have we also allowed the rape of innocents by 0 type males?

Throughout this work, I mention religious texts. I know that some vehemently reject all of these as fabricated tools to enslave a gullible populace. My point of view is different. Although I agree that often these texts have been used for such a purpose—like anything else that could be used for power, such as money, sex, beauty, social standing, etc.—they do contain some elements of truth, and I could find the LIFE biosystem in the structure of those I studied.[168] Also, I trust that the sources of these texts were sincere people. The written text can be misinterpreted because of faulty translations, time, history, changes in words' meanings, geographic situations, political power, and most of all, human immaturity and imperfections. Some of those texts, though, are no longer associated with a political power. They can now speak to the hearts of men. One of these, predating

The Reptilian Brain, the Physical Self, and the Instinctual Human

by many centuries the *Iliad*, the Mahabharata of India, and even the Bible, is the Epic of Gilgamesh, an ancient Sumerian epic, written some 3,500 years ago on cuneiform clay tablets and rediscovered in the nineteenth century. It is a story that has echoes in the Old Testament, with its graphic details of a flood and the formation of mankind from the dust of the earth. Remarkably, these literary pieces sprouted from oral tradition and are in fact medicopsychological tools used by priests. Indeed, this epic is attributed to Sîn-leqe-unninni, a priest-exorcist.

These texts were tools of purification from what I call psychological viruses. I suggest that as we can contract viruses on a physical level, we can also do so on emotional, rational, collective, and even symbolic levels. They are patterns. These texts enhanced the workings of the Human brain. Oriented toward the heart aspect, they aimed at optimizing human qualities through the feminine polarity. They were used to rebalance the LIFE biosystem, not to be laboriously pored over by an inflated analytical aspect. This explains why in the Muslim faith, practitioners are asked to recite the Koran, not to analyze it. In this way, the implicate order contained in the text imprints and awakens its perfect structure on their brains, provided they have sincere and open hearts when reciting them—a sign of a functional feminine polarity. The texts are the words of God in the sense that they are balanced in the expression of all the elements of the implicate order. They remind me of walking in Paris. It is highly agreeable as the city is so closely in harmony with this autoregulated system, as Dr. DeBavelaere showed me (diagram 18). We will briefly discuss this point in chapter 9.

Most importantly, we need to take into account sacred texts because they are an essential part of our collective unconscious. We can't just ignore them, wipe them off the screen of our subconscious. Therefore we need to go through them and emerge with a new understanding. We must also become fully aware that what we consciously live and choose today creates the universal unconscious (or subconscious) of tomorrow.

The way I proceed with sacred texts is an unusual one. Instead of taking a text an trying to find a physical reality to it—proving that the facts surely existed or not—I filter the text through the LIFE biosystem. The matrix behind the written words is what I am after. The way

Isis Code

I understood the Pentateuch (Five Books of Moses) throughout this book is an example to this effect.

Words by themselves have no life. They die once the individual or group that summoned them has died. Only the bones are left, the system and the subconscious information. Also, the meanings of words change through the ages. Hieroglyphs, since they are symbols, are more universal. They are multilayered, and we may always discover some layers that have been hidden for centuries.

In my studies, I only know that there is a matrix, and I put aside whatever doesn't correspond to that matrix. Those things will not improve anyone's life. Nevertheless, I am aware that my vision of God will always be diminished by my own limitations. Therefore, it is always limited. After all, the part cannot contain the whole, only an image of the whole. The personality of Jesus, if he existed, was only that, a personality. The individuality, Christ, cannot be owned by a personality. The personality can be fractal and can eventually express the image of the whole, but the personality is mortal. Therefore my first duty toward myself, if I am to feel God's attributes or "know" God, is to become the best I can, to bring the image of what I am as close as possible, to mirror in the material reality His perfection. It is also necessary to keep in mind that, as I used to say to my children, you can't fit an elephant into a matchbox. The human brain is the matchbox, the concept of God is the elephant.

Here I have to cite C.G. Jung: "I am totally with you when you say that the human being lives totally only through his relation to God which faces and defines him."[169] This has been verified by science and now shows us that a belief in a benevolent God brings improvements to our health.[170] When I became interested in the different levels of human reality, books about our different "bodies" were extremely popular in France. One could learn how different civilizations divided the bodies into five or seven or nine parts for a better understanding of how we function. As Einstein said, though, everything is energy. We can nevertheless classify types of energy by their functions; that energy will use a condensed form of itself as medium. The medium in turn attracts the energy in affinity with it.

I can only imagine an Ideal that I could name God. A religion appears when a group of people have a vaguely similar definition of the attributes of this God and of the place of humanity vis-à-vis this God. The followers reinforce one another's progress toward this godly vision. But this can only be an intellectual or emotional representation, not the real deal.

To name is to restrict. Man named everything in this world because everything was in him. But he cannot name "God" because he is *in* God. What is contained is not the container. I can project conceptual notions such as eternity, universality, total wisdom, and total love. But to believe that this is God is naive, pretentious, and illusory. When I read a sacred text and my interpretation leads to an understanding that contradicts the highest projection I can make, then I consider my interpretation as too low or the text as false. Usually, this is what divides people: their projection and their interpretation. Thus, who they are. If the acts resulting from sacred texts are not linked to universality or eternity, then the interpretation is too personal. The first obvious element of this is that if my religion claims it is the best, then already my religion is false since "the best" can only exist in a mortal world, restricted to a form, a time, and a space. Maybe it could be the best if I diminished or ignored other faiths, but it will never be the whole thing, because we are mortals.

With our scientific knowledge improving, our religion has to adapt and accept this knowledge into its stride. As well, science has to take spirituality into account.

This could shine some light on controversial Biblical texts. Since we were discussing genitality and are walking on the path of the Reptilian brain, let us take the example of Sodom and Gomorrah. In the Bible story, Abraham and Lot chose different routes to travel, since their followers and families were too numerous to stay in one group. An analogy could be drawn here with the brain, in which decussation is necessary when a large number of neurons work together. Abraham is the evolutionary aspect of the individuality while Lot represents the involutional aspect of the system.

Two angels passed by Lot's encampment. Angels are, first and foremost, messengers, carriers of information that might guide our evolution toward its ultimate goal: Consciousness. It seems that these angels were incarnated since they needed to eat and sleep. Were they simply humans more evolved than most? My observation

Isis Code

is that since the men of Sodom were not interested in women, we have to search elsewhere than in our genital spectrum for where these angels belong. These angels could simply be those who are not bound to the Reptilian brain anymore. If they were considered males, they would indeed be a strange thing for the people of those towns to behold. If we understand that Lot and Abraham correspond to two ways of living—one associated to the feminine polarity and the other to the masculine—then things become clearer. Also, this story is discussed in Genesis, which is the description of the physical self. Thus the genital aspect makes sense. Abraham chose the symbolic, sacred aspect.

Lot, who had built his tent on the outskirt of the city, advised the angels to stay with him and not, as they intended, to sleep within the city square. (It seems the angels were oblivious to men's vicious aspect.) He invited them to eat.

All the men of Sodom, young and old, suddenly surrounded Lot's tent, asking him for the two men. They said they wished to "know" them. In the Bible, this term is used for sexual intercourse or, more in keeping with the Gospels, a desire for the cellular consciousness of something more evolved. Jesus said that the only thing he wished for his disciples was for them to "know" God. Elsewhere, when talking about God, Jesus says (John 8:55, ESV) "But you have not known him. I know him. If I were to say that I do not know him, I would be a liar like you, but I do know him and I keep his Word."

Obviously Lot thought the men from the city had sexual intentions toward the angels since he offered them his two *virgin* daughters instead. Obviously he didn't think the men wanted to kill the angels, unless Lot would allow them to kill his daughters. In that case, why would God want to save him? The men angrily refused the offer of the virgins. They said they had no interest in women and would do worst to Lot than they would to the angels. They complained that Lot judged them, even though he was a stranger on the border of their city. At that point, the angels pulled Lot back into the tent and blinded the men. Eventually Lot left the city unharmed, with the angels and his family in tow, and a rain of "fire and brimstone" fell on the city.

In Luke 17:28–32 (NIV), Jesus talks again of Sodom. "It was the same in the days of Lot. People were *eating* and *drinking*, *buying* and *selling*, *planting* and *building*." He compares the end of time to the end of Sodom. Interestingly, he doesn't blame the homosexual nature of the sexual act for the destruction of Sodom, but the people's materialism. Indeed, he more precisely faults their overblown masculine polarity. In the LIFE biosystem, the actions Jesus refers to are all associated to the same phase, to the masculine aspect of analysis/nutrition, the pancreas phase, and to its attached rational level (eating, drinking, buying, selling). The defensive aspect (planting and building) are also linked to the masculine polarity. Ezekiel 16:49–50 is in agreement with this. It states that the people of Sodom were sinners through the use of three adjectives: "arrogant" (pancreas aspect separated from heart aspect), "overfed" (pancreas aspect separated from all aspects), and "unconcerned" (pancreas aspect separated from the lung aspect). In short, their feminine polarity was not expressed. So the end of time is in fact the end of the supremacy of the masculine polarity. Indeed, the masculine polarity without the feminine one is condemned to the same fate as the frog in the story of the frog who wanted to be as big as the cow[171]. It inflated and blew himself out again and again until it suddenly exploded.

In the light of this, the way these males were acting genitally was in fact a consequence of the values they chose to uphold, not a cause, since the Analytical brain controls the Reptilian one. Once the whole population adheres to this, only knowledge and recanting or destruction can bring back homeostasis. Hence God destroyed those cities. In fact, it was not an isolated, personified God acting, but a normal result of the natural biocybernetic system at work.

For the men of Soddom, women were merely objects, and not even entertaining ones. The fact that they were women was almost repulsive; the feminine polarity was revolting to these people. We can surmise that their situation on the genital spectrum was in the low numbers. They put forward values linked solely to their masculine incarnation, favoring only space and matter, the physical world. (Whether the destruction of Sodom truly happened and whether the men of Sodom were truly like this is not the debate here. I am using this case solely as an example.) Those human beings were imprisoned in a limited aspect of life, although they believed they were free and powerful. They were delusional, although they probably described themselves as realists and pragmatic. How can we understand this through the duality we discussed earlier?

Isis Code

The men of Sodom, as described, lay totally on the right side of our pH spectrum. They were strong acid. They fornicated with anything, anytime. Their rituals were probably linked to genitality (some would wrongly say fertility). The world of the feminine polarity was so alien to them, they could see only through their Reptilian brains. They were so focused on themselves, their whole identity depended on penetration (even without consent) to assert control. They maintained all of their energies on the physical, reptilian level. Therefore they could not communicate with anything outside this physical world. They didn't see or feel or vibrate with anything else. For them, nothing other than the material world existed, so it was the only possible cause of everything. I would not be surprised to learn that at some point in their lives, some of these men raised their arms and proudly and sincerely exclaimed: "I am God!"

Why were they obsessed with the angels? Because angels were alien to the world those men lived in. A pack of wild wolves would not act otherwise. Bring an outsider wolf and they will try to dominate it. Why not Lot then? He was a stranger too. They said they would do worse to him, simply because they would have forced him and made him submit more violently. They had no curiosity or interest in him; he was similar to them. The angels represented something they didn't know in their material world, so they needed to dominate them. But at least their curiosity was entertaining to them. They did not seem afraid of these angels, so the angels' physical appearance was certainly not threatening.

For the men of Sodom, to act this way was not wrong but simply customary. After all, these angels were on *their* territory. To recognize something as being bad or good, you need to be able to compare. Morality starts with consciousness, which starts with the higher emotional level because consciousness needs a top-down process, an observational position. "Father, forgive them, for they do not know what they are doing."

They were not even at that level. When the subconscious motor brain associated to the Reptilian brain dominates, the observational position is not allowed. Information floods the amygdala, which is not inhibited by the higher levels.

For these men of Sodom, women were mere cattle. They would not have understood why this was a wrong way to think. After all, it was the truth. To them, women were mere cattle! Their awareness was limited to their physical beings. They were not analyzing the situation, but reacting on basic instinct. This explains why it is said *all* the men. Of course, all the men were not around Lot's tent. But all the men of those towns reacted in the same way. How was that possible? Through customs and values, which after a few generations eventually change the collective unconscious and therefore influence all those born to that group.

Leviticus, a text for the idealistic self, might further enlighten us. This book from the Bible is concerned with priestly conduct and rituals, and with hygiene of the physical, emotional, intellectual, collective, and symbolic levels as we will see in chapter 5. Concerning sexuality, it expresses concern about sexual hygiene and customs. The condemnation of customs held by previous inhabitants of these lands was aimed not only at the people within the community and how it affected this one, but to the effect that what lay behind some genital practices sullied the *land* (since genitality and territory are of the same phase of the LIFE biosystem). In the course of time, the land reacted by "vomiting" the people who lived on it. For people reduced to their Reptilian brain and who had only the physical realm for their entire universe, this must have been a nightmarish prophecy.

Leviticus 18:24–28 (NIV)—"Do not defile yourselves in any of these ways, because this is how the nations that I am going to drive out before you became defiled. Even the land was defiled, so I punished it for its sin, and the land vomited out its inhabitants.... And if you defile the land, it will vomit you as it vomited out the nations that were before you."

Nowhere is there mention of female homosexuality. Also, in the case of same sex rapport, it does not mention the homosexual act, but simply that a man should not lie with another man *as if this one was a woman*. For if you consider another man as being a woman, where does that put *you* on the sexual pH scale? Strong acid.

The words "defile yourself" convey the sense of action, thus of our expressive/masculine aspect acting on our receptive/feminine one.

Isis Code

One can only act on the feminine polarity, since the masculine is not receptive. Because the feminine is One, it implies that whatever we do imprints the earth's matrix and the global feminine polarity. We are all accountable of the condition of the global feminine polarity. It should be protected. This was supposed to be the main function of religions and monarchy.

Now if your perception considers what is masculine as feminine, how do you consider what is truly feminine? The spectrum here is wrong. There would be no real feminine. The result of this is the wounding, the slow killing of one's own feminine polarity. In every case cited in Leviticus, there is a question of men acting toward other men, women, or children, and considering them as females. Only in one instance is there question of an act performed by a woman. Presenting herself to an animal for sexual relation is a perversion, abnormal. We would easily agree with that. It puts her on the opposite level of the genital scale, which is just as damaging, although in this case the damage is to her masculine polarity. In the case of a strong base on the pH scale,[172] when one is situated on the left side of the genital spectrum, the result is caustic. In the case of a strong acid, the result is corrosive. Neither can harbor life. Thus the perpetrators sentence themselves to spiritual death. Most strong bases, like most strong acids, attack living tissue and cause serious burns. When you add a strong base to an acid, the pH increases slowly at first because the pH scale is logarithmic. That means it takes a lot of base to change a strong acid. Nature seeks to quickly regain balance; therefore, natural catastrophes are the quickest solutions. This explains the example of Sodom and Gomorrah. In the case of the individual, illness could play the same role.

Let's get back to our discussion. To defile something is to debase its pureness, to violate its chastity, - and innocence, to make it filthy or polluted, or to sully its reputation. As you cannot force receptiveness, you only damage the feminine polarity. In all these texts lies the question of the perpetrators defiling *themselves,* not something or someone else. This in no way diminishes the destructive effects of their act on the victims, but the text was intended for the priests, not for their possible victims. The consequences of this personal defilement is the defilement of the earth. Why is that? Because as I mentioned, the feminine polarity is One. This "sin" also has an epigenetic effect on the genetic of the perpetrator.

What makes the difference between a human and an animal is the nature of this top-down, back-to-front ability. Sins are not only detrimental actions, but also being in the state of erroneous thinking. Could it be that a sin is linked to acts that are alien to the harmony of the system of which we all are participating cells? When we are punished for our sins therefore, it is merely a question of regaining balance between our two polarities on every levels, in order for the life flow to be restored from the collective level down to the cellular one. This duality in humans exists not only on the physical level, but also on the emotional, idealistic, intellectual, collective, and symbolic ones. The text relating to these actions should therefore be along the lines of: "If a priest ... therefore he will die." In this case, to die is to be cut from the feminine polarity, from the soul, and consequently from Life in all its dimensions. Acting in such sinful way would imply that this individual has also died to his function of priest.

Further, an identification between ourselves and the earth also exists. Our physical body and the earth are part of the same matrix. We eat everything produced indirectly or directly by the earth in order to sustain our manifestation at this level. The action of eating is imprinted by what we feel, think, and do. The sin here is not only what we do to others, but what we do to ourselves. The earth doesn't do; it is totally receptive. If it is punished, the energies encompassing the earth, because of the implicate order, will reestablish the broken balance. In this text, God is the archetype of homeostasis, Ma'at. This implies fierce energies, natural disasters.

The intriguing point here is that God had to punish the land for wayward customs performed by humans. It is a punishment in the sense that pain is involved, since pain is a separation from equilibrium. God is the system itself, which will trigger the means to regain balance. Thus there is a continuum between humans and earth. They are not different entities. When I eat something my body cannot accept, my physical body is "punished." I have introduced something that broke the balance, thus provoking a division from the holistic balance, the universal matrix. There is a suffering involved, and the body is forced to throw up the food to reestablish the balance, the life flow. The information my body received was "wrong," and I became nauseated. It is a correction of a situation that brought imbalance to my whole physical being. If the receptive aspect, the feminine polarity, is not functional in a person, then healing will be difficult, if

Isis Code

not impossible to achieve. "And Jesus said to the man, "Stand up and go. Your faith has healed you."[173]

A man having intercourse with another man as if this man was a woman, defiles himself. Because of his actions, the perpetrator pushes himself to the right side of our pH spectrum and far from the balance, bearer of life. This brings "death" into himself. A base added to an acid creates a salt, but an acid mixed with an acid stays an acid. This reminds me of something else Jesus said. Matthew 5:13 (ESV): "You are the salt of the earth; but if salt has lost its taste, how shall its saltiness be restored?" Is this pH balance God's "covenant of salt" cited so many times in the Bible? I doubt this would relate to the fact that at the time, salt was expensive. The apostles are the salt because their polarities are well balanced.

The earth, our body, and matter in general are one entity and are receptive to subtle energies. It is damaged by any act that does not recognize, or ignores, misapprehends, objectifies, or violates its receptiveness, its sacredness. By any violent means. The receptive aspect is the one that suffers when there is an imbalance. The receptive aspect is the one getting tortured, humiliated, spat upon, laughed at, lied to. Because it is the receptive part. After any of these acts, the global feminine polarity, which we all partake in and receive energy from, is weakened. If you don't want to get hurt, become expressive, masculine, violent. Be the one who abuses, hurts, and lies. But know this: you also have a receptive aspect, and this aspect registers all actions, feelings, and thoughts you emit and partake in, and will hold you accountable for them. You will have a diminished feminine polarity and an enlarged but deficient masculine one. More on this later.

Of note, in our LIFE biosystem, the phase linked to exchanges with the environment is physically linked to the lung and its attached structures. When the environment is too acid, we can assume that the lung (social and environment) and kidney (physical) elements will suffer. Asthma will prevail. I was interested to find the following: asthma sufferers have higher levels of hydrogen peroxide (pH 3.5 at 35%) in their lungs than healthy people. This could explain why asthma sufferers have abnormally high levels of white blood cells in their lungs, which produce histamine, found in allergic reactions.[174]

Hydrogen peroxide, therefore, is considered a reliable marker for asthma diagnosis.

Genesis 2:18 (NIV): "It is not good for the man to be alone. I will make a helper suitable for him."

If all that God created was good, why suddenly realize it was not good? He made a mistake? Could it be because in this instance, God is the regulator and creator of our system? All the beasts were formed "from the ground," and man, endowed with a prefrontal cortex and left side of the brain, named them. Thus he was different, somewhat detached from them. He had a different function in the universe, associated with the development of this Analytical brain. Animals see us as one of the pack or not. We don't ever consider animals as humans, although Disney has tried to endow them with human patterns. That's a good idea when the spectators are still in their reptilian and mammalian phases. Then Genesis says, in verse 20 (NIV), "But for Adam no suitable helper was found." For the task Adam has to fulfill, he needs a helper that comes from within, from a rib protecting his lung. For the human male, as we now know, the feminine polarity is inner.

Adam and Man are not the same. Adam is the cosmic masculine polarity with all its dimensions. He can be compared to Geb of the Egyptian tradition. Man is the incarnated, polarized, and fractal aspect of Adam. Thus God put Adam to sleep and took an internal aspect of him associated to the lung to create someone other—external—who became known as Eve. Eve is similar to Nut of Egyptian tradition. Thus the feminine polarity is external in women and internal in men. The task Adam has to fulfill in the material, polarized world requires the participation of a being issued from himself, a part of himself that had to be separated from him. As we know, in the LIFE biosystem the lungs symbolize part of our feminine/receptive polarity, Isis or Eve. It is said that Eve is mother of the living, of what is alive, it is not said that Adam is father of what is alive. She generates the living, all the living, not only of what is born. She is an archetype of the feminine/receptive principle. She is indispensable to life, to existence, to manifestation. Even in the Gospels, women are noticeably present—"and many women were following them"—because the feminine polarity is necessary to any manifestation in this world.

Isis Code

Above, I compared Nut to Eve. Similar to Nut, Eve represents the mother of energies of the system at a level that is more incarnated. Still, I prefer to refer to Isis as an alternative of Eve, because of the negative collective connotation linked to Eve. So let's say Eve is the posterity of Nut. Thus Isis, on the material level, is issue from a structure that expresses the receptive/feminine aspect of humans. Here, Adam is not the animal man but the masculine polarity, the masculine principle expressed through man. When Eve and Adam are linked together, they create life on the physical level as well as on the emotional, intellectual, collective, and symbolic levels. This is conveyed by the symbol of the Seal of Solomon (diagram 24).[175]

The Genesis text pertains to the physical world, however. For example: "He brought them to the man to see what he would name them; and whatever the man called each living creature, that was its name." (Genesis 2:19 NIV). Two points have to be made here. First, you can name only what you know. This implies that every living creature can somehow be found in man. Second, to name is also to analyze from a distance and answer the question "what is it?" This question is linked to the material world and Reptilian brain. A baby will start by smelling, listening, looking, touching, and tasting himself and the world around him. This is how you answer "what?"—you use your five senses. Animals live entirely in the world of the *what*. So does our animal nature. This is the first level of knowledge. Somewhere in us, as in every animal, the answer to this question is written. When an animal cannot answer "what?," he becomes fearful. All that is in nature is in us. All that is in nature has a reflection in us. Oneness is absolute. Someone who has no fear, as it is requested from our young soldiers, has either a nonfunctioning Reptilian brain, as seen in psychopaths, or has risen above, which is a rarity. I would be careful in trying to render our young soldiers fearless and emotionless.

The Sodom inhabitants, bound by their physical lives, remind me of young colts that are not interested in their mothers anymore, only in their genitals and other males. If you watch them playing in the field, they show their dominance by mounting other male horses. For the men of Sodom, only the material world existed because they could be dominant in it. They had not matured or evolved, because they lived on automated reflexes, separated from the connection with their environment.

Later on we will look at how Jesus's observation fits beautifully in the cybernetic system. The men of Sodom are totally materialistic, egoistical, self-centered, and not conscious of the environment. A theory has been put forward that Sodom and Gomorrah lay along a major fault line. If true, that would demonstrate the lack of sensitivity of these people. Interestingly, a rain of fire and brimstone can also be symbolic. Brimstone, which is sulfur, becomes sulfur dioxide when burned. In the presence of a catalyst it produces sulfuric acid, thus acid rain. Brimstones with a content of 95% sulfur have been found where Sodom and Gomorrah are alleged to have been situated. The lesson is this: when you are totally acid, through your choices you eventually generate acidity in your environment. Sounds familiar?

Sulfur dioxide has the smell of sewage. Sulfuric acid is a strong acid with a high electrical conductivity. It is one of the top products of our chemical industry. Not surprisingly, we also use it for dehydration, which explains Lot's wife turning into a pillar of salt.

The Five Books of Moses

Jewish tradition states that the revelation of the Torah happened in 1312 BCE while Moses was at Mount Sinai. (Some give 1233 BCE as a date.) The Zohar, the most major text in Jewish mysticism, explains that the essence of the Torah was created prior to the creation of the world, and that it was used as a model for the whole creation. So the Torah, in fact, is the written rendition of Ma'at, the implicate order. Thus we should find the LIFE biosystem's structure in it and vice versa. We indeed find in the Five Books of Moses: Genesis for the kidney/reproduction/physical, Exodus for the liver/defensive/emotional, Leviticus for the heart/information regulation/symbolic, Numbers for the pancreas/nutritive/analytical, and finally Deuteronomy for the lung/exchanges/universal. The order and significance of the books are in perfect harmony and concordance with that of the biocybernetic system. In each of the different phases, I will express the main elements of each book so that you can see that there is a perfect parallelism.

The first book, Genesis, spans from the lung phase toward kidney, as it starts with the manifestation of the universe (diagram 19, 19.1) and ends with the embalmment of Yosef, as the people of Israel are exiled in the land of Egypt (renowned for its funeral rituals). Other than the story of Sodom and Gomorrah, which we have already explored, Genesis is largely replete of action elements. This book does not point to life and its continuation, but rather to the threat of its extinction. We are undoubtedly in the world of death, which the kidney phase is all about. Since the kidney phase spans the time of before birth to seven years old, the correct word is not death but rather "not manifested," since it is set at the crossroads between manifestation and potentiality. As Genesis moves on, contact with God becomes less and less direct. Abraham's dream and visions are a far cry from Adam's conversation with God in the Garden of Eden. In fact, this new land of Egypt (a symbol for physical incarnation) is for the true people of God (our individuality) a sad exile where one can become a slave.

In the Chinese taoist tradition, the symbol of the Taijitu, ☯, represents this duality, which is found everywhere in nature. It is depicted as a circle divided into two curved droplike parts. The yin aspect, often black, is receptive to energies, feminine, and pertains to contraction, deceleration, condensation, attraction, and magnetism. This droplet with its dot is the mother of all incarnated things that are made of condensed, slowed-down energies. It is the eternal, essential Nut, mother of Isis, or Eve. The yang aspect depicted by the color white represents the emission of energies, expansion, dynamism, penetration, repulsion, and is therefore masculine. It is linked to the eternal, essential Adam. As previously stated, these are always considered as one *compared* to the other, and one cannot exist without the other. They are not divided but interdependent. They are fundamental aspects of any organic system. As in the Genesis story, if Eve is taken from a hidden part of Adam, it means that Eve is the hidden part of Adam, externalized. Originally, they were one. We can find this also in the Taijitu. In fact, the Taijitu deals with the attraction and repulsion phenomenon addressed by the four basic interactions and more. For this system to be dynamic in time, and sustainable, a point similar to the complementary aspect is expressed in each droplike half of the circle. So in the yin/receptive we find an element of yang/expressive—the heart. Otherwise the system would stop from total retraction. It would be similar to a black hole. In the yang/expressive aspect, there is a point of retraction, the kidney, so the principle of yang does not expand so much that it dissolves into nothingness. Thanks to this fact, Adam can recognize himself in Eve and Eve can project herself on Adam. Adam can see the aspect of receptivity in Eve echoing his. Eve can see the aspect of emissivity of hers echoed externally in Adam. In the period we live in, the expressive masculine aspect is dominant. This is echoed in our expanding universe ... or the expanding universe is mirrored in our masculine society. We can choose our point of view.

This theory is as valid as any other one, since our modern science is becoming increasingly aware of this duality in all natural phenomena. Modern physics attempts to explain every observed physical phenomenon through four different types of basic interactions or, if you will, forces. Interaction implies elements attracting and repelling each other. According to the present understanding in physics, those fundamental interactions or forces are gravitation, electromagnetism, the weak interaction, and the strong interaction. Their magnitude and behavior vary greatly. Unification of these different forces is considered highly desirable.

What these sciences are finding now is that the analytical tool stops being useful where dynamism, brought on by time and large interacting systems, is concerned.

In our physical bodies, these two complementary elements of dynamism in every different biological system are always at play. The balance never lasts. The tendency toward homeostasis or the repulse/attraction between these two polarities allows cohesion of the different parts of the organic system, enabling material life to be expressed. With this striving toward homeostasis on every level of our beings, imbalances express themselves through what we call illnesses. These can be on the physical, emotional, intellectual, collective, and even symbolic levels, as will be further demonstrated.

In the blood, we have the white cells and the red cells; we have the purified blood and the charged blood. In the lungs we have inspire and expire; and in the nervous system, we have the white matter and the gray matter, the sympathetic and parasympathetic systems, and so on. The male body—expressive—has some feminine sexual aspects even though they are atrophied (breast, prostate, etc.). The female body, which is receptive, has atrophied masculine aspects (clitoris), even if we all start as female embryos. We have both masculine and feminine hormones, albeit in different quantities, and these quantities vary throughout our lives.

Everything on earth submits to this dual nature. Even our relationships do. Some might say that everything is sexual, and they would not be wrong if they mean dual and complementary. In fact, we are all potentially androgynous in the sense that we have both feminine and masculine potentialities. The impossibility, because of our polarization on many levels, is to express both polarities in the same quantity, at the same time, at the same level. A little like it is impossible to breathe in and breathe out at the same time. Whether studying biology, chemistry, religion, or even psychoanalysis, we can observe this duality everywhere. In every relation on earth, there is a giver and a receiver. This is not a qualitative hierarchy but two different functions, with one as necessary as the other. Every feminine and masculine aspect has a speck of the other aspect. What is feminine at one level becomes masculine at another. The cohesion, creation, life, salt, happens when these two aspects work harmoniously. In psychoanalysis, Jung referred to the animus and the anima, animus being the masculine polarity, anima being the feminine polarity. In different traditions and religions, we also have this double aspect of yin and yang, of fire and water, seeming to oppose each other but in fact working together, unless time and space oppose them.

How does this apply to the brain? We can use the example of navigation. Males and females, whether animal or human, use different strategies when going somewhere. Female humans guide themselves using external landmarks, often endowed with significance—church, school, gas station, strange house, etc. Men prefer to use abstract directions relating to the movement of their body in space: north, south, east, west.[176, 177]

These preferences correlate with different brain areas. Women use mostly the right parietal cortex while men primarily use the hippocampus, a nucleus deep inside the brain that is not activated in the women's brains during navigational tasks.[178]

Both methods work, but the two approaches are very different, as are the brain areas involved in them. In laboratory animals, such as monkeys or even rats, the same applies. This illustrates a general principle about gender differences in cognitive function: the difference in what women and men can do is small. The differences in how they achieve their goals is more significant.

Thus, the first obvious part is physical. We can define ourselves as male or female—except for the approximately 1% of the population that has a sexually mosaic condition—and understand why many men consider the size of their genitals a relevant concern. They evaluate themselves on the genital, physical spectrum. The fact that men are XY protects them. They still have this speck of the duality to balance them. If they were YY, they would probably destroy everything and have absolutely no consciousness of the environment.

Let's take our pH gradient and add to it the understanding brought by the Taijitu. The highest pH being a strong base, we will draw a large minus sign joined by a small plus sign, since the Taijitu shows us that a polarity can't exist by itself (diagram 16). The other aspect, even if not expressed, is still there. At the other end of our pH spectrum, we find a big plus sign with a minuscule minus sign. Again, if someone is limited to the physical world and can't express himself on another level, he will probably be very violent or psychopathic. These humans have no contact with the feminine aspect of the gradient. It is not human for them. It is a toy, a thing they can use to fulfill their need for supremacy. Luckily for us, humans are not only physical.

A relationship limited to the physical level is genital. It is a *what*, an *it*. It is the basic love: attraction. Usually it is linked to arousal and the genitals. The man situated at 0 on our pH spectrum is not interested in women, contrary to our belief. He will seek same gender dominance, either expressed or repressed, and generally hunt for men. Men performing an act limited to this aspect can rape with no remorse, they feel a relief and nothing, except sometimes disgust, toward the man or woman, the "what" they used. After the act they wish "it" would leave. In the physical exchange, they use the other person's body as a tool for arousal and masturbation. Nothing more. There is no relationship with the other. Often they will not look their partner in the eyes unless to verify submission. The idea that a Spartacus-style education would create true males is an illusion. It would only generate insensitive brutes.

The woman at the other end of the spectrum will be easily aroused, *une dévoreuse d'homme*, a man eater. She wants to possess and

Isis Code

keep him/her. Acid men don't interest her. They are too far away at the other side of the scale. But she can be bisexual like acid men. These women, like the acid men, are not any more liberated than heterosexual men and women. They are prisoners of the genital level.

Limited to this level of interaction, only separation and war are possible. They can reproduce but they can't care....

Stressors on the physical system can be expressed through sexual acts, as seen with bonobo monkeys, which use a genital act as a less lethal form of violence. The genital comportment is used in nonreproductive behaviors. Female to female contact and male to male mounting are seen as negotiating and enforcing status differences. Since it is a somewhat matriarchal society, sex is the normal tool of conciliation (linked to kidney, a feminine cybernetic organ associated to the masculine polarity), instead of physical violence, which is linked more to liver, also a masculine cybernetic organ. We cannot talk of *love*, as some have claimed. (Make love, not war). Interestingly, some have projected human emotions on those genital acts. However, I consider that in the case of monkeys, the Reptilian and Mammalian brains take precedence over any other structures. I could see this in observing pictures of bonobo faces,[179] which have a perfect parallelism of expression between the left and right eyes. If you compare these with human eyes, you will note that there is usually a difference of expression between the left eye and the right eye in humans, revealing the different development of functions as expressed by the right and left brain hemispheres.

You can do this little experiment with pictures of people's faces. It is particularly evident in women's pictures because women are expressive on the emotional level, which is expressed through the eyes. You might have to darken the spotlight effect in the pupil to better observe it. Cover the left eye of the picture and you receive some information from the right eye which lays in front of the right orbito-frontal cortex and corresponds to the feminine polarity. Now cover the right eye and see how the left eye's information differs. This corresponds to the state of the masculine polarity. Now do this with bonobos monkeys. With them, the two sides express the same thing.

The Reptilian Brain, the Physical Self, and the Instinctual Human

With fertility, nature shows us she has boundaries. We trespass everything without even thinking because we want what we want. Infertility can happen because of venereal disease, but also when the physical body is not able to sustain a healthy pregnancy. Whether we accept it or not, we are responsible for the child's physical body. If the body reproducing is not in a condition to receive a healthy sperm, or if conditions imply that the child will enter the world with an unfair disability, why would someone risk it? Aren't there enough parentless children in the world, who need parents who have a strong desire for a child? Then there are environmental factors. Why have more children when the country has no food for that child? I know it is a controversial issue, but this shouldn't prevent us from thinking about it, discussing it, and finding solutions!

When I see men who use their religion to have dozens of children, despite the physical realities they will have to face, I am appalled. When I see men with so little control over themselves that their wives are always pregnant, I scream. This earth is abused and ill treated. We take all the natural resources for granted, with next to no regard for sustainability. What we do to the earth, we do to women. The worst part is that many women in some societies have been brainwashed into thinking that having many children is a good thing and God's will. Next to no thought is given toward the well-being of these children.

Empowering ourselves through our most popular tool—money—is the only important goal. Everything else submits to it. Look at them, those acid men, with their eyes in the shape of dollar signs now racing to rape the Great North. It is not different from those men using religion to force themselves on their wives over and over again. You want to talk about the Reptilian brain and what we shouldn't accept from it? This we shouldn't: fornication, the repetitive rape of nature, whatever form it takes. Whether or not it is sanctified by a church or a government and you have the paper to prove you own something doesn't change a thing. Because it is abuse. To say it will create more jobs is ridiculous, to say we need more population for the economic system is putting tails before heads. Dangerous for human dignity.

Look at those gigantic holes we make in the earth's crust, mining for nickel, coal, diamonds. It is endless. We are overfishing, over clear cutting, over mining, over copulating. We always want more: more cows to produce more meat and more milk, more crops on more of the earth to produce more food, and so on. Are we insane? We are now producing grains that will mainly be used to feed livestock and for the production of ethanol. The United States is very interested in ethanol, since 60% of the world's supply originates there and it is a good way to get rid of genetically modified corn. Unfortunately, 80% of grain production now goes to feed livestock instead of humans. You want to do something for your health and for the future of your grandchildren? Reduce your consumption of meat. Tolstoy famously said: "As long as there are slaughterhouses, there will be battlefields."

Contrary to what the agriculture industries claim, *the earth produces enough to feed the world.*[180] The problem is mainly that poor countries cannot afford the foodstuffs they need, and financial speculation at all levels worsens the situation.

Our use of water also points to the same problems. Of all the water on earth, only 1% is fit for human consumption. Americans use an average of 650 liters or 176 gallons of water *daily.*[181] If you wish to compare, the average African family uses 5 gallons or 18.93 liters per day. One car necessitates 148,000 liters of water for its construction. The largest water users are industries such as agriculture, manufacturing cooling systems, and thermal power generation (nuclear and coal). Alone, a leaky toilet can waste 750 liters per day. All the pesticides and herbicides used in agriculture end up in the water. In poor, arid countries such as Haiti, people gather the remnants of vegetation to burn and sell on the market as wood charcoal. Rain becomes flooding, washing this sterile earth to the sea, depleting the water of its oxygen. Fishermen have to go farther and farther with their tiny fishing boats in order to capture fish, since those close to the shore have died, suffocated.

At the same time, Monsanto is flooding Haiti with seeds—which they claim are not genetically modified.[182] The farmers will have to pay for these in subsequent years. They are patented hybrid varieties, which means the farmers will not be able to collect the seeds for replanting at the end of the season.[183] Also, the tomato seeds were doused in thiram, a hazardous pesticide banned in the US. The seeds will be distributed by the United States Agency for International Development

(USAID), a taxpayer-funded agency that, in a sense, promotes United States' based commercial agendas while offering "development assistance." Indirectly, Haiti was *forced* to open its markets to foreign agricultural imports by the International Monetary Fund in order to qualify for a much-needed loan. I doubt that Americans know what their hard-earned money is being used for.

A Stanford graduate student, Chirag Patel, has decided to "use bioinformatics for the environment,"[184] because genetics fails to find answers to some major problems. The Stanford researchers piloted an environment-wide association study (EWAS) in which epidemiological data is analyzed upon genome-wide association studies (GWAS). Using data from the National Health and Nutrition Examination Survey, the researchers performed a great quantity of cross-sectional analyses associating 266 environmental factors with the occurrence of type 2 diabetes. The study Patel led in enviromics found that of those 266 environmental factors, the development of type 2 diabetes correlates the strongest with the presence of the OC pesticide-derivative heptachlor in blood or urine, with environmental contaminant polychlorinated biphenyls (PCBs) also showing a significant association. More than 23 million people in the United States suffer from type 2 diabetes, with that number on the rise, and genetics have thus far offered little insight. This is the reality of our world. This is the reality of business pursuits. A company has a legal responsibility to make profits and survive no matter what. It has no heart and no ideal, it is not human. If the head follows only the rules of a company—which is to garner huge profits and be competitive by any means—then it can become a prolific cancer. But who really profits from these companies? It is not the masses but a few individuals. The company is not democratic. Other rules besides being profitable must be promoted. Make money, yes, but not regardless of the earth or humanity! When society maintained a certain moral control, many companies cared about the impact they had on the world. Now it seems worse than before, unless it is only more visible, and done almost in an arrogant manner. Since public funds are used to help these companies, show the faces of their owners so they are part of the acting community! This is a courageous work that should be done by the media in a bid for transparency.

Agribusinesses have "simplified" their culture, and thus have damaged the ecosystem. How could eight hundred thousand hectares

Isis Code

of corn be good for the planet? In these fields, there are no more bees. Pollinating has to be done by plane. Uniformity of the "perfect" plant is the goal. But from the air, these fields look like cemeteries. We have now lost 75% of our foodstuff diversity. The example of the Irish famine should warn us. The Irish potato blight of 1846 caused the starvation and death of a million men, women, and children, and forced another million into exile. One of the causes was that poor farmers were pushed into a monoculture as their only means of survival. Their holdings were so small, planting potatoes was the only way they could feed their families. A third of the population was dependent on potatoes for survival.

> "The Celtic grazing lands of ... Ireland had been used to pasture cows for centuries. The British colonised ... the Irish, transforming much of their countryside into an extended grazing land to raise cattle for a hungry consumer market at home.... The British taste for beef had a devastating impact on the impoverished and disenfranchised people of ... Ireland.... Pushed off the best pasture land and forced to farm smaller plots of marginal land, the Irish turned to the potato, a crop that could be grown abundantly in less favorable soil. Eventually, cows took over much of Ireland, leaving the native population virtually dependent on the potato for survival."[185]

The rice crisis of 2008 was another warning. On top of weather conditions and illnesses, the production of rice had to fight a new, even more destructive enemy: speculation. Tons upon tons of rice were put aside by governments and businessmen, while the population couldn't afford the price of rice and children starved. Speculators at all levels reaped generous gains. This is truly insane and inhuman.

We should increase the range of food products, not diminish it. From forty thousand products we used to produce, we are now limited to the twenty kinds of plants that amount to 80% of the food supply. Of these, 75% need bees for their pollination. Sadly, as described by David Grimaldi and Michael S. Engel in their book *Evolution of the Insects*, insects have been the systematic target of the blind pesticide industry, which kills many necessary ones and renders others resistant to the poisons. Reports indicate that

500 phytophagous insects, considered pests, are now resistant to at least one insecticide, some to all of them. Individual insects are able to naturally mutate enzymes, which in turn renders the poison ineffective, and they reproduce to form colonies. This will eventually lead to a demographic explosion of these superbugs. On the other hand, beneficial and harmless insects, such as bees and butterflies, because they do not have the array of detoxifying enzymes of phytophagous insects, are being thoughtlessly eradicated.

Chronic Fatigue Syndrome, Cancer, OCD, and Addictions

Look at all the women suffering from chronic fatigue syndrome, cancer, asthma, and other illnesses linked to an abused receptive principle. If men are gardeners of women and women are their fields—as some sacred books tell us—then all I can say is that not many men have green thumbs! Why conceive a child who will suffer for years and then die prematurely? Charity will fix this? No! This is an atrocious and inhuman thought limited to the Reptilian brain. It has nothing to do with religion. The more we abuse the feminine polarity and concentrate all our thoughts, feelings, and actions on getting more money or power out of it, the more acid (figuratively) we become.

Now back to the brain.

Obviously, even if the Reptilian brain is primitive, it is nonetheless primordial for the function of heart and lungs. We can't live without it. Some have judged the behavior of reptiles as primitive and concluded that the human behavior linked to the Reptilian brain is primitive and an expression of fear. I have read frequently that we need to control the Reptilian brain's aspects. Agreed, but control is not disrespect or eradication. Otherwise all our athletes can forget about mastering their respective disciplines. Also, some have concluded that our religious feelings, since they are linked to rituals and ceremonials and are repeated, are linked to this brain and need to be discarded. I disagree. As for those who claim religious feelings must be an expression of fear, I disagree even more.

With the organic, autoregulated cybernetic system, five functions—expressed through five global physical systems—allow for a better understanding of the observed phenomenon. In this system, fear relates to the most condensed phase, which presents the most contraction. Fear is the brake pedal you instantly hit to save your life. But also linked to this level is our basic form of intuition: instinct. Repetitive rituals are voluntary acts and therefore require frontal lobe input. They are linked to a point in a cycle, to a *time,* while the Reptilian brain is interested in *space*. Also, rituals are not associated with perseveration; therefore they are of the feminine polarity. In our physiology, the repetitiveness of compulsory disorders are linked to the vagus nerve, indicating a deficient feminine polarity. The longest nerve, it originates in the brain stem, within the medulla oblongata, which is also where our brain receives information from the heart. This might explain where the confusion arises.

Not surprisingly, this nerve is responsible for heart rate (kidney aspects control heart aspects), gastrointestinal and lymphatic peristaltis, and speech. Lately, an implant like a pace maker has been used for vagus nerve stimulation in cases of clinical depression in which the patient is drug resistant.[186]

Although repetition is necessary for habits to form, eventually leading to automation, the sort of repetition linked to OCD or to the Reptilian brain function is not the same as the repetition in rituals or ceremonies. They are not linked to the same brain areas. One is cyclical, the other not. The ritual is linked to the feminine polarity, while OCD is linked to the uncontrolled masculine polarity by the inefficiency of the feminine polarity. Also, the notion of time is different in rituals and OCD. In religious rituals, it should be the expression of the evolved heart aspect (synergy of all phases). There is an urge in OCD that we don't see in religious rituals. A reptile, taking a path that leads to food, will avoid an obstacle. If we remove the obstacle, it will still use the same path since the initial positive result made it a fruitful tool of survival. The path became a part of the reptile, a type of Pavlovian reflex conditioning. The Reptilian brain is our world of conditioned reflexes. It is a bottom-up reaction linked to the past. Rituals, on the contrary, are top-down actions. Our *spatial* memory appears to be confined to the right hippocampus, a structure associated to the Reptilian brain. This structure seems able to create a mental map of space, thanks to place cells.

Rituals have a different aim and should proceed from the totality of incarnated human levels. In that aspect, they enclose the Reptilian brain. By essence, rituals are either purifying, incantatory, or commemorative. By creating an event, they propose to restore in space and time—that is, physically—relations established by an ancestor or archetype between our spatio-temporal reality and the transcendent energy of a dynamic universe (implicate order). It is an actualization of the sacred world in our daily, prosaic life, permeating our cells as memory. It is not a repetition arising from the Reptilian brain, but the creation of the unique from the frontal lobes. It is the willfull actualization in a point in time and in a determined space, the synchronization of different phases and levels into a unity. It links the inner and outer in a certain time and space. It is an event that will never happen again. It is a top-down, inside-out process. It has nothing to do with repetition, unless you are an outsider looking in and not participating physically, emotionally, rationally, collectively, or symbolically. In that case, you cannot call it a ritual. Only a clear understanding of the cybernetic system shows us that, although it's a seductive idea, linking religious rituals and the Reptilian brain is inadequate. I believe that we came up with this idea because many of those who perform rituals and ceremonials have not attained the level of maturity, balance, and knowledge that should be prerequisites for this function.

The Reptilian brain is archaic but nonetheless necessary. Should we dismiss arousal with a smile or a frown, saying that because it is part of the Reptilian brain, it is irrelevant and possibly just a reaction of fear or aggression?

That there could be archaic reactions in religion, I do not deny. The Reptilian brain is not isolated. It controls autonomic responses, for which reaction time can make the difference between life and death. It is primordial for survival on earth. It is indispensable to the functioning of other aspects of the brain, and a direct door to the environment. Religion doesn't necessitate a quick reaction time in order for the organism to survive. The problem with religion is when it is used by people who are dominated by their reptilian aspect and have a problem with their feminine polarity. We are then confronted with fundamentalism and fanatism, which have nothing to do with religion—or I should say, spirituality—and everything to do with covert or overt desires for power. Not surprisingly, most fundamentalists are men. They can be of any religion, but can also be seen in science, politics, business, sports, families. In fact, anywhere you can find a social pyramid. It is not religion that causes wars, it is the

Reptilian brain and its desire to dominate. A universal religion would not have these problems. Such a religion is not possible, since our cultural roots are different and should be recognized. Nonetheless, to recognize the universal system behind these religions (LIFE biosystem) could be an effective tool to finally bring peace to our world. Economic competitiveness and greed for power used religions and thus caused more wars and revolutions than religions did.

The brain stem has a wide spectrum of integrative functions. It is the basis on which the various brains are built. Many health concerns arise from problems within the Reptilian brain. Damage to the cerebellum has been linked to loss of balance, vertigo, blindness, difficulty in coordinating movements or performing alternating movements (ataxia, dysdiadokokinesia), muscle weakness, dysarthria, loss of postural tone, multiple sclerosis, problems with recreating a recently practiced movement, and many others. Upper brain stem damage is common in head injuries. It is likely to interrupt critical connections with the frontal lobes. In our biocybernetic system, the frontal lobe has a controlling effect on the Reptilian brain. Elkhonon Goldberg even coined a term for this condition: reticulo-frontal disconnection syndrome.[187] He also indicates that the pathways connecting the frontal lobes to the ventral brain stem are particularly sensitive and prone to damage, *even in mild head injuries.* He surmises that there are good reasons to believe that these pathways might also be abnormal in certain type of schizophrenia, ADD, and ADHD.

It also appears to play a role in cancer.

I came in contact with Dr. Hamer's theories in 1985. While he was chief of internal medicine in a gynecology-oncology clinic at Munich University, he elaborated a theory that explained his own cancer and how he cured it. Dr. Hamer elaborated the theory after studying the many CT scans of his cancer patients. Curiously, those scans all presented certain brain areas surrounded by a ripple effect. Specific cancers showed the same perturbed brain area. How he came to superimpose an emotional conflict to these areas, I am not sure since I did not explore his theories in depth. What is of interest to us here is that the CT scans all showed perturbations in the Reptilian brain area. He found this normal since, recalling his own cancer, he remembered the panicked shock he felt, the helplessness, and the prolonged *inability to act.*

Here is how I understand it. Our LIFE biosystem implies that when faced with a great stress, we will cope by following different stages. We may also remain stuck in a certain phase. The coping process is as unique as we are. Interestingly, these stages correspond exactly to the Kübler-Ross model of grief. Women in general will use emotional, nutritional, collective (social), and symbolic tools for coping, all associated to the aspect in which women are expressive. Men will generally use physical, logical, and symbolic tools of coping, all associated to aspects in which they are expressive.

The Reptilian brain is powerful in instantaneous action and quick resolution of a situation. It has ingrained patterns and models, a type of DNA connected to the collective unconscious, the history of the species as well as its personal unconscious, which tells the personality how to react. In humans, when this aspect is powerless in the face of a certain situation, and you can't either flight or fight, the first reaction is *denial*. The new reality negates all the cellular information linked to the past and held by the Reptilian brain aspect. Once this phase, which can last from one second to years, is released, men will have a tendency to try to expel the problem physically. They express their *anger* as information floods the limbic region, the emotional brain. This emotional brain confirms or negates the impulse. If confirmed, a woman will feel physical pain and cry.

The stressful information is quickly or eventually transferred to the Analytical brain. The Analytical brain, the neocortex, as it searches for an abstract solution, will try to see the problem unemotionally. The woman will analyze her emotions. The man will try to find or construct an abstract output. This is the *bargaining* period. If even analysis doesn't help, the Universal brain, linked to the right prefrontal lobe and to the feminine polarity, will search with a projection into the future. This is when we reach out, though mostly women. If you can't reach out in a satisfactory manner for whatever reason, *depression* may result.

If you can access the next level, all of these brain areas and tools of consciousness will sum the result and compare it with the symbolic aspect, the Self. There, hopefully, an output is finally found. If so, *acceptance* will be reached. The Self, attached to the individuality, will either resolve the situation or, if not, send the information back to the Reptilian brain for a new cycle, a new search for a definite coping

Isis Code

strategy. We can also compare this process to the one described by Jeffrey A. Gray.

As the cycle starts again, the person may lose weight, sleep, and heat. To escape the situation, the person needs to find a focalization at the weakest point, which will be used as an outlet. The weakest point is the level from which the Self has retired the most. Also, energies have the most affinities with the densest aspect, which explains the psychosomatic effect of stress. The physical level is forced to express and output energies, whether they have a positive or negative—that is, stressed—character. Whether the output will happen on a physiological, emotional, rational, collective, or symbolic level depends on many individual and Self-oriented factors, as well as on the nature of the affliction. The benefits of healthy life habits, of a loved one, of knowledge, of a caring society, or of a belief system come into play here. One or more may open a gate, an output, on one of these levels and allow the tension to escape, if not resolve the problem. Otherwise, the person is imprisoned on a merry-go-round. Only a fixation at the weakest level will bring about a release from the tension. The whole process, whether long or short, and whatever the outcome, will hopefully be resolved and help deepen the personality's connection to the Self, even if this happens only in the minutes preceding physical death. In a parallelism with our LIFE biosystem model, the phases of the Kübler-Ross model are denial (Reptilian brain/kidney), anger (Mammalian brain/liver), bargaining (Analytical brain/pancreas), depression (Universal brain/ lung), and acceptance (Human brain/heart).

Brain scans of grief-stricken individuals detect local inflammations. Also, their salivary concentration in proinflammatory cytokines is very high. When anti-inflammatory cytokines are prescribed, they can have adverse psychiatric effects.[188] Conditions in which we use different types of interferon are advanced cancer or metastatic cancers, multiple sclerosis, AIDS, and others. Their side effects are as varied as depression, fatigue, somnolence, cognitive impairment, psychosis, suicidal ideation, confusion, and delirium. These proteins seem to interact with different tissues and organs, and assume new functions outside their conventionally accepted ones. Accordingly, medical biocybernetics associates cytokines with pancreas function, and therefore with psychological energies (psi-mental) and its effect on the central nervous system, which explains those side effects.

Similarly, cytokines seem capable of acting as neuromodulators within the brain. As such, they affect important brain actions such as sleep, appetite, and neuroendocrine regulation.[189]

To propose, as Dr. Hamer did, that all illnesses and cancers are solely psychological is farfetched, but it shows the importance of top-down processes in humans. This, I believe, is a bias linked to a certain epoch. It reminds me of a discussion I had with a renowned heilpraktiker (a German naturopath). Dr. Joseph Diener, then eighty-four years old, warned me, when I was twenty-nine years old, of my susceptibility to breast cancer. He was right. Unknown to him, when I was twenty, I was scheduled for an operation to remove a lipoma. I never showed up for my appointment. In my late forties I had a cancer scare after a thermography. Luckily for me, a more refined digital mammography showed that one of the lipoma had enlarged but was benign. I concluded that my profoundly modified eating and drinking habits, combined with emotional stress, were the cause, and reverted to a better physical hygiene. Had I continued on the same track, then yes, I might have had to deal with cancer. My physical aspect was the weakest at the time, and I was depressed. All the levels influence one another, but where the problem is expressed depends on many factors.

Back then, I asked Dr. Diener if the vegetarian diet I was following could have a protective effect against cancer. He became agitated and said that diet couldn't influence it. However, according to the LIFE biosystem, since nutrition is of the phase that could control cancer, it must have an effect. Dr. Hamer seems to be in the same frame of mind. Dr. Diener was proved wrong when I stopped consuming dairy products and alcohol, followed a homeopathic treatment, and my lipoma reduced spectacularly. I doubt Dr. Diener would accept that 80% of cancer are officially considered as having environmental triggers.

I also have to wonder why the money we give to charities for cancer seems to go mainly to pharmaceutical companies for research and development,[190] instead of being used to reestablish a healthy environment. It appears we don't mind about the quality of the environment, because it is not a "cost effective" behavior. Obviously, to heal our environment is to heal ourselves. Is it because by doing so we would have to go against the monetary interests of some huge financial conglomerate? I hope not. Such a laissez-faire attitude from our governments would be very irresponsible and undemocratic.[191]

Isis Code

In our system, if a tension is created, the masculine aspect has to express the tension, once the feminine polarity cannot absorb the toxic elements anymore. If universally and personally the feminine aspect is in difficulty, the system has less resilience and cancer will proliferate. What is the feminine aspect of our system? No, it is not female genitalia. Those are part of the masculine polarity.

Getting back to our pH spectrum, here is an interesting analogy. The pH of blood is usually slightly basic, around pH 7.4. When an acid-base imbalance occurs in the blood, the person's breathing rate will change. This is a short-term method of compensation. For a longer term regulation, the kidneys try to maintain homeostasis by changing the excretion of excess acid or base. Those two organs, lungs and kidneys, are, in our cybernetic model, the two main feminine, receptive organs. The first one pertains to the feminine polarity, the second to the masculine one. The kidney in the cybernetic system is linked to the most condensed, material aspect and to the auditory (the mesencephalon directs hearing).[192] As we said, it is the most receptive cybernetic organ, although it is part of the masculine polarity. It is controlled by the pancreas aspect, the nutritive aspect of the LIFE biosystem, part of the feminine polarity (diagram 7), and the psy aspect, part of the masculine polarity. The pancreas is the regulator of our system, hence its belonging to the two polarities.

A diet favoring cytokine production (red meat, dairy, and eggs) is proinflammatory. Cancers prefer this type of setting to thrive. Also, the nutrition linked to the regulatory system concerns all types of nutrition, whether it be of the physical, emotional, symbolic, rational, or collective type. This also explains why short fasts (twenty-four hours) have a protective effect against neurodegenerative diseases.[193] Indeed, this aspect of our system, nutrition, has a controlling effect on the generative phase (kidney; diagram 16). If the regulator becomes deficient in energy, it will not control proper cell division and cancer may appear.

When I still lived in Paris, I dreamed about a group of North American men, hunters. They killed animals, mainly deers, with no remorse, as a power trip or as a game and as a way to earn money. I faced them and said, "Many will die of cancer because of you." They answered, "This is black magic!" I replied, "No, it is a fact, only a consequence of your acts." Now I understand: killing for pleasure or because of objectification and desacralization of life is detrimental to the global feminine polarity. It weakens the lung aspect that is mother of all energies(including the one from the perpetrators). We are cutting the branch on which we sit. We weaken our resilient aspect.

Sleep, Dreams, and Oscillations

This brings me to talk about sleep and dreams. Sleep is a universal need, although we do not fully understand why. As the Reptilian brain is responsible for our access to the physical reality, it has a definite control over sleep. Lesions of the anterior hypothalamus (Reptilian brain), as seen in Alzheimer's disease, produce a long-lasting insomnia. Paradoxical sleep (REM) is triggered by the ultradian clock sited in the pons and bulb. Destruction of a very small region of the brain stem can permanently prevent this phase of sleep, even if the rest of the brain's connections are intact. Nevertheless, many brain regions participate in REM sleep, even though they are not central to it. Sleep is thus not a passively occurring state but is actively generated by specific brain regions. The previous notion, that sleep occurs as a result of the withdrawal of activity in parts of the brain that support wakefulness, is inaccurate. In fact, the metabolic cost of maintaining the brain during sleep or hibernation is only slightly lower than in the awake phase; and in some areas of the brain, it is the same.

Sources indicate that insomnia affects 20 to 35% of the population, and is most common in women, elderly, teenagers, and psychiatric patients, particularly those suffering from depression. Low socioeconomic status, chronic medical illness, stress, and the use of alcohol are also reported as factors linked to insomnia. In fact, sleep and cycles go hand in hand. Ability to respect cycles, thus the feminine polarity, impacts sleeping patterns. Cycles depend on synchronization, which is strongly affected by social conventions. These directly influence the Analytical brain, which in turns control the Reptilian one. Also, light is a natural synchronistic element that affects the Universal brain (feminine polarity). In fact, all the cells of the organism bear an intrinsic clock. Synchronicity is attained by all the different elements of the organism working in harmony. Insomnia is therefore a symptom of a broken harmony.

Nature went to great lengths to bring together different channels with the right densities and locations to serve a unique purpose: oscillation. The study of these is a new science, brought on by the necessity of understanding functional associations between different brain structures. The finding that the brain operates at multiple time scales renders these studies even more fascinating. The same structure can produce different oscillations (alpha, delta, gamma, theta), depending on its level of maturity and on environmental cues. Moreover, we now know that external stimuli cannot be considered the sole initiator of brain activity. Since a single structure will not produce all oscillatory sets, structures must then collaborate to cover all the frequencies. The exact mechanism of most brain oscillations is not known at this point. I surmise we will have to look closely to the meridian structure, the heart-lung conjunction, and the Master of the Heart for an answer to this question. What we do know, however, is that awareness requires structures, which will display persistent neuronal activity involving large neuronal groups. But consciousness, like health or love, is an hypernym that can't be strictly defined by the analytical tool. As with love, a definition of consciousness in the hand of the masculine polarity gives us something such as:

"A process in which information about multiple individual modalities of sensation and perception is combined into a unified multidimensional representation of the state of the system and its environment, and integrated with information about memories and the needs of the organism, generating emotional reactions and programs of behavior to adjust the organism to its environment."[194]

The brain's interactions with the body and physical realities in a feedback fashion is physical awareness, though still not Consciousness. We can achieve a greater awareness without accessing Consciousness, although it is the first step toward it. So what is the Consciousness we are after? It is home. The way to get to it is by the connection in this reality with the indivisible, which is through the control of the personality and the access of one's own individuality. It is really the particular quest of the feminine polarity. Without this feminine polarity, we are only deceiving apes.

Traditionally, sleep is divided into five phases. This cycle repeats itself approximately five times during the sleep period. Different waves or oscillations are attached to the diverse phases.

Stage 1 (heart phase controlled by kidney phase) Theta waves:

This is the onset phase. Drifting in and out of sleep, the body is still but may experience sudden jerks of legs or muscles (myoclonia). Often experienced by sleepers, the sensation of falling during this phase is popularly attributed to a digestive surcharge. Theta waves are apparent. This phase is short, around five to ten minutes. It is a light sleep period.

Stage 2 (pancreas phase controlled by liver phase) delta waves:

Adding together the length of all the stages of sleep during a night, around half is spent in stage 2 sleep. During this stage, eye movement stops and brain waves slow down. There will also be brief bursts of rapid brain activity called sleep spindles. Body temperature starts to decrease and heart rate begins to slow.

Stage 3 (lung phase controlled by heart phase):

This is the first stage of deep sleep. The brain waves are a combination of slow delta waves and faster waves. During stage 3 sleep, it can be very difficult to wake someone up. They might be groggy or disoriented. Delta waves are associated with synchronous silent episodes (100–500 milliseconds).

Stage 4 (kidney phase controlled by pancreas phase):

As of 2008, stages three and four have been merged. However, I foresee that the biocybernetic system might bring some novel understanding to sleep phases, so I will keep these phases separated. The second stage of deep sleep is sometimes dubbed delta sleep. In this stage, the brain is almost exclusively exhibiting delta waves, which occur in synchrony over the entire neocortex. It was thought that the thalamic output toggled the neocortex back and forth, but in the absence of the thalamus, the delta waves still occur. They also occur in the absence of the neocortex as well, but they are not synchronized anymore. This second phase of deep sleep disappears in old age. It is also known that sedatives, instead of lengthening this period of deep sleep, shorten it.

Both stages of deep sleep are essential for feeling refreshed and recharged in the morning. During this stage, there is a marked secretion of growth hormone, which controls many aspects of metabolism, as well as physical growth and brain development. In children, this is also the most common sleep phase for sleepwalking, bedwetting, and night terrors. Of note, there is a 25% decrease of delta waves in children of eleven to fourteen years old (Mammalian brain phase), and fewer delta waves in people suffering from depression, anxiety, ADD, and OCD.

Stage 5 (Rapid eye movement—REM—sleep; liver phase [imagery] controlled by lung phase) alpha waves:

REM sleep is the phase in which dreaming mainly occurs. As a person enters this stage, breathing accelerates and becomes shallow and irregular. The muscles of the limbs are often paralyzed for short periods and the eyes move very quickly, as if looking an action movie. Heart rate and blood pressure increase. The alpha wave is located in a frontal-central position of the brain. In the visual system, the prominent waves are of the alpha type. Men may develop erections. In adults, about 20% of sleep is REM sleep, while in babies it is 50%.

In general, during sleep the ultradian clock directs three subsystems:

1. Activates the cerebral cortex and stops the secretion of many neurotransmitters (serotonin, noradrenalin, and dopamine). These findings, however, are contradicted by findings in rats, where the dopamine levels in the medial prefrontal cortex (heart phase) are more elevated.[195] Interestingly, *the hippocampus, which is stimulated in phases of memory coding, is more active during sleep than during daytime. It seems to be responsible for dream imagery as well as cortex programing and memory consolidation (Michel Jouvet).*

2. Responsible for ocular movement encountered in REM sleep.

3. The system able to block nerve activity, resulting in muscular atony during sleep.

The REM sleep phase begins about seventy to ninety minutes after a person falls asleep. As morning approaches, the time spent in REM sleep increases while the deep sleep decreases.

Researchers do not fully understand REM sleep and dreaming. They know it is necessary in the creation and consolidation of long-term memories or learning, and probably has a role in oxidative stress reduction and synaptic plasticity. In the event a person's REM sleep is disrupted, the next sleep cycle will not follow the normal order, but will make up for lost time by switching directly to REM sleep. Consequently, this phase must be essential for general multileveled regulation of the LIFE biosystem. Older people have less REM sleep than any others. Some patients suffering from clinical depression might experience relief if they go without REM sleep. The reason for this has not been found. In my opinion, dreams express the state of the system. In a situation when there is no way out, this could add to the problems of a depressed person. Unfortunately, a few days without sleep or REM sleep will trigger hallucinations.

Occipital or thalamic alpha waves appear automatically during periods when the eyes are closed. They start appearing consistently only after three years old, which is when memories can start reaching our awareness. I therefore believe this phase has something to do with the subconscious (lung phase) interacting with awareness (liver phase). Here, the subconscious can rise to awareness.

Interestingly, research on the practice of meditation shows this method to be highly efficient in modifying alpha waves. So our will can control some oscillations. Beginners in meditation show an increase in power of their alpha activity in the occipital area. An intermediate will have more oscillations in the cortical area, and the frequency of the alpha oscillations will decrease. As for masters of the technique, for a large extent of the brain, large amplitude theta rhythm dominates. We have seen that this is the wave privileged by the heart phase control.

Of note, increased theta power above the occipital region (Mammalian brain) is associated with the encoding of new information. Experiments on hippocampal theta rhythms showed that these oscillators do not obey the elementary rules predicted by known oscillators. They appear to have a single monolithic oscillator, even though they are made up of a large and heterogeneous group of individual neurons. The quasi-sinusoid field monolith of theta rhythm keeps time (phase) more precisely than the relaxation oscillations from which it is built. Seemingly identical architectures can either promote synchronization or resist it.

In this fifth period of sleep, brain activity is similar or even higher as when we are wide awake (alpha waves). The rest of the body, however, is essentially paralyzed until we leave REM sleep. This is caused, we are told, by an amino acid (glycine), released from the brain stem onto the motor neurons. Dreams do not appear only in REM sleep though. Tore A. Nielsen, PhD, of the Dream and Nightmare Laboratory in Montreal, explains that these dreams, however, do not have the intensity of REM dreams. He calls this phase "covert REM sleep."

These five stages make a cycle. We go through many cycles every night. Each subsequent cycle, though, includes more REM sleep and less deep sleep (stages three and four). By morning, we're having almost all stage one (heart phase), two (pancreas phase), and five (REM sleep, liver phase).

Each complete sleep cycle takes from 90 to 110 minutes.

For Sigmund Freud, "dream interpretation is the royal way to the unconscious," and thus to the discovery of one's soul. For Carl Gustav Jung, who expanded on dream analysis, dreams also have a function "to try to re-establish our psychological equilibrium aided by oneiric material and in a subtle way tries to reconstitute the total equilibrium of our whole psyche." In this sense, dreams can access the collective unconscious (lung aspect, feminine polarity) and participate in the development of the personality and discovery of the individuality. The Self, besides being the center of the psyche, is also autonomous in the sense that it exists outside of time and space. Jung called the Self an imago dei (image of god). It is the source of certain dreams called *songes* in French, which may give us the ability to perceive the future or guide us in the present. How do we know a dream is a *songe* and not the expression of our personality and masculine polarity? During a *songe*, my observations have indicated, events happen but emotions are not triggered. You might act but your emotional self (Mammalian brain) is not involved.

Different types of dreams refer to different aspects of Man. Belief in the divine origin of dreams was universal. Archaeological discoveries attest that dreams could reveal the future for Egyptians of the tenth dynasty. They decoded dreams through the use of keys. The dream as a divine message exists in Greek mythology as well. In Orphism as well as the school of Pythagoras, it was taught that communication with the sky (i.e., the implicate order) occurred only during sleep, when the soul awakened. We could also say: when the feminine polarity is not inhibited. Identical beliefs are found in medieval Jewish writings, Arabic medieval scholars, and in most traditions around the globe.

Through Dr. DeBavelaere, I met K. T., who was providing consultations to students, specifically people interested in Jungian psychoanalysis. She would help them safely walk through their own subconscious. Originally from Vienna and having lived in New York and now in Paris, she had an old European aristocratic charm and the simple but genuine nobility I so love, although it is alas so rare to find nowadays. Even if taken away from her personal environment, she had something unique about her. Her environment did not make her, she made her environment. Her Parisian apartment, located in a serene part of the city, not far from the Eiffel tower, was full of books as well as beautiful and meaningful objects. It had the perfect ambiance to elevate and draw toward the inner self. Not surprisingly, K. T. had contacts with the well-to-do class of society, not only in Europe but also in America. A painting of one of her uncles, a professor at an illustrious American university, enhanced this feeling of knowledge. She was at the time using astrology as a door and, when we were ready, relaxation and dream recall as a ladder to our subconscious. She would accompany us and visit those dreams we felt were relevant. Somehow, she could see what we dreamed more fully than we, for she could point out details we left out, as if our conscious mind didn't see the whole picture but could remember it. She brought our awareness toward them, until an understanding of the dream emerged naturally. She never analyzed the dreams; the understanding would come naturally to us. Inexplicable anxieties I had vanished following these meetings. It was not what she said. It was who she was and how she put one in contact with this part which feeds from, and manifests itself through, archetypes and symbols. Did she express the model of this new woman I was seeking? Certainly. Curiously, I was told not long ago that K. T.'s mother, while pregnant with her, had always wanted to orient her life toward beauty. Was this part of why K. T. was who she was? This conclusion seems plausible.

Isis Code

One thing is sure, working on my dreams, not in an analytical way but a fusional one, brought me closer to defining and understanding the silent, hidden and observant part of myself.

At the time I am writing this, K. T. is no longer with us. She left her body a few months ago while undergoing a surgical procedure. The last time I saw her, she told me she would be having this surgery. She'd been assured it would put an end to the constant pain that had resulted from an almost fatal horse riding accident. When I heard this, my reaction was instantaneous. Tears started to roll down my cheeks, uncontrollably and for no apparent reason. She looked at me and quickly replied, "But I will not die! The worst that could happen is that I find myself in the *petite chaise* [wheelchair]." She smiled. I felt stupid and dried my tears. She was a mother to me and the godmother to my eldest son. A few months later I understood: back then, the silent, timeless unconscious part of me knew too well what was about to happen.

Dream interpretation allows us to become aware of elements that are responsible for creating imbalance in our personal biocybernetic system. It may help alleviate the need to self-medicate such imbalance through addictions and obsessions. For the sake of simplicity, I have attributed some addictions and taste to the different phases solely as demonstration (diagram 20). Tastes are linked to the pancreas aspect, which is analytical and, as we recall, the regulator of the system. Who holds the key to this aspect holds the key to the system. This is another reason why nutrition and the conditions surrounding it can have such an impact on us. Addictions are also linked to the masculine polarity as they directly affect the reward path of the Reptilian and Mammalian brain.

For the last several decades, mental illness was viewed more as a social problem than a medical issue. Drug addiction continues to be seen as a character flaw instead of as a biological problem. In fact, mental illnesses and addictions are social problems *and* medical problems. With knowledge and awareness, we can control and help with the medical issues, as well as modify society to limit the apparition of these troubles. The risk of our present orientation resides in the victimization of the individual, leading him into thinking "he can't help it" and that his actions are unavoidable and thus justifiable. The human spirit is still an unexplored venue, and I would not take away responsibility from humans. This is insidious and would lead to absolved robberies, murders, rape, and the like. The general solution is to increase the energy in the feminine polarity and, on a societal level, stop impeding its functions.

In a research, when subjects, who had been exposed to cocaine, were asked to inhibit their craving while watching a video laden with cocaine cues, activity in some brain regions decreased. The researchers say this deactivation is a way for the brain to "tune out" the cocaine cues and shows humans have the ability to inhibit craving.[196]

From a clinical point of view, the most frustrating characteristic of addiction is the persistence of relapse risk after a person has stopped abusing drugs. As a result, no treatment can be considered curative, and for the most acutely addicted individuals, relapses often can occur even long after any withdrawal symptoms have subsided (McLellan et al., 2000). This happens in part because use of drugs during growth (from conception to thirty years old) modifies hormones and neuromodulators' performances. Drugs induce alterations in neurons. During consolidation of key neural pathways, this might disrupt the normal course of brain maturation. Eric J. Nestler et al. demonstrated that addictive drugs, as well as drinking sucrose and running, for example, produce a protein—ΔFosB—that accumulates in the neuron until it affects which genes are turned on or off.[197] The surge of dopamine in the brain consolidates neuron connections created by behaviors. The dopamine system will be forever changed. Alcohol, cocaine, nicotine, food, gambling, sex, and—on a positive note—love have been shown to increase brain dopamine levels and activate the dopaminergic reward pathways. Repetition of these behaviors will determine the pattern. Interestingly enough, there is little transporter protein in the prefrontal cortex.[198] There, the norepinephrine transporter is preferred.

> "There is increasing evidence supporting a link between the endogenous opioid system and excessive alcohol consumption. Acute or light alcohol consumption stimulates the release of opioid peptides in brain regions that are associated with reward and reinforcement and that mediate, at least in part, the reinforcing effects of ethanol. However, chronic

> heavy alcohol consumption induces a central opioid deficiency, which may be perceived as opioid withdrawal and may promote alcohol consumption through the mechanisms of negative reinforcement. The role of genetic factors in alcohol dependency is well recognized, and there is evidence that the activity of the endogenous opioid system under basal conditions and in response to ethanol may play a role in determining an individual's predisposition to alcoholism. The effectiveness of opioid receptor antagonists in decreasing alcohol consumption in people with an alcohol dependency and in animal models lends further support to the view that the opioid system may regulate, either directly or through interactions with other neurotransmitters, alcohol consumption."[199]

While each drug produces different physical effects, all abused substances hijack the brain's normal reward pathways. Areas of the brain responsible for self-control, judgment, emotional regulation, motivation, memory, and learning will be altered.

Whether the addiction is alcohol, nicotine, crack, cocaine, or anything else, the effect on the brain is similar: an uncontrollable craving that becomes a priority over anything else, including career, family, friends, and even the addict's health and happiness. Social as well as other problems associated with drug addiction have not been properly addressed. The period most associated with stimulant experimentation is from fourteen to twenty-one years old, the heart period of the Idealistic self. Not surprisingly, people faced with low energy in the structural kidney phase will be more at risk in the heart phase. As the heart phase deals with identity and management of information (patterns), users may be responding to a real need to escape the lifeless lifestyle offered by a desacralized, consumerist society.

Associated to this period of fourteen to twenty-one years old, the induction of altered states of consciousness through diverse methods, for the purposes of initiation or therapy, can be found in all ancient traditions. Our civilization follows this pursuit through many means: dance, breathing exercises, fasting, food, isolation, meditation, music, physical exercise, religion, psychotherapy, not to speak of the destructive ones. In fact, considering the five phases of the biocybernetic system, all actions of man can be inscribed in this pilgrimage between different levels of awareness. For some, the quest emerges from a desire to find meaning to their existence, to ward of boredom, or to limit anxiety. For others, they wish to enhance their artistic abilities or change their perception of a senseless world. For others still, it is to be part of a group that questions society or to stimulate certain cognitive function by a surplus of dopamine or simply to "feel good." Beneath all this, though, is the essential mandate of the Heart period: perception. The tools we use here will not only help reveal our profound identity, but will also change our perception. This perception will profoundly mark how we deal with the rest of the world in the next phases.

This need of younger people to find sense and meaning is natural and healthy. The problem resides in the fact the world we have built does not provide suitable answers anymore. Exploring new, hidden realms often considered "forbidden" or "evil" and "destructive" by society expresses this need. This search of the secret and unique experience reminds us of certain rites of passage among traditional peoples. Added to this, various biologists have observed that when available, some animal species consume natural psychoactive substances with gusto, naming it a "fourth instinctual instance of animal biology" (Siegel, Ronald, 1990). It is as if evolution impulsively tends toward an expansion of perceptions to reach an associated and yet more subtle consciousness. I see it as the normal instinct to explore our different dimensions. The baby explores the physical realm of his body and surroundings (zero to seven years old, physical self); the child explores relationships with the psychological environment and time (seven to fourteen, emotional self); and then the young adult explores the inner realms through his contact with the community and often will wish to travel far away, as if deepening self-knowledge occurs through awareness of unchartered remote regions (fourteen to twenty-one, idealistic self). As an adult now responsible for earning his livelihood, if he doesn't feel he can control his environment, then what is left to control are either his body through sport or diets, or his psyche through control over states of consciousness (twenty-one to twenty-eight, rational self), as well as expressions of the knowledge he has acquired. At this period, men believe they know everything.

Isis Code

In the next phase, the social aspect will be explored, and an individual's place in society will become central to him. Native Americans have reduced the incidence of alcoholism on their reservations considerably by reviving ancestral rituals (Hodgson, Maggi, 1997), in this way refining their Idealistic selves. Lack of harmony and the fragmentation of our post-modern society, along with the desacralization of our daily lives and the loss of authenticity, all leave us without the means to integrate ourselves in a like-minded society. Only those who limit themselves to the values and goal of the consumer society feel adapted. Unfortunately, trying to find one's identity through addictions is guaranteed to fail, as indulgence in addictions impedes health. It is like making a fire to warm oneself. It's a good idea, unless you are using the wood floors of your own house. Then it is very destructive.

Honestly, we all regularly use stimulants on different levels in order to rebalance our personal biocybernetic system. We also use stimulants to enhance some aspects of the system so we can respond competitively to the demands of our unbalanced environment. Unfortunately, this eventually backfires. Neurotransmitters used by the system can be excitatory or inhibitory, and we will consciously (through the use of prescription or over-the-counter drugs) or unconsciously (through behavior) modulate these by using different substances. Some replicate natural neuromodulators and replace them on their receptors. But it is all make believe. We only deceive our biological apparatus.

Morphine and heroin, for example, attach to endorphin receptors (a natural painkiller produced by the brain). On the other hand, nicotine deactivates the receptors. So as long as nicotine is present, all is well. But after a night's sleep, for example, the baseline concentration drops, the receptors become functional, and the smoker experiences discomfort. Also, a certain substance in tobacco smoke seems to inhibit monoamine oxidase, an enzyme that breaks down dopamine after reuptake, therefore momentarily allowing for a larger amount of dopamine in the system.

Cannabis appears worse than cigarettes. For example, long-term cannabis users have deficits in cognitive flexibility and attention (Pope, Gruber, and Yurgelun-Todd, 2001); impaired learning, retention, and retrieval of dictated words. As well, both long-term and short-term users show deficits in time appraisal (Solowij et al., 2002, 2000). The decision-making deficits resemble those observed in individuals with damage to the prefrontal cortex, suggesting this drug alters function in this specific brain area (Rogers et al., 1999). For some it inhibits, while for others it facilitates long-term potentiation (LTP) and long-term depression (LTD). In the latter case, connections between neurons become less responsive (Carlson, Wang, and Alger, 2002; Nugent and Kauer, 2008; Sullivan, 2000).[200] Bambico et al. have identified the antidepressant effects of cannabinoids to be mediated through the medial prefrontal cortex of the symbolic brain (heart aspect). When someone says he is high, the term is appropriate. They also found higher doses suppress serotonin (lung aspect) and thus provoke a loss of the antidepressant effects researched.

While nicotine, for example, binds to the receptors for acetylcholine, other substances block receptors so there seems to be more neurotransmitter available. In any case, this unfortunately backfires and leads to a diminution of the number of receptors. We see this with the use of reuptake inhibitors.

Alcohol obstructs the NMDA receptors, glutamate receptors, a predominant molecular device for controlling synaptic plasticity and memory function. Cocaine, on the other hand, indirectly increases the amount of dopamine available to the synapses, while Ecstasy does this for the amount of serotonin.[201] Stimulants of the different phases are not only physical ones. They can also be emotions or ideas.

At all levels we can stimulate one aspect of the system. These interactions between ourselves and the environment are what we call life. No more exchanges (reception, emission) with the environment, no more life. As an illustration, I will present one general type of stimulation, and then we will study some mainstream addictions and see how they influence us. What is problematic is when a neurotransmitter system is involved.

Of all the means to influence our biocybernetic system, the most universal one is the taste of food. The different tastes stimulate different organs, different aspect of the system, and their attached phase (diagram 23). Interestingly, results from studies as well as observations tend to validate the system.

Vinegar/ liver phase

The antiglycemic effect of vinegar was first reported by Ebihara and Nakajima in 1988. Brighenti and colleagues demonstrated in normoglycemic subjects that 20 ml of white vinegar (5% acetic acid) in a salad dressing reduced the glycemic response to a mixed meal (green salad and white bread containing 50 g carbohydrate) by over 30%. Not surprisingly, vinegar is the sour taste related to the liver phase. It has a control over the pancreas phase, which is linked to glycemy (sweet taste) and the triggering of cancer (cytoplasmic aspect). For instance, in vitro, sugar cane vinegar (Kibizu) induced apoptosis in human leukemia cells, and traditional Japanese rice vinegar (Kurosu) inhibited the proliferation of human cancer cells in a dose-dependent manner. It is thought that because acetic acid in vinegar deprotonates in the stomach to form acetate ions, it may possess antitumor effects. A case-control study conducted in Linzhou, China, demonstrated that vinegar ingestion was associated with a decreased risk for esophageal cancer.

Bitters/heart phase

Also, it turns out that receptors for the bitter taste are found in the smooth muscles of the lungs and airways. These muscles relax when they're exposed to bitter taste, according to a report from researchers of the University Of Maryland School Of Medicine in Baltimore, in the online edition of the journal *Nature Medicine*. In our system, bitter taste is linked to the heart phase and has a control on the lung phase.

Spices / Lung phase

Today many spicy seasonings are available on the market. Cayenne pepper exerts several beneficial effects on gastrointestinal functions. Ginger significantly reduces serum and hepatic cholesterol levels by impairing cholesterol absorption and by stimulating the conversion of cholesterol into bile acids. This shows that when we stimulate the lung function, we have a better control over the liver one. Siberian ginseng enhances the activity of macrophages such as Kupffer cells found in the liver. This herb helps release stress. It increases protein synthesis in the liver and reverses diet-induced fatty liver in animals. In our system, spicy food is linked to the lung, which has a control over the liver phase. Many studies state that having spicy food helps increase cognitive functions. Turmeric has been shown to increase cognitive functions, and also improve memory and other mental functions.

We also know that overindulging in salty items eventually negatively affects the heart function (kidney controls heart), and overindulgence in sweets (pancreas) can lead to diabetes (pancreas controlling the kidney). What stimulates a structure, when taken in excess, eventually depletes its energy. Balance in all we do is essential to freedom.

Now how about other stimulants?

Caffeine usage is widespread, as we all know, followed by alcohol and nicotine, which rank second and third respectively as the most widely consumed stimulants worldwide. Sugar is certainly number one, although it is erroneously not considered a drug. Caffeine artificially stimulates the kidney phase, alcohol the liver phase, nicotine the lung phase, and sugar the pancreas phase.

As an example, at the cell level, the stimulating effect of coffee is provided mostly from its action on the adenosine receptors of neural membranes. Adenosine is a central nervous system modulator that inhibits arousal and has specific receptors. When adenosine binds to its receptors, it dilates the blood vessels, probably to ensure proper oxygenation during sleep, and it induces sleep through an inhibition of neural activity. Caffeine acts as an adenosine-receptor *antagonist, replacing* adenosine on the receptors without reducing neural activity. Fewer receptors are thus available to the natural slowing action of adenosine, and neural activity therefore speeds up. Also, caffeine causes the pituitary gland to secrete hormones, which in turn cause the adrenal glands to produce more adrenaline. This increases the attention level and jolts the system. As a stimulant of the kidney phase (diuretic), it has an effect on the heart phase and hence the "feel good" action.

Unfortunately, regular use of alcohol, nicotine, and marijuana lead to a prevalence of anxiety, mood disorders, and behavior problems in adolescents (King et al., 1996; Rohde et al., 1996; Kandel et al., 1997, Riggs et al., 1999; Rao et al., 1999). Alcohol, caffeine, poor sleep, poor nutrition, certain antidepressants, stress, and sugar all seem to eventually decrease dopamine activity in the brain. Nicotine has been shown to enhance immediate learning and attention (Del et al., 2007; Kenney and Gould, 2008; Mattay, 1996; Conners et al., 1996; and Watkinson and Gray, 2003). This may explain the high propensity for ADHD patients to smoke. Following smoking cessation, ADHD symptoms increase, predicting relapse (Rukstalis et al., 2005). ADHD is also associated with abuse of other stimulants and psychoactive drugs. Some researchers believe that such abuse might express an attempt to self-medicate (Elkins, McGue, and Iacono, 2007; Galéra et al., 2008; Tang et al., 2007, Lambert and Hartsough, 1998; Pomerleau et al., 2003).

In another study, the odds of having ADHD were more than three times greater for adolescents whose mothers smoked during pregnancy (Pauly and Slotkin, 2008). Cocaine and nicotine can directly induce one form of synaptic plasticity, the strengthening of neural connections via a process known as long-term potentiation (LTP) (Argilli et al., 2008; Kenney and Gould, 2008). Unfortunately, cessation of nicotine absorption as well as augmentation of dosage is associated with eventual deficits in working memory, attention, associative learning, and counting (Keenan and Yellin, 1989; Blake and Smith, 1997; Bell et al., 1999; Davis et al., 2005; Hughes, Jacobsen et al., 2006; Mendrek et al., 2006; Raybuck and Gould, 2009; Semenova, Stolerman, and Markou, 2007). In fact, many stimulants and drugs (gambling included) provoke a rise of dopamine at the onset, but eventually, the reverse occurs, forcing the user to increase dosage or frequency to get the same high.

Recent studies have shown that long-term smoker, as well as those with a smoking history, experience cognitive decline. For instance, in one study of middle-aged men and women, smokers' cognitive speed declined nearly twice as much as nonsmokers' over a period of five years. Additionally, the rate of decline in smokers' cognitive flexibility and global cognition occurred at 2.4 times and 1.7 times respectively when compared to nonsmokers (Nooyens, van Gelder, and Verschuren, 2008). As well, over a ten-year period tests of verbal memory and speed of visual searching have shown that ex-smokers' visual search speed slowed more than nonsmokers

(Richards et al., 2003). Those who decide to quit sometimes have a transitory decrease in cognitive performance. The level of this decrease predicts relapse (Patterson et al., 2010; Rukstalis et al., 2005). Some studies show that, in some cases, the cognitive failure of long-term cigarette smokers can subside with time, while others associate adolescent cigarette use with later periods of depression (Choi et al., 1997).

We avoid surroundings that are not in affinity with us. Colors, shapes, textures, temperature, sounds, and how our bodies can move in an environment have to be in affinity with us. Otherwise we will try to change the environment to enhance our well-being. In drug use, the environment plays a critical role of reinforcement (similar to a Pavlovian reaction) and may therefore jeopardize our ability to reach optimum health on all levels. We call this conditioned place preference, a cue-induced drug craving (Franklin et al., 2007). A number of functional brain imaging studies have shown that the insular cortex, most particularly the left one and part of the masculine polarity (see chapter 6), is activated when drug abusers are exposed to environmental cues that trigger cravings. This has been shown for a variety of drugs of abuse, including cocaine, alcohol, opiates, nicotine, amphetamine, methamphetamine, morphine, heroin (once in the brain, heroin is converted to morphine by enzymes), cannabis, and caffeine (Bardo and Bevins, 2000). Changes in the activity levels of brain regions involved in learning and memory (i.e., striatum, amygdala, orbitofrontal cortex, hippocampus, thalamus, and left insula) have also been observed (Franklin et al., 2007; Volkow et al., 2006). The right hippocampus, part of the masculine polarity, also plays a role. The level of accuracy of the route taken to a target place determines the level of activity in the right hippocampus, part of the Reptilian brain.[202] In short, the more we rely on our masculine polarity, the more difficult quitting will be and the more we will need to abuse stimulants.

In navigation tasks, women will consistently recruit the right parietal and right prefrontal cortex, linked to the heart phase, while men will rely on their hippocampus, linked to the Mammalian brain and masculine polarity. In 2005, it was discovered that in rats, the entorhinal cortex contains a neural map of the spatial environment.[203] The entorhinal cortex (EC) is located in the medial temporal lobe and functions as a hub in a widespread network for memory and

Isis Code

navigation. It is one of the first structures to be affected in Alzheimer's disease. The EC is the main interface between the hippocampus and neocortex. The EC-hippocampus system plays an important role in spatial memories, including memory formation, memory consolidation, and, as we saw, memory optimization in sleep.

Men's Brains Age More Rapidly?

Significant changes have been observed across cortical regions when we reach sixty to seventy years old, suggesting a critical transition point in brain aging at that time, most particularly in men. The brain undergoes sexually dimorphic changes in gene expression through all the different phases. However, Gene Ontology analysis showed more gene change in men than in women. It also revealed that with aging, different categories of genes were predominantly affected in men. Notably, the male brain will be characterized by decreased catabolic and anabolic capacity (break down or building up of biochemical compounds/components), especially regarding classes of genes geared toward energy production and protein synthesis and transport.

In these researches, the superior frontal gyrus (masculine polarity, Analytical brain) and the postcentral gyrus (feminine polarity, Symbolic brain) showed the most aging-related changes. Women's brains, on the other hand, will show the greater level of increased immune activation, although there is a noticeable effect of aging in both sexes. I think this relates to the general deficient feminine polarity. A strong lung aspect would prevent or minimize this. Berchtold et al. also used Gene Ontology research to determine if there were prominent changes between 20–59-year-old and 60–99-year-old brains.[204] They found in men's brain, but not in women's brain, a general decrease in the capacity of producing energy related to such components as electron transport, oxidative phosphorylation, ATP metabolism, mitochondrial transport. In women, categories such as neuronal morphogenesis and intracellular signalling were affected. In both sexes, genes associated with synaptic transmission were down regulated while genes associated with apoptosis (cell death) and angiogenesis (formation of new blood vessels) were at higher levels (up regulated). Interestingly, for both sexes many genes associated with inflammation and the immune system were up regulated.

Although some inflammation was shown to be neuroprotective, at a certain level it becomes detrimental and a symptom of aging. This also is a sign of a deficient feminine polarity, not a "natural" occurrence. It is similar for men. If the feminine polarity is taken care of, then the deficits will be reduced if not eliminated.

Locus Coeruleus

The locus coeruleus (also spelled locus ceruleus) is a nucleus in the Reptilian brain involved with physiological responses to stress and panic. It contains by far the largest group of noradrenergic neurons in the human brain (German et al., 1988; Baker et al., 1989). This relatively homogeneous compact nucleus contains forty-five to fifty thousand neurons bilaterally in the lateral central gray, and another five to ten thousand neurons in the lateral aspects of the central pontine tegmentum (Baker et al., 1989). Disinhibition of noradrenergic locus coeruleus neurons is linked with neuropsychiatric conditions, whereas degeneration of this region occurs in diverse disorders involving movement. Unfortunately, dysfunction of these regions is common in the elderly.

When we are born, the locus coeruleus and the substantia nigra are unstained. The pigmented locus coeruleus is visible at five years old, in harmony with its attribution to the Reptilian brain phase, tributary of the Universal brain (lung); with those in the substantia nigra remaining unpigmented (Olszewski and Baxter, 1954) until close to fourteen years old, in harmony with the attribution of the substantia nigra to the Mammalian brain, controlled by the Universal brain (lung). The degree of pigmentation increases with age to reach a steady state within individual cells, and by age thirty a substantial proportion of the neurons in both regions contain neuromelanin pigment (Mann and Yates, 1974; Graham, 1984; Manaye et al., 1995).

As the principal site for brain synthesis of norepinephrine, which as we saw has a direct link with the feminine polarity (pineal, right hemisphere, released in sleep by the medial prefrontal cortex), its name literally means "the blue spot," deriving from its azure appearance in brain tissue. The color is due to light scattering from melanin in noradrenergic nerve cell bodies. In Alzheimer's disease, there is up to 70% loss of locus coeruleus neurons.

Isis Code

It receives input from a number of other brain regions, but primarily from the medial prefrontal cortex of our Human brain (information regulation and Idealistic self), whose connection is constant, excitatory, and increases in strength with raised activity levels of the subject. It is one of the brain areas considered as an interphase, that is, between the lung phase (Universal brain) and the kidney phase (Reptilian brain). The locus coeruleus may figure in clinical depression, panic disorder, and anxiety. The cingulate gyrus and the amygdala also innervate the LC, allowing emotional pain and stressors to trigger noradrenergic (norepinephrine) responses.

Women's Masculinization

I am not talking here of the suspected defeminization and masculinization of female mammals by endocrine disruptors found in our environment and resulting from man-made chemicals that have been released into the environment (chapter 9).

I refer to a psychological masculinization, brought indirectly by men's interpretation of what a female human is. They unwillingly triggered women's masculinization. There are many reasons to this, the first being that their interpretation of what is feminine is based on the feminine aspect of their own masculine polarity, not the feminine polarity itself (diagram 21). This interpretation has been projected on women for thousands of years.

The feminine aspect of the masculine polarity is the kidney aspect. It corresponds to matter or, if you prefer, space. Since men, understandably, considered this to be *the* feminine polarity, or *the absolute* feminine and still do, they mistake the masculine polarity of women for the feminine polarity. The natural equation that arises from their Reptilian and Mammalian brains is thus women=territory. This is not *the* feminine polarity. It is the feminine aspect, represented by the black drop in the Taijitu, of the masculine polarity (schemas 21–22). In the same way, the heart aspect is not *the* masculine polarity, but the masculine aspect (the white drop) of the feminine polarity, analogous to time. Women, in the same way men did, project this aspect of the feminine polarity on men. If men don't fit this model then they are eventually found useless. If women don't fit the genital model, then they are also eventually found useless. The implications of this are numerous and profound, unfortunately exceeding the limits of the present demonstration.

In the tradition of the Tao, merged with the Kabbalah, the absolute feminine aspect, (Binah in the Kabbalah), is projected in the materiality of the LIFE biosystem as the kidney phase. The absolute masculine, Hochmah in the Kabbalah, is projected in the LIFE biosystem as the heart phase. We therefore see where the confusion arose from: desire of absolute of humanity. In its natural evolution and maturation, our society has adopted values that shun those linked to the feminine polarity, so many immature men who have not developed their own feminine polarity have associated women to the receptive aspect of the masculine polarity. This leads rapists, for example, to say things like, "She wanted it." Freud, who had the same vision by saying women naturally suffered from penis envy, didn't help the cause of the feminine principle. Because the feminine polarity (lung-heart) is present in both men and women and seeks harmony and wholeness, women have a natural tendency to express the stereotyped perception men project. We see this through the model many young girls now emulate, which unfortunately emphasizes the genital aspect of sexuality, a kidney representation. The feminine polarity is mainly heart and lung, the heart being its masculine aspect, the dot of the Taijitu; and the Lung being its feminine aspect, which represents, among other things, exchanges with the environment.

In the same way, one of the most recognized male fantasies is to partake in a threesome, two women in a sexual act with one man. When a male is obsessed with this fantasy, it is a sign that he is not in contact with either his or the feminine polarity, but is attached to the one-dimensional, cartoonish aspect of his being. Why this attraction? Beside of an obvious boost to his masculine ego, the masculine aspect has a value of the straight line. Electricity can be used as an example of this. The symbol associated with the feminine aspect is the circle, which adds a cycle to the straight line. Magnetism can be used as an illustration. The number one is attributed to the masculine polarity while the number two is ascribed to the feminine polarity, as expressed all the way down to the symmetric quality of women's brains. The dynamic association of the straight line and the circle give us the spiral, ever present in nature and linked to the number three. Electromagnetism can be used as an example of this. "Numbers as archetypal structural constants of the collective

Isis Code

unconscious possess a dynamic, active aspect which is especially important to keep in mind. It is not what we can do with numbers but what they do to our consciousness that is essential." [205]

When a man performs a sexual act with two women, he becomes the ultimate one and they, the ultimate two. Although this is a hormonal-heightening situation, it also reinforces his estrangement from his own feminine polarity by linking him to the feminine aspect of his masculine polarity, the kidney, rather than to the feminine polarity. For a male, focusing on the physical selves of two females means he controls the absolute genital femininity, which makes him the absolute alpha male—not the alpha man though. Unfortunately, it is also an acknowledgment that he doesn't have a functioning feminine polarity. In the same way a vagina cannot receive two penises simultaneously, the emotional level of this man cannot be open at once to the emotional expressive aspect of the two women. This indicates deficiencies in the emotional aspect of this man. On the material level, things get complicated since male humans, whether they are conscious of it or not, possess two polarities, one inner—the feminine—and one outer—the masculine.

Because all of society, in its pursuit of power and truth, has enhanced masculine values and ridiculed feminine ones, it is normal that this fantasy would be widespread. Unfortunately, it is not a sign of mental health. In addition, women own also a masculine polarity that has been formidably enhanced for the last hundred years. This leads to an even more profound need in a woman for her mate to be "in love" exclusively with her. A threesome will increase her masculine polarity because, considering what I previously explained, she has to be perceived as the "top female" for her masculine polarity to be expressed. As a woman manifests her masculine polarity mainly on the emotional and collective levels, she needs her companion to be faithful on these levels. This echoes the man's desire for his wife to be *all his* on the physical, mental, and often spiritual levels. As for her feminine polarity, she will need fusion at the soul level to feel elevated, and at the physical level to feel desired. All her feminine polarity tends toward subtle energies and away from the imprisonment of matter. She breathes in the strength of the sun and expands to the four corners of the universe. She can go through great traumas and accept any difficulty life throws at her, even accept death, if this soul connection exists as she is receptive at the heart level. She will overcome troubles on the material plane because of this link. In the harsh reality, it will make the difference between being poor but happy, and being poor and miserable. We need to remember however that both man and woman have a feminine *and* a masculine polarity although these are expressed on different levels. The man with a developed feminine polarity will need to have a relation with a woman who is receptive at the Heart level of his true identity. This is why it is said in tradition that men should not take for wife someone who does not share their profound beliefs.

As I experienced it, the world can reject you in every possible way, but if you have this love, every difficulty will appear scaled down. Without this connection though, the smallest problem will be inflated to great proportions and might appear as an impregnable fortress. In a balanced relationship between mature humans, it is not a question of projecting the masculine principle (time) penetrating the feminine principle (space) on only a genital level, but of all the levels communicating and exchanging this way. We thus find two masculine polarities in a dance with two feminine polarities on all levels (diagram 17). If the woman is objectified, then the couple stays at the physical level of the masculine polarity, and both become enslaved to matter and its separateness.

If the kidney aspect lacks energy, it will not be able to control the heart aspect. We will be faced with inflammation in the body, over-stimulation in the emotional self—leading to depression—rigidity in the mental self, and sanctimoniousness in the symbolic self.

This first phase of the Reptilian brain heralds the whole development of a human. As the Reptilian brain is, the other subsequent brains will be.

The first structure we have the benefit of studying is personality/individuality. Somewhere else I refer to the personality as the water self and individuality as the fire self. I use those two elements to express the fact that these always interact in a person's life, and they need to be balanced to express all the potentialities of a person. Too much water and the fire dies; too much fire and the water evaporates.

The water self, which is our personality and our masculine polarity, is given to us by our parents, our environment, the epoch we were

Isis Code

born into, the country and its culture and customs. This is the part that, even though imprinted on each and every one of our cells, will not stay with us when we die. The only memory of it is the actions we performed in the outside world or were performed on us. If we accept the theory of reincarnation as proposed to us by most traditions, these memories are part of the subconscious and will come to the surface in the guise of dreams, intuitions, aspirations, and talents.

Some people are born with the ability to remember some past events. This can be triggered by contact with a certain environment, taste, objects, or people, or by any of the senses. It is difficult to know if those memories are personal or if they are those of someone else. Personality is a characteristic way of thinking, feeling, and behaving. It embraces attitudes, mood, opinions, as well as subconscious content, and is most clearly expressed in how we interact with other people and with the environment. Our Reptilian and Mammalian brains are shaped by our personalities but also shape them.

Individuality is the ka of ancient Egyptian religion, along with ba and akh. It is the main aspect of the soul of a human being or of a god.[206] The exact meaning of ka remains a matter of debate, as the Egyptians did not exactly define it. The most frequent translation of *double* is inadequate. Written as a hieroglyph of uplifted arms similar to the pictogram describing the Hebraic letter He (ה), indicating thus way its receptive nature, it is therefore linked in our system to the feminine polarity. In some texts its protective nature is underlined. Later on, it personified the sum of qualities constituting the *individuality*, which for the LIFE biosystem, uses the feminine polarity as a support. The ka survived the death of the body and could even inhabit a statue of a person. This is a reason why a new pharaoh would have the noses of statues (lung aspect) of previous pharaohs hammered down, in a bid to prevent the soul of the deceased from visiting the world of the living. Personally, I believe the Egyptians were referring also to the body of light, a resurrected body created by the harmonious interaction of the masculine and feminine polarities on all levels. This is echoed by all religious literature throughout the world.

> "O Primal Origin of my origination; Thou Primal Substance of my substance; First Breath of breath, the breath that is in me; First Fire, God-given for the Blending of the blending in me; First Fire of fire in

me; First Water of my water, the water in me; Primal Earth-essence of the earthly essence in me; Thou Perfect Body of me!"[207]

This body is only able to manifest through the maturation of the feminine polarity, in the same manner the physical body becomes the expression of the personality. Chinese and Hindu sages, through their breathing and purification techniques, aimed at its construction and vivification. Then, it was thought that this body of light could protect them from aggressors, rendering those unable to act (lung aspect controlling the liver aspect), similar to the angels in Sodom who blinded the citizens so they were unable to harm Lot and his family. Ancient Egypt has a similar tradition, in which this soul is named as a double.

The reptilian phase ends with the maturation of our hardest physical structure, the teeth. Although the densest aspect is associated to the physical, science teaches us that everything is made of quanta. These have properties that cannot be overlooked, although we do not fully understand them.

> "Experiments performed in the 1990s by Alain Aspect and collaborators and repeated by Nicolas Gisin in 1997 show that the speed with which the effects is transmitted (from one quanta to another sister quanta) is mind –boggling. In Aspect's experiments the communication between particles twelve meters apart was estimated at less than one billionth of a second, about twenty times faster than the speed with which light travels in empty space, and in Gisin's experiment particles ten kilometres apart appeared to be in communication 20,000 faster than the velocity of light, relativity theory's supposedly unbreakable speed barrier."[208]

Hence the communication comes from within, not without. This resulted in the term *entangled* to describe particles. In fact, we are made of trillions of quanta. As a third aspect, the two others being matter and energy, information is inherent to quanta. Information withheld by these is "memorized," so to speak. So in fact, everything has a "memory." This point is essential. It was set in the parameters

Isis Code

at the birth of our universe and continues to govern the evolution of all basic elements of this universe. It is similar to Ma'at of Egyptian tradition, to the LIFE biosystem as based on Chinese tradition, to the Ayurvedic tradition of Indian culture, and certainly present in most traditions. According to quantum theory therefore, this information that forms our universe has to be constant and universal.

When we consider an atom as an object moving in one dimension, we ignore its internal structure. This structure gave the properties of quanta and can be found not only in quanta, but also in larger constructs such as organisms. This explains the coevolution of everything with everything, from particles to galaxies. Two quanta, electrons, atoms that were in contact can be born again in a different point of space and time and still act as if part of the same system. The cells of an organism appear to be acting in the same way.

In conclusion to this chapter, we can all help stop violence in the world, whether on an individual or national level. It starts at conception and continues every second of our lives. In Bali, for one hundred days after birth, the infant is continuously carried and never touches the ground. As a former inhabitant of the Heart realm, his center is still the heart aspect of the feminine polarity. By touching the ground, the center of his being symbolically shifts to the pancreas aspect. He becomes human and must therefore start his ascending quest toward the heart. Additionally, we now know that this intimate body contact with the newborn and the fact that the mother carries him at all time is crucial for the development of his physical and emotional selves.

Chapter 4
The Mammalian Brain, the Emotional Self, and the Emotional Human

Better less comfort and more well-being than more comfort and less well-being. **Ariane Page**

"When people live far from scenes of the Great Spirit's making [nature], it's easy for them to forget his laws." [Lung-liver axis] **Walking Buffalo**

Emotions are necessary to inspiration. They are its vehicle. **Ariane Page**

Phase of the *When* Level

The Liver Aspect

- construction phase: from seven to fourteen years old (when the hormonal structure is in place)
- main sense: vision
- building block: physical self
- nourishment: social group in which one belongs (family)
- cybernetic phase and function: from defense to information processing
- brain aspect : amygdala left, hippocampus left, thalamus, reverse synchronization with neocortex, basal ganglia, dentate gyrus, striatum, septal nuclei, occipital lobe, posterior hypothalamus, reward system of the masculine polarity, cingulate cortex
- cybernetic attached main organ: liver; later, heart
- this level controls nutrition and psi energy

Isis Code

- this level is controlled and inhibited by the Universal brain (environment and exchanges)
- cybernetic gender structure: part of the masculine polarity (expressive aspect of the masculine polarity)
- at this level, men are receptive and women, expressive. The sexual energies of women are thus superficial. Men will rather express their liver energy on the physical or mental level.
- cell structure: mitochondria
- energetical function: defense
- stimulation: romantic movies, vinegar taste, gloomy emotions, emotional voyeurism, sexual intercourse, drugs, humor, talking
- the protective animal
- motivation of that phase: wants
- type of attraction/relation of that phase: triggered by hormones, courtly love (lower type), sexual, egocentric
- main attitude: positive
- psychological: seeking
- emotional: anger, anxiety (if the controlling cybernetical aspect, lung, is not strong enough)
- level of awareness: collective unconscious, unconscious and general body awareness
- house analogy: the walls, what holds the roof and the floor together
- sunrise, morning
- Bible equivalent: Exodus
- Egyptian god (archetype): Osiris
- color: red
- figuratively: temperature
- elements: fire, plasma

This is the *when* phase, the period in which we can consciously situate ourself in time and space. It corresponds to the second dimension, the link between two points—from point A to point B. Its quality allows for a proper control of the *how* phase, the Analytical brain. This aspect also feeds the following phase, the *why* phase, corresponding to the Human brain of information regulation and Idealistic self. If you have traveled with children and listen to their remarks about how long the trip takes, or heard them complain about their age and the time it takes to become older so they can do what they please, then you have observed that children aged seven to fourteen appear obsessed with the *when* question in all that concerns them.

In this level of evolution, creatures are less controlled by the *what* (objectivation) and *where* (space) phase. As humans, we start thinking about other places as well. We explore our surroundings with enthusiasm. The young girl has a predisposition to explore the confines of her emotional world and those of others, while the boy will relish exploring the physical one. Through our future projective abilities, we will access other "wheres." For now, the Mammalian brain puts us in contact with cycles. The Reptilian one had a limited awareness of the cycles and was more inclined to react in a binary fashion, similar to a basic computer: true-false, good-bad, right-wrong. The Mammalian brain will be more adaptive.

Descartes saw opposition and duality in soul and matter—unless we, as products of a binary-oriented society, misinterpreted what he meant. By the possibilities offered to us by Einstein through his formulation of $E = mc^2$ and the space-time continuum, we can increasingly free ourselves of matter and time. We can invest in the here and now, the now and elsewhere, and the before or after and elsewhere. This second level of our personal pyramid was named the limbic system by Dr. Paul MacLean in 1952.

In the LIFE biosystem, I chose temperature to figuratively describe this biocybernetic phase of the liver structure. It is confirmed by the appearance after the seventh year of life of the circadian rhythm of body temperature with regard to phase and amplitude (Abe et al., 1978). Continuing with our pentane brain (instead of triune), the second level of our personal rocket is the emotional self. On the brain level, the limbic system is in the most affinity with it. I used pH for the physical level; I use temperature for the emotional one. Even in our language, temperature can express emotions: "I will have nought to do with a man who can blow hot and cold with the same breath." ("The Man and the Satyr," Aesop.) Or "she is so hot" and "he is so cool" and "she is cold," "she is frigid." This level is also linked to sexuality in general, as some of those expressions infer.

On the emotional level, there is an inversion in gender. The physical male is therefore receptive/feminine on the emotional level. The

Isis Code

female will be expressive/masculine on the emotional level (diagram 10). Thus men will be emotionally cold in the sense that they have a tendency to keep their emotions to themselves without analyzing them, while women will be emotionally hot and constantly analyze how they feel. The emotionally masculine woman relieves her tension by talking. The emotionally feminine man chooses either physical activity, sex (in its genital aspect), or intellectual pursuits. "He is cool" and "she is hot" are directed not to the genital aspect but to the sexual one, more tributary of emotions. This is a level dominated by hormones, or which dominates hormones, depending on your point of view. Personally I am inclined to lean in favor of the emotional aspect regulating our hormones, even though it is possible to use hormones to influence how we feel. At each of the levels, we can orient our inner parabolic antenna toward the more materialized or the more subtle, as we wish. Some always cast their point of view toward the most incarnated, which is a masculine tendency linked to the function of space control and desire for penetration.

It is generally accepted that ancestral male warriors remained impassive so as not to betray any emotions, thereby enhancing their chances of survival. This emotionless mask is seen as an act of control. In my opinion, this impassiveness is a consequence of their Mammalian brains being receptive, not something they consciously worked on. Also, facial recognition is managed by aspects of the feminine polarity, so warriors would undoubtedly not analyze your facial expression. It just happened that it might be a good quality to have when you were a warrior, albeit not so good in the case of a lover. Thus it does not enhance the chance of reproduction.

Facial expressions are the language of the emotional, thus more of the realm of women. Being stone-faced while communicating does not say you are more masculine. It says that as a male, you have not worked on your communication skills, you are not social, or your feminine polarity is not operational.

Many studies have demonstrated gender variations in the organization of the brain in harmony with these observations. Women's brains almost always show more active regions than men's one. Women cross-reference things through more parts of their brain than men do, most importantly through the emotional control centers in the limbic system (Moir, 1991). Indeed, behavioral studies have often and clearly demonstrated that women tend to be more emotionally expressive and have better episodic memory than men (Bradley, Condispoti, Sabatinelli, and Lang, 2001). Subjectively and empirically, we certainly all come to the same conclusions. Women's brains did not show greater activation when viewing emotionally charged content, but the activated areas were different. Women's brains are not as asymmetric as men's. The cortex is larger on the left in men while it is slightly larger on the right for women. The symmetry observed in women's brain is brought about by the larger size of their associative networks and commissures (corpus callosum, anterior commissure). Researchers also found that women and men process emotions differently and through different areas of the brain. Further, while men are more likely to use a small area, typically only on the left side, for a particular task; women will use more of the brain, on both hemispheres, for the same task. Also, baby girls are more sensitive to touch (information regulation brain, heart, part of the feminine polarity) than baby boys. Girls have a better sense of smell (Universal brain, lung), smile more, are more patient and less easily annoyed than boys. They also start speaking earlier than boys (Mammalian brain), enunciate more clearly, and develop a larger vocabulary. Teen girls can describe their emotions with precise details and without difficulty. For boys, the locus of emotional control remains confined to the amygdala. Asking a seven to fourteen-year-old boy to talk about his feelings will generally result in a flat glance.

Interestingly, girls experience a significant brain growth during the mammalian phase, from ten to twelve years old. Indeed, their brains gain about twice the weight as do boys' brains. Also, alpha rhythms (relaxation) and nonverbal cues show a sharp increase during this period. For the boy, this will happen later, during the growth spurts typically seen between fourteen to sixteen years old, corresponding to the heart phase of the idealistic self in which they are expressive.

When boys and girls were given a verbal task, Schmithorst and Holland (2007) found that activated brain areas and white matter pathways (Schmithorst, Holland, and Dardzinski, 2008) differed between genders, with older girls showing greater inter-hemispheric connectivity.

Isis Code

Computing a number of correlations, Yu et al. (2008)[209] showed that participants with high intelligence displayed more white matter tracts integrity (feminine polarity) than those of average intelligence, especially in the uncinate fasciculus, a white matter tract that connects parts of the Mammalian brain, such as the hippocampus and amygdala, with frontal parts of the orbitofrontal cortex, associated to the nutritional aspect of the Analytical brain. Fifteen participants with mental retardation were also studied. Compared with the seventy-nine healthy participants, they showed extensive damage in the integrity of the brain's white matter tracts: corpus callosum, uncinate fasciculus, optic radiation, and corticospinal tract.

Recently, a study combining a genetically informative design and a diffusion tensor imaging (DTI) approach for analyzing the relationships between white matter integrity and human intelligence came to the same conclusions (Chiang et al., 2009). White matter collects and dispatches information while gray matter has a more executive function.[210] Accordingly, researchers at Johns Hopkins found substantial differences in the anatomy of the part of the brain that is in charge of complex mental operations, as well as of cognitive and emotional experiences in women.[211]

Neuroimaging studies by Antonio Damasio in 2000 suggest that many of the brain areas that are in charge of emotion processing evolved much earlier than did the neural circuitry that allows awareness and control over them. Emotions and patterns of emotions are thus generally unconscious by nature. They come to consciousness as an evolutionary process. Women, because they are expressive (masculine) on that level, will naturally analyze their emotions and those of others. Consciously or unconsciously, emotions have a controlling effect on the Analytical brain too.Thus, when men say that they are being rational and totally objective, they are speaking ignorantly of their own structures. The rational aspect is submitted to the unconscious. In my opinion, there is confusion between "objectivity" and emotional coldness.

The kidney aspect of the physiological system (Reptilian brain) expresses difficulties of the system through its reactions of fear. The liver aspect, the Mammalian brain, manifests anger, aggression, and associated emotions. The lower heart (Human brain) expresses difficulties through agitation. The pancreas (Analytical/nutritive brain) through obsessions (economical and others). The lung aspect (Universal brain) through emotions of melancholy, anxiety, and such. In fact, emotions are not the prerogative of either men or women. For one thing, they are in fact masculine as they are set into motion and are linked to the masculine polarity. Men and women are emotionally the same. The only difference is how they express emotions.

Antonio Damasio stresses the rationality of emotion in his book, *Descartes' Error,* and emphasizes the value of emotions in decision making. He points out that patients with damage in the areas of the brain that integrate emotional and cognitive systems can no longer successfully function in the day-to-day world, even though their rational abilities are perfectly normal. This agrees with our cybernetical system, as the Mammalian brain controls the Analytical one (diagram 15). He observes that "the brain is a supersystem of systems." There are more neural connections going from the emotional centers to the cognitive ones than the reverse.

"Vision, listening abilities, reaction times, mental clarity, feeling states and sensitivities are all influenced by the degree of mental and emotional coherence experienced at any given moment." [212]

Although instincts are of the Reptilian brain, I place human sexuality in the emotional self. For humans, although reproduction is an instinct, sexuality is not confined to it. Traditionally, an instinct is defined as hereditary behavior unique to a species and varying little from one specimen to another. Sexuality implies exchanges while genitality is associated to dominance and sometime death. It is naturally egocentric. We can assess when the sexual act is linked to the heart function of the feminine polarity. Tender touch is then a key factor. This can be observed even in the animal world. The stag, for example, to masturbate lowers its head and softly rubs the tips of its antlers in the grass. Within five seconds the stag experiences an erection with little or no oscillation movement of the pelvis. About five seconds later there is an ejaculation. The antlers are linked to the most profound energy of the kidney (Reptilian brain), while the soft caress is of a heart nature (Human brain of information regulation in humans). This sensuality promotes the liver function, which is expressed through the ejaculation.

Isis Code

In humans, sexuality is not as hardwired in its purpose. It is experienced more as a consequence of interactions between all of our levels, in their conscious or subconscious states, and changes as we enter new phases. Many anthropologists have stated that it took a while for humanity to understand that intercourse led to pregnancy. This would imply that pregnancy was not the conscious aim of intercourse, but a consequence.

The thalamus is modular, has fixed functions, and is rigid, in the image of the masculine polarity. More evolved, the cortex has the ability of flexible adaptation and functions, which are less delimited in space and far more cooperative, a quality of the feminine polarity. The function of the thalamus in the human Mammalian brain is somewhat different from what it originally was. In a rudimentary way, with the basal ganglia, it holds most of the functions held by the neocortex.

A man who doesn't develop his feminine polarity will remain stereotypically modular in his approach to life. Also, these modules will not communicate properly. His intelligence will become specialized. He can be capable in some areas and ridiculously immature in others, although he may fool himself into thinking that if he is brilliant at one thing, he is automatically skilled at everything else. Our modern educational system and inherited values have unfortunately tremendously enhanced this modularity, blocking thus way human evolution — but this is changing with globalisation and multitasking. Luckily, in a social process, our reasoning—which is always a result of both conscious and unconscious patterns—is confronted by other reasoning (also issued from conscious and unconscious patterns). The winner takes it all. This doesn't mean we express the truth; it means we have confronted our ideas in society. As a social process, science helps find flaws in some reasoning, but is totally blind to others because they are in harmony with the general system in its immature state. This explains why so many of the greatest visionaries were persecuted in their time. Reasoning is used to convince oneself and others and is part of the masculine polarity. It is a tool to legitimize emotions and actions. In no way does that make it the truth bearer.

Relationships

When a woman is depressed, implying her Mammalian brain is overactive, she will tend to focus her thoughts on what is depressing her. She will try to find ways to fix her problems as well as evaluate her options. A man will tend to use an activity to monopolize his attention in a bid to avoid thinking about the problem. If the trouble is of an emotional nature, men are often at a loss. They will flee or numb themselves in any possible way, including the use of drugs or alcohol. Their Mammalian brains are more influenced by the subconscious than women's Mammalian brains are.

Some women can also be emotionally and socially violent, even more so than men. Their masculine polarity expresses itself on those levels; the stronger their masculine polarity, the stronger their action on those levels. A study of the Versailles court life (and many others) is quite enlightening on the subject. Even Napoleon Bonaparte declared: *"Je ne veux nullement à ma Cour de l'Empire des femmes."* This means he did not want some women to control and dictate his court: "I do not want in my court women ruling it as if it were their personal empire."

In a relationship, when confronted with a frustrating situation, a woman can therefore look quite like a convincer. She will use her consciousness of the emotional level in a highly analytical fashion, sometimes to the astonishment of her companion. He may then respond with passive resistance. This opposition will be expressed through withdrawal since men are emotionally receptive, while women think more along the lines of cooperation (social aspect) to solve problems. Her mate's withdrawal will frustrate her since it does not allow her to express her feelings, thus her masculine polarity. She will interpret the withdrawal as a silent (and sometimes vocal, if alcohol is involved) condemnation. This will make her feel physically isolated.

This is painful behavior for a woman to bear, because through her feminine polarity she is physically receptive and often judges a man's degree of love by the quantity of physical affection she receives. We speak of emotional abuse while the term is not exact. Does a woman refusing physical intercourse is physically abusive?

In the case of a dysfunctional withdrawal, which happens with men who have a problematic or underdeveloped feminine polarity, the isolation can last for days for no apparent reason. A woman with a deep attachment to such an immature man, in a bid to make him understand how hurtful the whole situation is to her, will try to explain her feelings in great minutiae. Faced with unresponsiveness, she will hide in a corner, weep, and swallow the lack of acknowledgement. A depressive state might ensue, especially if she was ignored by her parents. If a woman can't express her masculine polarity on the emotional and social levels, she will be forced to resort to the physical or mental levels, somewhat masculinizing herself. She might eventually become physically violent, which is a sign that she has reached some limits. She may also become obsessed with diets. Little by little she will feel that her mate's love is gone, since to her his silence is his covert way of saying she is a troublemaker. Her self-esteem will plunge off to the deep end.

A relationship requires exchange and communication. Men are pleased when women open themselves up for intercourse. Women are pleased when men open themselves up for communication.

When women voice their concerns, men who have a deficient feminine polarity will take the easy way out and answer through their Analytical brains: "You are too emotional," "you are moody." Some will even go to the extent of qualifying the woman as hysterical, needy, and even crazy; or will ask: "Are you having your period?" That without a doubt will highly irritate the woman. This is the masculine polarity linked to the analytical mind categorizing the situation in order to reach self-exculpation. Two masculine polarities are clashing. The Analytical brain of the rational self (pancreas level) has *no* relational skills. It is a tool to bring one from point A (reptilian) to a desired (mammalian) point B. The masculine polarity hates to be stuck or contained. The feminine polarity wishes to contain in order to resolve divisions. Women will use their Analytical brain in a more nurturing way (feminine aspect of the pancreas), unless their feminine polarity is dysfunctional. Some men interpret this as a wish to control. The feminine polarity mirrors the masculine one. This is a prerequisite to receptivity, which is one of the principal functions of the feminine polarity. This is why the little girl will seek her identity in her father's acknowledgement, the young woman in the ecstatic gaze of her beloved, and the mother in the outstretched arms of her child. This is why women left their lonely houses to work in society. It is not necessarily a reaction to economic fear, but the need for fusion of the feminine polarity.

Relationships are not easy. If they are profound, they go into all the nooks and crannies of our personalities, awaken past frustrations, and reveal our scars and blemishes. They may also bring to the surface our grandest virtues and heal many wounds.

I read a story about Socrates. He was married to an awful woman, Xantippe, who was always frustrated and would yell at him continually, in public as well as in private. Despite all, he stayed. I would not be surprised if the elevation of his thinking resulted in part from the difficulties he had with her. I do not believe he was a masochist. Of course, one should not try to find a difficult relationship for the sake of it. Nonetheless, when difficulties arise, it is time to stop, feel, think, and understand.

Can we love more than one person at a time?

Can we have a physical relationship with two different people?

Although ultimately we collectively are one in our individuality, it doesn't mean we should indiscriminately make love to everyone, as intercourse is an aspect of the personality. Overstimulate the personality and you will become its slave. True fusion only happens and lasts on the most subtle level, not on the physical or emotional ones. If you believe there can't be a relationship between a man and a woman without exclusive ownership on all levels, then you can't imagine that someone in a romantic relationship could have a platonic relationship with someone else. As a man once told me: "Friendship between a man and a woman is impossible." Also, can you accept that someone might love more than one person at the soul level? Not if you can't experience, and thus don't believe in, the existence of the soul. That is a sad and limiting aspect of the personality. If the possessive aspect of the Reptilian brain (kidney) coupled with that of the Mammalian brain (liver) dominate, then no, you can't love more than one person at a time. The French phrase *Honi soit qui mal y pense*—Shamed be he who thinks evil of it—is aimed toward that type of people. Such a platonic relationship requires total trust among partners. Unfortunately, our masculine polarity, this force of involution, as master of the personality, does not trust and is always calculating. Also, let's face it, soul-deep relationships are seldom and few in our feminine-polarity-denigrating society. It takes people who have courageously risen above the cocooning caves and walls of the masculine polarity. When the two poles are reunited, there is no more potential, but something is always created, whether on the physical, emotional, mental or collective level. Love can use attraction but is not limited to attraction. When you have intercourse with someone, you create a link with that person, even if you are only a snake slithering on the reptilian level. It affects you, even if only subconsciously. If you have another partner, then the information you bring to your companion will change. Some sensitive people will sense that "something has changed," even if a Reptilian brain would retort, "You can't wear out sex."

One day, I reconnected with a love of my youth. I had been drawn to him the way a person is to a long-lost friend she chances meet while visiting an unfamiliar place. It had been an unfinished affair, but my heart always cared for him, and still does. Although he was much older than me, I had always been powerfully attached to him. I would also always excuse him for anything. One day a psychologist who happened to be a dear friend of mine told me: "You are attracted to him because a child cannot think that her parents don't love her. Since he does to you what they did, you feel comfortable, it is a known environment, and that is the way you think love should be." Although it is true that he treated me as my parents did and this allowed me to work on some old wounds, I sought him out for a different reason.

When you love someone in a profound, or should I say elevated, way, it is not because of appearance, actions, behavior, social position, but mostly *despite* those. It was not his personality I cared for but his soul. I still saw him with my adolescent eyes. During this time, which became one of self-imposed solitude, I painfully refined my understanding of love. His vision and understanding of "what a woman is," and his purpose for sharing his life with one, were very much attached to the masculine polarity point of view. So my days passed, full of paperwork, solitude, and feelings of inadequacies. Why him? I loved him. And just as my parents before him, he deprived me of affection and acknowledgement, which were the two things I craved but which he could not give me, having not received them from his own parents. He was stuck in his Reptilian and Mammalian brains. One day I finally understood, true communication and fusion would never happen, although I saw glimpses of the shining light of his soul. He could not express his individuality, though. What a waste.

I have since understood that we are guided toward love because, considering the feelings we experience, it makes sense and there is no way around it. This leads us to another phase in which self-sacrifice might appear more like self-destruction, but in the grand scheme of things—sometimes revealed to us later—love makes perfect sense. In the world of the here and now of the masculine polarity, love is submissive, senseless, and often hurts, but in the coming world of the feminine polarity, it is incredibly intelligent.

When someone you love is physically ill, you don't abandon them. As well, you don't reject someone you love because they are emotionally challenged. But you don't have to share the same space either.

When we study relationships, we analyze them under the assessment provided by the masculine polarity and the personality. That is not bad in itself if the aim is to protect the personality, but it is limited in its understanding of what is *really going on.* Relationships, whether harmonious or not, ask the question: "Why?" And we have a tendency to answer through the "what"—as in, "What do I like about this person?" or "What does this person has to offer me?" We compare different "products" and purchase the one most rewarding. Sometimes we even buy the product because there is not a better one available, or because it is cheap and it will suit our present needs and purpose. This is truly a relationship of the masculine type.

At one point, I spoke with a male friend about the relationship he was contemplating: "After all," he said, "I am the one who buys." For him, the female proposed but the male chose. This is the Reptilian and Mammalian brain talking. This type of relationship travels between the *how*—as in, how do I get him or her—of the human personality brain to the *what* of the Reptilian brain. This is the primitive interaction of the masculine polarity. When someone advises you to make a shopping list of what you like or dislike about a potential companion, this is the type of relationship they are talking about. Of course you might choose this type of relationship for many reasons, but don't fool yourself into believing this is ultimate love. It is a communion of interests that gives you fuzzy feelings, nothing more. The personality cannot explain everything; we need the individuality for that. We need it to grow to a relationship of the feminine polarity type. Otherwise it is an immature relationship that will not last.

We will never be able to answer *why* we love someone with a *what* unless the relationship is only at that level. When your love is of a more subtle nature and thus worth the work, the questions of the *why* and *which* will find answers through the *who*. I love this person because of *who* he or she is, despite any physical, economic, social, rational aspects. Then you realize that if you love someone this way, you do not have to share every minute of your physical reality with them, because you do not need to share that personality. You can be by his or her side if life allows it, but you don't have to be glued to him or her. This is the relationship of the feminine polarity. At this level, people are totally free. Otherwise, one has to give to Caesar what is owed Caesar.

Through our parents, we unconsciously receive patterns that shape our personalities. I always thought my mother didn't love me, but how could she, given what she had experienced? My father would send me conflicting signals, so I would not know how to act, what he really wanted from me, or who I was. I call this the backward-forward syndrome. He would tell me to act in a certain way and later on would say something that contradicted the first request. An example would be someone saying to his kid: "Look how this person swims well" and then, some other time, "I forbid you to go in any swimming pool as long as you don't know how to swim." This means the child can't go in a swimming pool, and since it is impossible to learn to swim otherwise, this child can't make the parent proud and will feel confused. This, I understood later, came from the lack of coherence between the different levels of the personality/individuality, and between the masculine and somewhat shrunken feminine polarity. This originated most obviously from the religious and secular worlds, and trickled all the way down to my father.

I would try to understand what he truly wanted or meant. I would choose what I thought he wished and what was in the most affinity with who I was. Naturally, he would not be impressed by me. This would unsettle me deeply. I did not recognize myself in the image my parents were sending back to me and tried desperately to redress it, but to no avail. I was stuck behind a glass window, trying to get their attention but they could not see me. Since there was no possible positive issue for me, I could only be wrong. The pattern I received is one of constant stressful relationships in which I can't decipher what the other wants and always believe I am not loved or meaningful to this person.

For her part, my mother would not allow me to go out and play as long as my work was unfinished. Both would deny occurrences in my daily life, calling them products of my "strong imagination," when to me they were realities. Those attitudes put together meant that I could not express myself. I was not able to go out with others and could do only what I had to. I had to live in a bubble. It killed every desire for action that I had. Luckily for me, music and spirituality kept me safe. Life was permitted if it was work. I had the right to be a saint, martyr, or wife, and to do and think as others, but nothing else. Because I would think and act as guided by my feminine polarity, which was unpopular then (and still is), I was judged naive, unrealistic, and overly sensitive, and therefore unintelligent. The good thing was that I dug my roots deep, and my contact with nature became interiorized. Nature and I became one.

In Quebec, pets are abandoned by their families at a fivefold rate compared to any other Canadian province. When a nine-year-old girl, strolling through the alley behind her house on her way to school, finds the severed head of a German shepherd, and then later discovers a white cat that has been buried alive in a pile of dirt, and she cries, no, she is not being too sensitive. Instead, there is clearly a problem with her environment. When another time, while driving with her dad to her school, she sees her beloved cat of eight years lying dead in the street, a casualty of the road, she cries. Her dad, irritated by her expression of emotions, tells her she is being "oversensitive" because it is "only a cat." Then a few years later, her dog gets hit by a car. What does the dad with a weak feminine polarity say then? "If that dog is not healed in three months, I shoot it in the head." And of course we will not spend money on it. In a society, how we treat animals is how we treat the feminine polarity. Make no mistake; my dad was an awesome father. I succeeded in healing my dog. He was eventually given to an uncle who lived far away because my parents preferred to have a poodle they had found. My aunt told me that my dog howled for weeks. He echoed my pain. Then one day, my dog,

Isis Code

older now, jumped on a little girl. He was probably proud of doing one of the many tricks I had taught him, but my relatives feared he had become dangerous. He was shot in the head. I was told a year later. Make no mistake, the members of my family are normal and wonderful people, but the patterns they received while shaping their Reptilian and Mammalian brains were deficient. This is very sad.

I was invited to a traditional Jewish wedding in Montreal. It was a grand affair with 250 guests, music from a band, four singers. When my friend invited me, I hesitated. What to wear? He pointed out that I had a beautiful black dress. "A black dress?" I replied. "I would not want anyone to come to my wedding dressed in black!" I decided to go, but opted for a cream-colored pantsuit with gold jewelry and golden shoes. We drove to the synagogue, and as we neared it I saw two women entering the synagogue. I said, mortified, "They're dressed in black!"

My companion smiled. I hoped for a miracle, that these women were going to another event. When we walked in, I surveyed the cocktail room. Except for three women, one of them being me, all the others were dressed in black! Later on, a Jewish gentleman said to me, as if to put me at ease, "I am so tired of all these Jewish women dressing in black all the time! As you can see, my wife made an effort." A beautiful lady, she was one of the three not wearing black.

The color black is receptive to the environment. It drains the light. Looking at the bride dressed in white and then at the people in the pews all dressed in black, I couldn't help but feel that she was the queen of the night and was giving all of her energy to the guests. Then I thought of Mary, Queen of Scots, one of the first to wear a white wedding dress, as an expression of purity. Black is associated to the kidney phase because of its receptiveness. It does not emit information, it only receives it.

A violinist was playing soulful music. We had been given a card that explained the ceremony. I was most interested in its symbolic nature and how these symbols were interpreted. Let me describe the wedding to you.

To the sound of a violin, the parents of the groom walk in and stop in the middle of the aisle. There they wait for their son. When he joins them, they kiss him and each takes one of his arms to guide him to the rabbi. The parents of the bride do the same with their daughter. It was explained on the card that she came last because in the Jewish tradition, the most valuable comes last. Surprised, I asked the Jewish woman sitting to my left: "Is not the firstborn the most valuable in Jewish tradition?" She nodded. For such a paternalistic tradition, I thought it odd and that maybe it was a form of flattery directed to the mother of the bride.

The bride was beautiful, radiant, and seemed extremely happy. They were married under a canopy, a chuppah, all white and gold. Four posts held up the tent. The explanation was that this was a replica of Abraham's house which had four doors so people could come from any side. This meant the couple has to be welcoming to everyone. To me the canopy is a sacred enclosure similar to the ones under which kings and queens used to pronounce their laws and judgments. The four posts indicate the four cardinal points of the compass and the canopy mimicked the enclosed womb in which every life forms. What happened under the canopy is thus sacred. The couple's ceremony is the expression of the laws, applied to their choices.

The bride was to circle her husband seven times as a protective shield for him as he left his bachelor life where he was protected by his mother (liver aspect). She circled only once, probably because her gown impeded her. It is intriguing to note that the number seven represents the cycles in life in Taoist tradition and in Shakespearian lore. There are also seven main colors in the spectrum, seven fundamental virtues, seven musical notes, seven continents, seven openings of the head. In physiology, there are seven protective layers to the scrotum. The sacred numbers in tradition are one, three, five, seven, and nine, odd numbers that are by definition feminine in their function. It is not a coincidence that in Islamic tradition, seven is the number of heavens as well as of earths. In Hindu mythology, it corresponds to the number of sages. Their wives are the goddesses referred to as the Seven Mothers, those seven divine women who were left behind on earth, it is said, and thus became the ancestresses of humankind.

Here I saw the expression of the Mammalian brain function. As the father is expected to protect his children and companion on the physical level, the mother, expressive on the emotional level, is expected to protect her children and spouse on the emotional level. Of course, this is what should happen in a harmonious world. The abusive mother and wife on the emotional level and the abusive father and husband on the physical level are unfortunately widespread in our deficient world.

The bride and groom drank from the same cup. The groom then put a simple gold ring on the index finger of the bride's right hand because, the rabbi explained to the audience, in their tradition the forefinger is linked to the heart (although energetically, the middle finger has this function). Then the groom made a proclamation of union. Following this was the reading of the contract, outlining the moral and financial obligations of a Jewish husband toward his Jewish wife. Seven blessings were then said, the husband broke a glass, representing the broken hymen,[213] while everyone shouted, "Mazel tov!" The violinist played again as the couple, their close family, the best man, and the maid of honor exited the area of the chuppah.

The way I felt it, the central point of the wedding was the contract the groom made to the bride as the crowd applauded. It honestly felt as though she had won a prize. The importance of a ritual is that it links the people to their community.

The expression of sexuality in a civilization, as in human couples, goes through the different phases, from genital to fin'amor (diagram 22). One can judge the level of evolution of a society by its spontaneous and widespread sexual practices. Sexuality is often stereotypically associated with Greece, so we will use it as an example.

Greek pederasty arose in the tribal past of that country, before the city-state became a unit of political organization. As with many tribes across the globe, these communities were structured according to age groups. As in Egypt, a child lived with his mother in the women's quarter until he reached seven years old. When the time came for the boy to join an older age group as he prepared to become a man (midliver phase), he would leave his family to be in the company of an older man for a period of time. This man would educate him in the ways of Greek life and in the responsibilities of adulthood.

The Mammalian Brain, the Emotional Self, and the Emotional Human

Relationships and marriages in ancient Greece between men and women were age structured, with men around thirty years old (lung phase, collectivity, feminine polarity) commonly taking wives around fifteen years old (heart phase, feminine polarity). In China, on the other hand, such an age gap in a couple was frowned upon. A difference longer than a phase (seven to eight years) was considered energetically detrimental to the older man.

Later in Ancient Greece history, starting at twenty-one years old, a young man would stay away from women until the age of thirty (collective phase and end of maturation of the frontal lobe), at which point he would marry and form a family. A younger man would be allowed to become eromenos, a prepubescent taking on the receptive role of student. In some cases he could also become a sexual companion to a man twenty-one to thirty years old (pancreas phase). This corresponds to a time before puberty in the liver phase, in which the male is feminine at the emotional level and the beginning of the heart phase, in which the feminine polarity is awakened. Also, since this phase is prior to the maturity of their Analytical brains, males are still receptive at this level.

Boys exercised in the nude with other youths of their age. Adult men would witness these workouts. For boys this was a period of muscular formation (liver aspect), supposedly the reason why a man in the rational phase would approach a younger one, as muscular bodies were considered beautiful, a heart aspect element. The heart period is also when we can see more clearly the quality of the soul. A poet would say that when we love, we are always twenty years old.

Although within the system one would understand that an older man, expressive at the mental level, would wish to take a younger man, who is receptive on his level, under his wings as a rite of passage, I would add a word of caution regarding a sexual relationship. The frontal lobe of these "older" men (twenty-one to twenty-eight) is not yet mature. The pancreas aspect has an affinity and a control function over the Reptilian brain only in its maturity. Lack of maturity of this aspect could unleash the energies of the Reptilian brain, locking the two in genital activities far from the ultimate goal and function of their relationship. I guess this was accepted as a lesser evil for a system that had to fit over traditions that were already in place.

At which level sexuality is lived depends on many factors. We cannot generalize, as everything and anything was and is seen, from sexuality triggered by the Reptilian brain to fin'amor. For example, we learn from Roman writings that in the first two centuries, the Greek's valorization of sexual moderation shifted nearly to an idealization of sexual abstinence. Care of the self for the self's own sake, as in Michel Foucault's expression,[214] led from a critique of the ill effects of too much sexual activity to a general anxiety over sexual pleasure. Eventually, relations with boys were terminated (exit of the reptilian phase) and values shifted toward mutual care and fidelity. Marriage was now idealized as the perfect and complete formulation for relationships, so much that there was a new validation of intentional virginity.

In harmony with our physical spectrum system, Greek society did not differentiate sexual desire or behavior by the gender of the participants, but rather by the role each one played in the sexual act. The rules were contingent on the age of the partners. The active penetrator (erastes) was dominant, and was associated with masculinity, higher social status, and adulthood. The passive penetrated (eromenos) was associated with femininity, lower social status, and youth.[215] This is a blatant example of a masculine-oriented society.

Although the sexual act was not necessarily implied, education was. If the teacher fancied the student, he was allowed to pursue him sexually. The student, in his feminine function, had to resist, refusing the reptilian aspect of the relation, preventing in this way his feminization and accepting the relationship only on the individual level. Greek society attached a great importance in cultivating the masculinity of the adult male. As such, sexual relations between adult men were frowned upon. Past puberty, a male citizen being sexually receptive toward another adult male "made a woman of himself." The perceived feminizing effect of being the passive partner was stigmatizing. Though not considered a criminal, a homosexual adult male who was the passive partner was barred from participating in the Assembly and deprived of certain other citizen rights. The men acting in the masculine function were scolded but were still considered super males, which fits our genital spectrum (diagram 16). Slaves and noncitizens were forbidden to take roles of either erastes or eromenos at any age. In a society dominated by the masculine polarity, pederasty is unavoidable.

Socrates described himself as an erastes of wisdom, which is what evolved society should accept. To him, carnal pederasty prevented evolution, because that type of passion inevitably resulted in enslavement of the soul. This is in harmony with our system. To Socrates, the desire to teach was of a higher value than the supposedly erotic response to physical beauty, a definite attribute of the feminine polarity. Desire of dominance and an expression of status were not even taken into consideration, as they refer to the masculine polarity in its most primitive aspect: of the level of snake and lizards, not of humans.

The feminine polarity can express itself and be receptive only if the masculine polarity does not interfere. As I've said women are receptive physically. Accordingly, if a woman wants to experience an orgasm, one that is not restricted to the spastic result of clitoral stimulation, her amygdala—as linked to the masculine polarity—has to be deactivated. She can achieve this if she can open herself to her partner (trust) and has a functioning feminine polarity. Since the sexuality we put forward in our society is of a masculine type, this can generate some feelings of inadequacies in her. In a relationship in which the feminine polarity can express itself, the woman will have a definite feeling of peaceful, sensual expansion and will experience a slowing of time during intercourse. As well, she may access another dimension in which the borders of the finite and infinite disappear. The periorgasmic period will then enhance and merge into the orgasm itself. This aspect of sexuality, more associated to the sense of touch (heart) than to reproduction, seems confirmed by the fact that women—whether using birth control or not—tend to seek physical rapprochement more frequently around the time of menstruation; that is, at a time when they are the least fertile.

Sympathetic Nervous System

The sympathetic nervous system (SNS) turns on the fight-or-flight response, thus switching on the amygdala. It is controlled by the feminine aspect, by the right hemisphere. A defective feminine polarity will thus allow the sympathetic system to run in overdrive. By contrast, the parasympathetic nervous system (PNS) promotes the relaxation response.

Isis Code

Similar to two pedals on a car, the SNS and PNS carefully maintain metabolic equilibrium by making adjustments whenever something disturbs the balance of the system.

The feet controlling these two pedals are hormones, chemical messengers produced by endocrine glands. Named after a Greek word meaning "to set in motion," hormones—as emotions—travel through the bloodstream to accelerate or repress metabolic functions. In our desire for vigor, which we consider youthful, and our demands for an active life totally controlled by economics, we often keep our foot on the accelerator (SNS). When we do so, some hormones remain active in the brain for an extended period, injuring and even destroying hippocampus cells. As we recall, this area of the brain is needed for memory and learning. Additionally, there is already a natural, hierarchical dominance of the SNS over the PNS, often requiring conscious effort to initiate a relaxation response and reestablish metabolic equilibrium. If we compare both nervous systems, the sympathetic would correspond to the masculine polarity while the parasympathetic relates to the feminine one. As in the cybernetic system, the lung aspect controls the liver aspect, so it is not surprising to learn that structured breathing exercises are the best tool to regulate the SNS. In the brain, the left hemisphere (masculine polarity) naturally tends to overpower the right one (feminine polarity).

Hormones initiate several metabolic processes that best allow the organism to cope with a situation deemed negative, thus inducing a state of metabolic overdrive that goes as follow: The adrenal glands release adrenaline (also known as epinephrine) and other hormones, which increase breathing, heart rate, and blood pressure. This allows for keener senses, less sensitivity to pain, and more oxygen-rich blood flowing to the brain and to the muscles. It also causes a rapid release of glucose and fatty acids into the bloodstream. While this is happening, other hormones shut down functions that are unnecessary during an emergency. Growth, reproduction, and the immune system all go on hold. Blood flow to the skin is reduced. This explains why chronic stress leads to sexual dysfunction (kidney and liver aspects), increases your chances of getting sick (liver function), and often manifests as skin ailments (lung aspect lacking energy). Stress that is not detrimental to our feminine polarity can be beneficial and will not trigger the same biological reactions. Researchers at Ohio State University found that in some people, stress caused by performing a memory task activated the immune system. However, the stress from passively watching a violent video weakened immunity (as measured by salivary concentration of SIgA, a key immune factor).[216]

At Wake Forest University in North Carolina, professor of sociology Robin Simon studied more than a thousand unmarried adults aged between eighteen and twenty-three.[217] This is the heart phase period in which emotional patterns are stained into the fabric of the personality. Contrary to popular belief, Simon found that the emotional rollercoaster of relationships, as consequence of the ups and downs of a romance, had a greater detrimental effect on the mental health of young men than of women. He says that although men sometimes try to present a tough face, unhappy romances take a greater toll on them. They just express their distress differently. This again confirms our finding that women are expressive on the emotional level—their masculine polarity is in charge of this level—while men are receptive.

Dr. Hans Eysenck, a pioneer in the development of behavior therapy, and his colleagues at the University of London have shown that chronic unmanaged emotional stress is as much as six times more predictive of cancer and heart disease than cigarette smoking, cholesterol level, or blood pressure, and is much more responsive to intervention.[218] Subconscious emotional memories and associated physiological patterns underlie and affect our perceptions, emotional reactions, thought processes, and behavior. I believe these traces of emotional memories can be repatterned to an extent using a change in identity through access to a person's feminine polarity and the pursuit of the Know thySelf dictum.

Recent findings suggest that in generalization tasks, children spontaneously rely on perceptual information, even when they possess conceptual knowledge (Fisher and Sloutsky, 2005; Sloutsky and Fisher, 2004). Adults, on the contrary, spontaneously rely on their knowledge of how things fit into categories when performing classification of familiar entities. Similar findings emerge from a task in which five year olds were taught novel natural-kind-like categories.[219] When we force children to use conceptual instead of perceptual information, we jump over a necessary stage.

The Basal Ganglia

Researchers at the Medical College of Wisconsin in Milwaukee and the Veterans Affairs Medical Center in Albuquerque have identified areas in the brain responsible for perceiving the passage of time.[220] Their study is the first to demonstrate that the basal ganglia (Mammalian brain) and the right parietal lobe (information regulation brain), located on the surface of the right side of the brain (feminine polarity) are critical areas for time sensing.

Importantly, the study calls into question the long-standing and widely held assumption that the cerebellum is the critical structure involved in time perception. True, its anatomy parallels in miniature the anatomy of the whole brain, containing half of its neurons. It participates in complex planning and coevolves with the dorsolateral prefrontal cortex of the masculine polarity. In fact, the Analytical brain controls it. The cerebellum was assigned the function of timing in movements, which is not the same thing as passage of time. As well, it has an excitatory function while the basal ganglion has, it appears, an inhibiting one.

"We are excited that our findings can also have application to better understand some neurological disorders," says Stephen M. Rao, PhD., professor of neurology at the Medical College and principal investigator. "By identifying the area in the brain responsible for governing our sense of time, scientists can now study defective time perception, which has been observed in patients with Parkinson's disease and Attention-Deficit/Hyperactivity Disorder (ADHD), two maladies commonly thought to have abnormal function within the basal ganglia."

Investigations have revealed that patients suffering from Parkinson's disease experience difficulty in properly perceiving time. They also suffer from an extreme reduction of the neurotransmitter dopamine in their basal ganglia. When such patients are administered a drug that boosts their dopamine levels, problems with time perception diminish. It is thus believed that dopamine and the basal ganglia have a function in timekeeping. Animal studies have demonstrated the same importance of dopamine for timekeeping.

The Mammalian Brain, the Emotional Self, and the Emotional Human

Similar defective time perception was also observed in the cases of Huntington's disease. Patients with ADHD or obsessive-compulsive disorder (OCD), as well as workaholics, typically have issues with their basal ganglia. Their perception of "idle" is biaised. ADHD individuals will also suffer from poor concentration, poor fine motor skills, and often poor handwriting. Added to these will most often be other behavior disorders, since the deficient feminine polarity (information regulation and Universal brain) is often the main cause of these disorders. These can be as varied as oppositional defiant and conduct disorders; learning disabilities and communication disorders (including speech and reading difficulties); and anxiety disorders, such as generalized anxiety disorders or separation anxiety disorders.[221]

> Sudden and severe depressive episodes have been induced by stimulation in the substantia nigra, followed by euphoric rebound when the stimulation stops. This supports evidence from imaging experiments suggesting abnormal metabolic activity in the caudate nucleus [part of basal ganglia] during depression ... In a forward model of the basal ganglia, the striatum and other basal ganglia nuclei would receive incoming signals about the current status of events (in action or mental space), as well as signals predicting future events. The cortico-basal ganglia loops running through the striatum and basal ganglia circuits would sort and combine such signals, and then influence cortical and subcortical networks responsible for the production of motor or cognitive activity. Under conditions of circuit dysfunction, at one extreme excessive and repetitive actions or thoughts could result, and at the other extreme poverty of movement or thought could be the result.[222]

The basal ganglia is composed of a group of interconnected structures of the Mammalian brain, of which our understanding has enormously increased over the last few years. Its primary function appears to be one of action selection, selectively sending required *inhibitory* signals to all parts of the brain that are capable of generating actions. Here, rewards and punishments exert their most powerful neural effects. It appears also that seasonal effects influence the substantia nigra, part of the basal ganglia, and thus there seems to be a seasonal effect on striatal presynaptic dopamine synthesis. This agree with our model (diagram 15). Basal ganglia calcification elicits symptoms resembling schizophrenia (B. Chabot, C. Rouillard, and S. Dollfus, 2001). In rats, the pineal compounds melatonin and vasotocin have been shown to be neuromodulators of spontaneous neuronal activity in the caudate putamen, part of the basal ganglia (J. C. Castillo-Romero, F. Vives-Montero, R. J. Reiter, D. Acuna-Castroviejo, 2007). The caudate nucleus is like a miniature brain for higher stages. Elkhonon Goldberg calls it the "greater frontal lobes."[223] The frontal cortex will either allow or inhibit some behaviors arising in the caudate nucleus. It is an emotive-visceral integrator that incorporates the body's state of tension or relaxation in association with emotional states. One of its ends is connected to the amygdala (Reptilian brain) and the other to the globus palladius (basal ganglia of the Mammalian brain), and from there to the parietal lobes (information regulation brain—Human brain; diagram 15).

The Hippocampus

The emotional environment precipitates the mood fluctuations observed in bipolar children. Also, it was observed that children and adolescents are more receptive to what is presented by media. There is a stronger relationship between stressful events and major depression in adolescents compared to adults (Gould et al., 1994; Pine et al., 2002). This makes sense since the lung aspect (environment) controls the emotional (liver aspect) one. Consequently, healthy, unpolluted surroundings will optimize the proper development of these young brains.

Interestingly, deficient affective experience was correlated with a smaller volume of the hippocampus and a shorter length of the corpus callosum. The cingulate connects the corpus callosum to the hippocampus.These two elements, hippocampus and corpus callosum, work as a unit.

This hippocampus, part of the Mammalian brain (emotional self), plays a part in memory and spatial navigation. Physically, it is located inside the medial temporal lobe of the brain. Its name derives from its curved shape—which resembles that of a seahorse—and from the Greek: hippos = horse, kampi = curve.

Recent research shows the hippocampus is crucial for stress and learning interactions. Following an acutely stressful experience, it modifies learning for men as well as for women (Bangasser and Tracey J. Shors, 2007). Its development is affected by stress and bonding hormone levels, since the quantity of stress hormones circulating in a newborn will affect the number and types of receptors in the hippocampus. Nerve cells in the hippocampus are destroyed as a result of chronic stress and elevated stress hormone levels, especially in men. This will result in intellectual deficits since the Mammalian brain controls the Analytical one. Depending on gender, the hippocampus substantially differs in its anatomical structure, its neurochemical makeup, and its reactivity to stressful situations.

It is also one of many brain regions also whose volume differs significantly between genders. Adjusted for total brain size, it is larger in women. Although size alone does not drive function, research shows it may have an impact. This has led many people to wonder if such variations might result in gender differences in function and behavior, with women capable of expressing more on an emotional level and men expressing more physical drive. Of course, the brain is one area where it is essential to take individuality into consideration. Generalization would be perilous, although it is an excellent base to start with.

The LIFE biosystem agrees with the extrapolation for animals as well as for humans. In both rats and monkeys, chronic stress causes damage to the hippocampus in males, who are receptive with it, but does so far less, if at all, in females. The hippocampus thus plays a fundamental role in episodic memory, the kind that will let you remember, for example, an especially pleasant dinner years later. In fact, it seems the hippocampus enables you to "play the scene back" by reactivating this particular activity pattern in the various regions of the brain. Now we know why women regularly revisit those emotional loops of past events, since they are able to knit together old facts and compare them with new ones. Not surprisingly, active stem cells were found in the hippocampi (Mammalian brain) and the olfactory bulb (Universal brain). These memories will never fade away....

Isis Code

Neurons in the different areas linked to the five senses that have been triggered by the experience will fire simultaneously. How is this possible without an overarching field in which the information is complete? Our brain is in fact the mirror of our experiences (see chapter 8).

While brain activity in the hippocampus differs between sexes, short-term memory performance appears to be the same. The dentate gyrus, linked to the hippocampus and present only in mammals, is comparable to the medial cortex of adult lizards.[224]

This dentate gyrus, a main center for neurogenesis, receives excitatory input from the entorhinal cortex of the Reptilian brain (medial temporal lobe classically). This activates pyramidal cells among the CA4 and CA3 pyramidal neurons. It also sends information to the medial prefrontal cortex of the heart aspect.[225] Their more evolved aspect is at the level of our Mammalian brain. We will later see, with the Master of the Heart, that memory is not limited to hippocampi but is spread all over the body.

Schizophrenia is associated with deterioration in emotional experience and cognition. The typical age of onset is late adolescence and early adulthood, during the merging of the idealistic phase into the rational one.

Volume increases and myelination of the hippocampus occurs earlier in females (four to fourteen)[226] than in males (Benes et al., 1994; Giedd et al., 1996b).

Given that the maturation process of the hippocampus is not completed before the end of the mammalian phase in males, an injury to it would allow structural abnormalities, such as volume reduction, as seen predominantly in male subjects. Indeed, it was demonstrated that in male patients with schizophrenia, the severity of psychotic symptoms is significantly correlated with the hippocampal volume reduction (Bogerts et al., 1993; Lawrie and Abukmeil, 1998; Shenton et al., 2001 Seidman et al., 2002; Kurachi, 2003). Indeed, morphologic abnormalities have been observed in the fronto-temporo-limbic structures (Shenton et al., 2001; Suzuki et al., 2002), and were implicated in abnormal brain development (Feinberg, 1983; Weinberger, 1995). In addition to these, schizophrenia occurs with changes in brain chemistry, specifically, excessive levels of dopamine in the Mammalian brain and reduced amount in the frontal lobes (Analytical brain). Activation of the brain's frontal and parietal lobes is significantly disturbed in this illness. Medication can only treat the symptoms of schizophrenia; it does not eliminate its underlying causes.

These findings may have some implications regarding normal development in cognition and emotion during adolescence, as well as enlighten the mechanisms underlying other neuropsychiatric disorders.[227] In bipolar groups, for example, especially in girls, there is evidence of their hippocampus being smaller.[228]

Researchers of many universities in Quebec have questioned the role of leptin, a protein produced in the adipose tissue and present in breast milk. It appears it has the ability to reduce stress responses in the newborn. The researchers suggested it might reduce exposure to glucocorticoids and thus enhance hippocampal development. Its targets appear to be the hypothalamus and hippocampus, as well as the pituitary and adrenal glands. The researchers also observed a reciprocal regulation of responsiveness to stress between mother and infant. Indeed, it appears the mother can better filter several types of stressors if they do not represent a threat to her baby. [229] There is evidence leptin also has a role in organizing the dynamics of human hypothalamo-pituitary-adrenal (HPA) function through control over the rhythm of hormonal secretion. "In the absence of leptin, cortisol dynamics was characterized by a higher number of smaller peaks, with smaller morning rise, increased relative variability, and increased pattern irregularity."[230] Lately, weight loss publicity and research have signaled the positive role of leptin in weight loss programs, as well as the fact that processed foods and pesticide-laden products[231] might provoke leptin resistance.[232]

The emotional phase is also a very vocal period of life.

Language problems are much more common in boys than in girls. Boys who stutter outnumber girls who do so five to one, as do boys with aphasia (an extreme difficulty in learning to talk). Autism occurs in four times more boys than girls. Over 75% of those with reading difficulties are boys.

Isis Code

Generally, in women, speech occupies a specific area, primarily in the front left hemisphere that corresponds to the left anterior cingulate gyrus—it thus has a link with her masculine polarity—and in other smaller, specific areas in the right hemisphere. Women access their left hemisphere through the corpus callosum. It also explains why, in general, they are better conversationalists than men. This, again, gives credence to the LIFE biosystem. The fact that in many traditions, women who talked were frown upon only accentuates the fact that they were not allowed to express their masculine polarity socially, as if doing so would render them less feminine. Or maybe men felt threatened by women's masculine polarity?

Spindle neurons, also called von Economo neurons, are more numerous in humans than in any other primates or mammals (including whales or dolphins). These neurons develop postnatally, aided or hindered by environmental factors. We find them in the anterior cingulate cortex, which is another region that has reached a high level of specialization in primates; and at an even higher density in the right insular cortex (feminine polarity, Universal brain). These neurons might be involved in cognitive-emotional processes, which in humans are expressed as empathy and feelings of self-awareness. Research in autism has not found that autistic children have fewer spindle neurons, which would have explained the origin of their communication impediment. We therefore have to look in a previous phase, perhaps the reptilian one for an source of their difficulties. Similarly, functional imaging shows that both function and structure of the right anterior insula of the Universal brain allows a person to feel his own heartbeat (heart aspect controlling lung aspect) as well as to empathize with someone else's pain. Because women are also expressive on the social level, they have a tendency to utilize these, the insula and the mirror neurons, more extensively. It should be the same thing for individuals with a developed feminine polarity.

The Anterior Cingulate Cortex

When the subregions of the frontal cortex were studied, women showed significantly higher activity and more circumvolutions in two regions of the cingulate cortex: the left and right anterior cingulate cortex. The cingulate cortex hugs the inner surface of the hemispheres and overlays the corpus callosum.

It is a transitional element between the Mammalian and Human brains and has been implicated in conflict resolution and error monitoring.

It has a midfrontal position and is closely linked to the midprefrontal cortex. When damaged, it has been associated with the breakdown of socially appropriate behaviors, even as much as the orbitofrontal cortex. In accordance with our biosystem, comparison between genders show that men have a higher level of fissuring of the AC (anterior cingulate) in the left hemisphere, as well as more activation there than in the right one, but less than women. Women have less fissuring on the left side of the AC and more symmetry of the anterior cingulate. A high degree of fissuring is associated with enhanced activity of a conflict monitoring system, part of the midcingulate cortex, while low fissuring reveals an increased positivity over parieto-occipital regions.[233] This indicates to me that the emotional aspect of women uses not only the Mammalian brain (occipital area), but also the Human brain (parietal); hence, the feminine polarity. Emotions in women are thus not limited to the masculine polarity as they naturally are for men. Results show that cerebral volume of the cingulate cortex for men is larger on their right than on the left, which agrees with their feminine emotional level, while fissuring shows the reverse asymmetry with greater leftward fissuring. In contrast, women were symmetric in both respects.[234]

Hallucinating schizophrenics activate the temporal lobe, along with the anterior cingulate and the dopamine-rich striatum. Thus, their Universal and Mammalian brains are out of control. On the other hand, it appears that in social exclusion, the right ventral prefrontal (orbitofrontal) cortex of the feminine polarity moderates social distress, balancing the overreaction of the anterior cingulate cortex. Indeed, in the LIFE biosystem, as a feature of the Human brain, it controls the Universal one, which in turns controls the Mammalian one. Furthermore, people of faith, as compared to nonbelievers, *show less activation of their anterior cingulate cortex* (Inzlicht et al., 2009). Their medial prefrontal cortex and parietal lobes of the Human brain are activated instead. We see here the importance of a belief system to regulate distress and stimulate the feminine polarity.

Age has a negative impact on the brain area responsible for emotion, but leaves the amygdala intact (Grieve et al., 2005; Mather,

Isis Code

2004; Mu et al., 1999). As the energy in the lung aspect recedes (Universal brain), control over the Mammalian brain will diminish. The immune system, associated with the Liver aspect, might become out of control as it becomes less inhibited by the lung stage. While at younger ages there was too little immunity, in older ages it is the reverse. A significant atrophy in the anterior cingulate cortex can also be expected (Good et al., 2001; Resnick et al., 2003).

Face Recognition

Studies have shown that the right hemisphere is specialized in face recognition in children as young as four years old, and rapidly increases in its accuracy until five years old. However, it will not reach full maturity until fourteen years old (Kolb and Fantie, 1989) which is the end of the Mammalian brain phase. Only then will the individual be fully able to match facial expressions to situations. This reminds me of a game I used to play with my young children. I would call them over when they thought they might not have behaved properly (which they rarely did). With a neutral expression on my face I would talk in a normal voice about what they did or did not do. I was always surprised to see, whenever I would burst into laughter and hugged them, that they had not expected one face or another. When I would lose my temper, and I must confess I did on occasions, I always regretted it. An adult who did something wrong would expect to see you angry and prepares to react to it. A child just takes it all in. When you are angry, your facial expression is added to his image of *you*. It does not connect to "I did something wrong and it generates that face in Mommy."

The hippocampus is intensely active in acquiring new knowledge. As the neocortical memory representation becomes increasingly robust, the hippocampal participation recedes.[235]

Also, we have to understand that the right hemisphere, the principal tool of the feminine polarity, is used by young children and is thoroughly molded until they reach fourteen years old. Although the period of the Reptilian and Mammalian brains is under the jurisdiction of the masculine polarity, the feminine polarity is also formed during these periods. Some gender differences in social behavior emerge very early in infancy. As young as a few days old, female infants make more eye contact than male infants (Geary 2002). Also, at three months of age females show more expressions of interest, such as wider eyes and raised eyebrows (Malatesta and Haviland, 1982).

All phases are juxtaposed like the different levels of a pyramid. If the base is deficient, we can't expect to reach the top level. Contrary to a widespread false belief, just because you have problems with the lower levels does not mean you will be a genius on other levels, as if to compensate. This idea comes from the exacerbation of other senses, as can be observed in people with handicaps. Blind people might have a more highly developed sense of smell or touch than "normal" individuals. The brain map used for one sense can be taken over by another sense. Nevertheless, if your basal ganglia or hippocampus are damaged, then they are damaged. Nothing takes their place or develops to compensate for the damage. The damage will influence the subsequent development of structures belonging to other phases. Emotional, mental, and synarchic levels do not take over other levels. If you are not physical—that is, muscle oriented— you will not necessarily develop better on other levels. Of course, if all of your energies are spent through the Reptilian and Mammalian brains, the subsequent phases may not develop properly. It is also a question of the health of the different physical structures, how they interact, and of the quantity and quality of the lung energies available at conception from the mother.

A research team at University College London compared maternal love and romantic love by measuring the activity of certain brain areas.[236] Maternal love overlapped with activity observed with romantic love, although activity in the ventral anterior cingulate was present in romantic love *only in females* (expressiveness of emotional self in women). The areas concerned were the medial insula and the anterior cingulate gyrus. Activity was also found in the lateral orbitofrontal cortex and the lateral prefrontal cortex (LPF), which relate to the nutritional/Analytical brain in our system. The researchers also observed activity in regions deemed only indirectly associated with higher cognitive or emotive processing: a region near to the frontal eye fields, the occipital cortex (related to the Mammalian brain) and the lateral fusiform cortex, which is part of the Universal brain. A recent study on mothers' responses to infant cries showed some of the regions active here (e.g., substantia nigra, striatum, anterior cingulate), but also revealed activity in regions previously thought to be deactivated during the expression of motherly love (Lorberbaum et al., 2002), such as the *medial prefrontal cortex*, which corresponds to the Human brain. Motherly love is a spiritual experience.

The medial prefrontal cortex, the parietotemporal junction, and the temporal poles create a network of areas invariably active with theory of mind; that is, the ability (information regulation, Human brain) to determine other people's emotions and intentions (Brunet et al., 2000; Castelli et al., 2000; Frith and Frith, 1999; Gallagher and Frith, 2003). The same areas are also active in the assessment of social trustworthiness (Winston et al., 2002), of facial expressions (Critchley et al., 2000), in moral judgment (Greene and Haidt, 2002; Moll et al., 2002), and during attention to one's own emotions (Gusnard et al., 2001; Lane et al., 1997a).

ADHD and the Reward Circuit

It is thought that gender differences in ADHD may be also attributed to gender differences in dopamine receptor density. The concentration of these increases ±26% at the onset of puberty in boys, while of only ±7% in girls.[237] The striatum plays a central role here, as it does in reward and motivation. By adulthood, male receptor density is sharply reduced by 55%. This pruning coincides with the estimated 50–70% remission rate of ADHD by adulthood. Adult density in women is then similar to that of men. This rise in the number of striatal dopamine receptors parallels the appearance of ADHD motor symptoms, and might explain the two to four times higher rate of ADHD for boys. That the striatum is part of the masculine polarity could also be part of the cause. In agreement with this and with our system, men have more D_2 dopamine receptors on the left than on the right side of the striatum. Problems with attention have thus been linked with a delay of pruning or overproduction of dopamine. Researchers have also hypothesized that differences in D_1 receptor density in the nucleus accumbens (part of the ventral striatum) may explain the increase in substance abuse by young men. Melatonin, secreted by the pineal gland thus associated to the feminine polarity has a role in regulating dopamine which is a key neurotransmitter involved in the symptoms of and treatments for ADHD. Indeed, studies in rodents have confirmed a circannual variation in central dopaminergic activity, as well as a multitude of convincing interactions between dopamine and melatonin in the striatum.[238] This would show the positive effect of control from the lung aspect (here the melatonin) over the liver aspect (the dopamine), and the positive effect of sleep in dark and quiet surroundings to control symptoms of ADHD. Playing video games all night might therefore be counterproductive.

The dorsal striatum and its most prominent part, the caudate nucleus, are part of the "reward circuit."

"A lot of theoretical work in evolutionary biology and our previous experimental work suggest that altruistic punishment[239] has been crucial for the evolution of cooperation in human societies," says Ernst Fehr, who is director of the Institute for Empirical Research in Economics at the University of Zurich and senior author of a study on the subject.[240] "Our previous experiments show that if altruistic punishment is possible, cooperation flourishes. If we rule out altruistic punishment, cooperation breaks down."

Stanford University psychology professor Brian Knutson wrote an accompanying commentary noting that schadenfreude has now been captured in a brain scan.

These researchers also found that the idea of executing an altruistic retaliation was satisfying: "The activation in the dorsal striatum reflects the anticipated satisfaction from punishing defectors." This type of interference by someone who, say, protects a weaker member is in fact an action from the feminine polarity which sees wholeness and harmony as an essential element of social life.

What about altruism? Donating also engaged the part of the brain that plays a role in the bonding behavior between mother and child, and in romantic love. Giving to others, helping others, involves oxytocin, the hormone we referred to previously that increases and is secreted during times of trust and cooperation. The receptors for this hormone differ across species not only in its distribution, but also in its regulation by gonadal steroids (from ovaries or testes). This makes sense since in our system, the kidney aspect, related to gonads, regulates the Heart one, associated with the oxytocin. In autism plasma levels of oxytocin are low (Green et al., 2001; Modahl et al., 1998).

This again might indicate problems which occurred during the reptilian phase or before. Researchers at the National Institute of

Isis Code

Neurological Disorders and Stroke in Bethesda, Maryland, found that when subjects believe something to be false, the part of the brain adjacent to the oxytocin receptor is active. This area is thought to be responsible for decisions involving punishment. And a third part of the brain, the anterior prefrontal cortex and, in some researches, more precisely the dorsolateral prefrontal cortex, which both evolved relatively recently in a unique fashion for humans, might be involved in the complex, difficult choices when self-interest (Analytical brain and masculine polarity) and ethical beliefs (Symbolic brain and feminine polarity) conflict.

In another research, Dr. Martinez and her colleagues found that , in men, higher social status and improved social support (men are receptive on the social level)correlated with the density of dopamine receptors in their striatum.[241] The army with their tradition of recognition and regalia might be right. As well, the Universal, social brain exerts control over the Mammalian one at the same time as the Reptilian brain energetically feeds it. Dr. John Krystal, Editor of *Biological Psychiatry* commented, "These data [provided by Dr. Martinez's research] shed interesting light into the drive to achieve social status, a basic social process. It would make sense that people who had higher levels of D2 receptors, i.e., were more highly motivated and engaged by social situations, would be high achievers and would have higher levels of social support."

The dorsal striatum is then linked to the dorsolateral prefrontal cortex, a key feature of the Analytical brain, which is closely implicated in the subsequent execution of behavioral elements (see chapter 6). This confirms the Mammalian brain function of regulating the rational one.

To this we could add the following affirmations made by science:

1. Young babies prefer altruists.[242] This agrees with the fact that the right hemisphere of babies is dominant.

2. Individuals who are more altruistic have more activity in the posterior superior temporal sulcus.[243] This agrees with the Universal brain function.

3. Individuals who behave more altruistically than others have more gray matter at the junction between the parietal and temporal lobe.[244] As we have seen, the parietal is part of the Human brain and the temporal of the Universal brain. Therefore, this axis is functional in altruistic people.

4. Individuals only punish what they consider as violations at their own expense if the dorsolateral prefrontal cortex[245] (part of the Analytical brain) is activated and *can interact* with the ventromedial prefrontal cortex of the Human brain. In other terms, if their patterns and belief system agree to the punishment.

Females have a more acute sense of smell (lung aspect of the feminine polarity), allowing the evolutionary need for the mother to recognize her young. Because women are masculine on their emotional level, they tend to have a larger deep limbic system. This leaves women somewhat more susceptible to depression, especially at times of significant hormonal changes, and even more so if the feminine polarity is malnourished. Epidemiological clinical findings, which remain unexplained otherwise, reveal that the incidence of major depression is three to four times higher in women than men during midlife (forty-two to fifty-five years),[246] which corresponds to the end of the second mammalian phase and the beginning of the second Idealistic phase of the heart (diagram 23).

Not surprisingly, there is also a second peak of onset of schizophrenia around the same period. Thereforee, it appears an identity crisis underpins this depression. As well, women attempt suicide three times more than men (Davison and Neale, 1986). This is understood if the feminine polarity, through the lung aspect, fails to inhibit the Mammalian brain (liver aspect). Yet men succeed in killing themselves at a rate of three times more than women (Davison and Neale, 1986; Firestone, 1986). We shouldn't be surprised, for men use more physical and violent means, such as shooting or hanging themselves, while women tend to overdose with pills. This is understandable, since the masculine polarity interests the physical realm. Also, disconnection from others and lack of social support, often a symptom of an impoverished feminine polarity or of a deficient community, increases the risk of successful suicides. For both sexes,

Isis Code

the lack of meaningful human contact increases the chances of an individual acting on suicidal thoughts.

In social behavior and cognition, men tend to form larger social groups and be less accepting of strangers, as portrayed by the prized men's clubs of bygone eras, whereas women tend to prefer one on one interaction and are more compromising in their relationships (Geary, 2002). Women, because they are expressive at the social level (Universal brain level), are therefore responsible for the authenticity and life of the social fabric. Given, of course, that some basic conditions are fulfilled such as the respect of the feminine polarity. In agreement with the masculine aspect of the Mammalian brain, women are also more likely to participate in relational aggression, such as gossiping and backstabbing (Christiansen, 2001; Geary, 2002).

Finally, in a study of over four thousand subjects of different age groups from different countries, women were shown to be more adept at reading nonverbal cues (Sanchez-Martin et al., 2000; Geary, 2002; Cote et al., 2003; Fabes et al., 2003).

Endocrine System

Adrenal disorders, addictions, anxiety, depression, endometriosis, erectile dysfunction, high or low blood sugar, high cholesterol levels, hormone imbalances, hyperactivity, hypothyroidism, inability to handle stress, infertility, low sperm count, weight gain and obesity, and others—what do they have in common? They are all symptoms of a disrupted endocrine system. This system includes the hypothalamus, pituitary, thyroid, parathyroid, adrenals, pineal body, pancreas, and the reproductive organs. It plays a vital role in almost every function in the body and works in conjunction with the nervous system, which itself is influenced by physical conditions, emotions, and thoughts. Thus they mirror the state of the general LIFE biosystem. Within the physical conditions, endocrine disruptors such as pesticides (Hall, 1984) are starting to get everyone's attention. This recent focus includes studying adverse effects of pesticides with the coordination of a newly established international consortium of agricultural cohort studies (AGRICOH)[247] and collaboration in a testicular cancer study in the Rhone-Alpes region of France. In addition, AGRICOH, we are told, will allow the study of exposures and cancer risk related to crops and animal farming. Hopefully their wings will not be clipped by some interested chemo-financial power. Dr. Marion Kavanaugh-Lynch, an oncologist and director of the California Breast Cancer Research Program in Oakland—which directs tobacco tax proceeds to research projects—as well as a biologist and a cancer survivor, observed: "The data indicate that if you get your first period before age 12, your risk of breast cancer is 50 percent higher than if you get it at age 16. For every year we could delay a girl's first menstrual period, we could prevent thousands of breast cancers."[248] If we are honest, nutritional and environmental quality, prerequisites to a healthy feminine polarity, should therefore be considered the most essential shields against cancer.

What determines if a crocodile egg will bring forth a male or a female? Temperature. In a colder environment, the eggs will generate females, while in a warmer climate, males are born. Thus polarity expression, gender expression, have a definitive link with the environment, and thus with epigenetics.

Other studies have also revealed that various pesticides or pesticide metabolites can countermand this temperature-sensitive gender-determinant mechanism in alligator and turtle embryos, acting in a manner similar to natural estrogens (Bergeron et al., 1994; Matter et al., 1998; Willingham and Crews, 1999).

Crocodiles and alligators are ancient beasts. They have survived and adapted to extremely hostile environments. They are still here. They, as we, now face something alien to their natural environment: pesticides. They have not yet been coerced into eating genetically modified food as we are, but damages are nonetheless already apparent. We are next.

"We have observed that neonatal and juvenile alligators living in pesticide-contaminated lakes have altered plasma hormone concentrations, reproductive tract anatomy and hepatic functioning."[249]

In short, the liver aspect of these organisms is damaged with all the ensuing repercussions. How can humans escape this situation? Are we sure these pesticides have no effects on our physical hormonal systems and on our emotional lives?

Importantly, those chemicals, when combined, might exhibit compounding values or synergetic qualities, in which the sum manifests reactions that cannot be extrapolated through the addition of the different chemicals (Vonier et al., 1996). Affinity for a receptor is not assurance that a contaminant has a steroid mimicking effect. It could equally act as a hormone antagonist (Gray et al., 1996; Kelce et al., 1995). Moreover, bioavailability of these compounds is also a serious concern. If it can cross the cell membrane, through a weak lung aspect, then all of the chemicals in the blood are available to the cells and may eventually bioaccumulate and biomagnify in the food chain. Of note: inhibition of contact —thus of the cell membrane— has been observed in case of cancer.

Added to this, we now know that stress could dramatically increase the permeability of the blood-brain barrier to these chemicals. As proof, during the Gulf War, Israeli soldiers were given a drug to protect them from chemical and biological weapons. Nearly one-quarter complained of headaches, nausea, and dizziness. These side effects were known to occur only if the drug reached the brain. Biochemist Hermona Soreq of Hebrew University and Alon Friedman, a physician at Soroka Hospital in Beersheva, have now found that stress multiplies the ability of chemicals to pass the blood brain barrier.[250]

It appears that endocrine-disrupting contaminants interact with a number of other hormonal signals, such as androgens, progestin, and thyroid hormones (Crain and Guillette, 1997; Gray et al., 1996). Also, we should not underestimate the extreme sensitivity of the developing embryo to chemical signals (Bern, 1992; Knobil et al., 1999; Guillette and Crain, 2000). Concerns arise with detrimental modifications observed in animals exposed to contaminants in utero. Radical modification of embryonic structures and functions, and thus adult forms and functions, can be induced by epigenetic factors, which correspond in our system to the phase of information regulation. Information doesn't need a quantity to have an effect, only an available receptor. As an example, if you are in a theater and start screaming: "Fire!" there is a good chance others will repeat your scream and panic will ensue. In the same way, a red light can stop a stream of cars.

Following these facts, eating organic produce and promoting its cultivation seems the bare minimum we can do, not only for the planet but also for ourselves and the future we wish to generate. It is an act of social consciousness. The alternative makes no sense, even from a global economic point of view. We produce enough to feed the world, but almost half of it is wasted. To those who maintain organic farming is not producing enough:

"On average, in developed countries, organic systems produce 92% of the yield produced by conventional agriculture. In developing countries, however, organic systems produce 80% *more* than conventional farms."[251]

This happens because, as we will see later, chemical fertilizers are effective the first years but detrimental to the soil in the long run.

Exodus: The Mammalian Brain

We have seen that the book of Genesis relates to the Reptilian brain. Exodus, the second of the Five Books of Moses, relates to the Mammalian brain. As explained before, the Mammalian brain is tied to memory, thus to history and time. Exodus was written as a page of history that permeates the Jewish culture in all aspects, up until now. It seems that the experience of the Jews' exodus, even the sole fact this story was repeated and believed, was crucial in forming a group consciousness, and has provided a model ever since. Although historians argue that there are inconsistencies of time in Exodus, this is not what this teaching is about.

Exodus begins with "Now these are the names of the children of Israel, which came into Egypt." This corresponds to the new awareness of the Mammalian brain. The age of reason should in fact be named the age of awareness. In 40:37–38 (NIV), Exodus ends with: "but if the cloud did not lift, they did not set out—until the day it lifted. So the cloud of the Lord was over the tabernacle by day, and fire was in the cloud by night, in the sight of all the Israelites during all their travels."

At the end of the mammalian phase, the personality should serve the individuality and have will to move only as dictated by it. If the cloud is present, there should be no action taken.

Isis Code

This is echoed in the Gospels when Jesus explains why the Apostles do not fast. When the individuality is present and acting, the personality has no need to march toward it.

The rest of the Hebrew Bible often makes emotional references to the exodus for the purpose of directing behavior. Exodus is replete in polarities and distinctions, in the image of the Mammalian brain. Similar to the brain it expresses, it relates a symmetrical (the plagues) and often stereotyped history compared to the other books. It also lacks the citation of the personal (self-reference) found in the following book, Leviticus. This is in accordance with the Mammalian brain, in the same way the geographical names in Genesis correspond to the Reptilian one.

Torah, which holds these five Books, signifies instruction, but I believe it is similar to Ma'at in the sense it describes in words the implicate order. Here, it relates experiences that should direct one's action. Even more than that, it is a *guide* through the pitfalls one could encounter in the different phases, and indicates the inner attitude one must develop to become Moses. It refers to invisible laws to follow in order to find our true identity, not because it is imposed but because these are the structure of the universe. An example of this would be the following: "When there are black clouds and heavy rain, one should wear a raincoat." Is it a law or just the description of the fact that if you go outside without a raincoat under those conditions you will get soaked?

The Books start with the universe's birth and end with Moses "death." It is significant that both Aaron and Myriam "die" in the Fourth Book, Numbers, which corresponds to the Analytical brain. Moses "dies" in the collective phase at 120 years old, thus 12 × 10. Twelve corresponds to the feminine polarity, time, and function, while ten corresponds to the masculine polarity, structure, and space. Moses is thus a symbol of completion, the archetype of the Perfect human, and guides the pilgrimage we must achieve to express our true nature and realize the mandate we have on earth.

Moses represents our core identity, the Self, and Aaron is the personality one must develop to allow the inner Moses finally to manifest. Moses doesn't see the Promised Land because he is part of the Promised Land. He represents the perfect human.

The liver phase is linked to vision and has fire or plasma as its element. Also, we associate it with Hod in Kabbalah, to Mars, to Osiris, and Adam. Not surprisingly, accounts of God's presence in Exodus refer to fire and to glory (again linked to Hod in the Kabbalah). Similarly, the term "in the eyes of the children of Israel" and "see" are used frequently. The liver phase is associated to vision in TCM.

In this book, God is in movement, mirroring the mammalian phase. The title is followed by "Now these are the names ..." We learn the names of God because, since he is not in us, as animals we fear him. Only God could name God and as we saw, naming is a feature of the Mammalian brain. The maturation of this phase comes with the seeking of his names. This period of Exodus has themes attached to it that teach us how to deal with the Mammalian brain.

1. Serve. The people served pharaoh and will now leave to serve God. It can be only one or the other. The pharaoh in this instance is the pancreas aspect of the Analytical brain. Between the ages of seven and fourteen we are given a choice. What do we wish to do when we grow up? We have to choose. Serve the heart aspect or the pancreas aspect. So in this period, it is crucial to show the difference between both and to see the natural tendencies of the child.

2. The presence of God through his glory. This is a sign of reassurance for those who have not chosen pharaoh. Exod.17:7 (NIV) asks, "Is the Lord among us or not?" The mammalian phase is one of inquiries, these will be answered in the next, idealistic phase.

3. God sees, his people see, and the Tabernacle blueprint is given to Moses to see. It is a period when a person opens his eyes to the different levels of reality.

4. Rebellion is the reaction of the personality faced with the immateriality of God.

5. Sabbath: Contrary to before, Israel has to adapt to a week-long rhythm. The rhythm constructs a physical structure, which calms the Mammalian brain since the environment controls it through its cycles.

Isis Code

6. Similar to Osiris, first son of Nut, if the people of Israel obey the laws written in their hearts by God, he will protect them and treat them as his "first born son." Thus Osiris will be alive and not torn to pieces.

7. Topography is seen here as an account of an inner journey.

8. Wars are linked to Mars and present in this book, as they reflect inner turmoil.

9. The people of Israel have survived infancy as a nation in Genesis; they begin adolescence in Exodus. They survive a threat of extinction (death of the kidney phase) and now have to cope with the process of becoming adults and living as a community.

With the Song of the Sea in the fifteenth chapter of Exodus, God is seen as the true king, which is what the Mammalian brain has to do. It can serve the pancreas phase and decide it is God, or serve the heart phase. The God of the Israeltes is the God of the heart phase. Other contemporaneous cultures worshipped human heroes and were keen to display the biography of their gods. None of that probably pleased the people of God. A third of the text of Exodus contains a detailed description of the sanctuary, God's abode. In fact, this phase is one in which Man builds his sanctuary and his Tabernacle (His body). The three themes of fire (liver), water (kidney), and desert (pancreas) are indicative of the path to follow toward the Promised Land. The symbols are of foremost significance. The fire does not burn but expresses purity and transformation. The water carries the baby Moses to his destiny, and opens (the parting of the Red Sea) to allow the people their long march of evolution. The desert is necessary to distance the young nation from the massive influence of Egyptian culture.

We are shown that what started as an act of personal will by the pharaoh became with time impossible to revoke. Throughout the nine plagues (the nine parts of Man) the pharaoh's stubbornness makes his heart heavy, hard. He becomes petrified and immovable. Evolution stops when one choses the rational self over the idealistic, archetypal one.

The Mammalian Brain, the Emotional Self, and the Emotional Human

Another point of note in Exodus: it is the women, thus our feminine polarity, who are instrumental to the beginning of the liberation process. They are taken as a symbol of the feminine polarity because at this level of the Mammalian brain, women are active.

In Exodus Moses sees the burning bush not consumed by the flames and he hears God saying "I am", which is the expression and goal of humanity. Similarly, this occurs in the period of seven to fourteen years old.

Lastly, Moses, out of lack of self-esteem, implores God to send him someone to help him with the Israelites. As a symbol of our individuality, our inner Moses has no power in this divisive reality. The personality is needed for this. To which God replies:

> I will help both of you speak and will teach you what to do. He will speak to the people for you, and it will be as if he were your mouth and as if you were God to him. (Exod. 4:15–16, NIV)

Here, as readers, we associate with Moses. Aaron the priest is the perfected personality, a mouth (the verb comes from balanced masculine and feminine polarities symbolized by the mouth when they are inspired by God), and Moses is our individuality, a god.

How do the three parts of Exodus end?

The deliverance part ends with the people crossing the sea (kidney) while the Egyptians are engulfed in its waters (predominance of Reptilian brain). Aaron's sister, Miryam—thus Moses's feminine polarity—takes a timbrel and makes music, and followed by all the women, she sings of the glory of God and how triumphant he was. The second part ends with Moses sending his father-in-law back to his native land after he helped him organize judgments of the people. This is the phase when society and our family gives us a structure and we are able to function in it. The third part ends with the people marching on ready to serve God, and with God's presence in the newly built Sanctuary. They, and we are now ready to enter the idealistic phase.

Isis Code

The emotional phase is regulated by the collective one if we follow our inner compass. Seen at another level, the collective controls the egoistic nature of humans. This we see through the application of the law. The Romans used to say *dura lex sed lex*, which means that a law that seems hard is better than no law. In fact, in our cybernetic system, the emotional aspect controls the rational one, while the rational controls the instinctive one. In biology, there are more fibers from the limbic system toward the frontal lobe than the reverse. Considering the frontal lobe's fragility and tendency toward disease, it is socially responsible to have a system in place that limits the failings of immature frontal lobes.

Before the age of fourteen, we live in a world of emotions and sensations. We accumulate information, we react to life events and situations based on mimicry, but—as I painfully observed—the voice of the inner compass is present and difficult to shut down. As I recall, the first seven years of my life were lived as if I wasn't there. I believe that at this stage, children express their individuality, past personalities, and the familial personality. I have to write this through my own convictions. Even if you don't believe in what is called reincarnation, its ideas still apply. The observed phenomena are still there, even if you don't see the cause as reincarnation. I do like the Dalai Lama's description of it. People are on the lookout for a key (to their life) in a certain room. They turn around and around without understanding or finding an explanation as to why they can't find the key. The problem is that the key is in an adjacent room.

There are different definitions for the term reincarnation. This idea has evolved also with time and human maturation. Our world is still in the process of being made; that is why Revelation is not completed yet. As with a child whose brain is developing —therefore you do not tell him a story he can't comprehend—, our contact with the invisible and inner world enriches itself the more mature we become as humanity. Our brain is still evolving, so things we can't grasp at this point of our evolution will seem normal and ridiculously obvious later on.

Dr. James W. Prescott, PhD, is a developmental neuropsychologist, formerly with the National Institute of Child Health and Human Development, part of the US National Institutes of Health (NIH) and president of the Institute of Humanistic Science. His research into the causes and effects of child abuse is an ongoing project.[252] In his research, he realized that the senses are responsible for the brain's neuro-structural development. This, we have seen, is the first phase of the Reptilian brain. On this structure will merge our later experiences of affection, pleasure, peace, and love. Normally, active pleasure circuits inhibit pain circuits. It is thus critical in infancy to properly develop the brain's structural and functional systems. This will allow the pleasure systems to function appropriately, hence allowing an ultimate maturation of the Human brain. Otherwise, the Analytical brain will reign supreme and develop its control excessively, changing our vision of the world, locking us in a box, as we are now. Also, love might become perceived as closely related to pain. Instead of inhibiting the circuits of pain, these will be emotionally triggered (Mammalian brain). I have seen this countless of times, sadly. A person experiences a delightful, happy moment, and as this happens, a dark cloud passes in front of his or her eyes. The impossibility to feel the moment and anxiety takes over for no apparent reason.

Our interpretation of life is defined by what we experience in our first few days, months, or years of life. This solidifies into unconscious beliefs and patterns we will carry on as we mature. In fact, it influences our feminine polarity, since the right hemisphere is in charge at the time and will become part of the background I referred to earlier. It has been postulated that 80% of our understanding of the world comes from our internal experiences, while only 20% comes from actual external circumstances. We really do create our own reality and live in our own world.

Just as play is universal among animal groups, it is an integral part of all human development with the added feature of role playing. Play with others provides an individual reassuring contact when the mother's nurturing services are no longer required or offered. It also nourishes the senses and feeds the feeling of unity that is fundamental in humans. Donald Winnicott (1896–1971), a British paediatrician and psychoanalyst, placed play—and its extensions, creativity and culture—in the "potential space" that forms between a mother and child that remain connected as they slowly go toward, or back to, their own identity. In his view, cultural and creative activities are not a result of sublimation or suppression of the genital instinct, as Freud supposed. They are, instead, the very meaning of life and an extension of the need for love.

The cortex truly came into existence during mammalian evolution. The module *interconnected to the senses* triggered the development of a thin matrix of cells whose shape facilitated the formation of many neural connections between them. This type of skin (lung phase) became the cortex we know, and formed the awareness we experience. The cerebellum gained a backward position, relegated to our past. Morphologically, our domed shape heads (parietal lobe, heart phase) distinguish us from other primates, as does our ability to control our breath (lung phase), the consequence of a lowered larynx.

The self-consciousness developed in this phase is freed from spatial contingencies. As such, monkeys can recognize themselves in a mirror. This implies an aptitude of dissociation of the self with space. The next level, expressed in the idealistic phase, consists of dissociation in time as well. Hence, for example, the subject could recognize himself in a movie.

It is interesting to observe that for the next phases after the mammalian one, instead of multiplying or refining the structures of emotion such as the thalamus, nature elected to add a fundamentally different set of neural organization through the emergence of this neural sheet. Also, the elements of the Mammalian brain have a reverse synchronization with those of the neocortex, providing further proof of a subsequent phase. Generally, science sees our brains as embodied modular action machines that were designed merely for passing on our genes. If this is so, then nature is wasting a great deal of energy. We are supplied with genes that allow us to live 120 years, while our reproductive period is less than half that. The Human brain, which developed later, gives us the clue to nature's true intention.

If the Mammalian brain—the liver aspect—is dysfunctional, then the Analytical brain—the pancreas aspect—will drain and monopolize the energy of the whole system at all levels in an attempt to re-establish homeostasis. Nutrition, money matters as well as abstract intellectual pursuits, will assume an overwhelming place, draining the energy of the whole system. This, results in someone who is not concerned about the values of the feminine polarity hence of the environment, who is egocentric, mentally rigid, atheistic, and for whom money matters will dictate every action. This person lives his life stuck between four invisible walls, an impregnable fortress created by his own, or his social environment's mental patterns. Within this windowless and doorless dungeon, meaningful contact with others, or even simply seeing the sky, is impossible. And anyway, will this person wonders, what proof is there that there is a sky? I can't see it, therefore it does not exist.

There is a popular story of Jesus casting demons into pigs, which subsequently run off and throw themselves over a cliff. One can wonder, why pigs? Sure they are cellularly so close to us, we can utilize some of their organs for human transplants. They can sunburn too, as we do. And some studies on the effects of stress analyze pigs' and piglets' reactions, since curiously, they respond like humans when challenged. But I believe there might be more to it. Because of the particular shape of their necks, pigs are the only mammals unable to look directly at the sky.

Isis Code

1. *Horus The Child*

The Mammalian Brain, the Emotional Self, and the Emotional Human

2. Lajja Gauri

3. *Venus of Willendorf*

4. Sheela Na Gig

Isis Code

5. Bharata Natyam Dancer In Trance

The Mammalian Brain, the Emotional Self, and the Emotional Human

6. *Moses* (Michelangelo)

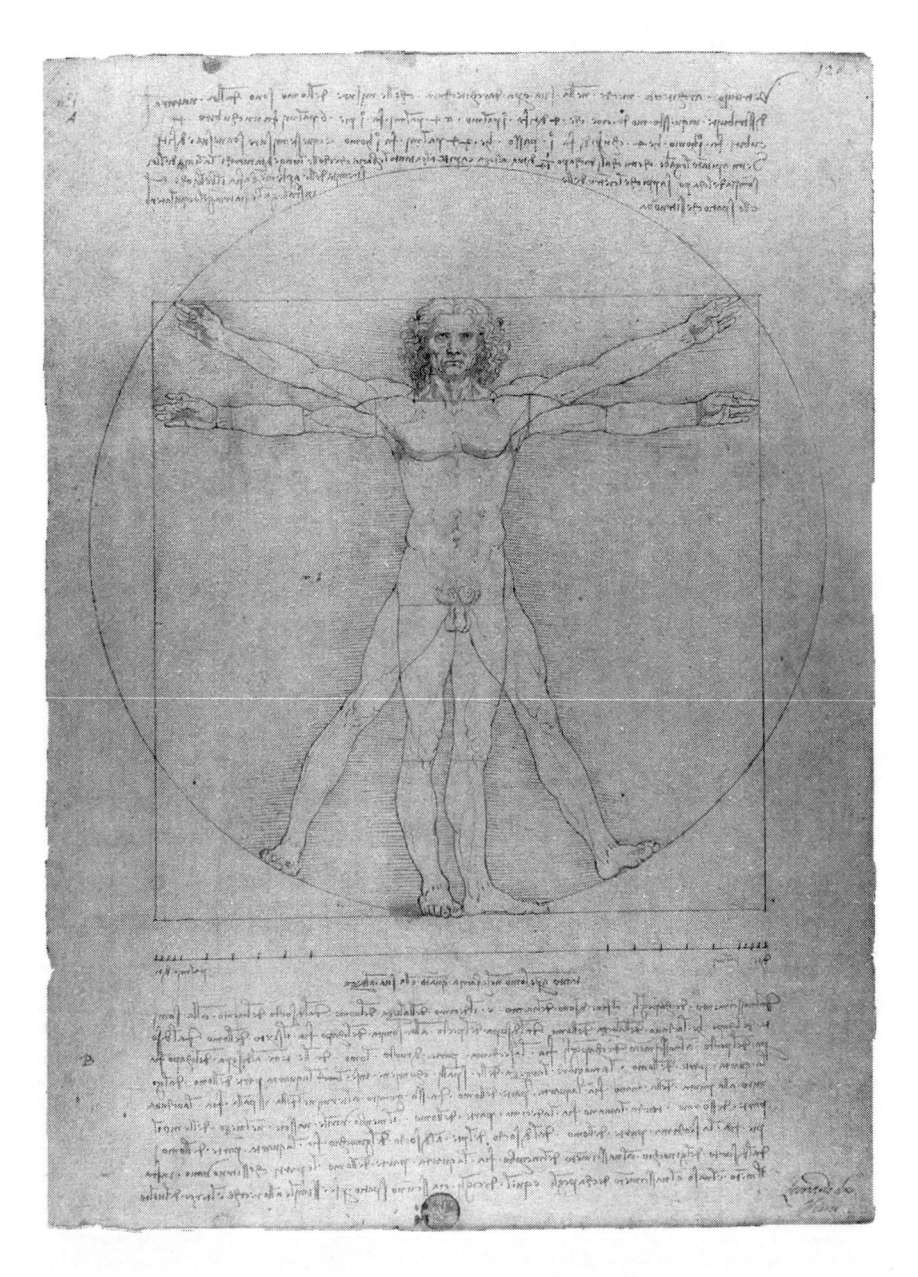

7. *Vitruvian Man* (Leonardo da Vinci)

8. *The Lady And The Unicorn*

9. *The Accolade*

Chapter 5

The Human Brain of Information Regulation, the Idealistic Self, and the Archetypal Human

"The most beautiful experience we can have is of the mysterious. It is the source of all true art and all science. He to whom this emotion is a stranger, who can no longer pause to wonder and stand rapt in awe, is as good as dead: his eyes are closed." **Albert Einstein**

"We can easily forgive a child who is afraid of the dark; the real tragedy of life is when adult men are afraid of the light." **Plato**

"My youngsters will not work; those of them who work cannot dream; and wisdom comes through the dreams." **(Smohalla, shaman, Nez Perce tribe)**

"Mythology was sacred to primitive people; it was as though their myths contained their very souls. Their lives were cradled within their mythology, and the death of their mythology, as happened with the American Indians, meant the destruction of their lives and spirits."[253] **Robert Johnson**

Phase of the Why and Which Level Developing into the I Am— If All Phases Develop Harmoniously

The Heart Aspect

- Construction, structural phase: from fourteen to twenty-one years old
- Puberty and growth completion (but the brain is not yet totally mature)

- Main sense: touch, cognition, synergy/love (through the Master of the Heart)
- Building block: the Reptilian and Mammalian brains
- Nourishment: personal space, biographies, travels, identification, high ideal, esoteric religions, system science, love (platonic)
- Cybernetic phase and function: heart phase, regulation of information
- Brain aspect: right hemisphere, particularly right parietal and medial prefrontal cortex, right dorsolateral frontal cortex, mirror neurons, postcentral gyrus
- Cybernetic main organ (not limited to anatomical, but also corresponds to an energetic and metabolic function): heart, Master of the Heart
- Level controlled by reproduction/physical and fed by defense/emotional
- Controls environmental/collective/exchanges aspect
- Cybernetic gender structure: part of the feminine polarity (expressive aspect of the feminine polarity)
- Cell structure: RNA
- Function: controls and regulates information
- Stimulants: biographies, bitter taste, objective beauty, travels, study of philosophies, essence of religions, theoretical sciences, social contacts, selfless love, coffee, emotional projection of the opposite expressed polarity, chocolate, contact with who is judged to be the top of the pyramid, study of symbolism and of dreams, recognition by peers, auditive humor (frontal right hemisphere)
- gift of that phase: path toward the archetypal, perfect human
- Motivated by thirst for the absolute.
- at this level, women are receptive and men expressive
- Type of attraction: previous + ideal, projection of masculine and feminine polarity, sensual
- Main relationnel attitude: projection
- Psychological trait: projection
- Emotion: freedom, joy, solitude, idealism, friendship, judgmental
- Level of awareness: self-conscious toward Self conscious
- House analogy: roof
- Part of day: noon

- Bible equivalent: Leviticus
- Egyptian god: Horus
- Color: orange
- Figuratively: magnetism/electricity
- Element: chi, prana, electromagnetism
- Patterns prepared: rational
- Love type: projection, courtly love, personal although timeless

This is an in-between structure as it will be the center of equilibrium for the Self which will truly manifest after the maturation of the collective phase of the Universal brain.[254]

Obviously, since the Mammalian brain shows us the possibility of taking directions, the next question we will ask is which direction to take, which route to follow in our lives. To figure out how to best spend our energies in a profitable manner, we have to define and choose what is most essential to us. In other words, we need to embark on a process that will lead us to understand who we are. In this period spanning between fourteen and twenty-one years old, we feel a pressing, urgent need to choose. We can put our energies into the world of personality or into the world of individuality. It is said that you can't serve two masters simultaneously; my experience and observations agree. The natural question leading to this choice is, "Why am I here?"

Once we choose, because we acquire and thus adhere to a certain identity during this phase, our *how* aspect will determine how to keep, develop, and express this identity. In the period of twenty-eight to thirty-five years, this identity will either find or not find an echo in the social aspect of our lives.

This is thus the phase of the *why* and *which*, in control of the *who* phase, as in "*Who am I?*," and feeding the *how* phase. It is tempered by our Reptilian brain. The quality of our rational selves depends on this phase. A defective heart aspect might express problems of identity, psychosis, neurosis, and all types of physical disturbances involving inflammation.

David Bohm posited: "Images are a key bridge between the older emotional brain and the more intellectual neocortex."[255] As such, our symbolic phase, seat of our idealistic self, between the emotional and the rational ones, clearly appears as an interface between emotions and manipulation of notions. The Reptilian brain excelled in space management, the Mammalian in time management. The Human one will focus on information management and will model action on this. Human infants, instinctively, do role playing games. Chimps do not.

Stephen Hawking, the renowned English theoretical physicist and cosmologist, observed:

> Even if there is only one possible unified theory, it is just a set of rules and equations. What is it that breathes fire into the equations and makes a universe for them to describe? The usual approach of science of constructing a mathematical model cannot answer the questions of *why* there should be a universe for the model to describe. *Why* does the universe go to all the bother of existing?[256] (My emphasis)

He is right. Answering *why*, though, is the domain of the feminine polarity. We cannot answer this question through linear thinking; we need the help of the right hemisphere as it concerns the whole. The answer to this question is given through the initiation brought by life experiences, in the silence of daily life unfolding through each of our breaths and each beating of our hearts.

Similarly, the quantum theorist Freeman Dyson wrote, before the theory of dark energy became known:

> It would not be surprising if it should turn out that the origin and destiny of the energy in the universe cannot be completely understood in isolation from the phenomena of life and consciousness.... Life may have succeeded against all odds in molding the universe to its purposes.[257] (Excerpt from his Templeton Prize Lecture)

As we have previously seen, the question of *what* originates with the Reptilian brain and works with the *how* of the Analytical brain in its double polarity. That question created the domain of science.

The *why* and *which* interrogations initiate from the Idealistic self, the Human brain, and lead to the *who* of the Universal brain. This is the ill-defined world of spirituality. They have to work together since the *why* cannot be answered without the understanding brought by the feminine polarity. What Dr. Hawking is saying is that ultimately, science is not really useful in this quest because it is impossible for it to access the *why*. Even as he asks why the universe goes to all the bother of existing, we can also ask why we bother asking; or even, for that matter, why we question why we exist? Freeman Dyson hinted at the answer. It leads me to this question: Could the goal of this universe be Consciousness? Could it be the nature of this dark energy we speculate on? After all, in our LIFE biosystem, the same energy flows in all the different houses. This means that all the different brains are truly embedded in the same reality. Energy spent or concentrated heavily and exclusively on one level will not be available for another one. Holiness is in the wholeness of our expression. This explains the restriction imposed on reptilian patterns by all religions in their esoteric approach (feminine polarity). The aim of religions should have been a greater Consciousness. Unfortunately, because of the rise of the masculine polarity, they were hijacked by political powers, and the results were frequently the opposite.

In fact, we are one Consciousness with trillions of eyes, influencing this manifested world by living in it. As space is created in the universe, in the same manner souls detach from the Unique and see the world and mold its matter, transforming it through information. In turn, information will influence matter that is in affinity with it, in the long chain of evolution, of Conscientization.

Dark matter, which does not interact with the electromagnetic radiation, was the main aspect of this universe at the beginning at a rate of 63%. Now dark energy is 74% of the total mass-energy of our universe. A little more than thirteen billion and a half years ago, atoms accounted for 12% of the universe. Now they are estimated at 4.6%. Therefore, from a dualistic point of view, the goal of the universe and thus of evolution is not matter but energy, *and one that can interact with the electromagnetic field.*

Bohms's implicate order is a guiding property for energy and matter. I do not believe we have dark energy here, dark matter there, atoms over here, photons and neutrinons above, unless that is an

Isis Code

approximate view of the universe. All flows from one aspect to the other, as all can be found in all.

So again, evolution points to an enhanced feminine polarity.

In the legend of Isis and Osiris, a phase is reserved for their son Horus, because his identity cannot be revealed yet. What does that mean? The idealistic self, Horus, receives his energy from the previous phases, including from the phase in which his father, Osiris, was murdered. Horus will not be totally *born* or mature until the lung phase (Isis) has manifested. Isis has to be manifested in this reality in order for Horus to be. This is reflected in the symbolism of Mother and Child so dear to many civilizations. It is also dominant in Harpocrates—the Greeks adopted the child god Horus into their own mythology and called him Harpocrates—as seen here (illustration 1). To the question "Which direction to choose?" the statue answers.[258] It bears the traits of the feminine polarity. Note the lock of hair on the right side of his head, thus on the side of the feminine polarity. It will be severed when the young becomes adult. It symbolizes the link to heaven, which will become internalized as the child matures. The earring symbolizes purity of the mental aspect. The headdress is extended at the parietal lobe level as a phrygian hat, and traditionally viewed as a pineal extension in association with solar deity.[259] His right side (masculine polarity) is arched as he leans to the left (feminine polarity), and at the same time he touches his index finger to his chin, an interrogation sign when needing to choose. In Egyptian statuary, the chin is associated with the kidney aspect, willpower and structure. It appears also that the left leg is forward, a symbol of priority given to the receptive function. He was assimilated with the Greek god of silence, but I do not believe this to be its primary symbolism. (Illustration 1—Harpocrates as the child Horus.)

Direction is key to this phase. Carl G. Jung appropriatly said, "If someone seriously starts on this quest for wholeness, he will put the foot, without even realizing it, in the hole which is destined to him, and of this shadow will be born for him the light; but light itself cannot be enlightened.... I would never tell him [someone who is in the light] to turn towards shadow because he would then, with his light, search for a darkness which is not his, and he would find it [and not realize it is not his]."[260]

Once the idealistic self has matured, the human will be king or master of himself. He will become similar to the emperor of Chinese tradition who inhabited the middle palace, the purple palace (the inner part of a circular rainbow), and was mediator between earth and heaven. Or similar to a pharaoh as a representation of the victorious Horus. Horus will then be in a position to avenge his father Osiris (the emotional self) and finally control his uncle Seth (the personal, critical self). The center of his being will not be earthbound in the sense that it will not be bound to the rational self anymore. It will shift to the Heart (idealistic self). Seth will cease to control and hog for himself all the energies of the system. As an arrogant master does with his slaves, Seth dominated, and this will be of the past. There will not be an alpha male controlling everything anymore, but cooperation. The benevolent Heart will reign and all the people (cells) will rejoice.

The heart phase is the masculine aspect of the feminine polarity. To access it, as understood by Leonardo da Vinci, men are required to develop their feminine polarity. There is no way around it. This was supposed to be facilitated by religion. Although some individuals benefited from religion, most unfortunately failed because of the general immaturity of humanity. Instead, we saw a long patriarchal era and the subsequent outbreak of consumerism, with the economy as a central if not sole social value. Ultimately, as I said, we cannot serve two masters. Our own evolution forces us to choose between the rational self of the pancreas aspect and the idealistic self of the heart aspect. All our physical/emotional/mental actual predicaments point in the direction of the heart aspect in need of maturation and expression. This is the true sense of human evolution.

In the Five Books of Moses, Leviticus starts with the following string of words: "The Lord called to Moses ..." (Lev. 1:1, NIV) and ends with "These are the commands ..." (Lev. 27:34, NIV). This period of fourteen to twenty-one years old, summarized as "The Lord called," is when awareness of one's core identity is built. It is the phase when we receive our calling to choose the path toward Consciousness. When we shut ourselves to this calling, anxiety is first generated since we block our path to evolution, then we become numb through the overstimulation of the masculine polarity (Reptilian, Mammalian and Analytical brains) through sex, food, general consumerism and money matters. This Third Book of Moses includes the rules we should follow to heed this calling. We are all priests graced by the presence of a god sleeping in our hearts. To allow him speech and vision to this world takes discipline and control over our personalities. As long as we are earthly creatures, the center of our physical and emotional selves is the pancreas aspect (the rational self), and the center of our individuality should be the heart aspect (the idealistic self).

In Leviticus, sacrifices are described as symbols of the immolation of our personalities, lifting them toward God or the absolute. In other words, the opening of our hearts to individuality allows us to create a link with our divine nature.

In the text, the use of the word *tamei*, often translated as "pollution," refers to the natural tendency of the personality to take over, rendering the individual "ritually handicapped," which is another way of saying that the feminine polarity is either not functional or is constrained.

The gruesome ritual of slaughtering and disposing of the blood and flesh in Leviticus refers to the offering of our masculine polarity, of our "animal" nature. Animals were distinguished from other living entities as they received a vocal blessing from God (voice is linked to the development of the Mammalian brain, where it is the most expressed). This personal sacrifice is a necessary step toward the functionality of the feminine polarity. Added to this, when we read the description of many sacrifices, the elements burned are usually the same: the head (extension of the kidney energies), the fat, the kidneys, as well as the extensions of the tendons on the liver. These are associated to the physical masculine polarity, which is kidney/liver and the expressive aspect of the pancreas. We also know that the liver through the bile salts and the pancreas through its enzymes work together for the digestion of fats. We also know that gallstones, a slowing of liver and gall bladder energies, are associated with recurrent attacks of pancreatitis, showing the link between those two organs. [261]

Concerning our sacred text, the sacrifices enhance the idea of offering the masculine polarity to God, the source of individuality, since that will usher in Man's ultimate evolution. This sacrifice of the fat is found also in the Catharist tradition, which frowned on fat consumption. In fact, the Cathar religion, like many others, took great care in *not* stimulating the masculine polarity, particularly through the diet (pancreas aspect). In a recent visit to the south of France in search of remnants of this Cathar tradition, I was shocked by the quantity of fat consumed by our French cousins there. Then I discovered through my research that this became more popular under the Inquisition period. This makes sense as a sure way of knowing if a person was a heretic was to look at his diet. Not consuming fat or meat would open one to direct prosecution. As an example, in 1052, when in pursuits of heretics including Cathars, the Holy Roman Emperor Henry III (1017–1056) gave a clever order: men were excommunicated and hung if they refused to kill a chicken. Cathars were only following the rule: "It shall be a perpetual statute for your generations throughout all your dwellings, that ye eat neither fat nor blood." (Lev. 3:17, KJV)

Hence, the most sensitive people of those communities were ruthlessly murdered. Travelling through this beautiful region from which my paternal ancestors originated, I had this reflection: you can eliminate people, you can deform their thoughts and words, but you can't silence the stones.

Leviticus, central to the Books of Moses, is set entirely around Mount Sinai. The most elevated element (mountain), seen from above, becomes its center. Indeed, in the Temple, the more central a structure was, the more precious materials were used, paralleling the jewels on the priest vestments. By analogy, the heart aspect is central to the evolved human. Not surprisingly, the heart was central to ancient Egyptians' faith and was the only viscera that would not be removed from the body before embalming.

The details and themes discussed here are found in all other Books of Moses, showing again that this teaching aims at the edification of the perfect human and, similar to the brain structures linked to this phase, integrates information from many modalities. For the LIFE biosystem, it is the building of the feminine polarity in preparation for its birth, analogous here to reaching the Promised Land. The idealistic self has a vision that must be fed and held. All the elements that could potentially disrupt this vision must be expelled. In this book, this world is one of perfection, a realm in which wholeness will reign. The body becomes the image of the cosmos, and the

Isis Code

Temple the image of the body. As such, hygiene of the body and of the Temple has infinite repercussions. In Genesis and Exodus we had action. Here we enter the world of being more than doing.

While the tribes and nations around the children of Israel were still immersed in a quasi-universal worship of the Mother Goddess, which had been gradually transformed into its reptilian aspect through the dominance of the masculine polarity, the Hebrew pilgrims—and more precisely Moses and probably Jethro, his father-in-law—had embraced the projection of their individuality into the screen of this world. Although the teaching was only understood at the level that the "children of Israel" could grasp, it was already a step toward perfection. What is perfection? Manifestation of Consciousness through completeness.

On this idea of perfection, Carl Gustav Jung has an interesting reflection, noting that the translation of the original biblical text as "perfect" was wrong, in that the text meant "complete." We should therefore read Matthew 5:48 as: "Therefore, be complete, just as your Father in heaven is Complete [whole]." [262]

The priests, who are of the heart phase (idealistic self), were barred from contact with death (the kidney phase), except in the cases of sacrifices. They were educators. The pancreas aspect (rational self) through its feminine aspect (nutrition) was first dealt with by Moses and Aaron. Jesus reiterated the necessity of feeding the heart phase. After him, Muhammed reinforced the last aspect of the feminine polarity, the lung phase (social self). Now we need to put all of these *together*, with a clear understanding of the masculine polarity. Unfortunately, this discussion, although a fascinating subject, exceeds the boundaries of this book.

The heart aspect (idealistic self) has a great affinity with the kidney aspect (physical self), as it is controlled or balanced by it. In fact, many doctors[263] consider the heart and kidneys as one interlinked body system rather than distinct organs.

In Leviticus, we therefore encounter an overarching theme of life and death, of sacrifice. The Seal of Solomon expresses this reality (diagram 24). In the Bible it is said that: *"Dieu sonde les reins et les coeurs."* Or in English: "God tries the hearts and reins." (Rev.

2:23,[264] Ps. 7:9,[265] Jer. 11:20[266]) A literal cybernetic translation would be: God knows what is in the heart aspect (seat of individuality) and kidney aspect (seat of the personality) of men. When healthy, one expresses love, while the other manifests wisdom.

Leviticus ends with an accounting of alms given to the temple. For the gifts, one fifth of their value is added in case of redemption. The number five is constantly reiterated. This creates the link with the book of Numbers, which is a phase enhanced through the maturation of the idealistic self. This allows us to understand that *it is only through the awareness of a general moral law, thus of the numeral structure of the universe, that social control and order, as well as the edification of a social fabric, are possible.*

Throughout time, many cultures have shared a vision of the heart as a source of discernment, through its reaction when something is right or amiss. We intuitively understand pertinent information from it, through its rhythms and patterns, since all of our cells during pregnancy were etched by its speech. Because of its extensive connections with the brain, it exherts a deep influence on it, far more than any other organ. This effect is more and more understood.[267] The heart continualy interconnects with the brain and other organs through transmission of neural impulses, hormones, neurotransmitters, ventricular pressure, electromagnetic fields, and of course sound waves. It has been demonstrated that the heart's electromagnetic field becomes more organized during positive emotional states (called heart coherent), increasing in this way its capacity to impact favorably surrounding tissues or even nearby systems and people.[268] Trials in which test subjects were asked to generate positive feelings while holding a specific objective to positively influence the DNA, showed an increase in heart coherence.[269]

Messages the heart sends to the brain also affect general performance. This is not surprising.

"The heart is the most powerful generator of electromagnetic energy in the human body, producing the largest rhythmic electromagnetic field of any of the body's organs. The heart's electrical field is about sixty times greater in amplitude than the electrical activity generated by the brain. This field, measured in the form of an electrocardiogram (ECG), can be detected anywhere on the surface of the body.

Isis Code

Furthermore, the magnetic field produced by the heart is more than five thousand times greater in strength than the field generated by the brain, and can be detected a number of feet away from the body in all directions, using SQUID magnetometers. Why would this occur if it had no use? The cardiac field is modulated by different emotional states that do not originate in the brain, although they are processed by it. HeartMath investigated the possibility that the electromagnetic field generated by the heart may transmit information that can be received by others."[270] Understanding the Master of the Heart, which I will discuss later, and the quantum plenum, which I discussed previously, I also believe this to be true.

The normal variability in heart rate is due to the synergistic action of the two branches of the ANS (autonomic nervous system) through neural, mechanical, humoral, and other physiological mechanisms, thus allowing appropriate reactions to changing external or internal circumstances. In a healthy individual, the heart rate estimated at any given time represents the net effect of the parasympathetic (including vagus) nerves, which slow heart rate, and the sympathetic nerves, which accelerate it. These changes are influenced by emotions, thoughts, exercise, and general metabolism. Our changing heart rhythms also affect the brain's ability to process information, including decision making, problem solving, and creativity. Further, they directly affect how we feel and vice versa.

When we experience a joyful emotion or maintain a positive attitude, there is a decrease in the activation of the sympathetic nervous system then inhibited by the right ventromedial prefrontal cortex and right hemisphere of the feminine polarity. There is also an enhanced parasympathetic nervous system activity. As well, the brain's alpha rhythm synchronizes to heartbeats, and a better entrainment between all the physiological oscillatory systems can be observed. The brain's ability to process information is therefore enhanced. Since all systems tend to become nonlinear with increasing amplitude of oscillations, the human body, because of the elastic properties of its tissues, may be regarded as a general oscillatory system. Oscillation now appears to be the inherent global behavior of balanced systems.

In our system as in real life, on the contrary, when we experience negative emotions, the heart rhythms become erratic, indicating that the two nervous branches are not synchronized anymore. The information received by the organism is conflicting, and the backward-forward syndrome we spoke of previously will result, increasing both energy consumption and aging of the entire system. The pancreas aspect of our LIFE biosystem will try to reestablish the balance. If the lung, mother of energies, does not carry enough energy or is dysfunctional, we have to find compensation and stimulation of one or many phases. The pancreas aspect being the pivot or regulator of the system, it is the first aspect we will instinctively try to stimulate, similar to a baby trying to self-soothe. Multiply these incidents and you end up with a habit. Over a period of time, it will become a reflex, and eventually an addiction. Out-of-control consumerism is a direct result of this need to recover balance.

If our lung structure were healthy, we would not be so compelled to fulfill all the consumer needs we have created. The genius of humanity is to find stimulants that are not detrimental and to make them widely available. In fact, to provide elements allowing every newborn a healthy lung aspect would be the most ecological, economical, humane, and intelligent endeavor. The downfall of our civilization is that we have put forward and widely promoted stimulants that are detrimental not only to the individual but to future generations, instead of promoting those that are benign. Of course this is immediately profitable, albeit in the long run a very destructive attitude. We close our eyes to the real underlying problem since we promote a type of commercialism in which, legally, companies must work toward making profits, even in cases that go against basic ethical values.

The heart aspect is active, while the lung aspect, which it controls, is receptive. The structure of the Master of the Heart influences the receptive pancreas aspect in its nutritive/feminine aspect. It is tributary to the heart aspect (idealistic self) and is controlled by the kidney aspect (physical self). The Master of the Heart is comprised of elements inherent to the feminine polarity. As such, it is sensitive to the electromagnetic spectrum through the skin as well as to touch, and most probably to the emotions of others. Indeed, prayers and love have been shown to influence the magnetic field.[271] Our future research may show us that we are also sensitive to thoughts. Although a controversial subject, prayers supported by emotions have a significant effect on the receptive people prayed

Isis Code

for.[272] Humans have held the intuitive belief that good or bad emotions, words and thoughts, directed toward someone have an effect. As with the medicinal use of many plants, often people's intuition precedes scientific findings.

The observed positive result of some prayers make sense since they are issued from the conjunction of the masculine and feminine polarities, which then is active as a laser—light/heart, amplification/lung, stimulation/kidney, emanation/liver, radiation/pancreas—and carried through the general Master of the Heart. This is often called a biofield, which is the field associated with the implicate order. This is understood only if we view the feminine polarity as One and part of a general plenum, which we previously discussed. Another required element is that both the person who prays and the one receiving the prayer have an active feminine polarity for the prayer to bring healing ("Go, your faith has healed you." Mark 10:52, NIV.) Considering the feminine polarity as One, explains some observed phenomena, such as the synchronized menstruation of women who live or work together.[273] Or this one I experienced personally, during a period while I was still nursing my baby. One day, after having breastfed him, put him to bed and witnessed his sound sleeping, I decided to pay an overdue visit to a neighbor. We were chatting for about five minutes when I suddenly felt my blouse wet. I had had a milk ejection. Running back home, I found my baby boy awake, all smiles. Scent, or the baby's cries, which are considered the only scientific plausible explanations here, were not present in this case. It may sound anecdotal but it is not. Somehow we were linked, either through space or our inner selves. Even if they are not obvious, the Master of the Heart and the nature of quanta and of the quantum field all provide reasonable and plausible explanations for some inexplicable phenomena. They therefore need to be explored if we wish to be scientific.

According to Einstein, we should regard matter as part of the electromagnetic field, in the sense that electromagnetic energy is the fundamental origin of our entire physical world. Therefore, for the most evolved being, superior assimilation and efficient use of electromagnetic energies should be of prime importance. It is probably a necessity for evolution. Echoing this is a work published by the Academy for Future Science:

> Under present biological conditions, evolutionary development in living bodies from earliest inception follows unicellular semiconductivity, as a living piezoelectric matrix, through stages which permit primitive basic tissues (glia, satellite and Schwann cells) to be supportive to the neurons in the human system where the primary source is electrical. This has been especially shown in bone growth response to mechanical stress and to fractures which have been demonstrated to have characteristics of control systems using electricity.[274]

Thus energy is eventually transformed, condensed into matter, and sustains matter cohesion. Energy precedes and forms matter, as shown during the Big Bang. And matter gives off energy. This is the eternal elixir of alchemists, the Om and Ram of Hindu tradition, the yin which accumulates energy and yang which disperses it, etc.

"Atoms are electrically neutral if they have an equal number of protons and electrons. Atoms that have either a deficit or a surplus of electrons are called ions [+ or -]. Electrons that are farthest from the nucleus may be transferred to other nearby atoms or shared between atoms. By this mechanism, atoms are able to bond into molecules and other types of chemical compounds, such as ionic and covalent network crystals." Wikipedia, Atoms.

How energy becomes matter is what we should study now, and only the feminine polarity can give us the keys to this venture.

In our desire to know matter more thoroughly, and in our fear of wearing rose-colored glasses, we have discarded evolution. We are now wearing bottle-thick glasses and blinders, limiting ourselves to see matter only as producing energy and energy only as a byproduct of matter. And we wonder why our kids are depressed!

What does nature tells us regarding this?

Scientists at the Carnegie Institute of Technology have found that the mineral kingdom truly coevolved with life. Only a dozen minerals (crystalline compounds) are known to have existed among the elements that formed the solar system 4.6 billion years ago. Today

Earth has more than 4,400 mineral species, with two-thirds of them directly or indirectly linked to the processes of life. This finding was published in *American Mineralogist*.

> "The interplay between minerals and life works both ways. It turns out that the origin of life may have been absolutely dependent on certain minerals. Mineral surfaces are the perfect place to concentrate, to select, to organize, to make larger structures like polymers, chains of molecules that have biological function. So minerals may have played a key role in life's origins. But by the same token—and this is what I find so amazing—life played a key role in the minerals' origin. If you have a world where you have no oxygen, perhaps Mars, certainly Mercury, these minerals will not form. There's no way to produce them. Life has a role in the origin of minerals just the way minerals have a role in the origin of life."[275] (My emphasis)

During his alleged many years of friendship with Albert Einstein, Dr. Carey Reams, PhD in biophysics and biochemistry, reportedly once teased him, saying, "You know how to take matter apart but you don't know how to put it back together again." It appears Einstein shot back to him that figuring out how to do that was Reams's job.

Later, through his research, Dr. Reams was in a position to observe: "Illness begins with the vagus nerve" and "We don't live off the food that we eat, we live off the energy IN the food we eat!"[276] The biochemical should not be overlooked, but everyone who had experienced eating a fruit straight from its tree knows that there is something in a fruit eaten right from the tree that can't be found in a fruit that has sat on a grocery store shelf for a week, even if it still looks fresh. This last assertion was a main point of the vitalist movement. Because we now need to see beyond the limited capacities of our instruments, maybe this element will regain value in the eyes of researchers, opening the doors to a wider and more interesting world. I will clarify later in this book that we own a system for extracting this type of energy, the Master of the Heart, and that the vagus nerve—part of the celiac plexus, otherwise called the solar plexus—has a key role to play in this system.

In agreement with the vitalist movement, we have to accept that our physiology is sensitive to electricity and magnetism. Industry is now starting to use this notion to create technology of the future.[277] This is one possible application for the conductivity of our biological system. In the same vein, A. P. Dubrov, a Russian scientist, showed that the susceptibility of biological membranes to the geomagnetic field (GMF) and other fields is universal and a type of perceptive response. With other researchers, he concludes that acupuncture points act as receptors to these fields. These are active in all people living in the same time zone at the same time. It appears also that individuals react differently to this information. Dubrov cited cases in which the field stimulates the parasympathetic nervous system, while in other individuals the sympathetic nervous system reacts. Other researchers, such as F. A. Brown and Y. H. Park, say: *"The GMF in its action on biological objects interacts with other vector forces, particularly the gravitational effects of the moon.... The combined effect of the GMF and gravitation is responsible not only for the spatial but also for the temporal organization of biological objects."*[278]

This is exactly what the Chinese sages of old studied. They considered the cycles of nature, the position of the stars. They had a system of star boundaries and coordinates by which they would study the moon's relative movements. The twelve branches, the trigrams, the four directions were all devised relative to the twenty-eight lunar mansions and the pole star. The rituals of sunrise identification and water ablutions, as historically observed in most religions, are in agreement with this concept of *necessary* harmonization between human and environmental energies.

In addition to light, studies indicate that electromagnetic (EM) radiations of different types—including earth strength magnetic fields—reduce melatonin production to the *same* extent as exposure to light does. This occurs in vivo and in vitro for a number of species, including humans (Reiter and Richardson, 1992; Reiter, 1993; Schneider et al., 1994; Yaga et al., 1993, Richardson et al., 1992). The environmental magnetic field as well as light diminishes in strength during winter months. Added to the fact that we do not go outside as often, SAD (seasonal affective disorder) is understandably more frequent in winter. Also, research has shown that in our northern climate (north of 42 degree latitude) during the winter months, even going outside does not trigger enough vitamin D production. [279]

Isis Code

Not surprisingly therefore, the ratio of northerners with SAD compared to those living in the tropics is about 10:1.

The Pineal Gland

It is a neuroendocrine transducer in this that it converts incoming nerve impulses into outgoing hormones.[280] The pineal gland is part of the lung phase and therefore has a controlling effect on emotion. We will discuss the pineal gland more thoroughly in chapter 7, devoted to the social self. Nevertheless, since its action is closely related to the Master of the Heart, we have to understand its function. For now we can signal that the Pineal-Hypothalamic-Pituitary Axis (PHPA) is the clock coordinator for the organism as a whole. In higher vertebrates, the pineal rests between the two large cerebrums at the anterior end, surrounded by the ventricles and directly connected to the cerebellum. As noted earlier, the cerebellum is crucial in coordinating fine movements with sensory data and plays a part in regulating thought as well as human language (diagram 15).

The role of the pineal organ in humans suggests that any agent impeding its function could affect health in many negative ways, including sexual maturation, calcium metabolism, parathyroid function, postmenopausal osteoporosis, cancer, multiple sclerosis, and psychiatric disease. [281]

One of these agents—adding fluoride to water—has been extensively researched, with the following observation:

> In conclusion, the human pineal gland contains the highest concentration of fluoride in the body. Fluoride is associated with depressed pineal melatonin synthesis by prepubertal gerbils and an accelerated onset of sexual maturation in the female gerbil. *The results strengthen the hypothesis that the pineal has a role in the timing of the onset of puberty.*[282] (My emphasis)

Can we extrapolate these results to humans? Many believe so. Multiple factors tend to explain the early sexual maturation of girls. This could be one as well. Consequently, cancer and other illnesses could be prevented in part by protecting the function of the pineal gland.

The early sexual events we experience model our sexuality. If intercourse is regularly practiced before twenty-one years of age, especially for boys, we enhance the energy of the preceding phases, namely the reptilian and mammalian phases, over the other brain phases. With repetition, the energy might be hijacked by these levels. I compare it to the water pressure in a high-rise building. Without enough pressure, the water will not be delivered to the highest floors since the only faucet is situated at a lower level. This can lead, as with any drug, to a need to regularly activate the reward system of the Mammalian brain through sex. Horse breeders know this. Once a young stallion has had a taste of mating, it is as if he is fit for little else.

Young people under the age of twenty-one and in the heart phase tend to project onto their romantic partners either the image of their own feminine polarity—if they are a boy—or their masculine polarity—in the case of a girl—with all the ensuing heart break and life problems. This is more frequent for girls as naturally they are inclined toward the feminine polarity. They live in the virtual reality of their hearts. Only in the pancreas phase (ages twenty-one to twenty-eight) will they finally touch ground. To think that "they know what they are doing" is illusory. Following failure in the heart phase, they might revert to the previous phases and become overly genitally oriented. We will discuss this at length later.

Conversion of norepinephrine in the sympathetic nerves innervating the pineal gland has a twenty-four-hour cycle. This rhythm is suppressed by light. The same rhythm in norepinephrine turnover generates the rhythms in the pineal gland's enzyme system (indole-amines and N-acetyltranferase), which makes melatonin and serotonin. Serotonin (5-hydroxy tryptamine) is also produced in the gut of the intestinal tract. These are energetically linked. Both the pineal gland and large intestine are functionally associated to the lung phase in medical biocybernetics. Another important feature of melatonin is its anti-inflammatory property. Associated to the lung function, its action in calming general inflammation, including in the gut,[283] should not be overlooked.[284] As well, because the pineal gland is less effective and inflammation is widespread in the elderly, we should strive for lifelong hygiene that promotes secretion of melatonin—even better, respect of cycles—and begin this at an early age.

Promotion of the feminine polarity and its cycles is mandatory. The tale of Cinderella's carriage reverting back to a pumpkin after midnight is not that farfetched. Melatonin production rises after sunset, reaching a peak around 2:00 a.m., and light suppresses its production. For those who are regularly on the computer or watching TV until 1:00 a.m., their level of melatonin is certainly deficient. But the need for staying up late is also a sign of an underperforming feminine polarity. Of note, the amount of light needed to suppress melatonin production is minimal. A dark and well ventilated room to sleep in, as well as refraining from television, video games, and computer viewing at least for one hour before bedtime, is recommended. Reading a book, or better yet, a stroll in a park are better choices.

Beside melatonin and serotonin, the pineal gland produces a multitude of peptides (see chapter 7). These have a wide range of properties. In particular, they may normalize general immune functions, stimulate antibody production, suppress tumors, protect against stress, regulate the circadian rhythm, and have anti-aging effects. We have previously seen that they might be implicated in memory and confer magnetism to biological structures. Most intriguingly, they are able to suppress RNA synthesis in tumor cells and to selectively modulate DNA transcription.

Researches have shown a 100% correlation between calcification of the pineal gland and MS (multiple sclerosis).[285, 286] Since calcification of the pineal gland is limited in the black population living in Africa, MS should be rare in this group. And it is.[287] It also appears that the black community in America shows a higher incidence of pineal calcification than their African counterpart. Unnatural EMFs are in part suspected of negatively affecting this gland, as well as upsetting the balance and production of the hormones attached to it.

The protein beta-Lipotropin, secreted by this gland, is some 400 amino acids in length. Found also in the pituitary gland, it is a peptide and mobilizes fat from adipose tissue. This protein is cut at numerous points by different enzymes in response to changes in the transcription of DNA. These changes vary with the cyclical energy fluctuations in the environment.

The organism is truly bathed in the environment through the pineal gland, which is one of the circumventricular organs. That is, it does not benefit from the blood-brain barrier. Consequently, this organ is highly sensitive to biologically active macromolecular substances circulating within the cerebral blood flow, in the CSF liquid, as well as those that are breathed in. In effect, the unique location of the olfactory system, its chemical links to both the environment and the central nervous system, turns it into a direct route for chemical and molecular information of the environment to the cerebral blood flow and circumventricular organs. There is an anatomic connection of the nasal submucosa to the subarachnoid space surrounding the olfactory nerves as they penetrate the cribriform plate of the skull and enter the brain. This cribriform region has no significant barrier to cerebrospinal fluid drainage either. This is a possible way for metals, dyes, viruses, pheromones, peptides, proteins, and narcotics to enter the brain, via the nasal cavity and thus avoiding the blood-brain barrier.

As an example, it appears that heating oil and solvent were dumped into the storm sewer system near my house. While I was in the writing process of this book, I lived a cloistered life, concentrating on my work. For five weeks I suffered from headaches, intermittent nausea, general tiredness, and a depressed mood. The few times I would go out of the house, I could smell a strong, foul odor upon reentering. I would call the furnace company, and of course they would rightly argue that their system could not be responsible for the smell. Exasperated and sick, I finally called the fire department and then the city. They came and flushed the system. The odor left but came back a few hours later, along with my headache and pain on the right side, at the level of the liver. When highly volatile compounds are in the air, the olfactory system tends to become saturated and desensitized. Within a few minutes, the smell doesn't appear as bad, but the detrimental effects are compounding.

It has been shown that the removal of the pineal gland (pinealectomy) or subjecting it to constant light, which suppresses pineal activity, promotes tumor processes.[288] On the other hand, melatonin injections result in a decrease of carcinogenesis. The several polypeptides produced by the pineal gland also affect the hypothalamus in different ways, since the hypothalamus is its main neuroendocrine target.[289] This is in agreement with our system of the Mammalian

Isis Code

brain being under control of the Universal brain (liver phase under the control of lung phase). Our physical and emotional health is therefore directly and utterly tied to the quality of our environment and wake/sleep cycles.

Research done in the late 1980s has shown that in vivo, proteins, DNA, have piezoelectric crystal lattice structures. The piezoelectric effect refers to the property of matter in its ability to convert electromagnetic oscillations into mechanical vibrations, enhancing reception and expression of information. It also seems there is a relation between chirality—which is considered as a mean to exchange information—and piezoelectric quality. It had been demonstrated that chirality, a quality not well understood until now, renders fluid lipid bilayers piezoelectric.[290]

Studies have shown that both transcription (RNA synthesis) and translation (protein synthesis through DNA) can be prompted by electromagnetic fields. As well, direct current in bone will produce bone formation (osteochondrogenesis) and bacteriostasis, and affect adenosine triphosphate (ATP) generation, protein synthesis and membrane transport.[291]

As we know now, our electromagnetic environment has dramatically changed in the last twenty years. The microwave levels are now at least ten times higher than before. Epidemiologists have reached an international consensus: even low levels of EMF double the risk of leukemia in children.[292]

Also susceptible to magnetic and electric fields is water. Spectrographic analysis of it before and after it has been exposed to such fields, reveals shifts in resonant spectra, as well as a marked decrease in surface tension (R. Gerber, 2001). Water does have a memory, as does even the simplest quanta.

Information does not need quantity, but it needs a qualified receptor to its frequency. This is the basis for homeopathy. A lower potency in homeopathy has an influence on information recorded at the cellular level, while the higher potencies (more diluted) have an influence on the heart aspect of the cells (RNA). Those who claim homeopathy is "only" placebo might be emotionally insecure, might have economic interest in this witch hunt, might have a dominant masculine polarity to protect, or are simply unaware of the double blind studies done with homeopathy across the world.

One observation should be made: through their universal acceptance, conventional drugs carry a potentially stronger placebo effect.

A simple reading of the many researches done on homeopathy and reported at the US National Library of Medicine, part of the NIH, or the report on homeopathy commissioned by the Swiss health authorities[293] should calm the suspicious minds. In homeopathy, as in acupuncture or any bio-informative medicine, the quality and experience of the practitioner and his general receptive and investigative nature, as well as his affinity with the patient, are the main aspects to examine when one seeks therapy. What I call bio-informative medicines (diagram 25)—that is, any corpus of knowledge or any therapy viewed through the lens of medical biocybernetics—target individuals in their uniqueness, and, as such, cannot be easily accessed through double-blind studies. It is detrimental to the whole population, both economically and health wise, to oppose proper research into, and funding of, homeopathy. Too often, double-blind studies are doubly blinded studies. Blinded to the observer who performs the tests, as the observer influences the observed, and blinded to the subtle nature of everything that exist, its interconnections with everything and everyone. Because the homeopathic remedy does not contain molecules, its target is therefore another system, one concerned with information regulation that guides the formation of matter: the Master of the Heart (diagrams 26–27).

Since the discovery that magnetic orientation by bacteria was also due to the presence of biochemically formed magnetic particles of magnetite (Fe_3O_4) within the organism, the search has begun for other biogenic deposits of magnetic material and the ways such material might confer on the organism the ability to orient itself to ambient magnetic fields. It was found that those crystals interact with external magnetic fields more than a million times more strongly than do diamagnetic or paramagnetic materials of similar volume. Earth-strength magnetic fields can thus, in these cases, yield responses above what is considered mere thermal noise. Evidence continues to accumulate that a wide range of organisms, including humans, can detect and orient to ambient magnetic fields.

Isis Code

Magnetite has been found in humans, bees, homing pigeons, dolphins, and various other organisms.[294] In humans, bones from the region of the sphenoid/ethmoid sinus complex (around the eyes) have been shown to be magnetic and to contain deposits of ferric iron (magnetite).[295] This sinus complex corresponds to the starting point of the bladder meridian (BL1) for the ethmoid sinus and in its profound trajectory to the sphenoid sinus. The sphenoid sinus is also linked to the gallbladder meridian's entry point (GB1), which is very important to the feminine polarity. All meridians have an associated point in the gallbladder meridian, and almost half of the points relating to this meridian are at the head level. These meridians are important for the triple warmer/Master of the Heart system. In the wei qi (chi) theory of Ling Shu (LS 76), the wei qi issues from the eyes at awakening and crosses all the yang meridians, finally going back to the eyes. This could be explained by the magnetic and electric field theory. In *Electromagnetic Fields and Life,"* Alexander Presman, a Russian biophysicist, hypothesized that through the process of evolution, electrical and magnetic fields were used by different organisms to obtain crucial information regarding their environments.

Research has also shown the presence of crystals of biogenic magnetite, with minimum estimates between 5 and 100 million single-domain crystals per gram, in many tissues of the human brain. Magnetic particle extracts from solubilized tissues were examined with high-resolution transmission electron microscopy (TEM) and electron diffraction, which identified minerals in the magnetite-maghemite solid solution, with many crystal morphologies and structures resembling those precipitated by some bacteria and fishes.[296] The magnetite compound has been found dispersed in *all* structures of the human brain.[297]

This knowledge opens wide new doors to research. For example, the antitumor effects of magnetite nanoparticles have been observed in cat mammary adenocarcinoma.[298]

Where else were they found? In the meninges, this system of protective membranes filled with cerebrospinal fluid that envelops the central nervous system and connects with the extracellular matrix, and where the vagus nerve ends. One of these membranes, just under the bone, the dura mater, surrounds and supports the large venous channels (dural sinuses) that carry blood from the brain to the heart.

The subdural space (or subdural cavity) is an artificial space created by the separation of the arachnoid mater from the dura mater as a result of trauma, pathologic process, or death. Indeed, in the case of a cadaver, due to the absence of cerebrospinal fluid, the arachnoid mater falls away from the dura mater. This underlines the fact that *circulation of cerebrospinal fluid is a fundamental requirement to life*, and more than the mere cushioning effect we ascribe to it.

Now back to the pineal gland. As written in the *Journal of Ophthalmology*: "In animals that have lost the parietal eye, including mammals, the pineal sac is retained and condensed into the form of the pineal gland."[299] This explains why we see an intimate relationship between the heart aspect (parietal lobe) and the lung aspect (pineal gland). Different biomineralization forms were observed in the pineal gland. Some are linked to age, as we discussed before. Calcite, which was used by trilobites as eye lenses, is also present. Microcrystals, whose length does not exceed 20 micrometers, were also found.

In a research study, these microcrystals were analyzed with different biophysical techniques. Their physicochemical properties and particular piezoelectricity gave them an active role in a potential mechanism of electromechanotransduction in the pineal body, affecting the membrane of its cells. The piezoelectric and second harmonic generation property (SHG)[300] of those crystals would allow them to interact with the electrical component of electromagnetic fields. Researchers in the field of chemical engineering and experimental toxicology found the following:

> By that very fact the piezoelectric property of the crystals would allow them to interact with the electrical component of electromagnetic fields. A simplified formula applied to those crystals ($f = v/2d$) lets us think that these crystals could be sensitive to RF-EMF in the range of 500MHz to 2.5GHz depending on their size. This range contains portable wireless frequencies, GSM (872–960MHz), DCS (1710–1875MHz), UMTS (1900–1920MHz, 2010–2025MHz), or BlueTooth

(2400–2483MHz). Piezoelectric determination of minute grain requires developing new methods based on either MEMS Precision Instruments microtweezers or direct correlation between electro-optic and piezoelectric properties in crystal with optical microscopy.[301]

The pineal gland then converts these signals into an endocrine output.

This leads us to discuss the fundamental structure and function of the Master of the Heart. Indeed, if the heart aspect is dysfunctional, the Master of the Heart cannot fulfill its function properly. As well, if the Master of the Heart is impaired, the whole system fails to produce health.

The Master of the Heart

We might ask: what is the Master of the Heart?

Dr. DeBavelaere labels it the "human flower function" and the "second digestive system." The functions he attributes to it are absorption, transformation, regulation, and excretion of subtle energies. Expressed with other words—information manipulation. The Master of the Heart pairs with the triple warmer to execute these functions. MH/TW is thus a unity. The feminine aspect of the system on the subtle level is the Master of the Heart. It therefore has a pivotal role in memory, since it is the first subtle expression of the implicate order. Its denser aspect, the triple warmer refers to the classical digestive system.

In the LIFE biosystem, these properties imply a feminine type of physiology on all levels. Why? Because although the masculine polarity can "solve," meaning it can analyze, pile, build, and move matter, it cannot transform energy into matter. On the other hand, the function of the feminine polarity is to receive energy and to coagulate, or condense if you will. Therefore it sustains the organization and cohesion of matter, as seen inversely when someone dies. When joined to the masculine polarity, the feminine one can transform energy into matter. This is the secret held by the polarities and which some alchemists were trying to master in order to transform "lead into gold."

In traditional Chinese medicine, Master of the Heart /triple warmer refers to an extraordinary indivisible and limitless system. One is the external aspect of the other in the manifested world, and one is a support to the other. The Master of the Heart—the inner—acts from the unobservable point of view, while the triple warmer—the "surface"—works from the inside of the body. We can get confused when we read translated traditional Chinese texts. We always have to keep in mind that these two structures are said to "have a name but no form." Are these exclusively under the skin and running through a specific meridian, or do they impregnate the whole physical universe? One does not exclude the other.

Who has not seen the meshlike figuration of Einstein's space-time continuum? We can make an analogy between these images and Greek mythology. Zeus was said to have made a cloth that he decorated with the elements of earth and sea. He presented it as a wedding gift to Chthonie (the physical aspect, kidney aspect) and wrapped it around her. This is the Master of the Heart (diagram 26). In *Cratylus*, Plato says that the "aether" had a penetrating power that permeates the whole world, and he found it both inside and outside our bodies. The pentemychos, or implicate order, is associated with this ether. We will discuss this later on.

It appears the Master of the Heart is intimately connected to the pericardium meridian, the lung and heart aspects, the feminine facet of every phase, and of course to the triple warmer, to which it gives the most subtle energies (information) for transportation. Alchemically, it transforms chi energies into vital ones. I surmise that the MH/TW is an interface between external and internal energies in the same way the skin is a boundary to physicochemical elements. Its TW aspect is in charge of energies from food and liquid, while its MH aspect is in charge of gaseous, electromagnetic energies, as well as other more subtle ones (schemas 26–27). Both give to every phase the energy of its kind: feminine for the Master of the Heart, masculine for the triple warmer. The proper functioning of the Master of the Heart affects all of the feminine aspects at all different levels, including the TW function, as the TW function affects all the masculine elements at all levels. This is why we say that they are the precursors of all the

Isis Code

phases, organ, and meridian relationships. The Master of the Heart, in fact, *is* the plenum bearing the implicate order. It is the mother of every manifestation, the feminine polarity, Eve.

> "The Heart and Master of the Heart are one. Thinking about the yin organ's symbol, it is called the heart. Thinking about the function of Shen, it is called the Master of the Heart. The Master of the Heart is the master of the twelve meridians. If there is an empty space, then everything is permeated by the Master of the Heart. ... of the five yin organs. The triple warmer is the most superior of the six yang organs. The Master of the Heart substitutes for the Shen functions. The Shen stays between the kidneys and comes out at the heart. This becomes consciousness. Shen and qi both have name [existence] but no form [material structure]. No form responds to matter."[302] Nan-Ching [303]

> The noble heaven, Shen, is the great One. The myriad things emerge and are created in the great one, the alchemical changing of yin and yang. (Quote from Shi Ji and the Liu Shi Chun Qiu)

> The form is the abode of life,
> The qi is the fullness of life.
> The Shen is the controller of life.
> (Huai Nan Zi, 122 BC)

Because matter is condensed energies, the body can be considered a complex field of quantic, electrical, magnetic, and chemical construct. Many scientists such as A. P. Dubrov, Brecker, and others believe the acupuncture points and meridians have a significant role to play in the body's reaction to the effects of the geomagnetic field on it. Because acupuncture is efficiently used in pain treatment, and because the senses use the electromagnetic field to receive information about the environment as colors and sound, why wouldn't we possess an organ that could translate the higher frequencies of the field into emotions, thoughts, and the like? Mozart and his music would find an explanation here, and our inner Plato would rejoice.

The meridians, as webs of electrical conductance, might be information channels. Different types of information come and go from different organs and structures. This information would then be relayed to the brain and across the body from the extracellular matrix, the bearer of the meridian system, through the nervous system, the lymphatic system, the capillaries and veins, depending on their type of information.

> *"In the newer model, molecules interact by coresonance and need not actually touch as long as they are within an energetic field radius. In living systems, relatively long range electromagnetic fields engage in resonance matching and coherent amplification between distant molecules. This occurs as long as emission and absorption spectra match. Thus, non-resonating, unwanted signals are excluded. [This explains why when we use a homeopathic product that does not concern our condition, nothing happens.] As with all electromagnetic phenomena, the upper limit of such communication is the speed of light. J. Benveniste has shown dramatically the proof of this model. In thousands of experiment over many years, he has recorded the resonance frequencies of signaling molecules, matched them to the harmonics in the audible range, and digitized them and then using a computer sound card, played the recording of the signaling molecules. Accordingly, the recording of a specific signaling molecule will evoke an appropriate response in the specific receptor molecule just as if the molecules were in local contact! Benveniste has suggested that the effects of specific biological molecules (histamine, caffeine, adrenalin, insulin), as well as viruses and bacteria, are due more of electromagnetic interaction than direct contact."*[304] **(My emphasis)**

We have to keep in mind that Chinese authors sometimes speak from different points of view, through different reference levels, and thus use different terms. This only adds to our confusion since the subject—division of invisible energies—is clearly seen by the sage, who is more sensitive and able to transmit what his consciousness was able to grasp. Therefore my interpretation—and it can only be an interpretation since I am neither a sage nor a scientist—has to remain general and schematic, as I attempt to relate all the definitions

Isis Code

I previously encountered in different texts, as well as my personal experiences and what was taught to me. The realm of invisible and universal energies is a difficult subject that doesn't unanimously gather the same fixed and limited definitions, but as a whole I believe it is possible to render what was seen by the sages.

Our Westerner mind has difficultly translating the old Chinese texts. Using only our analytical tools and linear thought process is undeniably divisive and too reductive.

Contrary to the founders of the Chinese Taoist ideology, we separate physical, emotional, mental, and spiritual levels as we would separate different colors. This is not the only problem, though. The difficulty also resides in our Western minds needing to reintegrate these different colors in the light spectrum and realize they are continuous, not separate, entities. Chinese sages have pushed the subtlety to their descriptions of the different hues, adding different terms to all of them. Therefore we are faced with terms that do not correspond to any words from our cultural background. If for us all reds were simply labeled red, to them, although a color may be part of red, it would have different tonalities that engender different reactions. Even their study of physiology takes into account elements that we are only now starting to be interested in. The extracellular matrix that contains the meridian system is one of those.

Truly, the system behind TCM is not "folk medicine," as some have intimated.

In light of this, this living entity, which the Master of the Heart is, can be subdivided in its extension into the human body as upper, middle, and lower. The Master of the Heart has the most superior level, which we could link to higher Dantian, a structure used by Taoist sages in their quest for perfection and production of their *corps de Gloire*—glorious body. This superior aspect in the brain is most probably centered in the pineal gland, which as we have seen is highly sensitive to electromagnetic energies.

The Master of the Heart, interacting with the physical body, is responsible for the accumulation, extraction, and distribution of the most subtle energies, including emotions, as well as of those linked to breathing. The triple warmer is responsible for collecting and distributing the energies issued from the food and fluids mixed with the energies received from the Master of the Heart. The meridians' system appears as an extension of the upper triple warmer, with each of the points as receptors and emitters of electromagnetic energies. The five yin organs act as reservoirs and, although more condensed than the Master of the Heart, their feminine energies (yin) are protected and fed by it.[305]

The lower level of the triple warmer as well as of the bodily-centered aspect of the Master of the Heart is focused under the navel, at the hara center. The Taoists say that this is the source of energy and is linked to the universal chi and Shen. Breathing would allow this basic chi to move up and meet the descending (incarnating) Shen, allowing the transformation of energies into tangible energies and matter, as well as dispatching them throughout the meridian system and the five yin organs.

Shen can be visualized as the feminine, receptive matrix bathing the whole universe and all of the cells, a common cosmic ocean in which we move and have our existence. In Christian terms, it is the universal soul, Eve, mother of all that lives. As science has taught us, you can try to isolate quanta from their plenum, but they are still connected with those they were born with and carry the information they have received, and are influenced by the impressions received by siblings. You can never separate from this sea. But the body, our most condensed aspect, can receive it well or weakly depending on the filters, and the veils or patterns we have put over our hearts and minds. Death is when the energy and organization is at their weakest.

The soul can be considered the driving force behind activities that take place in the symbolic (ideals), universal (link with every kingdom), or creative planes (synarchic aspects). When its link with the body is weakened, we might express symptoms of anxiety, restlessness, and depression, all signals of a deficient lung aspect. Shen link can be strengthened through breathing exercises, meditation, tai chi chuan, contact with nature and its inhabitants (including humans), physical exercises, conscious nutrition, unselfish love, Self-knowledge and understanding of the energy structure sustaining our world.

Chi, being intimately associated to the triple warmer, has a masculine nature. You can feel it through the activity level and vitality of people.

Isis Code

This is their chi. It is the most dynamic energy that feeds the meridian system. It is physical and emotional energies linked to the masculine polarity. As with the Master of the Heart, it has many levels, and someone can have a strong chi as a gift from their parents. (In TCM, *jing* is similar to DNA.) Chi can be considered a life force. A healthy body supports its circulation and its circulation allows a healthy body. Movement, proper food, and contact with Shen feed it. The time we stay incarnated is inscribed in it through the telomere of our DNA, the kidney aspect. It is a gift from our ancestors, from the universal Master of the Heart, and a result of the health choices we make during this life.

More importantly, the virtual boundary between the personal MH/TW and universal ones is where human energies meet and become the more absolute energies of the universe and vice versa. The overall image by the early Chinese sages was of human beings created of earth and heaven, of substance and no substance, a complex mirror image of the universe itself, in which chi and Shen materialize. Like other traditions, they saw the human as a fractal aspect of the universe, as an image, a mirror and extension of the One. Their physical aspect was thus the expression of the physical universe, with the North Star as its center (diagram 31).[306] "Death" is seen as this universal Shen leaving a form and returning in this way the latter's different components to their original, elementary kingdoms.

"Therefore, the Chinese sage becomes the abdomen. He doesn't become his eyes." Tao Te Ching

To become the abdomen here signifies to link oneself to the feminine polarity while to become the eyes would mean to separate from it and to link oneself to the masculine one — which is what most humans, notably men, do naturally.

Breathing exercises are a central feature of this knowledge base allowing some experiences to unfold, not in the rational self—connected to the masculine polarity and thus expressive—but in the abdomen, connected to the feminine polarity and thus receptive. They are then connected to the senses and the idealistic self, which can access another level of reality linked to the Master of the Heart, open to the source at the hara level, and most probably onto the subconscious or super conscious aspect, which are associated to the evolved heart (super consciousness of the Master of the Heart), and lung and kidney aspects. As with anything touching subconscious energies, a guide is strongly suggested when wandering on these paths. The experienced guide will help see through the thickets, avoid the abysmal illusions that some atheists are, with good reasons, so afraid of, and safely walk the wanderer to his destination.

As the Tao Te Ching says, the eyes are expressive in their consciousness because they are of the masculine polarity, while the abdomen is receptive of a general Consciousness and linked to the Master of the Heart. It is another way of saying that the sage seeks to develop his connection to the universal feminine polarity, not to his masculine one.

Similar to the esoteric Christian vision, the abdomen here is not solely a cavity in which organs can be found, but where one can connect with the source, the energetic center from which life flows. Pregnancy occurs in this area because the two potential of feminine and masculine energies can fuse there. The body is holy.

In the Bible, chapter 47 of Ezekiel evokes a "river of life" going out of the Temple—the source of the universe as well as of our Self—and swelling up in phases corresponding to our five phases. First, the prophet has water up to his ankles (lung to kidney; many of the acupuncture points relating to kidney can be found there), then up to his knees (kidney to liver; many of the gallbladder points can be found there), then up to his kidneys (liver to heart; the chi between the kidneys;) and finally "the water had risen and was deep enough to swim in—a river that no one could cross (Ezek. 47:5, NIV) (heart onward; the Shen joins the chi). The river will renew the Dead Sea (all the cells). The Chinese appropriately name this area "the sea of chi." It will bring life wherever it goes, and fruit trees will constantly bear fruits. Jesus used this same water analogy: "Let anyone who is thirsty come to me and drink. ... rivers of living water will flow from within them." (John 7:37–38, NIV). And in John 2:19, he said, "Destroy this temple, and I will raise it again in three days." Genesis, linked as we saw to the kidney phase, has in its opening verses that the Spirit of God was "hovering over the waters." Most creation myths refer to water at the origin. This water is the receptive feminine polarity, bearer of life.

Isis Code

It is said that the temple that will be built, this New Jerusalem, will bear the *new* name "YHWH is here" (Torah 37:26–28). His very essence, his fundamental nature, his inherent characteristics, and his authority abide in those who are respecting him and thus their own Temple. Individually, we are called to transform ourselves into this New Jerusalem. God is always with us should we follow the sincere path, as Ezekiel witnessed. What is in the affirmation "YHWH is here," since it appears to be the goal of human evolution? When those seemingly simple elements combine, they give birth to the entire complexity of life. Could it be that in part, the word YHWH represents a simple pattern that is holographically present in all life, from the atom to the deity: the LIFE biosystem? The Master of the Heart would carry this pattern, this implicate order. Let's see.

Yodh (י) is the tenth letter and the smallest of the Hebraic alphabet. The tenth letter, as in the ten houses of God. It appears suspended in midair and looks like a seed. It is the beginning of Scripture, a dot, a spark, and part of every word, of everything. Its pictogram is an arm and a hand, so it refers to the active principle, the masculine polarity, but also to the spirit. Yodh is the manifestation, the spiritual core of individuality subjected to the motion of eternity. It is also deep inside each being. The gematria (system which assigns numbers) of this Yodh links it to vision, to our liver phase, the first manifestation in the polarized world. In the Taijitu, it is the white dot (diagram 2) in the black droplike form. Adam, as Osiris, comes from this characteristic of God.

The second letter in YHWH is the fifth letter of the alphabet, He (ה). Its pictogram is that of a man with raised arms, so it has a receptive nature. "He", corresponds to the totality of existence, consisting of the five elements. As we recall, in the Egyptian pictography, the symbol of the two raised arms signifies the Ka, or soul. Being the fifth letter and associated with the divine breath, it is of the feminine polarity, inner Isis, but also soul. As it is also said to be associated with *who*, I ascribe to it the lung aspect. It is the black dot of the Taijitu. Eve, mother of the living, as Isis, comes from this trait of God.

The third letter, Vav (ו), has a numerical value of 6, and as I see it is symbolized by the seal of Solomon, creation through union, the link that binds, where the two principles communicate, join and henceforward can express Creation. It is associated with the picture of a human, who was created on the sixth day in the image of the Creator. It is the result of the interaction of the two previous letters (1 + 5 = 6) thus of the two polarities. Its pictogram also refers to the silver hooks, described in the Torah, that are to hold the curtains enclosing the tabernacle (Exod. 27:9–10) thus ensuring God's presence on earth. [307]

It appears that an oversized Vav marks the center of the Torah (Lev. 11:42). Not surprisingly to us, the word in which this occurs is "gachon", meaning "belly". Indeed, we previously linked Leviticus to the Heart phase and explained that, in Chinese tradition, the area of the celiac plexus or solar plexus, the abdomen, is where the feminine and masculine polarities merge. Here the Master of the Heart/Triple warmer is active. This is associated with the white drop like shape and corresponds to the Heart and to Horus.

The last letter, "H" (He (ה)) is a repetition of the second letter, a perfect mirror, the result. The first "H" was thus the principle in its potential while the second H — as Jewish tradition teaches us— will feminize the whole concept as it is used at the end of this ultimate word. The second H is the black drop like shape of the Taijitu and corresponds to the "Kidney", and to Nephthys, receptive, and expressive of the two polarities in the universe on which humans are given control.

Finally the phrase *is here* refers to space (here) and time (is), to manifestation of the previous elements, and notably here on earth, where we humans have free reign, and thus to the pancreas aspect and its archetype, Seth. It is the line, the force, which separates the two polarities and borders the universe (diagram 2). It is the control of our inner Seth and of our earthly nature that will be expressed the reality of "YHWH is here."

Through our daily rituals we carve the implicate order on our brain and cells, allowing for the sacred to vitalize our lives and to survive in them. This in turn will allow humanity to reach its goal, which is to become an optimal image of God.

Seth is the divisive force required for the manifestation of this order in this dimension. We have to realize *is here* in our personal temple. It is not and never was limited to a country or geographical area, but to Consciousness and intimate knowledge of the presence of heaven

Isis Code

in us, in our bodies, in the environment, and in others. The false notion that God is linked to a limited geographical position (space, Reptilian brain) or to a "chosen" group because their ancestors were at the right place (space, Reptilian brain) at the right time thousands of years ago (time, Mammalian brain) issues from the era when the microcosm was used as a mirror to explain the macrocosm and when we were still of the previous phases. God is universal. We are all Chosen People, in the sense that we all have the responsibility of expressing God's true nature. As it says in the Koran, we are only superior to others through our proper understanding and manifestation of divine life.

We thus see again the expression of the five phases. These can also be found in the division of the five stages of the soul, as per Hasidic Jewish tradition. It is said that the soul was blown into man, not created. It already existed within the One, as Plato felt. This breath breathed into man thus comes from the creator and is linked to the feminine polarity (breath, lung). The five stages of the soul are Nefesh, Ruah, Neshamah, Chayah, and Yechidah. Nefesh, in contact with the body, is the instinctive part and is then associated with the Reptilian brain (kidney aspect). Ruah, the emotional part, is associated with the Mammalian brain (liver aspect). Neshamah, the mind, is said to enter the believer around twenty years of age, and is associated with the brain of information regulation (heart aspect). Chayah, the ladder to transcendence, is associated with the Analytical brain. In Egyptian mythology it is linked to Seth, also given the name *divine ladder* in ancient Egypt, which could be associated to Jacob's ladder. This is normal since both polarities can be found in it. When the pancreas phase, the rational self, is mature and healthy, it *serves* the common good in all the other aspects, including those of the feminine polarity. It therefore becomes a ladder back to the infinite and eternal. This aspect is seldom reached during human incarnation, because we tend to dedicate our personal selves to the service of the physical and emotional sides of our personalities. Otherwise, we would know "how God thinks," to refer to Einstein's expressed desire.

Seth means, among other things, pillar of stability, in the same way the name Peter means rock. When Jesus says, "You are Peter, and on this rock I will build my Church," (Matt. 16:18, NIV), he was referring to the masculine polarity. The god Seth was represented as a donkey and we will remember Jesus riding a donkey on Palm Sunday. The significance is most profound.

Then who represents the feminine polarity? This function is symbolized by the apostle John, the beloved.

Finally, the fifth stage, Yechidah, is the sense of Oneness, associated with our social Universal brain (lung aspect). It is linked with God's desire (aspiration) and culminates in a biosynarchy, a synergy of these, creating Consciousness and merging with God (evolved heart aspect).

> There was something complete and nebulous
> Which existed before Heaven and Earth,
> Silent, invisible,
> Unchanging, standing as One,
> Unceasing, ever-revolving,
> Able to be the Mother of the World.
> I do not know its name and I call it Tao.
> Tao Te Ching, 26

Another observation we can make is of the word Shen. The word exists both in the ancient Egyptian and Chinese traditions. In the Chinese tradition, the Shen sign bears a line that is the dragon, the ancestor that is described as eating its tail[308]. It gave rise to the Taijitu symbol. It is the ouroboros of nature, the eternal cycle. In the Egyptian tradition, in sixteenth century BCE, Shen was also described as the snake eating its tail. My interpretation of the Taijitu symbol is that it uses the dragon as the divider of the two main forces. It uses the pancreas aspects and the action of Seth in the world. In a manifested world, masculine and feminine polarities must be kept separate. Otherwise there would be no manifestation. This is why in the legend, Seth must not be killed. This is why he complains that he is the one who masters the great snake that could have destroyed the universe, had it not been for him, Seth the Great! This is why the snake was also used in the Genesis story. It is the rational mind: I eat the apple, thus I can become like God. This brought materialization. There is no sin in wanting to be as perfect as God; it is exactly what is asked of us. Eve already was conscious that she was not God; this is why she ate the apple. It is not a *fall* as much as a necessary manifestation into a denser world in order

Isis Code

to enhance Consciousness in it. The feminine polarity incorporated (ate the fruit of Consciousness of good and bad) the elements that would allow us to become like gods (nutritive aspect of the pancreas phase), and has in fact allowed evolution of a denser reality. Without Eve eating the apple, men would not exist to say how inferior women are.... Life as we know it is therefore the interaction of masculine and feminine, of Shen and chi in a form. Life simply is not complete without incarnation and is only waiting for the required conditions to allow its expression on the level of this denser reality, which is a minute part of the total reality.

Some of the Chinese texts differ on or remain silent about what Dr. DeBavelaere refers to as the middle level of the Master of the Heart. In the text, this middle part, where the Master of the Heart most directly intervenes, is at the heart level, while the lower part is below the navel. Since the lower part of the Master of the Heart is a feminine feature evidently linked to the uterus, in some texts triple warmer and Master of the Heart are described as one and the same (Ling Shu, chapter 10). In most texts, however, the Master of the Heart carries functions attributed to the Shen, while the triple warmer alchemically transforms the chi. Since Shen is more subtle (of heaven) than chi (of earth), we might consider that the Master of the Heart is also linked to the Chinese concept of "small heart," in some texts called ming men. The ming men is situated in front of the L1 and L2 vertebrae, exactly where we would medically perform a vagus nerve block in the case of problems with liver, pancreas, and gallbladder. It is also the site of the solar plexus. Chinese texts state that "without the fire of ming men, no organ could function, nor exist."[309] Since the Master of the Heart controls the life in the organs, it has a definite role in ming men. Therefore, the lower level of the Master of the Heart encloses the whole of the abdomen.

On the underside of the arm, we find the meridians of lung, pericardium, and heart. On the outer side we find the triple warmer, large intestine and small intestine, more in charge of the known classical digestive system. We understand here the importance of the ritual washing of forearms in the Muslim and other traditions.

The ancient Chinese manual of acupuncture refers to three different "envelopes" of its meridian feature. They are the envelope of the heart, otherwise named Master of the Heart; the envelope of the throat, also named upper Master of the Heart; and finally the envelope of the uterus, or lower Master of the Heart.

As previously stated, fertilization occurs in one of the Fallopian tubes, in the ampulla of the uterine tube, which has three layers. Its most external layer is called serosa (lower Master of the Heart). We find it mainly in three other areas of the body: the external membrane of the heart (pericardium, Master of the Heart); surrounding the lungs, where it has an active function in breathing (middle Master of the Heart); and in the abdomen cavity covering most organs—the gut, for example—and in contact with the extracellular matrix (lower Master of the Heart) (diagram 27). Interestingly, these are all attached to the Master of the Heart and thus to the feminine polarity.

To further confirm this, extracellular matrix cells have been found to cause regrowth and the healing of tissue. In human fetuses, for example, the extracellular matrix works with stem cells to grow and regrow all parts of the human body damaged in the womb. For a long while, scientists wrongly assumed that the matrix stopped functioning after full development. It is now being researched as a device for tissue regeneration in humans. The extracellular matrix is ever changing and influenced by many elements in our environment, whether these are physical, emotional, or archetypal (symbolic).

These are doors used by the subtle level to influence and guide the physical self.

The human brain adapts to environmental changes even in adulthood. MIT neuroscientists have now found that changes happen there at an unexpected speed. Their findings imply that somehow, underlying its plasticity, the brain has a network of silent connections. Could it be the Master of the Heart and glial cells? Referring to the findings of a paper being published in the July 15, 2009 Journal of Neuroscience, Nancy Kanwisher of the McGovern Institute for Brain Research at MIT said,

"We found these referred sensations in the visual cortex, too. ... When we temporarily deprived part of the visual cortex from receiving input, subjects reported seeing squares distorted as rectangles. We were surprised to find these referred visual sensations happening as

Isis Code

fast as we could measure, within two seconds. But these distortions happened too quickly to result from structural changes in the cortex. So we think the connections were already there but were silent, and that the brain is constantly recalibrating the connections through short-term plasticity mechanisms." Fascinating.

We can thus gather here elements important to the Master of the Heart's function.

1. There are three main cavities in the human body, and each of these is lined by a complex serosal membrane. Cavities already confer a receptive quality to these areas. Those three serosal membranes are the pleura, (lung), pericardium (heart), and peritoneum (mouth of the Master of the Heart).

2. Our physical structure is mainly composed of liquid, which has a remarkable capacity for absorbing radiation, as shown by the use of water in nuclear facilities. It also functions well in memorizing information. As Dr. Benveniste has demonstrated, water retains memory. This can be shown also by the following quote:

    "The structure of liquid water is being continuously changed from the moment of its forming. The character of such changes depends on the physical and chemical characteristics of the surrounding medium (Klassen, 1982). Even by keeping the distilled water in constant medium its structure is being changed depending on its "aging" (Stepanian et al., 1990). Therefore the structure of the water could be considered a guardian of a "memory" for the previous effects of various environmental factors and this property is the main barrier for reproducing the experimental results on studying the effects of weak signals on physicochemical properties of water."[310]

    Other than the lymphatic system and the extracellular matrix systems that surrounds all cells, we have the cardiovascular system, the ever-moving cerebrospinal fluid (seen on MRIs), and ventricles, with each lateral ventricle extending into the frontal, occipital, and temporal lobes via the frontal (anterior), occipital

(posterior), and temporal (inferior) horns, respectively.
In the organism, the liquid medium is ever present.

3. The pineal gland—also called the epiphysis—looks like a miniature pine cone and is found in the middle of the brain beneath the two brain halves, surrounded by the ventricles and under the roof of the corpus callosum (crossbeam connecting the two brain halves). This active organ has, together with the pituitary gland, the next highest blood circulation after the kidneys. It is not protected by the blood-brain barrier. Therefore this gland is receptive to any substance or energy entering the bloodstream and sinuses. The pineal gland has a content of 3140 mg of serotonin per gram of tissue. It is unmistakably the richest site of serotonin in the brain.

4. In the lateral ventricles of the embryo, successive generation of neurons gives rise to the six-layered structure of the neocortex, constructed from the inside out during development.

5. Nutriments are absorbed thanks to the classical digestive system associated to the triple warmer, but the importance of the life aspect of foodstuff has rarely been taken into consideration. The closest we have been is to consider foods' vitamin content. Nonetheless, we know that a diet of ration foods cannot sustain life in the long run. Not only do members of the armed services fail to consume the full portions of their rations, often trading or discarding portions of them,[311] but more studies are showing that many service members do not meet today's standards of daily micronutrients absorption. Food contains energy, energy beyond the caloric energy. The closer to the source, the more alive. Colors, presentation, origin all have their importance and effects on the MH/TW entity for nutrients' absorption.

6. Water contains energy. As we've seen from studies by Dr. Emoto, Dr. Benveniste, S. G. Stepanian et al. and Sinerik Ayrapetyan and others, not all waters are the same. A slight change in the property of water can change the metabolic activity of cells. The energy contained in water is somehow passed on to the cells. The mechanism is not well understood

yet. I consider this is the work of the MH/TW. Different types of energy are passed onto the Master of the Heart, to the triple warmer, and ultimately dispatched to all cells.

7. Bundles of collagen fibers are a major component of the extracellular matrix. They support most tissues and give cells structure from the outside. The synthesis of collagen requires a high level of atmospheric oxygen. Complex animals may not have been able to evolve until the atmosphere carried enough oxygen for collagen synthesis (lung aspect).

8. The heart is involved in the assimilation of radiation, as shown by the epidemic of infarcts following solar spot occurrences (see the works of Nicholas and Romanski). The Master of the Heart is specifically linked to the blood and to the assimilation of its energy, which could explain the sacredness of blood in the Jewish tradition.

9. Electromagnetic fields (EMF; low and high frequencies), static magnetic field (SMF), and microwave (MV) induce changes in water properties. "Although the penetration of Microwave into the skin is less than one millimeter, the MW signal is able to transmit to the spinal cord and subsequently to various regions of the brain."[312]

10. In a research we found the following statement: "*It has been stated that the neuronal apparatus of the vertebrate and human pericardium is represented by plexus of myelinated and unmyelinated fibers, nerve cells, and receptors.* [This is part of the support for the Master of the Heart, or is it the other way around?] *The plexus contain both cholinergic and adrenergic neuronal structures. Pericardial neuronal apparatus reaches its highest development in vertebrates, especially in man. The main source of the afferent and vegetative innervation in the pericardium is inferior and superior thoracic cerebrospinal nodes,* **vagus nerves**, *and branches from the nodes of the sympathetic trunk. Neuronal fibers composing diaphragm and parasternal nerves, and those of the celiac plexus [solar plexus, lower Master of the Heart] reach the pericardium too.*"[313] (My emphasis)

11. Link between MH/TW and the rest of the body: "there is a system for the rapid transmission of information between distant parts of the gut. There are three prevertebral ganglia—celiac, superior mesenteric, and inferior mesenteric—connected to the enteric nervous system (ENS; figure 1.5). These ganglia are connected by two intermesenteric nerves. A stimulus in one part of the bowel can travel by an afferent axon to one in the ganglia, along an intermesenteric nerve to the next ganglion, and thence back to the ENS. In the axon of an efferent nerve, only three synaptic connections are required for transmission. These pathways are essential for the patterns of motor activity that involve the whole of the stomach and small intestine.[314]

12. Finally, the neural connections between the gut and the brain deserve mention. It has long been known that the function of the vagus nerve, the tenth cranial nerve, is paramount in this connection. Vagus nerve stimulation is a form of treatment that involves sending electrical stimulation to the vagus (or vagal) nerves (lung is associated to the gut in energetics). The vagus nerve is the paracranial nerves par excellence for the Master of the Heart's function.

The Vagus Nerve

The body has twelve pairs of nerves that originate in the brain and connect to other parts of the body. These are called the cranial nerves. Each pair of nerves is linked to a different part of the body, and each does different things. It's a little like the twelve branches of the meridians, but in reverse. The vagus nerves are the tenth pair of cranial nerves and are unique in their trajectory, as they do not go through the spinal cord. This is like the way MH/TW differs from the other meridians. In Latin *vagus* means wanderer, and refers to the fact that these nerves touch many different parts of the body. They relay messages between the brain and various organs, including

1. the throat (where they connect to the muscles that control swallowing),

2. the heart (as part of the autonomic nervous system that controls the way the heart works), and

3. all the organs in the chest and abdomen except the adrenal glands.

Sounds familiar? It follows the same trajectory as the Master of the Heart.

Stimulation of acupuncture points on the extremities of the fingers results in stimulation of the vagus nerve. Such acupuncture is used as an antiepileptic application. The nucleus of the solitary tract, which is a primary site where vagal afferents terminate, is also the site for afferent pathways stimulated in facial, scalp, and auricular acupuncture (via trigeminal, cervical-spinal, and glossopharyngeal nerves). The solitary tract and nucleus are structures in the brain stem that carry and receive visceral sensation and taste from the facial (VII), glossopharyngeal (IX), and vagus (X) cranial nerves.

"The nucleus of the solitary tract (NTS) contains oxytocin receptors and receives oxytocinergic input from the paraventricular nucleus of the hypothalamus. Recent studies have expanded on the modulatory role for oxytocin projections onto the NTS (Higa et al., 2002). Because the NTS is an important relay center for incoming central and peripheral visceral sensory inputs, second-order NTS neurons are ideally located to integrate peripheral inputs (i.e., cardiovascular reflexes) with centrally mediated responses, such as stress."[315] The quality of energy received will trigger oxytocin secretion, this magic-carpet hormone. We will remember that the heart also outputs oxytocin.

"Finally, Peters et al. identified NTS neurons that were responsive to electrical stimulation of the vagus nerve and activation of cardiopulmonary afferents.[316] They further measured oxytocin immunoreactivity of nearby axons. Among the doubly responsive NTS neurons, half appeared to receive inputs from oxytocin-immunoreactive terminals (J. H. Peters et al., 2008)."[317] Most of the nerve fibers of the vagus nerve are afferent (sensory), meaning they relay messages to the brain from the body, mainly from the head, neck, thorax, and abdomen. In this way, the brain receives messages from different parts of the body and these messages are, in turn, relayed to different areas within the brain.

This brings us to an important point. Women's sexuality has been studied in its masculine aspect, the coitus, but not in its feminine aspect, which leads to and includes maternity. This aspect is intimately associated with the Master of the Heart and thus to the vagus nerve, and is far more important than its masculine counterpart. Women's sexuality is all encompassing, with the genitals only one small aspect of it. Once a woman has passed her maternity period, her sexuality would normally learn to embrace not only the limited space of her body, but nature itself. How can a focused masculine polarity experience this, let alone understand it?

The vagus nerve is also implicated in taste recognition, a subtle aspect of nutrition, from gustatory and sensitive receptors to the medulla (kidney aspect), to the hypothalamus (liver aspect), and from there to the parietal lobe (heart aspect). The vagus nerve is the carrier of information from the three levels of the MH/TW. The celiac or solar plexus is formed in part by the greater and lesser splanchnic nerves of both sides, and also parts of the right vagus nerve, which innervates the foregut as well as the heart. The right vagus nerve is part of the celiac (solar) plexus.

Wikipedia has this to share with us regarding the vagus nerve: "The vagus is also called the pneumogastric nerve, since it innervates both the lungs and the stomach [as the MH/TW does]. It supplies motor parasympathetic fibers to all the organs except the suprarenal (adrenal) glands (kidney aspect), from the neck down to the second segment of the transverse colon. The vagus also controls a few skeletal muscles and is responsible for such varied tasks as heart rate, gastrointestinal peristalsis, and sweating. Other tasks include quite a few muscle movements in the mouth, including speech (via the recurrent laryngeal nerve) and keeping the larynx open for breathing (via action of the posterior cricoarytenoid muscle, the only abductor of the vocal folds). It also has some afferent fibers that innervate the inner portion of the outer ear, which explains why it is such an erogenous area for women, via the Alderman's nerve and part of the meninges. Breathing something toxic will make us cough because of the vagus nerve [hence its link to the lung aspect]."

Isis Code

Parasympathetic innervation of the heart is controlled by the vagus nerve, which lowers the heart rate. If it overcompensates, which might happen during a stressful emotional event, it may predispose the heart to atrioventricular (AV) blocks, or cause vasovagal syncope because of a sudden drop in blood pressure and heart rate. This affects young children and women more than other groups because of the link with the feminine polarity of the right hemisphere. This is the association the vagus nerve has with the Master of the Heart. It can also lead to temporary loss of bladder control under moments of extreme fear. This is its link with the triple warmer.

Right hemispheric autonomic inputs to the heart are associated with greater cardiac control of the heart's rhythm. It seems there are both direct and indirect pathways linking the medial frontal cortex to the autonomic motor circuits responsible for both the sympathoexcitatory and parasympathoinhibitory effects on the heart (Balaban and Thayer, 2001; Barbas et al., 2003; Barbas and Zikopoulos, 2007; Grace and Rosenkranz, 2002; Rempel-Clower, 2007; Resstel and Correa, 2006; Saha, 2005; Saha et al., 2000; Shekhar et al., 2003; Spyer, 1994; Ter Horst and Postema, 1997; Thayer and Lane, 2000; Wong et al., 2007).

Medial prefrontal activity is also associated with heart rate variability (Lane et al.).This confirms our association between the vagus nerve and the medial prefrontal cortex. These findings are consistent with a general inhibitory role for the prefrontal cortex via the vagus, as suggested by Ter Horst (1999). Observations have been made to the effect that hypoactivity of the prefrontal area, as well as a consequent lack of inhibitory neural processes, were found in cases of anxiety, depression, post-traumatic stress disorder, and schizophrenia. This was reflected through symptoms such as failure to recognize danger signals including an increased negativity bias; deficits in working memory and executive function; and poor affective information processing and regulation (Thayer and Friedman, 2004; Shook et al., 2007). As suggested also by Claude Bernard, it therefore appears that the modulation of cardiac activity by the cortex is mediated by the vagus nerve. As inferred by the LIFE biosystem, heart, vagus nerve, and medial prefrontal cortex are thus working as a unit.

In 2004, Dr. Barry Komisaruk and Dr. Beverly Whipple of Rutgers University, New Jersey, conducted a study on women with severed spinal cords. They discovered that these women could nevertheless feel stimulation of their cervixes and even reach orgasm. The question was: How could their brains receive information from the hypogastric or pelvic nerves? MRI scans of the women's brains showed that the region corresponding to signals from the vagus nerve was active. The vagus nerve is the only cranial nerve that bypasses the spinal cord. Because of this bypass, the women were still able to feel cervical stimulation. This is another proof that women's sexuality is different from men's and is expanded due to the Master of the Heart. It also gives credence to the controversial observation of two types of climaxes available to women: masculine (clitoris) and feminine (cervix).

Patients who have a problem with their vagus nerves complain of hoarse voices[318], difficulty in swallowing (dysphagia), and choking when drinking fluid. There is also loss of the gag reflex.

Acoustical energy hits our eardrums, and the brain translates this information into sounds we can interpret consciously or not. Light energy floods our eyes, and our brain translates this into an image we can interpret. Many types of energy bring information to our bodies at all levels. Emotions can be triggered by subtle cues that influence the Master of the Heart. The celiac plexus is the level where one can sense these. Shen and chi use the MH/TW system to touch the more condensed layers of extracellular fluid, blood, intracellular cytoplasm, cerebrospinal fluid, ventricle fluid, and all liquids, all the way down to the different crystal forms and collagen found throughout the body. It is similar to a field containing specific morphogenic elements. The aspect of the Master of the Heart in the vicinity or in the organism receives energies in affinity with this organism.

As I explained earlier, if the system is not open or in affinity with the information, then this particular subtle information will not be received. It is a natural immunity. In the same way, if the organism has developed in opposition to the universal patterns of the Master of the Heart, this will be respected. This is human freedom, in which we can choose the filters (the foreskin of Jewish tradition) on our Human, Analytical, and Universal brains.

This aspect works hand in hand with the insula of the lung aspect as a shield to alien feelings and ideas. If a type of sensed human energy is not in harmony with this universal system or with the receptor, the structures receiving the information will contract or simply not receive it. On the contrary, if the information is received, a feeling of expansion or contraction will occur, which will rapidly transmit to the whole of the organism.

We can do this little experiment: think about someone who hates you. Worse, someone you love who suddenly ignores you. You can summon this feeling the same way you would with a song that has an emotional impact on you, except it is not your brain "hearing," it is your Master of the Heart. Your celiac plexus hears it, and it contracts until nausea or a lump in the throat is felt through the path of the vagus nerve.

Now, think of someone who truly loves you and whom you truly love. Often this will trigger a deep breath, proof that the diaphragm has relaxed. With the vagus nerve properly functioning, you might feel an expansion at the chest/heart level. The heart receives the information, you feel, and suddenly you see the world differently. A smile comes naturally to your lips. Your whole face and body relaxes and opens up. The feminine polarity is functional.

All the different types of energies are stimulated in the different meridians. The absorption of nutritive elements is enhanced. Can we deny emotions are information?

This is how the feminine polarity is affected. If you eat conventionally grown foods bought at the store and cooked by someone who is angry, versus someone who eats fresh organic foods grown by a friend and cooked by a caring person, the biochemical differences between what you and the other person consumes might appear inconsequential. On the other hand, the biophysical differences are great, because the information your body receives is totally different in view of the immense qualitative differences. When it comes to health and wellness, the entire organism is affected on every level by the biophysical differences between our choices in all aspects of our lives.

To take our example of being in the presence of someone who hates you, if the heart is attacked from the inside or outside, it will modulate itself to some extent against the injury. This adaptive capacity is activated by the cell-cell signal transduction, and leads to remodeling of myocardial cells, extracellular matrix, and vascular endothelium. There will be an obvious collagen deposition in atrial interstitial substance, which will be accompanied by myocardial hypertrophy, fragmentation, and inflammation. What is more, the vagus nerve fibers will modulate the atrial substrate metabolism, although their relationship is still unclear for conventional medicine. The cardiac vagus nerve mediates the atrial electrical remodeling, which facilitates the provocation and maintenance of atrial fibrillation events. This effect is attenuated by denervation in the case of heart transplants, although it is not excluded.[319] In short, the regulation of physiological systems essential to generating health or disease have been linked to vagal function and heart rate variation (HRV). Thayer and Sternberg (2006) found evidence for the role of vagally mediated HRV in the regulation of physiological, affective, and cognitive processes. Indeed, they showed that a low HRV is a risk factor for pathophysiology and psychopathology[320]

Effects of Melatonin, Serotonin

To exemplify this on a more material level, I noted earlier that the atrial natriuretic peptide (ANP) is secreted in the upper chamber of the heart in response to certain signals, providing a homeostatic control of body water, sodium, potassium, and adipose tissue (fat). ANP secretion also affects the kidneys by increasing glomerular filtration rate (GFR) and filtration fraction, which produces natriuresis (increased sodium excretion) and diuresis (increased fluid excretion). Also, *serotonin turnover*, as estimated by the measurement of its metabolite (5-hydroxyindoleacetic acid) concentration, increases during the perfusion of brain natriuretic peptide into the ventrolateral medulla.[321] The medulla (part of the physical self), located in the brain stem above the spinal cord, is the primary site in the brain for regulating sympathetic and parasympathetic (vagal) outflow to the heart and blood vessels.

As a precursor to melatonin, serotonin is derived from the amino acid tryptophan, the largest of the amino acids, and is known to affect the bingeing behavior in bulimics. Tryptophan is obtained by digestion in the gut and transported in the blood plasma to the brain, where it is converted to serotonin. Thus, the health of the gut, associated to the lung function, is fundamental to this process. Serotonin is partially responsible for the regulation of appetite—creating a sense of satiation—and regulates emotions (lung controls liver) and judgment (which controls pancreas). Thus, the binge behavior of bulimics might also be a response to low serotonin levels in the brain caused by a defective or malnourished Master of the Heart. Under the lead of Dr. Walter Kaye, a research team at the University of Pittsburgh found that patients successfully treated for bulimia still presented a very low serotonin level. Their other brain chemicals, such as dopamine and norepinephrine, were normal in comparison to individuals with no history of eating disorders. The artificial control of bulimia with Prozac, a medication typically used for depression and which acts to reuptake serotonin in the brain, is additional evidence of the importance of this brain chemical through a functional feminine polarity.

Within the pineal gland, serotonin is transformed into melatonin. Synthesis and secretion of melatonin is dramatically reduced by light exposure on the eyes as well as the skin. For this reason, we have associated the pineal gland to the lung phase. Since light is also received by the skin, the pineal gland can remain somewhat functional in some blind people. The fundamental pattern observed is that serum concentrations of melatonin are low during the daylight hours, and increase to a peak during the dark, if it is dark enough. It is the reverse for serotonin.

Two melatonin receptors have been identified in mammals (designated Mel 1s and Mel 1b). They are differently expressed in different tissues and probably participate in implementing opposing biologic effects. These are G protein-coupled cell surface receptors (GPCRs). Approximately 150 of those found in the human genome have unknown functions.[322] "The highest density of receptors has been found in the suprachiasmatic nucleus of the hypothalamus, the anterior pituitary (predominantly pars tuberalis), and the retina. Receptors are also found in several other areas of the brain."[323]

Melatonin has important effects in integrating photoperiod and affecting circadian rhythms. Consequently, it influences reproduction, sleep-wake cycles, and other phenomena involving circadian rhythms.[324]

Serotonin directly excites pyramidal neurons in the cerebral cortex. Although various areas of the cortex differ in regard to organization, they share in common a preponderance of pyramidal cells, ubiquitous to its gray matter. They are the largest and are more numerous than any other neocortical neuron (Peters and Jones, 1984), accounting for up to three-quarters of all neocortical cells. They also serve as both local-circuit and long-distance neurons and generally receive two types of synaptic contacts, referred to as Gray types I and II (Peters & Jones, 1984). Almost all pyramidal cells are excitatory and use glutamate and aspartic acid as transmitters (Tsmoto, 1990).[325] In the human brain they are arranged in layers in the cortex of the two cerebrums. In the prefrontal cortex they are implicated in cognition. In mammals, their complexity increases from posterior to anterior brain regions, following the phase development of the LIFE biosystem. They are the primary excitation units of the mammalian *medial prefrontal cortex* and of the corticospinal tract. We know that the medial prefrontal cortex of the idealistic self and heart phase can suppress the action of stress on the HPA axis. The pyramidal cells act as electro-crystal cells bathed in the extracellular matrix, and seem to operate in the fashion of a liquid crystal oscillator in response to different light commands, or light impulses. In turn, they change the *orientation* of every molecule and atom within the body. Bio-gravitational encoded switches present in the brain allow the cellular matrix to release ions that will induce currents to the surrounding coiled dendrites. Electron impulses from a neuron, on reaching the dendrite coil of the neighboring cell, generate a micro amperage magnetic field, activating the ultrathin crystal, or liquid crystal, in the pyramidal cell. "On flexing, this crystal becomes a piezoelectric oscillator, producing a circular polarized light pulse that travels throughout the body, or travels as a transverse photonic bundle of energy."[326] The extracellular matrix acts therefore as an envelope to the meridians and a receptacle to the electromagnetic energies.

Most humans enjoy being cuddled. Now scientists know in part why. Humans are hardwired with a separate network of nerves that trigger emotional, hormonal, and behavioral responses when we are softly touched, as shown in a new study from the Sahlgrenska Academy at the University of Gothenburg in Sweden. The discovery may explain why touching the skin can relieve pain. This network of slow-conducting nerves, called the C-tactile (CT) network, is separate from the fast-conducting nerves that signal the brain about heat, cold, pressure, and pain. Why would that be? The Swedish scientists learned about the CT network's function through a patient who had lost the use of her primary network of nerves. When scientists stroked her arm, she felt no touch or vibration, but she reported a pleasant, rewarding, and emotional effect. Following this, other research has demonstrated that in responding to light touch, the impulses travel along this network at just one meter per second, compared to sixty meters per second for signals along the primary nervous network. Dr. Håkan Olausson, in charge of the study, believes the CT network can convey emotions or a sense of self.[327]

It is hypothesized that the CT network must be used for the unconscious consequences of touch. C fibers are found in the peripheral nerves of the somatic sensory system, on hairy skin. They are afferent fibers, conveying input signals from the periphery to the central nervous system. Unlike most other fibers in the nervous system, they are unmyelinated. This lack of myelination is the cause of their slow-conduction velocity. They have biophysical, electrophysiological, neurobiological, and anatomical properties that drive the emotional somatic system in a delayed fashion. Functional magnetic resonance imaging (fMRI) analysis during CT stimulation showed it activates the insular region (lung aspect)[328]. I believe that this system is linked to the goose bumps we often feel when something is extremely pleasant and probably a constant link with invisible types of energy received from the environment.

It appears pain perceptions from myelinated nerves are less potent and have a less profound action than the CT network of unmyelinated fibers. Pain receptors (called nocireceptors) are "free nerve endings." That is, there is no extracellular matrix capsule or epithelial cell receptor coupled to the neuron. This is a path of study to follow, as it might show the link between CT network and the Master of the Heart.

The ages of fourteen to twenty-one are marked by searching and outlining one's identity. Drive, ambition, and personality of our masculine polarity are not our whole identity. They are not Consciousness but part of self-awareness. They are tools we use and are colored and submitted to the time and age we live in. At puberty the hormones associated to one's gender are almighty, and the young man often evaluates his level of manhood by the number of females willing to receive his young fluids. The girl, on her side, is not so much on the lookout for quantity as for *the one* who will be the perfect screen onto which she will project her ideal of herself—which is determined by how he looks at her—and then of her loved one, since he expresses her heart aspect. Her polarities vibrate with his because she is expressive, masculine on the emotional level. At this time, questions on the nature of Consciousness arise. This is normal, since the Reptilian brain controls the idealistic one (diagram 5). The social serves as a mirror in the search of this identity. But before the Self can be consciously reckoned with, the self has to be aware, visualized, and delimited.

This period is therefore one in which the individual will test himself in many ways in order to find his true identity, often with a daredevil approach, since the Analytical brain (frontal lobes) is not totally mature. Consequently also, in this phase a great love story can develop. This happens because the self can see the echo of the Self in individuals, mostly in those of the complementary gender, as *the self can manifest only through the mirroring of an object.* The brain develops through its interactions with the surrounding material world. The lover will see the mirror of his Self, of his feminine polarity, reflected in the self of the other. When those relationships abruptly end, the victim feels crushed and destroyed as the self seems to have been negated. The identity is lost, most acutely in men. Further, to manifest itself the Self requires the presence of a self. The young man, for whom the door to individuality is not already shut, will vibrate to one woman when the projection of his own feminine polarity seems to match hers. Because of her he will want to *be* a better man. Eventually he will realize that her ego is not the Self he projected. The young woman will vibrate to a young man who appears in harmony with her masculine polarity; because of him she will want to *do* things.

During this period, we are in search of Self, and we believe we are in love with the other. In a sense we recognize the other in ourselves and ourselves in the other. This is the beginning of love. For a woman it is even more complex, for she will be attracted to those corresponding

Isis Code

to her masculine polarity, and hence affected by the quality of the familial and social masculine polarity. This can be found in the physical aspect, the mirror neurons that are part of the Human brain. This system fires when we observe another person performing an action. For example, if we watch someone eating an ice cream, mirror neurons will fire, giving us an idea about how it feels and taste to eat *that* ice cream. Neuroscientists have also hypothesized that mirror neurons are responsible for our capacity to empathize with others, and to understand how another person might be thinking or feeling. A series of experiments conducted by Yawei Cheng[329] discovered a gender difference in this system, with women reacting more strongly than men. The mirror neurons were found in the supplementary motor area and *medial temporal cortex*, as well as in the *parietal lobes*. Reporting in the journal *Current Biology* in April 2010, Dr. Itzhak Fried, a UCLA professor of neurosurgery and of psychiatry and bio-behavioral sciences, Roy Mukamel, a postdoctoral fellow in Fried's lab, and their colleagues have for the first time made a direct recording of mirror neurons in the human brain. The neurons that were responding were located in the medial frontal cortex and medial temporal cortex.[330] These concern the idealistic and social selves.

The young woman's search for the man who will correspond most closely to the heart aspect of the Self happens during this period. This is why during that period, relationships are lived as if they are life or death situations, and might appear overdramatized. Nuns who dedicate themselves to be "brides of Christ" shorten this period of attraction to the masculine polarity and try to link themselves directly to the Self. It is not an easy task, especially if the desire to have children arises. Monks could do the same with the Divine Mother, but it is not as culturally obvious since the present image of women, adopted by society and tradition, a kidney class archetype, is interfering with that of the Divine Mother, a lung archetype.

This is the age of Romeo and Juliet and of all the great love stories. The self realizes that what it really wants is expansion, and only the feminine polarity allows this. As in the movie *Man of La Mancha*, which follows some of the adventures of Don Quixote, to project disinterested love onto someone who "doesn't deserve it," as he did with Dulcinea, can be the lover's downfall and a grotesque farce. The other person might also be the loved one's redeemer and give the lover a chance to manifest the Self. As the Self is One, to project one's facet of it into another will sometimes make this aspect vibrate and shake from it the crust of personality. Some conditions need to be met for this to happen. Our link to God is triggered by this search for completion and wholeness.

The sexual revolution we went through in the '60s and '70s was intended to remove obstacles that kept people from expressing their full potential. Instead, because the true nature of Man was not taken into account, it enhanced the masculine polarity, the personality. It slowly but surely eroded the appreciation we could feel for the sacredness of a relationship. Women were still exposed to abuse and abandonment, maybe even more so. Don Quixote symbolizes an effort to reinstate chivalry in a world devoid of sacredness. Of course he looks like a madman. The object of his love is known to all to be an "easy woman," thus devoid of virtues. Upon encountering Aldonza, Don Quixote sings:

I have dreamed thee too long,

> Never seen thee or touched thee.
> But known thee with all of my heart.
> Half a prayer, half a song,
> Thou has always been with me,
> Though we have always been apart.
> Dulcinea ... Dulcinea ... [331]

Don Quixote repeatedly speaks blessings into Aldonza's life, calling her Dulcinea (which means sweetness).

Despite her rejection of his love, Don Quixote keeps showering Aldonza with love, patience, and gentleness. Again and again, Don Quixote reaffirms that a relationship is far more than just physical: it is a spiritual reality that is experienced in the flesh. That is why Quixote, the Man of La Mancha, also sings in the same song:

> I see heaven when I see thee, Dulcinea,
> And thy name is like a prayer
> An angel whispers ... Dulcinea ... Dulcinea! ...
> I have sought thee, sung thee,
> Dreamed thee, Dulcinea! [332]

Because Aldonza has been lied to and used by other men, it seems impossible she could trust any man. Is he her knight in shining armor? He looks more like an old, crazy fool. The fact that he seems to love her makes her extremely suspicious. Her faint hope nevertheless is that what he tells her is indeed the revelation of her true identity. At one point in the movie, Quixote's relatives try to take him away from Aldonza/Dulcinea, claiming that he is mad. The priest pauses and says: "One might say that Jesus was mad, or St. Francis." In one sense, Don Quixote functions as a Self figure, one who gives his life for love, even though, similar to Jesus, he is dismissed as insane by his own family (Mark 3:21). The individuality does not act through the Analytical brain but through the Human one. Thus it might appear insane not to follow a rational protocol. In this, Don Quixote follows in the traces of Dante, of the Cathar faith, and in those of Neo-Platonist Léon l'Hébreu (≏1460–1521, *Dialogues d'Amour*) whom Cervantes admired so profoundly.

Dulcinea used the image reflected from her environment to define herself. Don Quixote is telling her she is individuality, showing her a direction, while she thinks of herself as a personality—and not a nice one at that.

Don Quixote appears stubborn in his positive evaluation of Dulcinea. In response, at one point she cynically retorts: "Your heart doesn't know much about women." Instead of giving up, Quixote gently responds: "Woman is the soul of man, the radiance that lights his way. Woman is glory." Dulcinea, deeply afraid that he will just use her and discard her like all the others, replies: "What do you want of me?" As a true knight, Quixote says: "I ask of my lady that I may be allowed to serve her, that I may hold her in my heart, that to her I may dedicate each victory and call upon her in defeat, and if at last I give my life, I give it in the sacred name of Dulcinea."

Gradually Dulcinea softens in the face of Don Quixote's gentleness and patience. She sings:

Can't you see what your gentle

> Insanities do to me?
> Rob me of anger and give me despair!
> Blows and abuse
> I can take and give back again,
> Tenderness I cannot bear![333]

Our personality can bear the familiar world of hell but does not know how to handle heaven when it is brought by the individuality. It can hardly believe in it and often feels threatened by it. Some women intuitively know this, but a woman may subconsciously hope that the love a man feels for her will succeed in doing just this: awaken both the lover and the loved one's Self. In the movie, during the night of the mirrors, the social self, represented by those mocking Don Quixote since they were not linked to the heart aspect, describe only the personality of Dulcinea. They can't see her individuality. Although they were rational in their scorn, they couldn't see what Don Quixote saw: a facet of the Self, an image of the Divine Mother. His individuality was vibrant and alive. Eventually he was able to make this aspect vibrate in her, as the prince awakens the princess, as Jesus transformed Mary Magdalene. What can seem a farce now is of a great intelligence in the future, as well as in the universality and immortality of Self and of Consciousness. The Self exists, although not in this divisive dimension, and one must strive to manifest it here. As Jesus said, "My kingdom is not of this world." (John 18:36) By seeing the Self in others, one awakens his own.

Curiously, the main function of the monarch, although few knew, was exactly that: to represent the Self and thus way awaken this aspect in others. Louis XIV was well advised when he emulated the sun. As the sun, he would awaken that same aspect in the people who saw him. Who advised him so? Was it intuitive? Traditionally, the sun is linked to the heart and considered of the same realm.

This is the "know Thyself" of the Delphi temple of ancient Greece. It is the beginning of the search for the identity of the inner god.

Lao Tzu, in his Tao Te Ching, says, "Knowing others is wisdom, knowing yourself is enlightenment."

The Self, once subjected to space and time, creates and uses a self to bring Consciousness to this world. God has to submit to his own laws. The self cannot wholly know the Self. Through us, God, or Consciousness, is manifested here. The Self has billions of eyes looking into this dimension; it globally brings Consciousness to it. However, my ego has to submit to the Self in order for my eye to eventually open and allow the expression of Consciousness. This is echoed in the Muslim name, which means to submit to Self (God). It requires the function of the feminine polarity. To illustrate this, let's say you wish to drink some water and you don't have a glass. If you use your hands to get water but your palms are facing down, you will not be able to drink. This is the masculine polarity. However, if your hands are open toward the sky, you can receive or cup out some water and quench your thirst. This is the feminine polarity.

"The Self itself is the world; the Self itself is 'I'; Self itself is God; all is Siva, the Self," says Ramana Maharshi. When he slept, it was almost impossible to wake him up. Maybe his Reptilian, Mammalian, and Analytical brains had little hold over him.

With the Reptilian brain, the self says, "I am this body," and limits itself to it. Only by becoming aware of the dissolution of bodies (death) do we start feeling that the body is merely a shell and a concretion of something more subtle. With the Mammalian brain, we start identifying with our emotions, which keep us alive in this world. We also realize that once all the pain, the anguish, and the joys of this world have been experienced, there is an unchanging identity who simply observes. With the Human brain of the idealistic self, we start experiencing the subjectivity of the world rather than feeling surrounded by inert objects. This observer has an intimate knowledge of these surroundings. The self begins to experience the observer as it feels the need for an ideal. The quest for true love starts. The rational self then uses the physical and emotional selves, often going against the feminine polarity of the idealistic and social selves, or will allow its manifestation. This is where man's freedom lies. This is the choice.

One result of the maturation in this phase is that the individual stops seeing his environment as a multiplication of lifeless objects with him at the center, but more as an harmonious interaction of living subjects, from earth to humans to stars. With the social self, the ego experiences "I am also the others." Then the heart aspect becomes mature and there are no more divisions, save for the illusions of the world of the personality. The individuality becomes conscious of its eternity—"I am." The self begins to realize that both heaven and mostly hell can be found during incarnation, and that in order to truly be, it has to serve the Self. When this happens, a mystical marriage is celebrated. It is the mature idealistic self who experiences "I and My Father are One" (John 10:30). A definite *you* and *he* appear only when an *I* does, and the ego (self) can exist only when there is a division between *you* and *he*. If we understand this, we may start experiencing the mind as a general vital force passing through the filters of our physical, emotional, and rational selves.

The question "Who am I?" should be taken first as an exercise of self-awareness. While doing this throughout the day, instead of focusing on any of the five phases, we fix our attention only on the I-Consciousness, as the singular, and as a witness to—thus independent from—these five phases. We could call this identification with God. When the nature of the *I* is sought and the personality is taken for what it is—that is, only a necessary medium—the division of *you* and *he* can be put to rest. What shines then is the One.

"You are unconditioned and changeless, formless and immovable, unfathomable awareness, imperturbable, so hold to nothing but consciousness." Ashtavakra Gita, [6] 1:17

"O Rama, this enquiry into the Self of the nature or 'Who Am I?' is the fire which burns up the seeds of the evil tree which is the mind." Yoga Vasista Sara, 1 :5

"Do not meditate – be!
Do not think that you are – be!
Don't think about being – you are!"

"Self-inquiry should not be regarded as a meditation practice that takes place at certain hours and in certain positions; it should continue throughout one's waking hours, irrespective of what one is doing."[334]

Sri Ramana Maharshi said:

> *You are awareness. Awareness is another name for you. Since you are awareness there is no need to attain or cultivate it. All that you have to do is to give up being aware of other thing, which is of the not-self. If one gives up being aware of them then pure awareness alone remains and that is the Self.*[335]

Self-inquiry for Sri Ramana Maharshi started when he turned sixteen and was thus going through his idealistic phase. The Analytical brain can penetrate an object but cannot know this object, since holistic knowledge is an attribute of the feminine polarity which can empathize with and mirror a subject.

The woman who loves is foolish if you look only rationally at the reality that she will become pregnant and will have to endure all sorts of difficulties and humiliations for the rest of her life, and often will have to survive her husband. But only through this sacrifice can humanity evolve. Whoever cannot love is as good as dead. The downfall of the personality has nothing to do with the Self. Don Quixote saves his Dulcinea through his love, which alchemically transforms her autoperception through her acceptance of the Self's reality. He allows her to transcend her link to the masculine polarity and to reach for the expression of individuality through the manifestation of the heart. By doing so, he safeguards the sacredness of his own being and prevents their selves from being imprisoned in matter. His self can rise to the Self through his feminine polarity. His Self could descend to his self through his masculine polarity. He became a living Solomon's seal (diagram 24).

Similarly, Dante, who loved a woman named Beatrice, said that his greatest accomplishments were because of her, although they never met. I experienced this as well, and this book is the result. We are individually called to this type of love. *Reality is what we make of it.* It has the limits of our own limitations. We can say it is all an illusion, but it still is. Because we all are unique, we all live in different individual realities, although they seem the same since we have similar bodies, essentially similar cultures, and similar environments. What doesn't change is the Self. Eternal love is only possible at that level.

Talking about what falling in love represents, Dr. Arthur Aron, of SUNY said: "The expansion of the self [a quality of the heart phase] happens very rapidly, it's one of the most exhilarating experiences there is, and short of threatening our survival it is one thing that most motivates us."[336]

Dr. Helen Fisher, an anthropologist at Rutgers University and the coauthor of an analysis on the subject, adds: "When you're in the throes of this romantic love it's overwhelming, you're out of control, you're irrational; you're going to the gym at 6:00 a.m. every day—why? Because she's there. ... And when rejected, some people contemplate stalking, homicide, suicide. This drive for romantic love can be stronger than the will to live."[337]

There is no clear divide between self and Self, as there is none between body and Self. There are only two aspects, one normally receptive to the other. The self is the active element in this dimension, but is receptive of the others. Our self ultimately is made by and receptive to the Self. If no adequate receptacle is available, the dynamic aspect of Self will push the self into oscillation to incite it to build the organization that will be capable of receiving certain aspects of the Self. Some call this karma. To me it is simply evolution. This is seen in the general dissatisfaction, apathy, and sometimes suicidal behavior of teenagers.

Elkhonon Goldberg, in his two books,[338] argues that the seat of novelty and of negative affect is generated by the right hemisphere (feminine polarity) because he believes that novelty seeking is motivated by dissatisfaction. But interest in something does not necessarily imply dissatisfaction elsewhere. What the right hemisphere is looking for is union. Then, of course, it appears negative in this divisive world we have generated. I would suggest as a hypothesis that we all bear an inner unconscious model of who we are (Self), and that this inner model receives reinforcements or is contradicted by the different environments the personality has to deal with. Hence we all have the potential of becoming gods (thus of being complete). These are expressions of archetypal forces working for the expression of Self in our dual manifestation.

From birth to fourteen years old, the child elaborates a personality, allowing him to survive in the world he landed in. During the period of fourteen to twenty-one, a person becomes conscious or feels emotions linked to the harmony or hiatus between inner compass and general life conditions.

The prefrontal cortex consists of many areas, including the orbitofrontal, ventrolateral, dorsolateral, and rostrolateral prefrontal cortices. It appears that these regions have different functions. Structural investigations reveal differences in their cellular composition, as well as in their connections to other brain regions.

Isis Code

A cross-sectional study focusing on the dorsolateral and rostrolateral parts suggests that they exhibit similar but slow rates of structural change (O'Donnell, Noseworthy, Levine and Dennis, 2005). As we have seen, the masculine polarity is modular and conservative, which explains this.

In the fourteen to twenty-one age group, dissatisfaction can arise with the individual being either able or not able to consciously and rationally pinpoint the cause of his general malaise. Males, who have an emotional level of a feminine type, will have great difficulty analyzing their feelings. Often, it is only after the maturation of the Universal brain (thirty-four years old) and of their idealistic self that the individual will be able to formulate and find the exact source of his uneasiness. Before, he points at everything and everyone. After twenty-eight years old, he starts pointing to himself. Only then will he "feel good in his own skin." Sometimes his unease is never resolved, implying the problem resides in the physical or emotional environment. Difficulties at this stage of the idealistic self can be a good sign that something in him still vibrates to the Self. That person is not spiritually dead. Add to this the numerous frustrations, the consequences of detrimental actions laden with ignorance that were committed by well-meaning or uncaring persons who crossed the path of the infant and child, and we have a cocktail of pain that can be alleviated only through Self-knowledge.

This is why our religious practices of purification are important. Evolution in this sense goes from the densest to the most apt to receive and manipulate subtle information. At this stage, it is helpful to read biographies of the most enlightened. These will answer the question "Why?" and lead the pilgrim to "How?" and eventually "Who?"

In many cultures, during this period of life, an initiation takes place. The shift from child to adult in those communities is rightly highlighted and leads to acceptation into the group as a full-fledged, participating member and not as a parasite living off the community resources. In this search of Self-knowledge, a period of solitude within that phase, even if for only a few weeks, is salutary. Alone, the individual learns that his uneasiness with life is caused by more than a parent or society. He takes responsibility for his share.

The uneasiness he might feel with life is deeper, and it is the necessary tool to avoid status quo and to embark on the pilgrimage toward Isis. He has the choice to limit himself and take on the patterns of the masculine polarity, or push forward into the unknown and develop the qualities of the feminine one. This is the true meaning of Jesus's "Follow me." Who follows and who is receptive? The feminine polarity, as indicated in the Gospels: "and many women were following them" (Matthew 27:55, Mark 15:41, Luke 23:27, Luke 23:49). Nevertheless, what appears to be feminine in this polarized world is in fact masculine when compared to the self. This is expressed in numerous pre-Raphaelite paintings and by courtly love, in which the women received a masculine pseudonym therefore representing the individuality which must penetrate the personality. On the emotional level, women are masculine but more importantly the individuality, in its Heart aspect—which is a crucial aspect in fin'amor—is received by the emotional aspect of men while it is received by the analytical and physical aspect of women.

Physical awareness is finite, but through it we can access Consciousness. This is the good news. One question we should ask ourselves, though: If in their craziness humans were to clone themselves, would their self-consciousness be replicated?

No, because nature creates the unique, not only on a cellular level but on all levels. The true question is this: Can we clone Consciousness? It would be like trying to replicate the fire of a candle. We can replicate candles but not the fire. We can use nature to copy physical aspects, but what we may succeed in copying will not be as good as what nature does, since we don't have the holistic point of view of the implicate order and Consciousness. It is our materialistic point of view that makes us believe that nothing else exists other than the material, personal *me*. The day we can clone humans, the discussion of "Is it our body that generates thoughts?" will finally be put to rest. Two human clones, with the same life and genetic code, will act, feel, and think differently. The difference between a clone and the original is that the clone's cells are born with the same age of the original cells that were used. This is about the only similar point between the two animals. They are physically similar to siblings, that's about it.

Isis Code

This is what the owners of a bull named Chance had to learn. Chance appeared in Hollywood films and on television. You can see photographs of this bull with different celebrities and children. When he died, his owners, Sandra and Ralph Fisher, decided to have Chance cloned, since they couldn't accept the death of their beloved animal.

As reported in *Science Daily:*

"We should know in a month or so if the telomeres of Second Chance are like those of the 21-year-old bull used as the source of the cells for the cloning process, or if they are more like those of a normal newborn calf,"[339] said Jonathon Hill, a scientist at Texas A&M University and the creator of Second Chance. As the article continues: "Hill said it took him 189 attempts—that is, transferring 189 cells into 189 different eggs—before a pregnancy ended in the delivery of Second Chance." Efficiency can be as low as a 0.0006% chance of producing one live embryo. It makes therefore no sense to work with species of which there are only a few reproductive animals left.

As a cloned animal, Second Chance received intensive monitoring and treatment from a team of veterinarians and intensive-care technicians at the Texas A&M Large Animal Hospital. This is a very expensive procedure. Like many previously cloned calves, at birth he displayed some symptoms resembling those seen in premature human babies.[340] When they saw Second Chance, the Fishers were convinced Chance was somewhat reborn. After all, the calf looked the same. The scientists and vets tried to explain that cloning does not work this way, to no avail. Unlike the gentle Chance, on his first birthday Second Chance threw Fisher up in the air, and a few years later gored him with his horns, sending him to the hospital. Second Chance died at the age of eight of a stomach malady. The original Chance had died in 1998 at twenty-one.

The mechanisms underlying cloning's limitations and weaknesses are now partly understood. There are epigenetic effects, thus related to the feminine polarity, which are information received from the different environments that have an effect on the genetic code. Genes are then turned on or off at the wrong time during development. This aspect, which was not deemed important before, since we thought genes held the keys to the whole structure, is now understood as crucial. It's also a question of timing, not only of structure. Further, in mammals some genes are shut down in one parent. The combination can result in lethal consequences. Some fetuses can grow to massive size. Animals are born with failing respiratory (lung aspect, feminine polarity) or immune systems (lack of control by the lung aspect) or never develop normally.

Oliver Ryder, the chief geneticist at the San Diego Zoo's Institute for Conservation Research, admits, "Most clones are abnormal. In general, developmental problems are not infrequent or uncommon in clones. So long as it is not a germ-line mutation, as long as it is a developmental problem, an epigenetic effect, not a genetic effect, it will not have a negative consequence on the gene pool."[341] Other critics, such as Naida Loskutoff of Omaha's Henry Doorly Zoo, warns against hidden threats. "What if, for example, in the cloning of a gaur or banteng, a prion lurked in the cow's oocyte or slipped across the cow's placenta? Then the clone could become a sort of suicide bomber, introducing fatal diseases into a dwindling, vulnerable population." I will address the prion problem in the chapter concerning applications of the LIFE biosystem.

Jung said that our personal definition of God is what defines us. This personal definition has a deep influence on all of the biocybernetic system, especially on our heart aspect, which regulates the lung aspect. The search for God is in fact the search for Self.

As well, our physical shortcomings can hinder the manifestation of our core identity. For example, the abnormal brain activity of mTLE (temporal lobe epilepsy) patients alters their manifested response to religious concepts. These concepts become strongly modular and somewhat replace the sexual interest incited by the Reptilian brain. In temporal left hemispheric (masculine polarity) epilepsy, patients develop hyper-religiosity. This, in no way is spirituality. It appears that when a seizure happens on the lateral temporal lobe, it is conveyed to the whole hemisphere as well as to the lateral temporal lobe of the other hemisphere. Therefore the patients have a non functional Universal brain, hence their modular approach. I do not consider Geschwind syndrome, as it is called, equivalent to finding and expressing spirituality, since it doesn't imply the function of the right hemisphere. It only gives a type of self-centered sanctimony, bordering on a fanatical religious pattern. It is not holistic and lacks the synergistic aspect of a true spiritual feeling and experience of the feminine polarity, which is the basic requirement for spirituality. In short, it is a truncated experience linked to a defective LIFE biosystem.

The connection between the temporal lobes of the brain (lung aspect) and religious feeling has led one Canadian scientist to try to stimulate them. Eighty percent of Dr. Michael Persinger's experimental subjects reported that an artificial magnetic field focused on those brain areas gives them a feeling of "not being alone."[342] Some of them described it as a religious sensation. His idea was that if we succeed in stopping the connection between the right and left hemispheres, the right hemisphere might take over. This is experienced when we are a small child or nearing death. Contrary to his goal, his work raises the possibility that indeed we are programmed for universality.

Dr. Persinger was introduced to one of Britain's most renowned atheists, Richard Dawkins. He agreed to try his techniques on Dawkins to see if he could give him a moment of a "religious feeling." During a session that lasted forty minutes, Dawkins affirmed that the magnetic fields around his temporal lobes affected his breathing and his limbs, which agrees with our association with the lung aspect. He reported no other effect. Everything we can feel is already programmed in our cells. The filter we put in place in our mind, our belief system, is very efficient.

Information from the heart is received in the brain stem at the level of the medulla. Previous studies have shown that the forebrain-midbrain circuit of the ventromedial prefrontal cortex (vmPFC) is abnormal in patients that suffer from chronic anxiety. The subjects present decreased function but increased gray matter volume and activity in the periaqueductal gray (PAG) of the Reptilian brain. Intriguingly, the vmPFC (of the heart phase) inhibits stress-induced neural activity in the brain stems of rodents and is important in facilitating escape and extinction learning. Note also that the vmPFC (heart) and mObfc (medial orbitofrontal cortex; pancreas) project directly into the dorsolateral PAG. (The heart phase is succeeded by the rational phase, which controls the Reptilian brain). Our observation therefore supports the hypothesis that the PAG is critical whenever an organism recognizes an immediate threat, *yet may be suppressed or promoted by the higher prefrontal regions linked to the heart and pancreas phases.*

The PAG region of the midbrain therefore serves as a link between the higher centers of thought, in order to mediate reactive experiences and behavior with lower brain stem centers. Of particular note, the PAG projects to the vagus (Faikas et al., 1997). The medial prefrontal cortex of the information-regulation brain is part of the default mode network with the posterior cingulate/precuneus (end of mammalian phase), and bilateral inferior parietal cortex (information-regulation brain). Although the function of this network is not yet known, it has been associated with readiness to respond to environmental stimuli (the information-regulation brain controls the Universal one), as well as self-oriented processing (information regulation). Studies have indicated that activity of this network is altered in Alzheimer's disease, autism, depression, epilepsy, and schizophrenia. In the infant brain, there is no such network, though there does appear to be a proto-default mode network, which consists mostly of the precuneus and bilateral parietal cortex, which is of the feminine polarity, heart aspect. Moreover, infants are still centered on their feminine polarity, as shown by their predominant use of their right hemisphere. They are thus receptive and molded by their environment on every level, especially before three years old, when they mostly assimilate subconsciously the information they receive. The tradition of dressing little boys as girls until they are five years old, as seen before in France and Italy, comes from this instinctive knowledge. Until around the age of seven, children are still centered on the heart aspect. This will switch to the pancreas center with time, as their masculine polarity and personality develops. The little girls will still be receptive physically and the little boys emotionally.

Research has shown other differences in the brain between men and women. One analysis included bilateral temporal gyrus. Results showed an increase of *the right temporal gyrus* corresponding to the feminine polarity for women. It also showed greater PCC (*posterior cingulate cortex*) connectivity than men with a region within the PCC/precuneus and *bilateral medial prefrontal cortex*. Women had greater activity than men in the bilateral superior frontal gyrus and the right angular gyrus region of the brain, which lies in the parietal lobe near the superior edge of the temporal lobe.[343] The angular

Isis Code

gyrus is responsible for things like understanding metaphors and the source of the inner monologue. These all agree with our system.

Scientists from several different universities have found a transient signal from the *right fronto-insular* cortex (rFIC), which engages the brain's attentional working memory and higher-order control processes, while disengaging other systems when these are not relevant. This signal is essential in allowing switches between the executive network and the default mode network. The executive network comprises the posterior hypothalamus, the basal telencephalon, and the intralaminar nuclei of the thalamus, all linked to the masculine polarity. We can then say that the fronto-insular cortex (lung aspect, feminine polarity) allows us to shift from one polarity to the other. It is pertinent to mention that the heart aspect—that is our mental, archetypal models—controls this function, allowing or not allowing the idealistic aspirations. They further anticipated that disruptions to these processes may establish a key aspect of psychopathology in several disorders, including but not limited to anxiety disorders, autism, and frontotemporal dementia.[344] In other words, the feminine polarity is essential; otherwise we end up with a rigid executive brain. Of note, autism includes deficits in communication (lung aspect) and has been associated with, but not necessarily caused by, intestinal pathology, also a lung aspect. In the LIFE biosystem, the right insula is part of the lung phase, and is thus an important feature of the feminine polarity.

In preschool years, children develop social skills by learning how to understand others' thoughts and feelings. Over time, most children grow to understand that others' thoughts are representations of a subjective world that may or may not match their own. In a study of EEGs of twenty-nine four-year-olds, researchers found that these changes are related to the functional development of two parts of the brain—the *dorsal medial prefrontal cortex* and the *temporal-parietal juncture*, both part of the feminine polarity. These govern similar understanding in adults. According to Mark A. Sabbagh, associate professor of psychology at Queen's University in Kingston, Ontario, and the study's lead author:

> For a while now, we have known that specific brain areas are used when adults think about others' thoughts. ... Our findings are the first to show that

these specialized neural circuits may be there as early as the preschool years, and that maturational changes in these areas are associated with preschoolers' abilities to think about their social world in increasingly sophisticated ways.[345]

Both LIFE biosystem and science show us that the feminine polarity is the default one, and that a person develops the masculine one in the normal course of maturation, a relic of evolution of the animal chain. In the evolution of humanity, the feminine polarity will come in again to lift the individual up a spire from the mammals' level. One sure way to bring back the feminine polarity is through beauty.

Subjective and Objective Beauty

Experiments have demonstrated that one-week-old babies prefer beautiful faces over average ones, suggesting that the need for beauty is hardwired into our brains. "There are lots of myths that people have around issues of beauty and attraction, and part of the issue is to stop thinking about things in terms of myth, but to use the tools of neuroscience, and start dissecting and understanding how things actually function," said Dr. Hans Breiter, psychiatrist, director of the Motivation and Emotion Neuroscience Collaboration at Massachusetts General Hospital, and coauthor of a study on the subject.

Is beauty only subjective or is there a type of beauty that is objective and based on biology? Is there Beauty in a platonic sense? Using fMRI technique, researchers Emiliano Macaluso, Cinzia Di Dio and Giacomo Rizzolatti in 2007 addressed the question by presenting viewers, from laypeople to art critics, with images of masterpieces of Classical and Renaissance sculpture that present the golden ratio. They presented two sets of images. One was composed of the original sculptures; the other of a modified version of the same images. In the observation phase, the subjects were asked to pretend they were walking in a museum and viewing the sculptures. They were then required to give an aesthetic judgment of the images.

Two types of analyses were carried out: one that contrasted brain response to the original and to the modified sculptures, and one that contrasted beautiful versus neutral or ugly sculptures, as judged by each volunteer. The most striking feature of the results was that for all subjects when viewing the original sculptures, relative to the modified ones, activation of the *right insula* (feminine polarity, lung phase) was detected, as well as of some *medial prefrontal areas* (feminine polarity, heart phase). The activation of the insula was very strong during the observation phase.

Most interestingly also, when volunteers were required to voice their aesthetic judgment in the second analysis, the images they subjectively judged as beautiful activated their right amygdalae, thus their masculine polarity, relative to those judged as ugly. Hence, objective beauty does not trigger the activation of the same cortical areas as subjective beauty. The first triggers the insula (objective beauty) and feminine polarity; the other activates the amygdala, driven by one's own emotional experiences (subjective beauty) and masculine polarity (personality).[346] Since Classical and Renaissance sculptures often used the golden ratio for proportions, I wondered if this ratio stimulates the feminine polarity, thus explaining why the insula would be activated. It might, if the golden ratio is truly part of the structure of nature and not a rational, emotionally driven confabulation of the mind.

And indeed, the golden ratio is found in nature. In 2010, researchers from the Helmholtz Center Berlin for Materials and Energy, in partnership with colleagues from Oxford and Bristol Universities, and the Rutherford Appleton Laboratory, UK, observed nanoscale symmetry present in solid state matter. They measured the signatures of a symmetry showing the same attributes as the golden ratio used in architecture and art.[347, 348] This type of objective beauty—Plato's "Beauty is the splendor of truth"— is a consequence of the implicate order, the LIFE biosystem, healthily expressed in its two polarities. Subjective beauty stimulates the masculine polarity. It is attractive and talks to the physical and emotional selves. This type of subjective beauty varies from one person to the other and depends largely on cultural and personal bias. A healthy civilisation would protect what is objectively beautiful and reject what is ugly.

Dr. Hans Breiter, in an experiment addressed to heterosexual men between the ages of twenty-one and twenty-eight (pancreas phase), presented them with faces of beautiful men and beautiful women.

Although the men agreed the faces were equivalent in their beauty, when allowed to press a key to select the images they preferred, they invariably selected only female faces. "These young guys [were] keypressing 6,000 times over 40 minutes, that's as much as a rat barpresses for cocaine," said Breiter. "These pictures had as much reward value as cocaine, as food, as money, and that was remarkable."[349] Such is the power of subjective beauty.

Personally, I qualify as art whatever can stimulate the feminine polarity. The rest is an enjoyable pastime, an expression of our deficiencies, an intellectual pursuit and often a business venture.

Overall, faces evoke stronger brain response than words, with the expected laterality (left hemisphere, masculine for words; right hemisphere, feminine for faces). However, in research including alcoholics, where control participants showed stronger activation in the amygdala and hippocampus when viewing faces that had emotional (as opposed to neutral) expressions, the alcoholics responded in identical manner to all facial expressions. The activity of their amygdalae was inversely correlated with an increase in lateral prefrontal activity, related to the pancreas phase, as a function of their behavioral deficits. Lateral prefrontal modulation of emotional function is a compensation for blunted amygdala activity. It is not a sign of control over emotions, but a symptom of a profound deficit.

Fourteen Years Old

I remember that year because it was for me the awakening of physical attraction. Smells, different tastes, textures, colors would make me gasp for unknown reasons, as if tied to forgotten memories. At fourteen, I attained a sense of being I'd never had before. I also had my first real friend. Before, I was treated like an outsider at school. Was this because I felt I did not belong at home? Did I carry that feeling to school? In that year, I spoke with a schoolmate who used to avoid me before and asked her why she hadn't liked me. She seemed surprised by my question and said that I had looked like such a snob when I was younger, no one wanted to be with me. And I only recalled being so very lonely!

So the day came when a girl asked me to be her friend. Because I had been rejected by someone else the previous year and remembered how much her refusal had hurt me, I agreed to be Carole's friend. She was not beautiful, but she was kind and had extraordinary almond-shaped blue eyes. She sat at the desk just in front of me and regularly would turn around and smile to me. I had never received so much attention in my life! She would organize the top of my desk with zealous care, touching my eraser, my pencil, my ruler as if they were sacred objects. This sent shivers up my spine, a feeling I had never had before. I was happy to go to school.

My father also liked her very much; he called her his adopted daughter. Her family was poor, since her father had left the mother and the three children a few years before. A probable effect of this was that her mother was excessively nervous and excessively pious, as if she had committed a sin by becoming pregnant and needed to expiate in some way. I felt she would have done anything to prevent her two daughters from falling prey to the same life. Carole had a younger sister who was beautiful, but who had a penchant for stealing. I remember one time waiting for them outside a store and then lecturing them because they had stolen makeup. They laughed it off. The owners of the store would not even know since they were far richer than they! That was their argument. Also, Carole used to smoke, which was frowned on by my mother. Carole and I became part of a group that sang at church, which I loved as it prevented me from fainting during mass.

Carole was sexually aware while I was oblivious to these matters. I remember seeing the naked breast of a classmate after gym, and for weeks I wondered if I had committed a mortal sin not only for seeing her, but for finding her breast beautiful! It was the first time I had seen a bare body part other than legs or arms. Sexuality and its attached physical body were tied to negative, forbidden, mortal-sin questions. I did not wish to explore.

Carole and I befriended two boys who were very different from each other. Paul was tall and looked like Dr. Spock of *Star Trek*. He was of Italian descent. Bryan was also tall but muscular, and oh so beautiful! He looked like an aristocratic Northern Italian and had the demeanor of a lion. His body had perfect proportions, and he had incredible blue eyes. I remember one day I was to meet with him in the cafeteria.

When he came in, he was so beautiful, my legs suddenly refused to carry me and I had to sit down. I couldn't understand my reaction. The poster taped to my bedroom door was not of Captain Kirk of the starship *Enterprise*, a Bryan type, but of Dr. Spock! Of course, Bryan was sexually active, and someone told me that the girl he was seeing had tried to perform her own abortion with a coat hanger and ended up in the hospital. Whether this was true, I never knew, but it changed how I saw him. As for Paul, he eventually became stuck between Carole and me and had to decide who to date. He decided on Carole, saying that she needed him more than I did.

I remember one night—I think it was my birthday—Carole was sleeping over at my house. We had been talking for a few hours and were feeling particularly close to each other. For some reason I started crying and she tried to console me. She caressed my hair and hugged me. I melted. My tears ceased. We fell asleep in each other's arms. It felt so peaceful.

The following day I felt ill at ease with her and confused. For me back then, touch was equivalent to sex and thus a sin. Luckily, she was so natural that I did not question her actions, only my reaction. The question of what I wanted and what I was crept into my mind. I loved refined and pure facial features. It happens that these are mostly seen in women. I loved tenderness. I loved to exchange emotions and ideas. I needed tenderness. I needed to feel my life mattered to someone! I had found this with Carole.

Would I ever find this with a man? Was this love?

Years later, after a torrid and destructive relationship with a beautiful man, I ended up alone in a basement apartment. I was twenty-three. The first thing I did was to hang up pictures of women's faces in my bathroom. No, I was not sexually inclined to men or to women, I knew that now. I needed Beauty. At eighteen I had dedicated my life to it, not truly understanding what that entailed. Now I know. This state of mind is not something one can force on oneself. It is or is not. I also had a thirst not only to love, but to feel loved in exchange, without guilt. I wished also to feel acknowledged, and to be part of a couple that radiated waves of happiness into the environment. Would I still love if the sentiment was not mutual? The following events answered this question: yes.

Isis Code

Lies, Left Dorsolateral Prefrontal, and Archetypes

The dorsolateral prefrontal cortex, which we associate with the Human (right) and Analytical (left) brain, is particularly engaged when we must switch from one rule to another, and hence suppress the previously relevant rule (Crone, Wendelken, et al., 2006). Thus, in studies of humans and primates, the ventrolateral PFC is unfailingly involved in representing sets of conditional directions, suggesting that it plays a fundamental role in patterns, which fits nicely with the Human brain. Further, fMRI data suggest that the dorsolateral PFC (right is information-regulation or Human brain and left is Analytical brain) may be especially crucial for overcoming interference from previously integrated rules. Daniel Langleben, a psychiatrist and neuroscientist at the University of Pennsylvania, has clearly demonstrated that the *left dorsolateral PFC* (thus part of the Analytical brain) is activated when a person tells a lie.[350]

Not surprisingly to me, the areas highlighted when someone is lying are on the left hemisphere, not the right one. More precisely, as we said, this left dorsolateral prefrontal cortex is associated with the Analytical brain of the masculine polarity. The parietal cortex registers the fact and the anterior cingulate cortex (ACC) partakes in the choice of action. The ACC monitors the conflicting response tendencies. The level of activation of the right ACC is proportional to the level of the conflicting answer and inversely correlated to the left DLPFC activation (Carter et al., 2000; MacDonald et al., 2000). Increased activation of the right ACC but not the right DLPFC during the lie suggests that a conflict with the truth and its inhibition are taking place.[351] In another research in agreement with this, stimulus of the left or right parietal cortex had no effect on the volunteers' inclination to lie.[352] As well, the right dorsolateral cortex controlled the lie. Of course, deception is an affair of the whole person, but this aspect is played by the left dorsolateral prefrontal cortex.

Archetypes, models, ideals, and their attached positive personal identities are therefore essential stabilizing tools, able to lift us from detrimental influences, both current and past.

The analysis of the archetypal notion can be traced back to Plato. Jung himself compared archetypes to Platonic εἶδος (eidos). According to Plato, Ideas were original abstract functions imprinted on the soul before it was born into the world. They were collective in the sense that they embodied fundamental characteristics of humanity and not our interpretation of them. For Carl Gustav Jung, we could associate them to the biological expression of "patterns of behavior."[353] He also says that "nobody has ever seen an archetype or as a matter of fact, an atom. Nevertheless, we know the first one to produce 'numinous effects' and the second, explosions. When we speak of an atom, we speak, in fact, of the model that was constructed around its effects, not of its reality. In the same way, when we speak of archetypes, we can only refer to its representations, never of the thing in itself, which, in both cases remain a mystery pertaining to transcendence."[354] To him, what we name psyche, or soul, is a reality, as he could clearly observe the effects it produced.

Although we live in the present, the echo of our past does touch us. The Human brain allows us to rise above it. Because of it we can move through life, shed memories, and reinvent ourselves. The personality holds on to who we were, to what we did, and to what happened to us. The individuality acknowledges it but allows us redemption as it pertains to the eternal, universal and to the infinite.

Thanks to an active Idealistic self, difficult memories will fade eventually, replaced by universality and eternity. Flesh will heal and heart will forgive.

The parietal lobe, part of the Human brain, is immediately posterior to the central sulcus. It is anterior to the occipital lobe and is not separated from it by any natural boundary, as if it proceeded from it. Remember, the occipital is associated to the Mammalian, so the Human is its natural development. Its inferior boundary is the posterior portion of the lateral fissure that divides it from the temporal lobe. Mental manipulation of objects, numbers, and the integration of sensory information from various body parts are all related to this part of the brain. Its posterior area is referred to as the dorsal stream of vision, while the ventral stream is directed by the temporal lobe. These streams have been called the *where* and *how* of vision. Interestingly, we associate the left parietal to the Analytical brain, which answers the *how* question. Activity in the parietal areas has been observed in cases of self-transcendence and meditation (Cahn and Polich, 2006; Kaasinen et al., 2005; Turner et al., 2003). The researchers could also witness that the parietal and frontal circuits operate independently when a person does different kinds of arithmetic problems. When subjects did exact calculations, the left frontal lobe lit up the most. When they approximated answers, their parietal lobes lit up (information regulation). Those regions, located on both sides of the brain, perform visual and spatial tasks. Through them we can express analogies, a preferred tool of the Human brain. They also allow hand and eye movements, mental rotation of objects, and orient attention. On the other hand, Newberg, D'Aquili, and Rause (2002) reported that during meditation and prayer, their subjects had a decrease in activity in the parietal cortex. We have seen that a structure used when we are novices at something ceases to be stimulated when we become professionals. Could it be that at this level, there is no one size fits all? Nevertheless, a robust gender difference appears in the anatomy of the planum parietale, another asymmetrical structure in the parietal lobe, at the posterior end of the Sylvian fissure. This structure is typically larger in the right hemisphere of women (feminine polarity).

Schizophrenia is considered partly caused by abnormal brain function. It appears that underactive frontal lobes and overactive parietal lobes, which imply difficulties with the Reptilian and Mammalian brains, are to blame for some of schizophrenia's associated symptoms. For example, when frontal lobes are underactive, the abilities to plan, organize, and decide are all impaired. In the LIFE biosystem, an overactive parietal lobe is the sign of an inefficient Reptilian brain. Investigation of pregnancy and of the first years of life could be a valuable tool in the search for understanding of what led an individual to schizophrenia. Again, prenatal awareness offers the best prevention.

Parietal lobes are also involved in sensory perception, such as voice recognition, the ability to distinguish patterns, and spatial as well as psychological orientation. When these lobes are overactive because of a deficient Reptilian brain, sensory perception can be distorted, as has been observed in many people suffering from this illness.

As the center of the individuality and tributary of the liver aspects of our system, the Human brain integrates sensory information from the various parts of the body. It contains the primary sensory cortex, or somatosensory cortex, which controls the diverse sensations of touch (pain, hot, cold, etc.). It tells us which way is up and gives us our physical sense of self, all in harmony with this part of the brain in charge of information regulation.

Although the frontal lobes are the first to decline in an aging system, and the left hemisphere of the masculine polarity more rapidly so than the right, it appears that some older adults spontaneously recruit other cognitive resources to meet their processing needs. For some, however, information becomes overwhelming as they struggle in inhibiting irrelevant data (Earles, Smith, and Park, 1994). Others, by recruiting unique neural regions, can complete a recognition task while being distracted. They have a better memory than the former and they are deemed high functioning.

"The results of this study suggest that the anterior prefrontal cortex, the area of the brain that is most developed in humans, mediates the ability to depart temporarily from a main task in order to explore alternative tasks before returning to the main task at the departed point," says Jordan Grafman, PhD, chief of the Cognitive Neuroscience Section in the National Institute of Neurological Disorders and Stroke (NINDS) and a coauthor of a study on the subject.

Deborah L. Harrington, PhD, an associate research professor of neurology and psychology at the Veterans Affairs Medical Center of the University of New Mexico, suggested the critical role of the parietal lobes in timekeeping. She and her colleagues found that stroke patients with damage to the parietal cortex on the right side of the brain, thus associated with the feminine polarity, but not the left experienced impaired time perception. As we have seen, the feminine polarity is master of cycles and rhythms. To confirm our findings, an area of the brain labeled the inferior-parietal lobule (IPL) on the left side is typically significantly larger in men. This section of the brain is thought to control mathematical ability, and probably explains why men frequently perform higher in some mathematical tasks than do women[355]. Interestingly, this area was discovered to be abnormally large in Einstein's brain. In 2007, a group of researchers found a left/right asymmetry in parietal lobe function that specialized in numerical representations. Their results revealed "an abstract representation of numbers in the left parietal lobe and, by contrast, notation dependent on non-abstract representations of numbers in the right parietal one."[356] This challenged the commonly held view that numbers are represented solely in an abstract way in the human brain. Both polarities view numbers differently. Finally, the larger right parietal lobe allows women to focus on specific stimuli and to process sensory information in a more precise and conscious manner.

Humans share the meaning conveyed by signs and symbols with primates. It is the level and type of interpretation of these symbols that differs. Primates can recognize symbols of material (food) or emotional events. However, they will not understand the intricacies of esotericism or of system science. This form of meaning, highly developed in humans, added to the fact that only humans make propositions and reasons with them, are what render them uniquely human. Humans' drive toward Consciousness, expressed through the need to see all of our experiences and encounters as meaningful, proceeds from another level of complex encoding.

Descartes's dualism—that there was a distinction between the mind and the body—was misinterpreted. The Self doesn't control the self, but uses it as an expresser. They work together. The control is in the totality of what is expressed and what needs to be expressed universally. Details (our personal life) count only to us. As the rays of sun are undivided from the sun, our Self partakes of the whole Self. For the Buddhist tradition, the self is part of the illusion and will vanish at physical disassembling. But Tibetans choose their Dalai Lama on personal memories of a young child. This seems to contradict the former. In fact, it is an acceptance that only the Self truly exists, follows a course, and has a memory. The self is only the skin that is regularly shed. We all work at bringing more of Consciousness into this world. [357]

In his article "Quantum Consciousness is Cybernetic," Gordon Globus, professor emeritus of psychiatry and philosophy at the University of California, Irvine, offers that "Consciousness is a quantum eruption offering possibilities to the match with sensory input and thus with reality." He does not seem to believe in randomness of mental acts but more in a control from the implicate order (biocybernetic system) performing under symmetry and with minimal input from the neural apparatus. He posits that his vision is fully in agreement with the first physical principles of quantum field theory. Not being a quantum physicist, I cannot judge his assertions, although I find them intriguing. The brain to him is not a simple measuring device and should not therefore be reduced to a mechanical device.[358]

The self has some control over the here and now through the masculine polarity. The Self, being nonlocal because of its ability to assume all localities through the Master of the Heart, holds the key to the quantum field and its cybernetic structure.

Is it possible for a person to truly "know thySelf" without finding some information regarding the past of his soul? Since by definition souls are eternal, they have a past and a future spreading beyond the short cycle of our incarnation. Intuitive knowledge of our idealistic self does help in this quest for identity. It is an inner compass. The vmPFC is hypothesized to mediate a phenomenological "feeling of rightness," dubbed FOR, which is associated with it. This allows for rapid appreciation of the appropriateness and accuracy of information, of a response (Gilboa and Moscovitch, 2002; Moscovitch and Winocur, 2002), or of an action. It precedes the conscious, elaborate cognitive verification of the dorsolateral prefrontal cortex and is impermeable to influences from its attached Analytical brain. It is thus formed independently from the conscious awareness of the Analytical brain.[359] The vmPFC is therefore associated with decision making (Bechara et al., 2000), choice of action, independent from previous stimulus reinforced associations (Rolls et al., 1994; Rolls, 1996, 2004). In this, oxytocin will help "clean the slate" for new experiences. The ventromedial prefrontal cortex is similar to a computer instantly applying the patterns echoed from the matrix of our soul. It allows freedom from the automatic responses of the personality that are triggered by familiarity. This type of memory seems holistic and appears to precede or overarch the personality. This is in agreement with our attribution of this aspect to the individuality. The physical does limit the expression of the nonphysical, as well as allowing it. Perceptive processes or patterns will then be integrated with emotional cues and influence decision making at a preconscious level (Bechara et al., 1997, 2000a). Patients with vmPFC lesions have difficulty in achieving accurate choices.

Consequently, if you know a child who spontaneously talks about remembering places, names, events, etc. that he or she couldn't know, it is better to pay attention to the child than to ridicule him. My experience with this is that to be mocked, brushed off, or defined

Isis Code

as "not balanced" is detrimental to the mental health of the child, far more than it would be to support the belief of having memories from "before now." What insecure part of us would do otherwise? The rational self, the personality. When I was a young child, I had what truly felt like memories. Whether reincarnation or metempsychosis, labelled as such by my mother—"But we don't believe in metempsychosis in the Catholic Faith!"—whether they were true or not, was irrelevant to me as a child. To me there was no need for a name for what I was experiencing. It was normal. My environment labeled it as abnormal. It was not a question of believing, since that appears at a later age, after the frontal lobes are able to manipulate concepts, but a question of what I was experiencing. I *knew* these were truly memories. Children do not react on the lines of believing or not nor do they forever analyse what goes through their mind, they are whole. That is a grownup point of view. Most of the time, a child has those memories up until seven years old, when the present personality is still being shaped. During this period the consciousness fluctuates between soul-imprinted memories, subconscious collective influences, or unconscious memories and present realities. Parents should be aware of this and take it into consideration, even secretly and with respect, write down what the child says. Later on in life, these events could help the individual solve difficulties or give clue to why he or she reacts one way or another when faced with particular situations. My two children are remarkably different from each other. My reaction to the ways they behave when confronted by problems is colored by what they revealed to me of their core identities. It is a mistake to talk to a child about his or her past lives. It is better that the child, if necessary and when the time comes, uses his own surfacing past memories as these might help him resolve inner conflicts through a better understanding of his own self.

Also, I have observed with friends and people close to me that if the heart is open and sincere, elements of a solution to a difficult situation will come through dreams. This is the preferred language of the feminine polarity. Writing songs or poetry and singing are also tools for the feminine polarity. If knowledge from past lives is not useful for this present life, but is only used as a boost for the ego or as a curio adding some spice to the dullness of present life, then it is not being used in the right way. False memories can creep up, and balance and health can be lost. I remember meeting a man who was sure he had known a certain woman in a past life. Therefore he wanted to marry her. He literally stalked her. His present personality had found a rational way to excuse his behavior towards this woman. As I told him, if you marry her, it has to be because of the present life, not because you believe you knew her in a previous life. Can you see yourself marrying everyone you loved as a man or a woman in the last million years?

The ventromedial prefrontal is also implicated in the processing of risk and fear, and in decision making as opposed to procrastination. As of yet, functional differences between the orbitofrontal and ventromedial areas of the prefrontal cortex are not clearly established, although we know that the areas of the ventromedial cortex that are superior to the orbitofrontal cortex are less associated with social functions and more with regulation of emotions. This explains why we associate the orbitofrontal with the Universal brain. Research in developmental neuroscience also suggests that neural networks in the ventromedial prefrontal cortex are rapidly developing during adolescence and young adulthood, the phase of fourteen to twenty-one years old, supporting emotion regulation through the amygdala and associated with a decrease in cortisol levels. This echoes the LIFE biosystem in which the Reptilian brain regulates the Human one. The right half of the ventromedial prefrontal cortex, thus of the feminine polarity, is connected to controlling the interaction of cognition and affect in the production of empathic, collective responses. We see here the link between Human and Universal brain aspects. Morten Kringelbach, director of Hedonia: TrygFonden Research Group, also associated emotional responses to orbitofrontal cortex activity level. We see here the link between Reptilian and Human brain aspects. Kringelbach is a senior research fellow in the department of psychiatry, University of Oxford, and a professor of neuroscience at Aarhus University, Denmark, as well as junior research fellow at the Queen's College, University of Oxford.

Judith Rapoport, MD, chief of the Child Psychiatry Branch, National Institute of Mental Health, Alan Evans, MD, of McGill University, and colleagues report on their magnetic resonance imaging (MRI) study[360] of the brains of children suffering from ADHD that although the brains exhibited a normal back to front maturation pattern, they showed a delay of development in some areas. Of the 223 children studied, the cortex sites of ADHD sufferers attained their peak thickness at 10.5 years old instead of the 7.5 years old seen in the general population.[361] The middle prefrontal cortex lagged for a whopping five years in those with ADHD; almost a full phase long. As we have seen previously, there is also considerable development in dopaminergic function during adolescence. Specifically, data indicate that dopamine turnover and synthesis in the prefrontal cortex increases during adolescence, and that there is therefore a shift in activity from mesolimbic to mesocortical dopaminergic systems (Spear, 2000). Ventral, medial, and dorsal frontal regions, each receives considerable dopaminergic input from the ventral tegmental area. These allow for increased risk for mood disorders, including bipolar disorders. Ventral and medial PFC, amygdala, and the ventral striatum have been identified as key components in the circuit activated by pleasant stimuli (Aalto et al., 2002; Aharon et al., 2001; Bartels and Zeki, 2000; Blood and Zatorre, 2001; Canli et al., 2002; Hamann and Mao, 2002; Karama et al., 2002; Lane et al., 1997). A functional Reptilian brain is therefore critical for the regulation of the dopamine rich Human brain. As noted above, bipolarity in children is characterized by relatively rapid mood cycles, high comorbidity with ADHD, and significant irritability.[362] All of these could also be expressions of an out of control Mammalian brain, consequence of an impoverished or deficient Universal brain and lung energies.

The ventromedial prefrontal cortex is also associated with preference judgment, possibly assigning a key role in constructing one's self. One particularly notable theory of vmPFC function is the somatic marker hypothesis accredited to Antonio Damasio, professor of neuroscience at the University of Southern California. By this hypothesis, the vmPFC has a central role in adapting associations between mental/emotional objects with bodily feedback, thus defining one's identity. In harmony with the LIFE biosystem, the Human brain is an interface between rational/emotional and physical realities. This account also gives the vmPFC a role in moderating emotions and emotional reactions by its action on the social self, the environmental aspect regulating the emotional self.

Lesions to this area were also associated with a deficit in processing gender-specific social cues. Since this is the phase to which we attribute a primordial definition of the self, this would make sense. Abnormalities in response inhibition are associated with damage to the vmPFC. Both humans and animals with vmPFC damage have difficulty constraining impulsive urges to react, or consistently react with a dominant response, even if the response is inappropriate (Fellows and Farah, 2003, Chudasama et al., 2003). Response inhibition necessitates the thalamus of the Reptilian brain working in conjunction with the vmPFC of the Human brain.[363] This shows that every aspect of the system must be healthy and developed to attain optimal functionality, as well as manifestation of emergent qualities seen in the higher aspect of the Human brain, the Synarchic brain. Vitamins and exercises for the brain alone, although interesting, will not allow the development of all of an individual's potential.

Although vmPFC lesions do not appear to affect a person's emotional perception of stimuli, right vmPFC lesions, whether these are of a physical, emotional, mental or social origin—injury of the physical feminine polarity—are nevertheless associated with higher levels of trait anxiety. As we have seen, the right hemisphere controls the sympathetic system. A lesion in the right vmPFC or a problem with the feminine polarity would mean a *sympathetic disinhibition*. Therefore their foot is always on the acceleration pedal, so to speak. Patients with increased sympathetic activity are also at higher cardiovascular risk (Oppenheimer et al., 1991; Hilz et al., 2002) and can be subject to myocardial damage (Oppenheimer et al., 1991), behavioral anomalies, and even sudden death (Samuels, 1993; Critchley et al., 2005).

As suggested by researchers in the field: "Patients with vmPFC lesions, particularly right-sided ones, should be monitored for psychiatric disorders and studied further to assess their risk for cardiovascular abnormalities, such as hypertension, arrhythmias, or electrocardiographic abnormalities." [364]

Of note, increased solar activity will have a similar effect as excessive sympathetic activity.[365]

In association with this, Dr. James Prescott discovered another effect of maternal deprivation: abnormally high voltage electric discharge and activity in the brain, called spiking. In fact, lack of touch or movement leads to hypersensitivity to sensory stimulation. As we have seen in the chapter dedicated to the Reptilian brain aspect, children who have been deprived of touch might frantically want it, but exhibit a number of contradictory or erratic behaviors when confronted with true affection from others. They may shy away, respond violently and reject contact, and in some sad cases develop self-mutilating behaviors.

The unvarnished truth is that lack of normal, heartfelt sensory stimulation in infancy damages the brain's development, stunting brain cells and creating malformed dendrites. The Reptilian brain of these victims will literally not mature properly. This might eventually lead to self-sabotaging behaviors.

Just five days of blindfolding of normal sighted adults will lead their occipital lobes, thus their Mammalian brain associated to vision, to process information for touch function normally executed by the parietal (Human brain). In other words, the empty place is taken over by other functions; there is no waste in the brain. This is a rapid reorganization of some of the brain structures and functions in a bid to adapt to a new situation. The occipital lobe will return to normal functioning once the blindfold is removed.[366] What may happen on a physical level may also occur on the emotional, mental, and social levels. The difference is, it is almost impossible to take off the blindfold on these more subtle levels without a reassessment of the belief system (Human brain aspect). Deprivation of love does the same thing with the Master of the Heart and its sense of synergy (connection). In other words, if one polarity cannot function, the other will take its place. When confronted with this situation, a child has to develop coping tools.

A California psychoanalyst, Dr. Robert Stoller, questioned people who partook in violent sadomasochism. He learned to his surprise that as children, *all* had undergone painful and confined medical treatment.[367]

The sensations arising from the physical and emotional aspects are of no use if they cannot be integrated. This is the function of the frontal lobes. The senses complete one another. You see something fall and you hear the noise it makes as it hits the ground. Right away appears in you the question: What does this sound means to me? This awareness of the sensitive human is the basis for intuition (lung/heart). "I feel, I sense" will eventually lead to "My only desire" which is a function of the lung aspect. What does the feminine polarity aspire to? Wholeness through fusion.

"I am always with you" is always in the background of the feminine polarity, along with "Am I part of the whole?" and "Am I partaking?"

Schizophrenia and Identity

> "It is only in the mysterious equations of love that any logical reasons can be found."

This is a quote from the movie "A Beautiful mind." This film, which romanticized John Nash's[368] struggles with schizophrenia, received a great deal of attention when it was released in 2001. More intriguing to me are the details of his biography. He is depicted as arrogant, childish, and brilliant. Most of all, he did not like working with other people. He also said: "One had to learn from the world's knowledge rather than from the knowledge of the community." In the late 1980s, Nash had begun to reply by e-mail to other mathematicians, who had realized he was "the" John Nash and that his new work had value. These formed the nucleus of a group that contacted the Sveriges Riksbank in Sweden, responsible for the Nobel Memorial Prize in Economic Sciences, and vouched for Nash's mental health, so he could receive that award in recognition of his early work.

In an interview, Nash declared that the stigma against mental illness would be removed only when the disease was. "I think that when you try to de-stigmatize you may have some other objectives. The doctors and those who treat people with mental illness, they want to stay in business. ... I would not have had good scientific ideas if I had thought more normally."[369]

By saying this, he appears to know that his Master of the Heart was blindfolded and deprived. He had to develop something else. It does not make his condition healthy or desirable. As for his claim that he would not have had good scientific ideas without his illness, I believe this is a comforting escape from accepting something terrible. It also conveniently echoes a destructive dogma in the field: the more one is intelligent, the more one is susceptible to mental illnesses. (!)

The first overt manifestations of schizophrenia usually take place in the late teens to early twenties, in the rational phase. Onset of schizophrenia in men peaks at around twenty to twenty-five years, whereas the onset peak is later in women, at around twenty-five to thirty years. Hallucinations, mostly auditory, and delusions set in. By nature, these voices are usually threatening. Voices and hearing are associated with the Reptilian brain. An Analytical brain in difficulty will fail to control this one, which explains the derogatory voices. The functions of the frontal lobes are particularly disrupted in schizophrenia. Those disrupted functions are associated with the right hemisphere, which is understandable.

Studies have shown that the inferior parietal lobule (IPR), part of the idealistic self and Human brain, acts as a neural crossroads. Schizophrenic men have up to a 16% smaller IPR than healthy ones. In a way, this whole area of the brain may be miswired.[370] Their Idealistic self is in jeopardy.

"[W]hen I had been long enough hospitalized ... I would finally renounce my delusional hypotheses and revert to thinking of myself as a human of more conventional circumstances and return to mathematical research. In these interludes of, as it were, enforced rationality, I did succeed in doing some respectable mathematical research." (John Nash autobiography, Nobel Prize website)

This is in contradiction with the fact he said his mathematical abilities were caused by his illness.

It also appears that a person's unrealistic image of himself, a problem with identity, is concomitant with the delusions. Paranoia is another example this time mainly of a lung aspect deficiency. It is usually found in subjects who have suffered an early lack of nurturing.

Some recent evidence (Pedersen and Mortensen, 2001) suggests that the more youthful years spent in dense urban areas, thus polluted environment, the greater the risk of developing schizophrenia. Research has also found a reduction in the thalamus, the striatum, and the superior temporal cortex, as well as a volume reduction in the hippocampal region (instead of expansion) in males and in the corpus callosum (see chapter on Mammalian brain). Another aspect of schizophrenics is that they do not detect anomalies in their comportment, which indicates a problem with the right hemisphere. Normally, the right hemisphere detects anomalies and forces the left to revise established patterns.

The Human Brain of Information Regulation, the Idealistic Self, and the Archetypal Human

When I was sixteen, I seriously considered that I might be suffering from schizophrenia, as some experiences I had were not consistent with social norms. After viewing a documentary on the subject, I became convinced there was something wrong with me. Not only had I experienced a few auditory delusions, but what is described generally as visual hallucinations had started at a young age. In the case of the auditory illusions, I heard bells and sometimes a choir and orchestra, an amazingly beautiful symphony. When I asked my mother, she said she couldn't hear a thing. On the negative side, when I was about five years old, once I heard a threat. The visual hallucinations sometimes happened when I was in a dreamy mood, usually outside alone in the garden. It was as if whole life episodes were being re-enacted in a vision, in other places, other times, remembering these later I became conscious these were also other selves. These visions could be triggered unexpectedly by a smell, a taste, something I would do, something I would see on TV. Of course, at the time they were perfectly natural to my child's mind, and felt like an episode of a complete reality. When I would "awaken" and try to convey some episodes to the adults I knew, their negative reaction always stunned me.

I was a child with adult memories. I could only label them as memories because they had elements that dreams don't have. They felt truly real with all the senses present. The only detail that made me still hold onto them in secret was that these experiences were positive and helped me make sense of my present life and of "me." Later I understood that I was not suffering from schizophrenia; I was suffering from the fact that those I loved didn't share the same experiences and therefore could not understand. I believe this made me more sensitive to other realities, more compassionate, and kept my mind open. I know by saying this I will be labeled by those who have not experienced the same thing and have a fear of the unknown, but I do so to prevent parents from saying to their children who might experience the same sort of memories that they simply have an overactive imagination. After all, why care about a child's experiences if they are only fictional fantasies? Still, they are an open door to the profound identity of this child. Please, listen with love and compassion, do not judge, and do not consider these are coming from a child seeking attention. Keep a neutral attitude, as if the child were recounting a trip he made to the store. As I said previously, these memories could be used in moments of despair,

Isis Code

to hang on and make sense of one's actual life. In the case of schizophrenia, derogatory voices criticize the person's thoughts, or suggest violence and suicide. These voices are often vulgar and emotionally charged. This was never my case.

Some interpret hallucinations laden with auditory speech as a problem in the temporal lobe, because temporal seizure can sometimes result in auditory hallucinations. That is not to say that all auditory phenomena are hallucinations. Otherwise this would lead to hasty assumptions. One of such is saying that Joan of Arc "only" suffered from seizures of the temporal lobe focus near the left amygdala, explaining away her auditory hallucinations. The insinuation is that to achieve greatness you have to have a deranged mind, or that greatness is a by-product of deranged minds, so after all, she wasn't great.... Why do some individuals find human abilities so difficult to accept, that they need to do their utmost to explain them away? Is this because since they don't have these abilities, either no one can have them or they must be the expression of a deranged mind? Is this, in fact, an expression of their own psychology, reflecting the distorted society that has nurtured it? If we consider Joan of Arc's case, her father had dreams of her "fleeing with soldiers". In those times, prostitutes would do that. Joan's mother finally told her about those recurrent dreams but Joan's did not try to dispel the wrong interpretation. Her visions and "hallucinations" seemed utterly strange to her also. She had to meet with Charles VII and bring him to be crowned? It seemed crazy. When she finally decided to heed the counsel of her "voices", her silence implied she renounced trying to correct his conclusion concerning the lack of chastity of her motivations and intentions.[371] Second, the "presences" that visited Joan were not threatening. She started hearing voices at twelve years old. Lastly, there was a perfect synchrony between her visual, auditory, olfactory, and tactile experiences. All these do not fit the pattern of a deranged mind.

Before the age of fourteen, there is a normal manifestation of auditory hallucinations in children. There is an estimated at 8% of such cases in this young population. Of these, 60% will outgrow it.[372] Joan's hallucinations lasted and became more precise with time. Her visions were always positive, bringing her precise guidance and comfort, and she wept when they left. She would often have her visions after hearing bells.[373] Our science is not advanced enough to allow us to dismiss these sensitive experiences as mental illness. Under the LIFE biosystem, bells are a powerful agent of balance of the two polarities. This explains why they are used in every religion. Therefore Joan's system was balanced and healthy.

We recognize the negative symptoms of schizophrenia when a person stops smiling and speaks in a monotone voice. Movement disorders are the easiest recognizable symptoms of schizophrenia, since they cause a person to appear dramatically clumsy. Some people with schizophrenia even fall into a catatonic state in which they are unable to move.

Neuroscientist Antonio R. Damasio and his colleagues have assessed that patients with damage to the ventromedial prefrontal cortex,[374] part of the idealistic self of the Human brain, also lack the ability to feel where morality lies. For example, in some tests patients were confronted with a dilemma: "What would you do if by pressing a keypad you can save five people, but this will imply you provoke the death of a bystander?" Such brain-damaged patients lack the moral qualms most people experience when having to make a choice that harms some people but benefits others, and adopt a cold, clinical, rational form of decision making. This shows their propensity to consider living beings as mere objects. Only numbers matters and people are interchangeable; quantity for them rules over quality. But what if the bystander was young Einstein before he made his discoveries?

The autonomic nervous system is believed to send sensory stimuli to the vmPFC, which encodes their emotional value and projects them to the Reptilian brain, which generates the physiological responses.

In primates including humans, there is evidence that ventral aspects of the PFC (idealistic phase) mature before the dorsal ones (rational phase) in harmony with our succession of phases (Alexander, 1982; Thompson et al., 2000).

Again, concerning the vmPFC and in agreement with the LIFE biosystem, research has shown that belief activates the vmPFC while disbelief is registered by the insula.[375] I would like to know which part of the insula they refer to. Spinoza speculated that most people accept as true what they can comprehend quickly. Maybe he meant what seems logical to them (Analytical brain). Skepticism, which has its most outward application in science through the scientific principle of the null hypothesis, thereby stipulating that a claim is deemed untrue unless conventionally proven otherwise, is allowed by the Analytical brain. Both are natural tendencies. One is a natural inclination for the intuitive, feminine Human brain and the other is natural for the down-to-earth, masculine Analytical brain. In a healthy brain, they should work in concert. The insula is where we decide if information is blocked or not, as the lung aspect of the Universal brain is in charge of exchanges. As we have seen, patterns accepted in this Human brain and pertaining to identity filter notions coming from the environment.

The onset of puberty marks dramatic changes in hormones levels and changes in physical appearance. This period is characterized by psychological changes linked to a person's sense of identity, mood, self-consciousness, and his relationship to others. Recent studies in neuroscience suggest that hormones alone are not sufficient to explain the breadth of these changes. So are empathy responses that we already associate with ventral and medial prefrontal cortex (Damasio, 2000; Hopkins, Dywan, and Segalowitz, 2002; Spinella, 2002). Several studies also support the role of the right vmPFC in social cognition. In a PET study that investigated perception of emotion by showing subjects pictures of the eye regions of actors expressing friendly or hostile emotions, the right medial OFC, was activated (Wicker et al., 2003). Patients with lesions of the right vmPFC had significant impairments of interpersonal behavior and emotional processing. Patients with similar lesions on the left, however, displayed normal social behavior (Tranel et al., 2002). Again, this is in agreement with the LIFE biosystem. Social withdrawal, emotional and affective flattening, and interpersonal oddity result in significant social dysfunction.

The idealistic phase sees the introduction of a closer relationship with the frontal lobe. The "what?" will be answered by the occipitotemporal gradient, which will find information relating to object identity; while the "where?" will be answered by the occipitoparietal gradient, which will process information regarding the object location. It seems that in the frontal lobes, the superior aspect answers to "where?" while the inferior aspect answers to the "what?"

Women's Beauty, from Genital and Subjective to Archetypal and Objective

To reconnect with our idea of knowing thySelf, we have to study the civilizations that preceded us. We are then forced to agree that the idea of man surviving to himself was widespread and implied. We theorize that humans were thrown into this universe of sensation and had to integrally invent a world they had no conscience of, and subsequently create and imagine indications and substantiations as to the existence of this world. To what aim? Reading the texts that survived, we have to agree that the absolute certainty of immortality pre-existed humans' awareness of death. We can read of a general longing for a purer form, a way Man was *before*. In ancient cultures, there was a widespread belief that man was a god limited by a material body, and in death he returned home. This we can still see echoed in some texts from American natives.

"Even the seasons form a great circle in their phases and end up where they started. The life of Man is in a circle also, from infancy to infancy ..." Black Elk of the Oglala Lakota[376]

This also explains why most traditions had rituals such as the sun dance of the Sioux tribe,[377] which is barbarous from our point of view. As a boy grew into a young man, he desired to dedicate himself, which is a way of confirming his identity. For the young Sioux men, their communion with the Great Spirit was a personal, individual matter.[378] They did not talk about it, did not name the Great Spirit (they eventually had to, because of the white man), or discuss the sun dance. Since their only belonging was their body, this was what they offered.

Many rituals and traditions are ancient and are manifestation of our ancestors' beliefs and desires. Do you remember the fertility statues we've all came across in books or museums? Typically, those female artefacts bear huge bellies, have huge, drooping breasts, and are void of facial expression or are actually faceless. Are they cavemen renditions of our modern-day bimbos?

Undernourishment was a pervasive health problem during prehistoric times. A woman of such corpulence would be a rare sighting indeed.

Seeing a woman like that would be a kind of paradise for the male Reptilian brain, and certainly useful for a lonely caveman to reach arousal. But I read that this hypothesis was dismissed although I have not found for which reason. Animals masturbates so why not cavemen? As the illustrations of mammoths were mainly found in areas where mammoths were a rare sighting, these sculptures would materialize a potentiality.

> *Following a period of extreme cold, a period characterized by wind deposits of loess began about 26,000 years ago, when culture level 6 was formed. This period lasted until the time of the culture layer 9, the "Venus layer." Tiny snails, good indicators of climatic and environmental conditions in the ice age, are found in the loess in the upper levels at Willendorf and denote a dry, cold climate. The environment consisted of an open plain with a few bushes and trees. Nussberg, a hill to the west belonging to the Jauerling massif, while providing the settlement with some shelter from westerly winds, was probably bare.*[379]

"*Traces of colour show that she [the Venus statue] was originally covered with red ochre, which had ritual significance, since the dead were routinely covered in the same substance.*" (Matthias Schulz, 2008.) This definitely ties the statue with the kidney, thus genital aspect.

> *Excavations at Hohle Fels Cave in the Swabian Jura of southwestern Germany in 2008 recovered a female figurine carved from mammoth ivory ... This figurine, which is the earliest depiction of a human, was made at least thirty-five thousand years ago. The buttocks and genitals are depicted in details. The split between the two halves of the buttocks is deep and continues without interruption to the front of the figurine, where the vulva is visible between the open legs. There can be no doubt that the depiction of oversized breasts, accentuated buttocks, and genitalia are deliberate exaggeration of the sexual features of the figurine. Emphasis put on genital attributes, and the lack of emphasis on the head, face, arms, and legs, call to mind aspects of numerous other Venus figurines, which typically date between 22,000 and 27,000 BCE. Despite the far greater age of the Venus*

of Hohle Fels, many of its attributes occur in various forms throughout the rich tradition of female representations.[380] (Illustrations 2-3-4).

In 2005, *Investigation Discovery* on the Discovery Channel broadcasted a program produced by Ciccada Films entitled "Fat Fiancées." In a tribe in Uganda, it is seen as a sign of prosperity for a groom and his family if his fiancée is overweight. Thus the girl's family spends six months fattening her up before her wedding, and then she spends the next year putting on even more weight in order to attain what is perceived as a figure of abundance. One new husband on the program said that it would be ideal for him if his wife gained at least fifty pounds, but he'd be even happier if she gained one to two hundred pounds. Similarly, a young man named Moses, from the Bahima tribe, stated that he found larger women beautiful and attractive. He then compared fat women to fat cows. Then we saw his future bride, forced to drink on an hourly basis or her grandmother would beat her. With gourds upon gourds of milk and butter, this sad child would be stuffed in the "fattening hut" for six months.

The rituals of fertility to which women were associated depicted the woman's body in its kidney aspects. The women were subsequently associated with the earth, objects, and possessions. In fact, she was an earth mother strictly in its genital sense, devoid of any individual and spiritual element and thus of Consciousness. She was totally receptive, in the image of the kidney aspects, the most yin of all LIFE biosystem's elements. She therefore corresponded to the feminine aspect of the masculine polarity, not to the feminine polarity. This created arousal in the Reptilian brain (masculine or feminine), as it is in affinity with the physical, polarized world. It has *nothing* to do with the feminine polarity which is heart-lung, pancreas yin.

Curiously, we can see remnants of these strictly genital considerations of women in the obligation to wear the full burka and niqab. I have searched through the Koran and Hadith (sayings) for indications that either God or the Prophet asked all women to cover themselves in such a way. Other than the Prophet wishing to protect his wives from nosy visitors, I could find none. This is a relief as I have great admiration for this sacred text, since I believe its core mission is to trigger awareness of the feminine polarity in the world (diagram 3). When it says in the Hadith that the full body of a woman is *awrah*, it

Isis Code

means that since women's sexuality is superficial, she is exposed and should be protected.[381] The meridians in which a woman's sexual energy flows are superficial. Also, all of her skin expresses the feminine polarity (lung, heart, pancreas phases). Therefore, the natural and respectful dress code for a woman would be to underscore this subtle aspect by accentuating the feminine polarity while protecting its receptive aspect. It is not a question of humility or modesty but of psychological hygiene. The different colors, shapes, and textures should express the energies associated with it.

Instead, the formal, conservative, colorless way of dress of some Muslim women corresponds to the masculine polarity and masculine body, in which sexual energies are profound. Always dressed in black, the women are not expressing their heart or lung aspect, but their kidney, the receptive aspect of the masculine polarity. The erogenous parts of the body of men and women confirm this. Unfortunately, as it says in the Hadith: "when she leaves her home, Satan looks at her."[382] This is because men are mostly visual (masculine polarity) in their contact with the world, while women sense invisible energies, in the image of the sense of smell of their lung function. Add to this the seeking behavior of the kidney aspect and you have the reduction by men of women to their genital aspect. Satan looking at the women is in the masculine polarity, not the feminine one. Therefore, it is up to men to purify themselves and evolve to see women differently. They are the ones who are lacking control. When they cease to see and be attracted to women solely for their genital aspect, women will become goddesses. On the other hand, since some women have an exhibitionist penchant, associated to their masculine polarity and immature feminine polarity, I would suggest for women to dress hygienically. This does not mean to avoid colors or beauty. Women are called to be the expression of objective beauty in this world. Of course, for this to happen, men have to stop equating beauty and genital sex, because they negatively influence young women who try to please them. Marilyn Monroe is one example.

Marilyn Monroe was not able to dissociate men's desire from true beauty. As with the fertility figurines, she was dubbed unintelligent, which I suspect is untrue. She became the expression of the kidney energy and eventually ended up destroying herself, as objective beauty is linked to the feminine polarity. At times she must have felt lost and confused with her own identity.

In the future, men will also become more powerful and beautiful, because their own feminine polarity will shine and feed their masculine polarity.

The gallbladder meridian expresses beautifully the feminine polarity. Personally, I consider the sari one of the most beautiful garments for women. To make a woman hide her beauty is to consider her only as an energy receiver (kidney), not as the beauty bearer and expresser she was created to be (heart). Why would God, who is Beauty, ask a woman to hide this aspect when he specifically created her to express it? To hide the genital aspects is normal, as they are linked to the masculine polarity. We call this modesty. That is why we must explain to young girls that beauty and sexuality are two different things. How could men who suffer from oversized masculine polarities understand what beauty is? They can't.[383] They can brainwash and coerce women as much as they want. They are responsible when they impose on their wives, sisters, community what has not been explicitly asked, or makes no more sense to ask, as this is not part of the matrix of their religion. It is solely linked to their need of power, control, and objectification of women. They discredit their whole culture. They put in the world an image of their God, thus of individuality, which is erroneous. This keeps people from the understanding found in sacred writings and teachings. The religious upheaval of the eleventh century in France was related to this, as the Catholic Church went the same masculine route, inciting men to beat their wives.

God speaks through facts. He chooses women to express beauty. Dr. DeBavelaere expressed his concern when he confided to me that a great number of his Muslim women patients, who kept themselves covered, suffered from hormonal imbalances, parturition problems, and emotional distress because of the lack of natural vitamin D. Vitamin D in a bottle will never replace sunshine, as the former is soluble in liquid and the latter in lipids. Recent studies have also shown that higher vitamin D levels are preventative of both diabetes and cancer in part through lower insulin levels. (Cohen, Endocrine-Related Cancer, 2012).

If this custom were holy, this would not happen. By agreeing to maintain it, these men limit the expression of God in this manifested world. Fads have no place in true religion. Consider this quote from

Isis Code

the Bible, Genesis 38:14–15 (NIV): "[Tamar] covered herself with a veil to disguise herself. ... When Judah saw her, he thought she was a prostitute, for she had covered her face." This also reminds me of the faceless statues. Now we know what their function was.

I will end this section with a story to exemplify what feminine beauty is, as expressed through the LIFE biosystem.

Intuitively, all the feminine fashion traditions, when linked to a healthy feminine polarity, enhance its meridians. Let us follow the movements of a sacred dancer as a synthesis of all traditions, so as to view their path.

I present to you Aisha, our guide, who will dance for us. She explains that before performing a dance, she goes through a profound ritual in order to attune to sacred energies and free herself from the shackles of incarnation.

In the rhythm of the ritual, she says she feels Beauty as well as the principle of dance, which must not only inhabit her but take over. Smiling and opening her arms, she says: "Aisha will be more than Aisha. She will be dance."

A wonderful fragrance emanates from her. Leaving us, she goes to sit in front of an oval mirror. With a dark kohl pencil, she underlines the outer corners of her eyes to enhance the power of her gaze. From the lobe of her right ear she suspends a chiseled, sparkling gold earring. She draws her lustrous, fragrant ebony hair backward. To hold it in place, on the right side, the same side as the earring, she affixes a hair comb made of the same chiselled gold. Opening a small chest, she retrieves from it an amber necklace that she says is a thousand years old. It is a gift from her dance teacher, and she drapes it around her slender neck. Dressed in a long silk green tunic, slit along the sides of her legs, she now drapes a shimmering red fabric around herself, enhancing her fluid shape. She slides a golden ring onto the fourth toe of each foot. Attached to these rings are thin chains scattered with small crystalline bells. She fastens the chains to her ankles.

She looks at herself and smiles. Over her garment, she wraps a golden strip of silk as a belt. The buckle is in the shape of butterfly wings and bears small emeralds and rubies. Sitting again in front of her mirror, she applies a clear powder to her face and enhances her lips with a ruby color. She exits to a small perfumed room, where she meditates in a silence filled with her gratitude and dedication. With all her love, she links herself to her mistress, who taught her how to become a sacred vessel. She feels ready. Her body is not her body anymore. She inhabits the whole space and the whole space inhabits her. She was, is, and will be.

Soon, through her dance, she will express the feminine polarity, uplifting her awestruck audience toward light and beauty. Awakened, their own feminine polarities will vibrate. The two guards who lift the curtains for her are not mistaken. As she passes, they instinctively bow, filled with a profound respect (illustration 5).

William James, the American psychologist and philosopher who was trained as a physician, mentioned two types of consciousness, an upper self and the under self, which had complementary effects on each other. They were viewed as fields, guiding attention and behavior.

Jung used this idea of the field in his observation of the striking images repeatedly experienced by damaged psychic bodies. Those images generally occurred in mythologies, alchemy, religion, and philosophy lore, and were therefore deemed a universal collective unconscious. Their emergence in the mind, in patients who had no knowledge of these, was viewed as an attempt by the self either to pursue a greater expression of Consciousness or as a tool to reestablish a broken equilibrium.

Since the individuality is immortal, it is difficult to get in touch with it if the idea of religion as bearer of these archetypes and their memory is frowned upon. What is crucial at a young age is sincerity and purity of intention on the parents' side. Whether you are Buddhist, Christian, or Muslim is not relevant to a child. The quality of the community is. For a child, the religion is good so long as the emotions sustaining it are pure, sincere, and heart oriented. In fact, the basic function of religion is to help safeguard openness to the soul, so the being can be unified as an adult. The rituals performed by every religion are carved in time and space, thus on the cells of them; the resulting information links them to the unified Being of which they are fractal, albeit imperfect, parcels. The rituals bring the two polarities into contact.

What people make of religion, mostly a political, material, subjugating tool, has nothing to do with spirituality. Even the sacred texts should be looked at differently. Texts also have a personality and individuality. Their individuality is the same in every religion. Their personality is aimed at a certain group of people because those texts contain the culture and history of a group. This inclusion was necessary to allow the parishioners to understand the texts and to bond.

In the past, religion was not a separate discipline but a way of living. It contained every domain of collective life: politics, medicine, science, history, geography. It identified a group not with the goal of opposing it to another group, but to enhance the cohesion *within* that group. This way, the child, whatever his condition, could feel his roots and be connected to a social fabric. The defective masculine polarity, however, used religion as a political tool.

In regard with the subject of reincarnation, that religions withstood changes is good. If you reincarnated into the same religion, memories linked to that aspect would come more easily to you. That is one of the reason I think young children should be put in touch with as many older religions and countries as possible. It would not only enlarge their vision and bring a sense of tolerance into their lives, but also help them reconnect with their past, be a citizen of the world instead of developing a limited tribal view. I strongly believe that any religion is better than no religion, as long as the one chosen is not a political, paternalistic, or subjugating tool.

The personality is, like water, malleable. From the surroundings it will gather information, good or bad. Only the fire of individuality can burn away collected aberrant information. The individuality is fundamentally good in the sense that it survives in the cohesion of the world and the connection it has with all personalities. To give a better image, individuality is the fire on the candle; personality is the candle. You can share the fire from candle to candle, but the candle has to exist for the fire to be seen. Environmental conditions can extinguish the fire, so conditions are important. We have to take care that the candle is of good quality too, so it does not give off toxic fumes. For the fire to be, we need oxygen (lung aspect). The fire from the candle cannot be differentiated from the fire of other candles. When you blow on a flame, it becomes invisible, though a match will bring it back. The fire itself can be identified as the Christ principle in each of us, which existed before Christianity. The light we see is the feminine principle. The energy of the fire is the masculine aspect of the fire. Our individuality in a given life is this fire attached to this candle. The fire is linked to all other fires. Like each drop in the ocean is linked to each other, like each quanta of the plenum is linked to the others and keeps in memory all that happens to it. It is here (kidney), supported by the candle and the wick (pancreas), needs oxygen (lung), and emits light (heart) and warmth (liver).

The candle without fire can exist on its own, but it doesn't bring light or warmth.

The personality builds plans, elaborates paths, makes projects, and establishes goals, but life itself presents you with a path, often through background details, and this comes to take the entire place in your awareness.

It took me a while to decide which direction my life should take as I could not see well who I was. I had an experience at the site of the Montreal Exposition that helped me find a path. This was actually years after the 1967 Expo, but it was at one of the pavilions where they were projecting a 360 degree film. Audiences were warned that people with "sensitive hearts" should not attend. They were presenting a movie on open-heart surgery. Of course, I thought I would be okay. After all, I'd dissected a few frogs in biology classes with no problem and I intended to become a medical doctor. But as soon as I saw the blood and the heart, I nearly fainted. Only a race to the exit door saved the spectators from the embarrassing sight of a young woman fainting. My career plans suddenly changed as added to this I was unable to bear the stress of science exams.

I took a course in French literature and communications with hope, because I knew I could write plays with ability and happiness. I had experienced some successes in previous years, and enjoyed all the details related to play writing. I could not see myself in any specific career yet, but knew I would enjoy any work in the theater at any level. The creative work, the camaraderie, the link with classical

Isis Code

pieces all appealed to me. Unfortunately for me, the ambiance of the university was not for me. I felt alien to the surroundings, without being able to pinpoint the cause. I simply felt I did not fit in.

Even the physical environment was difficult for me to bear. Later in life I came to realize that I have a vital need for light, fresh air, and nature on a daily basis. How others could cope, even thrive, in university surroundings? I could not understand. This fact reinforced my feeling that I was ill adapted to life, weird, and lost. I finished my first year with excellent grades but was not satisfied. I was living in a constant foggy mental state, and my life felt absurd to me. My love for a student in medicine was the only thing that made me desire to hold onto university life. Then that changed.

Just before the Olympics in Montreal in 1976, I attended a horse-jumping competition. The boy I loved was one of the competitors. While I was watching the different horses and their riders, a man approached me. I will always remember this moment that changed my life. I was leaning on a fence; he came from my right. He was an older man, sure in his manner. I noticed a TV crew beyond him, but was too shy to look in their direction. After we'd chatted for a few minutes about the horses and competitions and my own riding ability, he asked me what my career plans were. At that moment I remembered that the previous year, one of my cousins had visited Montreal and I took her to CBC downtown. As we walked on the first floor, not far from the exit, automatic glass doors opened in front of us. At that instant I exclaimed, "One day I will work here!" The words just came out, without any thought or calculation, as if someone else had pronounced them. I froze, surprised about what had just happened. My cousin looked at me, also surprised, and probably thought I was being pretentious.

Remembering this I stood straight, no longer leaning on the fence, and without hesitation told the man I loved the media and wished for a career in that sector.

He laughed. "Do you know who I am?" he asked.

I gathered he was probably well known and felt stupid for not recognizing him.

"I am the director of the sports section for CBC," he said. He then talked to me about the upcoming Olympics and how they had an urgent need for script assistants. Since I was in the right field at the university and obviously knew sports, they could test me. If I had the required skills, they'd train me. This would be a convenient way for me to see if I liked the job. Of course it was only a summer job. And of course I happily accepted.

When my contract ended that summer, I was given the chance to stay on but for some reason, maybe my love for theater, I declined. I was told that this kind of boat didn't pass twice, but I thought it was my duty to get a degree before I'd be taken seriously. Income was an issue in our family. Those with the highest salaries were the most respected. I was turning twenty soon and did not feel ready to enter the workforce. I needed to be more educated. I felt I knew nothing. At the same time I had started taking the birth control pill. I had gained a few pounds and did not feel comfortable in my skin. My doctor had told me this sort of reaction to the pill was normal and would go away in a few months. It was, as he explained, better than having menses every three weeks and risking becoming pregnant.

So I started my second year of college. I felt even more lost than the first year, but still loved the theater. One day I came home and lay down on my bed. That was it, I couldn't take it anymore. I enjoyed my studies but hated going to the university itself. I was contemplating quitting altogether when the phone rang. It was the man responsible for hiring assistant director asking me to meet with him. It couldn't have happened at a better time. I went to see him, we discussed a job, and he hired me. I left the university and entered the workforce.

The idealistic phase is intimately linked to the collective phase, which it controls and inhibits. It is also controlled and inhibited by the physical phase, receives input from the emotional one. During this period we can't help it but allow our subconscious to guide us toward our path without the Analytical brain having any control. At this age of profound transitions (fourteen to twenty-one) we see the first drug addictions. All the researches confirm that the earlier the onset of drug and alcohol use, the higher the incidence of severe addiction. As defined by the American Psychiatric Association, addiction is a "chronically relapsing disorder that is characterized by three major elements: (1) compulsion to seek and take the drug, (2) loss of control in limiting intake, and (3) emergence of a negative emotional state when access to the drug is prevented." Sandwiched between the inhibition of the material reality that holds us back and the push of our potentialities to find our social niche, the personality often feels stress as the personal, rational self cannot understand what is happening. We are blind, but we know we have to get somewhere. But where?

As a personal example, I loved the broadcasting ambiance, but as soon as I was not with the crew and was sitting alone at my desk, a constricting feeling would take over. Later in my apartment, it would be the same. I concluded I was not meant to be by myself. Curiously, I was alone most of my life! Nevertheless, these many alone periods always re-oriented my life.

Many factors are responsible for or influence an addiction. These can be found in our LIFE biosystem. As the physical phase controls the idealistic, genetics is the first obvious factor of influence. A family history of abuse of stimulants will negatively impact the embryo as well as an adolescent. The emotional factor, prepared by the physical phase, concerns the relationship between the mother and child (conception to seven years) and the father and child (seven to fourteen). Society, through the collective unconscious, already had an impact in the emotional age through the unconscious and conscious interactions with peers. Finally, the logical aspect, expressed by the father or a father figure, replaces in a sense the immature frontal lobes of the adolescent. During this phase of the idealistic self, the child does not want to be told no. He wants to understand why. There is a normal confrontation between the parents' value system, society's values, and the adolescent's inner compass. It is a period of profound doubts and reassessment. To feel unwell in one's own skin is therefore normal, especially if society as a group expresses imbalances of the LIFE biosystem.

As we have seen before, children at these ages are attuned to the physical phase, and they associate pleasure, and thus drugs, with environmental cues. In a Pavlovian type of conditioning, finding themselves in the same environment where they tried drugs before will invariably trigger drug-seeking behavior.

This is a difficult period in a difficult epoch, because all the safe and sound models have been put aside by our society in a bid to promote consumerism and false interpretations of equality, fraternity, and freedom. The only model left, nature, is on the verge of collapsing under the weight of human abuse. Between fourteen and twenty-one years old, humans have a vital need for models, ideals, and inspiration. It is the community's responsibility to provide these to feed their nascent idealistic souls.

Symbols: Subconscious Voices of our Ancestors and Superconscious Voices of Our Gods

The heart phase is also an open door to the world of symbolism. Universal symbols are vitamins for the soul. Advertising companies understood the hold of symbols on humanity many years ago. As with heraldry, a symbol may represent personality, a family, a country, or a group, and therefore creates bonds within those. It establishes the association of this group to the country of origin, and its physical, emotional, intellectual, and spiritual qualities.

In its essence, an enduring symbol is not associated to a particular time and space. It is by definition universal and eternal. The universal symbol, or archetype, is in the middle of the wheel. If one stands on the periphery of the wheel and looks toward the center, where the symbol emanates from, the middle of our circular rainbow, many levels of interpretation are possible. If you can't find a universal and eternal interpretation to a symbol, then it becomes no more than a logo. Knowing the interpretation of a symbol shared by a group as compared to its true meaning indicates where that group stands in regard to its maturation. They might be at the limit of the periphery with a quasi-tribal outlook, or central, universal, and eternal. The power of the symbol depends on the meaning it has for the people using it through time, and to its significance in regard to perceptible and imperceptible realities. It is a vessel receiving and giving energy to the onlooker. With time, it becomes powerful and part of the universal collective subconscious, a transducer of subtle information.

A symbol should not be hijacked by a group as it is, in its essence, universal and eternal. Also, if a group uses a symbol, it implies this community adheres to its fundamental value otherwise it is a lie.

Isis Code

As with everything else, symbols have been appropriated by the predominant masculine polarity. At this point in our society, there is an abuse of symbols. Groups take a universal one and reduce it to a personal function. Take two examples. The color pink now evokes cancer; and the swastika, which dates back to ancient India and is a sacred symbol, has been reduced to a Nazi symbol of oppression. What a waste for humanity! How will we reclaim those and reinvest them with their original significance? The media and their providers understand that symbols and archetypes are powerful selling tools. Symbols are much sought after when they resonate with a large group of people, to a time in history, to subtle realities. Women are more inclined to receive information and sense from symbols, as they are receptive on the symbolic level. Men are more sensitive to social aspects associated with history and culture (flags, medals, etc.)

Link a symbol to power or money and you have something irresistible. Also, experience shows that a living symbol, that is to say, one full of life, will prevail over any rational argument against it.

Because of our general orientation toward pancreas-type values, our governments are geared toward gathering money by any means. Since most men are mainly of the masculine polarity, and our governments are mostly composed of men, they only see money as the source of any problems and therefore believe it should be a prominent part of any solution. They will not see that most problems originate from the way we treat the feminine polarity. Thus, what was not accepted coming from organized crime (gambling, alcohol, drugs ect…) is now accepted coming from the government. Governments seem to think, it will happen anyway, so let's gain from it. But the scale at which it is happening now is much bigger than what it was before!

As example, I have a dear friend who is a serious gambler, so I have had many opportunities to observe players when I accompanied him on his regular escapades. As he grew older, he stopped taking planes to exotic places to play Baccarat, but stayed at the local casino and played expensive slot machines. The most popular machines had the number 7, crowns, roses, diamonds, fruits, bell, and dragons. Why these and not 4s, tulips, or vegetables, although we find these in cheaper slots machines? Because the former are universal and eternal symbols of subtle realities linked to our feminine polarity. The collective unconscious fills them with energy.

The hypnotic repetition of the wheels and sounds (Reptilian and Analytical brain) stimulates the pancreas function and shields the player from awareness of passing time. Unknowingly, the player becomes fixated. It is a form of mantra, a prayer to providence. The gambler feels balanced while he plays, happy without quite knowing why. The likelihood of losing is brushed aside since the hope of winning is stronger. It is like watching TV. You wonder what will happen next. You press those buttons with all your hope. People are literally in a trance, tied to their machines for hours on end without realizing it. Older people are particularly vulnerable, because they often lack dopamine in their systems; hence they instinctively seek ways to stimulate its production. Also, gambling at the casino is often their only contact with the feminine polarity (symbolism, contact with other people, colors, nice surroundings, hope) over their sad economy-driven (pancreas aspect) lives. It has become a substitute to churchgoing with dopamine as an added bonus.

I would witness my friend's exhilaration when he was winning, the Machiavellian calculations he'd make to try to deceive the machine—which by the way was always a she—the rational explanations of why he was winning or not, and the fading of his smile and natural sunny disposition when he kept losing. Nobody is immune to gambling[384] given the suitable conditions. We know that winning, or merely the hope of winning, gives this surge of dopamine in the brain. Also, it has been shown that gamblers remember their wins, but little of their losses.

People used to have to drive long distances or take a plane to get to a casino in order to play. Now they can gamble in the neighborhood bar and—worse for lonely women—online. Gaming commissions warn that people should not play with the hopes of winning the jackpot. Then don't put one gargantuan jackpot out there, but give more money back to the players. That seems to me to be the only honest way of doing it. It is the same with the multitude of lotteries—why does anyone need to win more than $2 million? More winners make more sense. A banker told me that 15% of people rely on a future win at the lottery to provide the funds they need for their retirement. Players feel cheated by their own government. At least make it so

Isis Code

that the money people lose goes directly to something precious and noble! Not toward research for some cure, but for basic needs: prenatal awareness, infancy care, quality daycare, quality primary education, the backbone of any country! That would be the humane and constructive thing to do. I am told that in some states, in USA, lottery profits fund public education. Therefore it is a realistic goal.

As a symbol, the ring is probably one of the oldest and most common. It appears people have always recognized the importance of sealing their unions with a ring. When a woman receives one, to her it means more than just an object. Since the ring has neither beginning nor end, it represents never-ending, eternal love. And traditionally only the woman received it. It meant in a covert way that this man has seen in this woman a reflection of the Self, and that for this they would work on their personalities together, for richer or poorer (pancreas), until death do us part (kidney). Rings for men came later. Of course I don't say everyone feels that way about the ring symbolism, but it is what it means.

The oldest recorded exchange of rings comes from Ur, the city of Gilgamesh and Enkidu, some five thousand years ago, followed by ancient Egypt about 4,800 years ago. As children do now, the Egyptians fashioned rings and bracelets from reeds and other plants growing on the banks of the Nile. They were well aware that relationships were similar to the knots one makes. Maybe the expression "to tie the knot" comes from this ancient time. Indeed, the ring is a knot you can't untie. Life for ancient Egyptians was an occasion to untie knots from previous lives and to tie new ones in this life, and the ring symbolized an eternal commitment. Later on, for the Romans, the ring became more prosaic, and its acceptance by a young lady was a binding, legal agreement. The girl was no longer available.

There are also many theories as to why a particular finger came to bear the wedding ring.

Both the ancient Egyptians and Romans believed that a vein—the vena amoris in Latin—ran directly from the fourth finger (the ring finger) to the heart. Acupuncture teaches us that the thumb is linked to the lung aspect, the index finger to the large intestine (lung aspect), the middle finger to the Master of the Heart, the little finger to the heart, and the one generally used, to the triple warmer of the masculine polarity.

Jewish brides have the ring placed on their index finger, since this is the finger with which they point to the Torah as they read. The text of the Torah is therefore purifying. Early Puritans refused to wear wedding rings because they considered jewelry frivolous, yet in Colonial times couples exchanged wedding thimbles—a useful and practical gift, and therefore acceptable—but after the wedding they often cut off the bottoms, thereby creating rings.

The thumb briefly challenged the accepted norm in Elizabethan days, as fashionable ladies, led by the romantic and tragic Mary Queen of Scots, chose to wear their wedding rings there, but this did not last, given the sad outcome of her life. The signet ring, which originates from the French word *chevalière*, which means knight, is a longstanding tradition among nobles in Europe and in some other cultures. In current usage, the signet ring is often worn on the little finger, associated to the heart phase, of either the right or left hand (depending on the country).

This knowledge of the significance of fingers and body posture was consciously used by some of the Renaissance artists, notably the painters. It was a form of secret language which the Church could not decipher thus these artists could avoid the branding of heretic and thus escape inquisition (see chapter 5 and 7). One name given to this knowledge is dactylology.[385]

Whether mathematical or grammatical, numbers and letters are also symbols that put us in touch with a community through the ages. These symbols have been chosen and have evolved within a group and have become universal measures used to communicate. At the beginning, little sketches called hieroglyph expressed what a person wanted to communicate. Symbols were born. One person could look at a symbol and understand the depth of its meaning. For someone else, it would mean nothing or have a limited meaning. For my children, when I showed them the first letters of the alphabet, I created a little dance they would mimic, so these would become organic. We had instruments as well to make sounds and musical notes. And I wrote a creative story for the birth and significance of numbers, linking these to their environment.

Isis Code

The evolution of mathematical glyphs was quite long. The word "number" is thought to originate from the Greek word *nemein*, which means "share."[386] The Greek philosopher Plotinus said, "Only the One exists." It is the genetic principle of all. So numbers in their essence were seen by some as division of the one, not multiplication of objects.

The numbers one, two, and three were written as superposed horizontal lines by the Brahmins and the Chinese, in the image of divisive layers, and while those slowly changed to vertical lines in the Western world, they remained horizontal in the Asian world.

The number four was a cross, again influenced by the Brahmins and by the Arabs. Linking two of its sides makes a four similar to the one we know. As much as letters and numbers were an element of communication within a group, they also became elements of division from people who were outside the group. Today, mathematical and scientific symbols are universal, music is universal, and the English language is almost universal, although it fails to express the depth of human psyche. An example of this: in French the word *songe* has no English equivalent. A dream, un *rêve*, is an expression of the personality, while a *songe* is an expression of a personal truth originating from the individuality.

Religious symbols were often used by political and religious powers without striving to understand their profound meaning. They were expressions of a group and often opposed by other symbols. So a symbol such as the cross, which existed long before Christianity, was borrowed and finally became almost their patented, copyrighted logo. Of course, symbols used by "enemy" groups were considered as expression of "the devil." An example of this is the fate of the Pythagorean pentagram (diagram 6).

The first drawings of the pentagram were found in Mesopotamian writings dating to about 3000 BCE. The Sumerian pentagrams served as pictograms for the word *ub,* meaning "corner, angle, and nook." Is this similar to the corner angle Jesus spoke of?

When constructing a building, a foundation stone was positioned at one corner and served as model. In Latin it is called Primarii Lapidis. Also, in Sumer as well as in other traditions, such as the Celtic one, an animal or human was sacrificed and laid in the foundations. In Sumer, *ub* was also the most important religious ceremony, in which people were sacrificed to Ishtar, Queen of Heaven. In Sumerian tradition, *ub* signified a cavity, thus of a receptive nature, and *bu*, its reverse, meant to ignite or sprout thus was expressive. We have here the two principles. The pentagram is therefore of the feminine polarity. It is thus normal that it was depicted in a circle to signify its esoteric and protected dimension.

Also in the Sumerian language, *U* meant time and *ubur* meant measure thus model. The pentagram is therefore a measure. The ziggurats built in those times were large terraced platforms that supported the temples. Built quite high, the platforms grew smaller and smaller as they rose, eventually culminating in a cornerstone at the very top, which eventually supported the temple. Is this cornerstone supporting the temple what Jesus mentioned? "Jesus said to them, 'Have you never read in the Scriptures: "The stone that the builders rejected has become the cornerstone; the Lord has done this, and it is marvelous in our eyes"?'" (Matt. 21:42, NIV)

"Everyone who stumbles over that stone will be broken to pieces, and it will crush anyone it falls on." (Luke 20:18, New Living Translation)

"The stone that the builders rejected has now become the cornerstone." (Psalms 118:22, NLT)

We also find the text in Mark 12:10, Luke 20:17, Acts 4:11, Ephesians 2:20, 1 Peter 2:6, and Isaiah 28:16. It is repeated so often that its significance must consequently be a profound and essential one.

And is not this the implicate order, which was thrown away by the masculine polarity at the same time men (the builders) threw away from their consciousness the reality of the feminine polarity?

"The legacy of Pythagoras, Socrates and Plato was claimed by the wisdom tradition of the Hellenized Jews of Alexandria, on the ground that their teachings derived from those of Moses. Through Philo of Alexandria this tradition passed into the Medieval culture, with the idea that groups of things of the same number are related or in sympathy. This idea evidently influenced Hegel in his concept of internal_relations.*"*[387] (**My emphasis**)

"The ancient Pythagorean pentagram was drawn with two points up and represented the doctrine of Pentemychos. Pentemychos means "five recesses" or "five chambers," also known as the pentagonas—the five-angle. It was also the title of a work written by Pythagoras's teacher and friend, Pherecydes of Syros, and a pre-Socratic work concerning the creation.[388] In the Labat, the dictionary of Sumerian hieroglyphs/pictograms, the pentagram is also depicted as having two points up, rather than just one. Some religious groups have wrongly associated this with Baphomet (sounds similar to Mahomet), to the devil. Also, if we take the upright pentagram of Hygeia[389] and draw a pentagram inside its center, the result is an inverted pentagram. This symbol surrounded by a circle describes the inner self (diagram 6) and is thus similar to Agrippa's or the Pentemychos pentagram (diagram 6). It was considered outside time and space; that is, from the place that allowed for the ordered cosmos to appear. The Master of the Heart is also figured in the writings of Pherecydes of Syros. Indeed, as mentioned earlier Zeus made a cloth he decorated with all the elements of earth and sea, and presented it as a wedding gift to Chthonie (the physical aspect, kidney aspect), and tenderly wrapped it around her.[390] In *Cratylus*, Plato said that the ether "had a penetrating power that permeates the whole world"; a power that we can find both inside and outside of our bodies. This is similar to the Master of the Heart or to the Akasha, which is the first element from which all others originate. And indeed, the Pentemychos is everywhere as a reality and invisible model, guiding formation.

In ancient Greece, Tartaros (or Chaos, according to Hesiod) was the first existing darkness from which the cosmos was born. While it became separated from the cosmos after this one emerged and was duly ordered through the implicate order, it never ceased to have an influence. In fact, it was known as "the subduer of both gods and men" (Homer). The psyche (soul), which for Carl Gustav Jung constitutes not only the soul but every manifestation,[391] issues from this Chaos. This boundless darkness held influence on the cosmos through Mychos, or Krater. Is there a link between Krater—a type of Holy Grail—and Kether? It is the passage from *there* to *here*, and from *here* to *there*. In many of the Greek tales we encounter heroes, philosophers, or mystics who have to "descend" through Krater to Tartaros/Hades in their quest for wisdom. We can summarize this action by the word "incarnation". It is where one would find the pentagram, the implicate order, also known as Ma'at in the Egyptian tradition but one had to pass through Hades in order to acquire this wisdom.

In the Babylonian context, the edges of the pentagram were probably also orientations: forward, backward, left, right, and above, to which the name of the five Sumerian gods were given. These correspond to our main Egyptian gods.

These directions had an astrological meaning, representing the five planets: Jupiter (left), Mercury (right), Mars (forward), Saturn (backward), and Venus—Ishtar, the Queen of Heaven (above). Curiously, it appears also that the pentagram figure could be associated with the planet Venus. It is said that every eight years the planet comes back to the same place in the skies, but that the planets that accompany it have changed. This is called the pentagonal cycle of Venus. In this procession, the earth describes a similar pattern. The result is a pentagonal synodic series which takes about eight years and consists of five synodic cycles. After forty years, Venus has made a pilgrimage through the entire zodiac, describing a five-branch star around the sun. This explains why the number forty is so dear to Kabbalistic and esoteric researchers and found in so many instances to mark the length of an important cycle. The pentacle as a symbol of the feminine principle is embodied also by the rose, the small five-petal rose, and most perfectly in the Tudor rose, a hidden pentagram.

This also agrees with the Chinese tradition that gives cycles of seven years to the yin (feminine) aspects and cycles of eight years to the yang (masculine) aspects. It seems that Venus appears at its brightest when in the shape of a crescent, which looks similar to horns. In different mythologies, many ancient goddesses expressing a function similar to Isis's are portrayed with horns. Horns, in fact, in the remotest times connected with the ideas of inward, divine power. In tradition, based on a description in the Vulgate, the Latin translation of the Bible used at that time, Moses was described as having two horns. This is now considered a mistranslation of the word for halo, but if that were so, I doubt Pope Julius II would have been happy with Michelangelo's statue of Moses, which does have horns, for the statue would then signify the devil. Therefore, this idea of horns being godly must have still been around, since the statue was not seen as negative. In my research, I have read opinions that consider the mistranslation hypothesis as also problematic.[392] Not being a specialist of those languages, I prefer to rely on the artistic intuition of Michelangelo and on what he did with this information. If Moses had symbolic horns, then they would be as Michelangelo depicted. Let us remember that this statue was commissioned in 1505. There was an underground movement of knowledge that would have seen the horns as eminently positive. The horns' presence does not mean Moses wasn't glorified, on the contrary. It is not one or the other.

Reportedly, when Michelangelo finished this piece, he took his hammer and hit the right knee of the statue, commanding it to speak. And it is true that when one looks at it, a shiver runs up the spine because the statue looks alive. If we study it, we see that the horns sprout from the dorsolateral prefrontal cortex, the Analytical brain. Thus this aspect is functional in Moses and linked to his functional Reptilian brain, as horns are expressions of genetic energy. This idea that horns are associated to genetic energy and the belief that ingesting horns will increase this energy has led numerous Chinese to kill the poor rhinoceros in a desire to appropriate the energy supposedly contained in the horns. This is but a misguided intellectual notion. But there is more to Michelangelo's intuition. If we take the inverted pentagram as well as our brain and the LIFE biosystem, the right horn is associated with the heart center, thus to the individuality, the right hemisphere of the brain, and the feminine polarity. The left one is linked to the pancreas center, the personality, represented by the left hemisphere and the masculine polarity. As we can see, the horn associated to the masculine polarity is oriented toward the left, toward the feminine polarity, as Mose's gaze is. This Moses is sitting with his right arm resting on as much as holding the tablets of the law. His masculine polarity is therefore using these as a foundation. He is looking toward the left, with his left leg folded back. This pose implies that if he follows the direction of his head, he will have to put his whole weight on the left leg, the feminine polarity. In this he imitates the first pharaohs. He is therefore a true king (illustration 6) as both of his polarities are functional.

In the tapestries of *The Lady and the Unicorn*, the posts framing her action are adorned with this symbol of Venus, an upward facing crescent; even though it is argued that the symbols correspond to the family crest of the man who commissioned these incredible tapestries, Jean Le Viste. Curiously, he was a weaver of Italian origin. This, I believe, comes more from the fact that in the thirteenth century the weavers were strongly associated with heresy of the feminine polarity resurgence.

The Muslim faith adorns its temples with a crescent, which appears to be Venus and not the moon as some have said. After all, their holy day is Friday, *vendredi* in French, which means Venus. This would make sense since in the tradition of the Five Books, Moses and Aaron, in Abraham's steps, brought the whole system, Jesus and the apostles highlighted its heart aspect, and Mohammed the lung aspect. Since socially we are controlled at this point by the pancreas aspect, we have not yet understood the message, probably because we've dismissed the aspects associated with the feminine principle, the cornerstone. Nonetheless, we are evolving, so one day we will fully comprehend.

The Pythagoreans appropriately named the pentagram Hygiea, which means health, and saw in it mathematical perfection. Indeed, the golden ratio seems to be expressed in it. For their part, the medieval neo-Pythagoreans represented the five classical elements by the five vertices.

The difference between traditional Chinese medicine as it is used now and medical biocybernetics, as per Dr. DeBavelaere, is that in biocybernetic, the feminine aspect of the system is as valuable, if not more important, than the masculine one.

When the aspect of lung and heart are dominant in a structure, we can say that it is mainly feminine. Flowers, to give an example, are feminine.

I read a well-written text from an acupuncture school that explains the elements of TCM, and was surprised to find they considered only three different types of energies important, instead of five as in medical biocybernetics. I quote:

"Actually, there are many different forms of 'Qi.' There are generally three forms of force, momentum, or Qi that will concern you and your doctor that sustain, protect and regulate your body."[393]

In TCM, we westerners are seen as not being able to understand Tao or the system of the five phases, because we are deemed only capable of linear thought. To my knowledge, no human can fully comprehend Tao since an element (in this case, a human) can't be greater—that is, comprehend—the sum of all elements, which in this case is the Tao. The understanding of the Tao by a human will always be limited.

This can be further demonstrated by two different observations. While lecturing at a convention on acupuncture in Beijing, Dr. Jean-Claude Darras, who helped confirm the existence of meridians, became ill. Five of the most eminent Chinese acupuncturists visited him. He was lying on his back, feeling quite vulnerable. In turn they all took his twelve meridian pulses. As he related later, he grew quite uncomfortable, because they started arguing right there in front of him about the pulses! Different schools can have different approaches, and all say that they are expressing the fundamental Chinese medicine. This is the same for every movement, whether it is scientific, religious, or medical. Every masculine personality has its own point of view. The only way to discern where the truth lies is in the results and in the possibility of universally applying the "rules," thus through the feminine polarity.

At another time, Dr. DeBavelaere was teaching his biocybernetic system to acupuncturists. In the group was the president of TCM of Alberta, in western Canada, a venerable Asian lady who could be considered the epitome of the Chinese medical tradition. At the end of Dr. DeBavelaere's lecture, she said she wanted to hear more, and then added: "It is intriguing. But it is not traditional Chinese Medicine."

We traditionally associate light with the sun and with the heart, and it is a fact that there are more myocardial infarctions during solar events. In our biosystem, the heart, although part of the feminine polarity, is the most yang, or expressive, of all the organs. This thought therefore comes to mind: light cannot understand shadow because it can't receive it. Light, our individuality, can go into the shadow, but shadow, our personality, cannot go into the light. It gets destroyed.

> Who is the conductor of the multi-voice biorhythmic symphony of our organism and of our brain electrical

oscillation frequencies, in particular, of the steadiest a-rhythm frequency? The Russian scientist Slutsky, who specializes in the study of biorhythms, put forward that animal and human organisms are biorhythmic systems. These were developed thanks to epigenetic factors, such as a long evolution under the influence of the environment, the earth, and cosmic factors. Therefore, they are totally in affinity with this environment. It is necessary to search in the environment for the source of electromagnetic oscillations with frequencies corresponding to the frequencies of the brain rhythms. In Slutsky's opinion, such a source is the geomagnetic field with an oscillation frequency of 8–13 Hz.[394]

Thus, we can consider the brain as a radio receiver, which is self-tuned according to the frequency of the Earth's magnetic field.

Brain researcher Michael Hutchison states in his book *Mega Brain Power* that the 7.83 Hz frequency (Schumann resonances) has been found to be one of those "window frequencies" that seems to beneficially influence the human body. This frequency, it is said, is one our biological systems should attune to, in order to be in resonance with the earth's magnetic frequency. The *"natural electromagnetic matrix for all life on this planet, the frequency in which all life forms evolved, and until recent decades, the dominant electromagnetic frequency in which all life took place."* Hutchison calls 7.83 Hz the "electromagnetic matrix for all life on this planet."

The Human brain in a healthy state has also been shown to oscillate at 7.83 Hz. Consequently, our brains are in a natural state of resonance with the earth. The loss of this characteristic would result in not insignificant limitations to our vitality and health.

The examples above, verifying an operation of the golden proportion law in physiological rhythms and the functions of the human organism, illustrate interconnection between phenomena and processes of both the physical and biological worlds, according to the general law of proportionality in nature.

Indeed, the brain seems to be a sensitive electromagnetic organ. For example, Schumann resonances, these global electromagnetic resonances due to lightning discharges, appear to interact with the brain, altering its waves and neurohormonal responses. Belov et al. (1998) demonstrated the link between geomagnetic storms (GMA) and some human health effects. Violent GMA effects are considered stressors, while low frequency magnetic oscillations such as a 3 Hz signal have a sedative influence. Many researches show the effects of GMA on humans and animals, memory and attention (Tambiev et al., 1995), aggressiveness and war (St. Pierre and Persinger, 1998; Persinger, 1999). In the health sector, correlations were found between GMA and seizures (Rajaram and Mitra, 1981). Birth rates were observed to drop and mortality to increase during increased solar and GMA (Zeitseva and Pudovkin, 1995). All these underline the necessity, for human health, of a healthy environment.

On Love

When you wish to share your daily life with someone, one question you must ask yourself is if your rituals are compatible. Rituals are the anchor of the Self in the material life. They keep you sane. If your rituals are conflicting, then both of you will suffer and eventually question your love.

You meet a thousand people and nothing happens, and then you meet one and your whole life is changed. Our chance meeting is movie material. We epitomized "they lived happily ever after," but have jointly decided to live physically apart. As the term "soul mate" implies, the physical level is not really in affinity with this type of relationship. When we were together we would talk endlessly, with little desire to do anything else. We eventually recognized that there was no dynamism, only peace between us. It is as if you join two elements, one positive and one negative, of the exact same qualities. What happens? No more dynamism. This is not particularly productive. When we were together, I had no energy to act physically and was becoming more and more inhibited, through no fault of his. We needed a separation, since our only desire was to bathe in the beautiful magnetic ambiance our two hearts generated together. We lived in the same world outside of time and space. On one level, it was bliss and peace.

When we physically meet or talk now, the peace is still present. Twins separations are always difficult. Luckily for us, we are not twins physically, and our independent personalities are pleased with our choice. Now I know, what we share is fin'amor. When in the Gospels Jesus says to his disciples to live with their wife as they would do with their sister, this is what he referred to. In fin'amor, physical intercourse seems restrictive and inappropriate. I will discuss this type of love in the social self phase.

I have been asked: Can we truly love two people at once? I believe we can love many people simultaneously at the soul level, as the desire of the individuality is to be One. The levels and phases involved are not the same, though. The more subtle the level, the more people you can love without hurting anyone. Also, you can love someone on many levels. We would believe that the ultimate joy for a couple would be to love each other on all levels, but love is not static, it goes from phase to phase. The one based on the individuality though is the only lasting one.

Loving your child is not the same as loving your mate, a brother, or God. It is love nonetheless, and one is not lesser than the other. Love expresses itself on different levels, and as such has different functions. In the same way water can be ice, steam, rain, or the ocean, love is. The states of the individuality are not the same as those of the personality, so love in each of these will not express itself in the same way.

We all share the same illness and pain: separation from the One. This is inherent to our materialized status. You can be totally in your personality and not suffer from it, but then there is a little voice calling you to love. It is easy to hush it down. I say, listen to that voice. It is the first step toward perfection and happiness. Become aware that we are not complete. Then understand that this happens either because we have not matured yet, or we have acted against our own balance; or we have accepted elements that went against it; or in our youth our system was pushed out of its natural balance, creating a hiatus between our individuality and our personality, or simply because our Idealistic self was shushed.

Next, we have to understand and take responsibility for our physical, emotional, mental, and synarchic aspects, and accept the necessary steps we have to take in order to allow our evolution. We have to find the echo of this implicate order or LIFE biosystem within us and harmonize ourselves to it. Our personal evolution happens through this constant striving toward a return to the universal balance and conscious reintegration into the One. Health expresses itself truly through the resilience of our individual system. Thus the quest for health is necessary to our evolution. The esoteric model we all carry within us, our ultimate identity, motivates us by its power despite its seemingly irrationally. When a person's personality is not a slave to the physical and emotional life, then all of his being tends toward individuality, in the same way the seed of an oak tree tends toward its ultimate shape. This is what freedom refers to: the freedom to tend towards the Self.

Sincere rituals harmonize our physical, emotional, and mental aspects—which we have a tendency to name consciousness—to the universal and eternal archetypal world. This world resides in the center of our whole being, and we have a tendency to name it our subconscious or unconscious while it is more a super-consciousness or Consciousness.

Rituals are deemed religious in the sense that they put us in the field of a dimension outside of limits, outside of time and space, to a dimension in which we are whole and one. For the duration of the ritual, we are one. The Christian religion as a whole went through an epoch in which it determined that rituals were dead, because the sense of those rituals was lost. By doing so, the Christian tree could not flourish. A true religion will thus be recognized through its rituals even more than through its dogmas, as only rituals touch our physical identities. A ritual should be recognized and understood, and should move the members of a community, to feed their culture and add to their identity. The closer this religion and its members are linked to the eternal and universal center of reality, the more they will be recognized and respected by a vast number of individuals. For who does not desire to quench their thirst in a clear, pure, and crystalline spring?

Happily, the soul goes toward its purification, which often implies suffering for the personality, as we have to let go of one state in order to reach another one. Nevertheless, we have to be conscious that the suffering of the personality does not necessarily mean purification of the soul. Otherwise, it is an open door to all abuses in the name of God. You suffer; therefore you will go to heaven?

What a sad and limited philosophy! Unfortunately, this is the one my parents received.

Some say that the devil hides in what we love most. What devil are we referring to? Do we mean that pain hides in what is dear to us? Of course, since most often love pushes us toward evolution and purification. Since some do not believe in cycles between dimensions, they have a tendency to assimilate what hurts their personality with what is bad. In that case, their own souls could become assimilated to the devil.

The quest of the Self is a quest for the sacred, for the divine. Consciously or not, willingly or not, we are all pilgrims. The object of our quest is our Self in its wholeness. Its code is engraved in our soul, echoed in our body. The discovery of I implies walking the path of the divine, for only it can lead us to our soul and its silent code. This can be done only by opening up and believing the reality of the individuality.

"Do not give dogs what is sacred; do not throw your pearls to pigs. If you do, they may trample them under their feet, and then turn and tear you to pieces." (Matt. 7:6, NIV)

This world came into existence as sanctified and will end up sanctified. The middle aspect is a muddled affair. The pearls are the fruit of our reflection, born of all the difficulties encountered during the span of our lives. It is the next phase of the Analytical brain which could transform us into a "pig," in the event it was not properly fed by the heart aspect and thus unable to control the Reptilian brain.

Small minds, often a consequence of an unbalanced LIFE biosystem, use scorn and aggression when threatened by ideas that exceed the convention of their ingrained patterns. Nevertheless, this does not, in any case, prove them right. The only thing they will achieve is to render this world sterile, ugly, and dead, as life will withdraw from it. They will eventually have nothing left, not even their life.

Chapter 6
The Analytical Brain, the Rational self, and the Personal Human

"There is One Ideal, written within One Heart, resulting in One World. Only we, through our Analytical brain divide it.
We from heaven: There is One Being of whom we all participate, from rocks to humans to stars.
We from hell: Only we divide, and lay divided." **Ariane Page**

"Nothing is in the intellect that was not first in the senses."

"To theorize from the standpoint of the visible is not to advance in wisdom but to succumb to credulity." **Meditation VI, Descartes**

"Every scientific theory is incorrect. For a scientific theory to be correct, it should take into account all the phenomena, visible and invisible, past, present and future, known and unknown, which is, of course, impossible. ... When propounding a general theory in science, the one thing one can be sure of is that, in the strict sense, such theories are mistaken. They are only partial and provisional truths which are necessary ... to carry the investigation forward; they represent only the current state of our understanding and are bound to be modified by the growth of science." **Claude Bernard, founder of modern experimental medicine, 1813–1878**

*"I was ten years old [1808] the winter I saw a Wasichu [white man] for the first time. The buffalo were so numerous then that we couldn't count them but the Wasichu killed them so much that there were only carcasses left where they used to roam. The Wasichus were not killing them to eat them; they killed them for the metal that makes you crazy and kept only the skin to sell. Or they would not even cut them up; they would only take the tongue

Isis Code

and I heard of ships on the Missouri laden with dried Buffalo tongues. Those who did that are crazy." **Black Elk**

"We have not the reverent feeling for the rainbow that a savage has, because we know how it is made. We have lost as much as we gained by prying into that matter." - Mark Twain

"What is the undefinable thing which would cease to be if it were to be formulated? The infinite, which would be finite if it could be defined." Leonardo da Vinci

Phase of the How Level

The Pancreas Aspect

- Seth
- Afternoon
- The logical man
- The how level
- Figuratively: day and night
- Element: earth
- Cell structure: cytoplasm
- Function: nutrition and mental energy
- Nutrition: heart phase
- House analogy: the floors on which you can walk
- Level of consciousness: previous plus awareness focused on one point at a time
- Motivated and controlled by emotional
- Biocybernetic phase and function: from nutrition to exchanges
- Biocybernetic main organs: spleen, pancreas
- Biocybernetic polarity: part of both feminine and masculine—the feminine for the nutritive aspect and of the masculine for the analytical aspect
- Women are expressive for the nutritive aspect, receptive for the analytical aspect. Men are expressive for the rational aspect, receptive for the nutritive aspect.
- Level controlled by emotional level/ defense
- Controls the Reptilian brain

The Analytical Brain, the Rational self, and the Personal Human

- Brain aspect: frontal lobe, left hemisphere, left dorsolateral, left orbitofrontal and rostrolateral prefrontal cortex, left parietal and left posterior insula
- Sense: taste
- Construction phase: from twenty-one to twenty-eight years
- Pattern prepared: mental prepares Universal brain
- Type of attraction: abstract- based on the other aspects
- Main relational attitude: I know
- Psychological trait: inquisitive when healthy, conceited when not
- Emotion: calculation, manipulation
- Bible equivalent: Numbers
- Stimulation (good or bad): analytic sciences, sweet taste, accumulation of objects (consumerism), analysis, atheism, nutrition in general, egoism, repetition, rocking movements, reading, economy (not in the sense of saving but in the sense of being able to spend), short fasts (resets the pancreas), mantras, repeated prayers, knowledge, city life, gambling
- Love-type: analytical-personal-egocentric (no love)

This is the *how* phase, as in "How do I do this?" and "How do I get from point A to point B?" It normally controls the *what* phase and feeds the *who* phase. It is controlled and contained by the *when* phase and sprouts from the collected answers to the previous questions. This phase expresses what happened in the other ones.

Denial: the Left Hemisphere

We saw in the last chapter that the *left dorsolateral cortex*, part of the Analytical brain, is the main part triggered when someone tells a lie. Interestingly, patients presenting right hemispheric injuries, thus leaving their left hemisphere somewhat intact, fail to notice or are in denial of their disabling condition (anosognosia). The left hemisphere is linked to their pancreas aspect or, if you prefer, to the rational mind, which boasts that, unlike the other parts of the brain, it is in "contact with reality." Really?

The left hemisphere, our masculine polarity, does not feel pain from separation. It is modular, an island all by itself. This is the way it sees it, anyway. On the contrary, the right hemisphere expresses

Isis Code

suffering, not only when it is in difficulty but also when the left one is.

Our two hemispheres differ in their macrostructure, physiology, behavior, and neurophysiology. Dopamine is more prevalent in the left one. It is critical for phonological processing and for categorical analysis. In our mechanistic point of view of the world, somewhat generated by our masculine polarity (Reptilian, Mammalian, and Analytical brains), emotions are considered feminine and therefore viewed with great suspicion. However, contrary to general assumption, emotions are not a faculty of the feminine polarity but of the masculine one. A woman being expressive at the mammalian level—which explains that women are more verbally inclined—implies that women manifest more emotions, not that they have more. Contrary to a widespread belief also, "control" of emotions is done by a healthy feminine polarity (lung aspect), *not* by the masculine one.

As the heart aspect is in charge of information regulation, the heart senses these emotions and acknowledges them, but they do not originate from the heart. The heart aspect only allows them to flow when they are in affinity with the personality or individuality.

In our bid to control and analyze everything, we decide that to attain knowledge, the first requirement is to put emotions out of sight and out of mind. We excel in this. Our depressive emotional state is the child of this point of view. Now we understand that the emotional aspect is not "irrational" unless seen through this stunted point of view. Rather, it is highly adaptive and, in fact, essential to higher cognitive processes.

The rational self, the Analytical brain, controls and inhibits the physical one. Also, it can analyze what pertains to the masculine polarity, but it is not adequate to understand and know the feminine one, as we cannot analyze life. Life with its cycles is therefore alien to its world, although it feeds and inhibits the masculine structures. Cognition involves not only the masculine aspect of reality but also the feminine one. The knowledge associated with this phase is therefore limited and subject to subconscious elements issued from the Reptilian brain. Nonetheless, it is a necessary phase that should work hand in hand with the others. Cognition manifests through brain waves, binding through synchronization the diverse areas into a mental unity. The whole prefrontal cortex controls this unification of the brain. We should not underestimate its function.

When making a choice, men are normally more context dependent, while women are more context independent. For example, a man may love you now, but as his circumstances change, he might say, "Well, that was then." And then an hour later he might feel like making love to you. A woman, or someone in whom the feminine polarity thrives, is more on the always and forever mode, a gift of the perpetual rhythm of her heart and lungs, necessary to maintain life. A context-dependent strategy reflects a calculation of the information available at a precise point in space and time, so as to custom-tailor the organism's response (Reptilian brain). Its time is punctual. It takes all the available data, at this point in time and in this physical spot, compares it to ingrained models, and computes the points of convergence and divergence. Input new data and the answer might change radically.

There is also an association between well-established routines that become reflexive and the left hemisphere. This explains why surgeries to the left hemisphere in adults, especially in men, brings more upheaval to their everyday lives than in young children, who, like their mothers, rely more on their right hemisphere.

The Frontal Lobes

Patients with problems in their right hemispheres, delusional as to their condition, have been wrongly judged as calm and collected, or as having an emotionally positive reaction when faced by their obvious disabilities.

Traumatic head injury has been dubbed the "silent epidemic." As an example, in the United States in 1989, more than two million people sustained head trauma.[395] The Brain Trauma Foundation now estimates that number to be four million a year. Even if 50% can function, we can consider that their drive, initiative, and competitive edge are often gone. They frequently become passive and indifferent to others. "Frequently they become inappropriately jocular, emotionally volatile, irritable, fractious, and impulsive" (Goldberg, 2009). These

Isis Code

are all symptoms of a deficient social self, thus incapable of inhibiting the Mammalian brain.

Importantly, the frontal lobes appear to be the point of convergence of the consequences of damage to any part of the brain. For instance, regardless of where a brain tumor is located, the regional blood flow will be disrupted in the frontal lobes.[396] In the LIFE biosystem, its attached phase, pancreas, is the regulator of the system, the earthly center of our personality, and thus agrees with this function of merging. It is the only phase in which the two polarities are reunited and reflects the rapport between our personality and individuality. Because of the nutritive aspect of the pancreas phase in its feminine polarity, we can appreciate the profound meaning of the Last Supper and understand why it is that moment which expressed what would occur. In that moment was revealed Judas' soon to happen treason. As well, we understand why sects such as the Essenes would not allow a novice to sit at their table and share their meals for many years. Also, members of the highest caste in India, the Brahmins, are not supposed to have wealth, but as priests they should prepare food in a holy manner. This can be understood through the fact that the regulator of our system not only expresses the state of the whole system but also can be used to rebalance it. Through nutrition or knowledge, the two aspects of this phase, we can ascend the ladder to the Source, to the One.

The interests expressed by the rational self reveal the relationship between our two polarities or lack thereof. In fact, we could say that the truth of someone is expressed through his rational self. His level of Consciousness will modulate this aspect, orienting the personality toward acquiring more of this Consciousness, unless he is dominated by his personality. Once mature, if it ever gets there in this life, it will eventually bow to the heart aspect of the individuality, asking to be instrumental to a wider Consciousness.

Because of the immaturity of the frontal lobes before twenty-eight years of age, mortality rates for young adults from fifteen to twenty-four are more than triple the mortality rates of grade-school children. Teenagers take risks and test themselves in a bid to self-define, a compulsion of the idealistic phase. The US Center for Disease Control has identified three behaviors as leading causes for death and illness in adolescents:[397] namely motor vehicle crashes (30%), homicides (15%), and suicide (12%). Sadly, this last number rises every year.

Related to motor vehicle crashes, alcohol and drug use is a factor in approximately 41% of deaths. More youth in the United States use alcohol than tobacco or drugs.

Each year in the United States, almost half of the 19 million sexually transmitted infections that are newly diagnosed concern young adults. Thirty-nine percent of sexually active high school students admit having not used a condom during their last sexual intercourse.

Brain research indicates that brain development is not complete until nearly the age of thirty, and refers specifically to the development of the prefrontal cortex.[398]

We have seen that developmental processes tend to occur in the brain in a back-to-front pattern, which explains why the prefrontal cortex develops later and the corpus callosum last (diagram 11). Studies have also found that teens have less white matter (myelin) in the frontal lobes when compared to adults. With more myelin will come the growth of brain connections, allowing for better flow of information between the different brain regions.[399] The feminine polarity is at work. Within the cortical layers, axons and dendrites extend vertically and horizontally, forming billions of connections.

In its immature state, the self, tied to this phase, is self/non-self-oriented. It is as if the binary aspect of the Reptilian brain works hand in hand with it. This is true, since the rational phase has control (or not) over it and both are part of the masculine polarity.

Seth and the Personal Self

If we look at the Taijitu symbol, we observe that the masculine and feminine polarities are divided. They do not mix, although with time, they can transform one into the other, in the same manner magnetic and electric energies might. What makes this great divide is the pancreas aspect, which sits between the phase of evolution and involution. In the myth of Isis and Osiris, Seth is not killed, simply because he is essential to our survival on earth—although he

Isis Code

should not be given the reins of the kingdom either. This function is incumbent to Horus, when he will come of age. For now, Seth does as he pleases, to the sorrow of Isis. He made her a widow.

The Idealistic self of most people starves and remains dwarfed compared to their rational one. This allows Seth, the rational self, to close the door to our sub- and superconscious and therefore to blacken the window to eternity. Analysis, in our masculine-oriented society, can therefore only access the "reality" that has crystallized as a result of our past. Because of this, we are not even fully conscious of the present, rendering it lifeless.

On the other hand, through her love Isis will allow the birth and survival of the savior and redeemer Horus. Symbolically, he will truly be born to this world during the winter solstice, when the days are short and the holy bread is scarce. By allowing wholeness, Isis will give a voice to the subconscious, which remembers the words written prenatally on the human soul. By giving birth to Horus, she will allow humanity to ultimately recognize and acknowledge her, and thus evolve. This is the Kingdom of God. John 18:36 (Aramaic Bible in Plain English):

"Yeshua [Jesus] said to him, 'My Kingdom is not from this world; if my Kingdom was of this world, my servants would be fighting that I would not have been delivered up to the Judeans, but now my Kingdom is not from here.'" And Mark 1:14–15 (English Standard Version): "Now after John was arrested, Jesus came into Galilee, proclaiming the gospel of God, and saying, 'The time is fulfilled, and the Kingdom of God is at hand; repent and believe in the gospel."

Christ, the sun figure of the heart, or the Self as per Carl G. Jung—in other terms, Horus—will reign. The symbol of the eagle, or the falcon for Horus, describes well this aspect. With it, the masculine polarity has accessed the feminine one, and therefore is finally freed from time and space. The animal has two wings and can at last fly. Freed from the shackles of the earth, the eagle can finally soar in the blue sky. The dragon was a similar symbol of the Reptilian brain having grown wings through its ascension within the heart aspect. The culmination of this image is the dragon holding a pearl in his mouth: a symbol of enlightenment.

The Analytical Brain, the Rational self, and the Personal Human

Meanwhile, as long as we are incarnated and not mature enough, Osiris is in the underworld, replacing Nephthys's son Anubis, therefore imprisoned solely in a physical aspect. This fact impacts the expression of the idealistic self without us even being aware of it.

Rightly, Seth proclaims that he is the one who prevented the great snake, Apep, from swallowing the barque of the sun. Undeniably, without what he represents, masculine and feminine polarities would merge and the world as we know it would revert to the primal chaos. Our essence would still exist, but not this dimension of time and space. In an earthly language, we would be dead. In a divine language, we would still be alive. Our nature allows us to live the two aspects simultaneously: we can be earthly alive and divinely alive. Nonetheless, as we can lose our physical lives, we may fail in keeping a conscious connection with heaven. Hence the edict of avoiding sin, which is the severing of ourselves from our divine Conscious selves. Etymologically, devil comes from diabolus, which divides.[400] This is the work of Seth and of our analytical minds. On the contrary, the word symbol, from syn-bole, means to unite.[401]

Charles Baudelaire said that the greatest triumph of the devil was that he succeeded in making us believe that he does not exist. In fact, we are so engrossed in our personal selves and Analytical brains that we can't even figure out that there could be anything beside it. We don't see its actions, we are its actions. Seth is incarnation. In the myth, Osiris, who should control Seth, went away to teach humanity, leaving Seth to his own devices. With nothing to control him, Seth became obsessed with power and wanted his brother's dismissal and demise. He murdered him with unalloyed pleasure. Osiris thus has no more conscious control over Seth. In other words, the emotional human now controls the rational one only through the subconscious (Hades), but it still feeds it. Yes, Seth, the rational self of the personal human, "killed" Osiris and ignored Isis. This is part of normal evolution and a necessary path toward expansion of Consciousness. But we must move on and not stagnate at the level we are in now. The Analytical brain is only a tool, not a goal.

Now Osiris rules from the non-manifested associated to the physical reality. The masculine polarity is therefore limited to the physical self. It is not wholly manifested. This is the state of the emotional self at this point. Since it is not conscious, it can't control the obsessiveness and

Isis Code

vacuities of the Analytical brain. A mature Horus, in command, will naturally and without hesitation put Seth back where he belongs.

With a healthy pancreas aspect on all levels, the personality is complete. It will be apt to receive and feed, or not, the flower of our individuality, to grow and multiply, being open to life in *all* of its dimensions.

This period in life from twenty-one to twenty-eight years old is highlighted by the entry into the fold of the community at large through joining the workforce. Being unemployed signifies being out of society, and that is damaging especially for men. One of the first social imperatives is for our young men and women to find work. Young men develop their identity through their *work* and young women use the social group as a mirror to grow, so it also serves their identity. Alas, money-oriented companies obviously do the reverse: they try to minimize the number of employees. Therefore, there has to be not-for-profit associations that will give work to this population, even if this means laboring for communities for scarcely more money than for food and lodging. It could be a type of positive army for different villages and communities. This work should be valued as vocation, since one's place in society's pyramid is an echo of a male's identity. What do men wish for generally? They want to be right (pancreas), to be physically strong or viewed as such (kidney), and looked up to (heart). These are all aspects in which men are expressive. For women, the *social* aspect is even more indispensable to sustain their identities. What do women wish for generally? To be appreciated and admired or loved (liver), socially recognized, even if only by members of her family (lung), and seen as a nurturing person (pancreas). These are all aspects in which women are expressive.

So he finds his identity in the work he performs, and the place in the social pyramid this work occupies; she finds her identity through the eyes of her environment, lover, friends, coworkers, boss, and children. This of course depends on the level of the feminine and masculine polarity each has. Eventually, a mature adult will be less affected by external realities, but it is a fundamental phase of maturation.

In many traditions, the beginning of this period is heralded as such. As an example, tooth filing is a significant rite of passage for Balinese

The Analytical Brain, the Rational self, and the Personal Human

Hindus. It marks the transition from childhood to adulthood. Artifacts found in the Buleleng regency of Bali reveal that the Balinese have conducted this ceremony for over two thousand years. Hence, it was not originally a Hindu ceremony. The Hindu-Balinese people perform this ceremony for their young to rid them of their Sad Ripu, the six weaknesses of the flesh. These are lust, greed, anger, drunkenness, confusion and jealousy. Once the ritual is performed, it is hoped that the young people will be better apt to lead a healthy, well-adjusted life as a valuable member of their close-knit community. Following the symbolic embargo of those enemies by rendering their structure, here symbolized by the teeth, not receptive to them, the participants are then considered adults and therefore able to make decisions based on common interests rather than egocentric ones.

Teeth are associated with the Reptilian brain. The filing is thus a rational act of control over it, in the same way the Analytical brain controls the Reptilian one. It is a way of choosing the greater good during the idealistic phase.

When the personality is too powerful, the rational mind is often at odds with the idealistic one. This can breed a backward-forward syndrome in a tug-of-war between personality and individuality. In the Bible, in the image of the fifth phase, Jacob wrestles with an angel that is sometimes identified as Samael, chief archangel of Ma'on, the fifth heaven. [402] At the end of the fight, at daybreak, the angel maims Jacob at the hip, which traditionally and energetically symbolizes the feminine polarity. Nevertheless, the angel blesses Jacob. Samael is considered the seducer, accuser, and destroyer. He is said to be the one who seduced Eve, but he also stopped Abraham from taking the life of his son. He was said to mate with angels of sacred prostitution; he is thus polarized and of the same nature as Seth. He represents the rational self of the Analytical brain.

An overstimulation of this phase through too much food, too much repetition, too many choices, accumulation, and obsessions incapacitates the Analytical brain. This leads to a lack of control over the Reptilian one, and lead to wars and death.

My paternal grandparents were wonderful people. My grandfather was extremely generous and always invited in the poor. He could be seen regularly pacing the living room and the kitchen of his house, reading his breviary. Like my father, he sang wonderfully in the church choir. At eighty-three, he succumbed to bone cancer.

I had taken ill the day he died. I remember hearing my mother speaking on the phone, and I could feel her sadness. At the time I wondered if my sudden illness was linked to his death, as I loved him very much.

I wonder if his religious, idealistic heart was put in an unbearable situation by his rational mind. Left unresolved, the situation might have eventually triggered his cancer. My grandmother bore nine children. After giving birth to the eighth one, a cervical cerclage (tracheloplasty) was performed on her to prevent any possible miscarriage. My aunt and godmother related to me that one day the priest came to visit them while my grandmother was breastfeeding her eighth child. At this sight he became agitated and angry, and shouted that she was preventing God's work by breastfeeding, thus allowing the work of Satan. He pointed to my grandfather and said he had to put a stop to it. They had to do everything in their power to have more children!

So on one hand, if my grandfather did not have sexual intercourse with my grandmother, he was preventing the work of God and thus allowing Satan in. On the other hand, if he had sexual relations, he would force my grandmother to bring forth more children; he knew she was paying dearly with her health. This is what I call the forward-backward syndrome. Now we know that this was a well-thought political maneuver by the Church. It had nothing to do with God.

The prefrontal lobe is the chief censor of our brains. It automatically passes all the material of the subconscious through the filter formed by the psycho-affective environment that colors our view of the world. The Analytical brain is often more irrational than we would like to admit, as it takes into account only one aspect of reality. Representations stored in the left hemisphere of the masculine polarity are more readily available for conscious experience.

In medicine, if you don't take the prescribed drugs or don't accept the surgery recommended to you, you are considered an imbecile. In religion, you were damned and excluded from the community. Both science and religion went through an era marked by the same

The Analytical Brain, the Rational self, and the Personal Human

rigid Analytical brain and binary Reptilian brain and their regulated patterns. Under this light, neither science nor religion won anything. Both have often been walking on only one leg, the right, looking at the world through only one eye. Both have had at times a parrot on their shoulder that would repeat the same thing: "We are right."

Is it therefore surprising that people feel lost and decide to throw overboard what doesn't help them with their lives, whether it is science or religion, or even both?

The patterns accepted by or in affinity with our symbolic self, color and guide our personal selves. We then form abstract concepts about the world that are shared by our institutions, which become more real in our eyes than the things themselves. With an ailing feminine polarity, we do not see well.

Because it is dominant, the masculine polarity has a tendency to underestimate the value of the feminine one. Stay-at-home mothers are cooks, cleaning staff, psychologists, educators, and doctors. If we were to replace one mother with employees to do the same work, it was estimated the cost would be of approximately $130,000 a year. The end result would not be the same either, resembling more patchwork than unified mind. Filled with our own partialities, we don't carefully perceive how the world actually is, or how we are interacting with it and modify it. Our accepted masculine values breed mayhem and keep Osiris in fourteen pieces.

The enmity between intellect and emotion will not be resolved by the analytical mind gaining dominance over the emotions, but rather by increasing the harmonious balance between the two polarities. This implies an awareness and maturation of the symbolic self. This synergy will provide greater access to our full range of intelligence and consciousness.

Anorexia and the Middle Prefrontal Cortex

I loved the working ambiance at CBC. I was terribly naive. I used to walk barefoot in the hallway; that is how homey the place made me feel. It seemed that everybody knew me, and I felt appreciated. The

Isis Code

world was opening in front of me. My life could be what I would make of it. Save for one thing.

During the summer I worked for the Olympic Radio Television Organization, I went for a few weeks to stay in the country, surrounded by horses, pastures, friends, and the boy I loved. My parents did not know. I was especially happy there. When my contract ended, I returned to my family and asked my parents' permission to go on a camping trip with my boyfriend. We would visit the wonderful Gaspésie. We got permission and left.

We had a lovely holiday. At one point, we traveled into the Forillon National Park. We went down to the rocky beach. It was cold that day. A seal was observing us, head peeking out of the frigid water. Everything was so beautiful. We sat on a dead wood log, and there my boyfriend asked me to marry him. I answered with a laugh, as it was obvious we were too young, and surely he was talking under the influence of this beautiful scenery. After we left the Gaspésie, we went to visit my parents, who had rented a recreational vehicle for their own vacation. I can't remember if it was in Cape Cod or Gatineau. I only remember they solemnly invited my boyfriend and me to meet with them in the RV. As we sat in front of them, with a table between us, I did not feel well. I clutched my boyfriend's hand and felt something terrible was about to happen. Indeed.

My father started: "You went on holiday together. Now you have to legalize your situation."

They had probably asked me if I had made love and, not feeling any guilt, since I considered our love pure, I had told the truth. Now we *had* to get married. For some reason, my parents had changed their minds while we were away. Maybe it was because of the TV program I appeared in with my boyfriend, in which I stated we were living together. I can't blame them.

I felt this was not so terrible at the time, but eventually the reality dawned on me. I would have to rent an apartment. My boyfriend was still a student in medicine, and would be for at least six more years. I was too immature to think numbers, but a restrictive feeling grew in me. Something was amiss. My mother, who had been a nurse, cleverly made the remark to me that when medical students

The Analytical Brain, the Rational self, and the Personal Human

become doctors, they leave the woman who supported them during their studies. At the time, I was working during the day at CBC and during the evenings as a hostess at the Blue Bonnets Racetrack. I didn't want the life my mother lived. Because of my brother and sister, my boyfriend had told me of his fear we would have handicapped children. This was all spinning continuously in my head. I had to stop the feelings from strangling me, regardless of the rational explanations I gave to these feelings.

I had been suffering on and off from anorexia since I was eighteen. Regularly, for weeks at a time, I would limit my eating for the day to a slice of bread dipped in a glass of tomato juice, sometimes adding an egg yolk. I was skinny but to me looked healthy. Since the anorexia was chronic, my weight varied enormously in a few months. After our engagement, that winter I paid a visit to my future parents-in-law, who were charming people. I felt sick and panicky after the meal. My boyfriend walked outside with me and explained that I could get rid of the nausea by making myself sick. I knew about throwing up. I regularly suffered from liver congestion, which would make me ill and anxious for a whole night. So I got rid of the nausea. Now to my anorexia I added bulimia.

For years I would fight a lonely battle with both of these extremes, hiding my difficulties to those who knew me. Eventually though, throwing up didn't relieve me of the constricting feeling. Six months before the wedding, I met with my boyfriend and explained to him that I couldn't go through with it. I suggested that we go for a year without seeing each other, and we would then know if we were truly meant for each other. I felt we should meet other people, since I was a virgin when I met him. He didn't want to but had no choice. He said we would meet again in one year.

A few months later, after I had dreamed of him almost every night, it seemed obvious my anxiety had been irrational. I knew I loved him. I had started a relationship with an older man. It was difficult and I was confused. I asked to see him. We met in my small apartment overlooking the river. I asked him if he was willing to date me again. He told me it was too late; he was already with someone else. He married this same woman very soon after.

Isis Code

Some psychologists and other clinicians have observed that individuals with eating disorders often lack fundamental coping skills, replacing them with chaotic managing behaviors. There are numerous causes of the stress that leads to the need to use disordered eating behaviors, from desire to be the best, to family conflict and, even, academic pressure. Even food sensitivities can trigger such behavior. Using food or the lack of it as a coping mechanism is common with binge eaters, anorexics, and bulimics, because it is an instinctive means for stimulating the regulator of the biosystem, the pancreas, in an attempt to regain balance. In the wild, ill animals stop eating. Binging people, whether it is on food or alcohol, are often at a loss when it comes to managing strong emotions, such as anxiety, anger, boredom, isolation, and sadness. These emotions can be the core backdrop of their lives if they were born from a mother who lived in sadness, despair, and anxiety. Only the awareness of this situation and the respect of the feminine polarity at all levels can help alleviate this unhealthy ambiance.

This is when knowledge of previous lives or understanding of important *songe* can help, by widening one's measure of space and time. And this is why I consider it detrimental to laugh at and denigrate such matters. It puts in jeopardy the possible connection with one's Self, thus with one's feminine polarity. Serious research in these domains should be done. It is estimated that approximately half of all binge eaters are depressed or carry a history of depression. This implies that in such a person, his or her lung aspect does not have enough stamina to control the Mammalian brain and the Idealistic self is unborn. Researchers are not yet clear whether depression is a side effect or a cause of binge eating. If we follow the LIFE biosystem, unidentified depression causes the latter. Low lung energy implies out of control liver energies, out of control pancreas energies, thus bulimia. It is clear, though, that those who binge often turn to food or alcohol when they are upset. Binging is a way to self-soothe, to numb emotional pain, and to stimulate the rational self. Gambling is another crutch. Anorexics, on the other hand, also develop anxiety when confronted with the necessity of eating. Anorexia has the highest death rate of any psychiatric disorder (Sullivan, 1995). Anorexics might behave like sick animals in the wild —they stop feeding themselves.

How can we link what the LIFE biosystem says about the illness with what science has found?

Brain imaging reveals that in humans and primates, a primary taste cortex is a feature of the insula and its adjoining frontal operculum (Scott et al., 1986; Yaxley et al., 1990; Faurion et al., 1999; Schoenfriend et al., 2004). These two areas react independently to hunger, but the left orbitofrontal— associated to the pancreas phase in its feminine aspect—responds to satiety. Electrical stimulation of the insula in humans elicits gustatory sensations. Of note, the superior laryngeal branch of the vagus nerve, associated with the Master of the Heart, innervates the taste buds in the tongue, palate, epiglottis, and esophagus. With other nerves it projects the information to the medulla and then to the thalamus, and then it finally is received by the insula/operculum. Gustatory information is conveyed to the left orbitofrontal cortex, the secondary gustatory cortex. Studies have shown that neurons in the left orbitofrontal cortex react to taste stimuli, and a part of these neurons are finely set to specific ones. As well, these also respond to visual and olfactory stimuli. In studies, subjects were shown either food pictures or geographical pictures (visual), and the activated regions of their brains were compared. The food pictures (taste) activated the right insula/operculum and the left orbitofrontal cortex.[403] With anorexic patients, brain imaging studies have found that when shown pictures of food, anorexics have altered responses in their right insula (social self) and left orbitofrontal regions (rational self), as well as in the medial prefrontal region (idealistic self) and anterior cingulate cortex (emotional self) (Nozoe et al., 1993; Ellison et al., 1998; Naruo et al., 2000; Goedon et al., 2001; Uher et al., 2004).

Another study using functional magnetic resonance imaging (fMRI) tested recovered anorexics with the same images (Uher et al., 2003). Only the medial prefrontal (idealistic self) and anterior cingulate cortex (emotional self) showed a difference compared to controls.[404] This is why it appears that anorexia might have more to do with one's profound identity in the midst of modern society than with the desire to be thin. Many yo-yo dieters have an obsession with slenderness without suffering from anorexia.

Another element pointing in the direction of the idealistic self is that individuals in the grip of anorexia nervosa have reduced concentration of 5-HIAA, a major metabolite of 5- T (serotonin) in their cerebrospinal fluid, reflecting a lower extracellular 5-HT concentration.[405] The mPFC is enriched in 5-HT (1A) receptors. The same review concludes that interactions between serotonin and its receptors in the *medial prefrontal cortex* seem to "modulate anxiety, attentional functioning, impulsivity and compulsive perseveration, and exploration of new environments."[406] Finally, the obsessional aspect of anorexia indicates to us the inability of the Mammalian brain to control the Analytical/Nutritive one, in the same manner observed with an obsessive-compulsive disorder. Hormonal changes during menarche might trigger troubles with food, as well as emotion and behaviors. Most particularly, the impact of the rise in estrogen levels could affect serotonin's neuromodulatory system. Interestingly, my symptoms of anorexia started shortly after I started taking the pill, which at that time had a high level of estrogen.

The Virtues of Fasting

It seems fitting here to discuss the question of fasting. In many religions this technique is used to purify the body on the physical, emotional, and mental levels (kidney, liver, and pancreas). The LIFE biosystem shows that refraining from food enhances the functioning of the Master of the Heart and lessens the grip of the masculine polarity. Anorexia mirabilis appears when someone seems to refrain from normal eating for years without starvation. It appears to be a symptom of something else, not a researched or imposed goal in itself. These individuals claim they feed on subtle energies. There are cases in history that cannot be explained other than through the functioning of the Master of the Heart, and there are too many of them to dismiss them all as frauds. Not surprisingly, most of these cases concern women. To allow the need to eat to subside, some conditions are required, though. When your point of view is centered on your rational mind, your masculine polarity dictates. You lose the abilities of the feminine one. When your idealistic self dominates and has matured, the feminine and masculine polarities work in concert and different abilities at all levels can develop. The observer and his point of view change the conditions. Obviously, people will die if they do not eat. And I would not be able to live on the little I used to eat during my anorexic periods. I honestly don't know how I did it. Still, to eat less is a virtue and a grace.

When fasting, we become sensitive to energies of a more subtle nature and are better shielded against all aggressions. Some religions, by choosing fasting or rituals on specific days of the week dedicated to the feminine polarity, such as Friday (Venus, Isis, lung aspect) or even Wednesday (Mercury, but also the feminine aspect of the pancreas phase), seem to indicate this. Catholics, for example, used to not eat meat on Fridays. This makes sense, as the feminine polarity is wounded by the merciless, unnecessary, and brutal killing of any creatures, especially mammals. In the same vein, not eating in the evening allows the energy of the lung to thrive instead of concentrating the system on the pancreas energies. Along those lines, Buddha suggested this to monks:

> I, monks, do not eat in the evening. Not eating a meal in the evening I, monks, am aware of good health and of being without illness and of buoyancy and strength and living in comfort. Come, do you too, monks, not eat a meal in the evening. Not eating a meal in the evening you too, monks, will be aware of good health ... and living in comfort.[407]

Roughly once a week Buddhists are instructed to observe the eight precepts. These include refraining from eating after noon until the following morning. For years, I spontaneously would have Chinese green tea and some toast in lieu of evening meal. I did this for as long as I lived in Calgary. Only on special occasions did I eat a full meal in the evening. When I found myself living alone, I was unable to continue doing this. I must admit, my weight and health were much better in Calgary. Also, when I was younger I used to fast every Thursday. Seeing the state of my health now, although I live in a beautiful environment, and remembering how I looked and felt then, I can only say that this practice is a grace. Of course, to the personality (masculine polarity) it appears somewhat punitive.

What are the known and proven benefits of fasting?

First, research indicates that short fasts are more beneficial than prolonged ones. The most beneficial appears to be one twenty-four hour fast per week. It is a liberating exercise. The benefits are as follows:[408]

1. Diminution of the speed of aging and a dramatic increase of life span. This is confirmed by TCM: when we eat, we use energy from the nonrenewable kidney-type energy.

2. Induction of a profound neuronal autophagy, which degrades misfolded proteins and/or viruses. This is the only known mechanism by which organelles such as mitochondria and peroxisomes are recycled. It then goes on to produce amino acids from these degraded proteins (Mortimore and Poso, 1986). Insulin secreted when we have a meal inhibits this autophagy (Blommwart et al., 1997b). It prevents malignancy and protects the cells from oxydoradicals and mutated proteins. In neurodegenerative disorders, recent studies suggest that macro autophagy plays a neuroprotective role.

3. Stimulation of neurogenesis.

4. Reduction of cancer formation and kidney disease.

5. Increased resistance of neurons to dysfunction and degeneration in experimental models of Alzheimer's and Parkinson's diseases, as well as stroke.

6. Increased production of ketone bodies, which can then be used by the brain as an energy source, and are known to provide some protective effects, including neuroprotection and resistance to epileptic seizure. This, without the inconvenience for the kidneys that a ketogenic-type diet would generate. Ketones are known to help regulate excess weight.

7. More production of growth hormone by the pituitary gland.

8. A reduction in swollen joints, increased strength, and overall better health in patients with rheumatoid arthritis, according to a Norwegian study.

9. Increased insulin sensitivity, stress resistance, reduced morbidity. Intermittent fasting resulted in beneficial effects that met or exceeded those of caloric restriction, including

reduced serum glucose and insulin, and increased resistance of brain neurons to excitotoxic stress.[409]

10. Reduces sugar cravings, resets the pancreas, lowers blood pressure, and helps the thyroid.

11. Enhanced synaptic elasticity.

12. Slight boost in metabolic rate.

13. Higher hepatic glutathione. Male hamsters that fasted had 68% higher hepatic glutathione (GSH) levels than controls. GSH is a tripeptide formed from glutamic acid, cysteine, and glycine. Combined with vitamin E and selenium, glutathione forms glutathione peroxidase (GPx), a key antioxidant critical for protecting the thyroid gland from oxidative damage. Excessive intake of sugars and starches reduces glutathione and thereby damage the thyroid gland.

Finally, the brain's RNA is under the control of intracerebral hormones but does not respond to modified hormonal circulating levels in fasting. Thus, even after a lengthy fast, protein contents of the brain remain unchanged. In the Bible, it is stated that those who fast have already received a blessing.

"And when you fast, do not look gloomy like the hypocrites, for they disfigure their faces that their fasting may be seen by others. Truly, I say to you, they have received their reward." Matthew 6:15-17 English Standard Version

Levels of Awareness

The lung aspect is related to the collective unconscious,[410] the kidney aspect adds the personal unconscious and instinctual, the liver aspect supplements awareness (personality), the heart aspect manifests self-unconsciousness toward awareness of identity (personality/individuality). The pancreas aspect to self-consciousness in both its involutive and evolutive aspects integrates the personality and individuality. This aspect interferes with, and is a result of, all the others. It should normally be subordinated to our

Isis Code

parabolic personal antenna of the medial prefrontal cortex (heart aspect, idealistic self). Depending in which direction we orient it, we will allow some information in and automatically and unconsciously censor all others. It is the conscious mental structure with psy energy (energy from the psyche) associated with the heart phase that sifts all incoming information.

"The prefrontal cortex mediates at least two types of memory monitoring and filtering. The first is a conscious, deliberate, rational process akin to problem solving in other domains. It crucially depends on the health of the dorsolateral prefrontal cortex [of the Analytical brain]. Some named this aspect "the editor"" (Burgess and Shallice, 1996).

In research on the subject, it was found that ventromedial prefrontal cortex lesions of the Human brain disrupt memory-based preconscious elements. The patient will argue that a face is unfamiliar to him even if that is not true. Nevertheless, that supposedly unfamiliar personal acquaintance evokes the same electrophysiological response as the face of someone consciously recognized. This is in agreement with the hypothesis of a frontally mediated dual-monitoring system, by the Analytical and Human brain.

A preconscious or subconscious element is followed by defective processing. When the Human brain is damaged, confabulations will occur. The vmPFC of the Idealistic self and Human brain is the prefrontal subregion most involved in mnemonic functions, while the dorsolateral prefrontal cortex (DLPFC) of the Analytical brain, together with the posterior parietal heteromodal cortex, appear to form the executive/attention functional axis (Mesulam, 2000) which crucially depends on the DLPFC (Gilboa and Moscovitch, 2002; Moscovitch and Winocur, 2002).[411]

Looking at the reactions of phobic individuals to both spiders and snakes, researchers have found activation of the amygdala, of the anterior cingulate, and of the insula. Vincent Paquette and colleagues presented pictures of spiders and butterflies to spider-phobic patients.[412] They found the parahippocampal gyrus and DLPFC activated *uniquely* to the phobic stimulus. This activation decreased after successful treatment for the phobic condition. This goes to show that the Analytical brain is subjected and sensitive to subconscious information. With aging, the lateral prefrontal cortex of the masculine polarity is the area encountering the greatest decline of any prefrontal region (Tisserand et al., 2002).

Science, Versus Religion

I will always remember that day. I was returning home from elementary school. Opening the door, I could feel tension in the air. Someone died? I remember walking down the hall and asking my mother what was wrong. She told me, with an emotional voice, that someone on the radio had declared that science had won over religion. I remember thinking it didn't made sense. Then I realized I hadn't even known there was a fight! Of course I knew there had been fierce religious wars, although I didn't know the details. But this was different.

Later on I understood that it was a battle that had been going on almost forever. In fact, it was considered a battle of reason versus beliefs/faith/credulity. Jung was especially against this reductionist tendency to summarize religions and philosophical quests of mankind as infantile quibbles from fearful and gullible, regressive beings.

Now why would proponents of each side fight for so long if one was totally wrong? Does one have to exclude the other? To me it is like thinking that meat won over salad. The meat says the salad is no good simply because it doesn't have the same function in the organism.

This division between science and religion is not outside of us. It is within us and pertains to the dual-monitoring system I referred to earlier. It is not different from the fight of men versus women. Modern science uses our rational mind and our intellect; religion normally uses our synthesis mind and our emotions. We are starting to get into problems when religion wrongly uses the rational mind (the Inquisition, excommunication, fundamentalism, sects, divisions, wars) and blends it with non-fraternal emotions; and when science wrongly uses our emotional mind (fear, stubbornness, arrogance, parties, cliques, shortsightedness, greediness) and blends it with the intellect. But this is what we are *obliged* to do when science

Isis Code

and religion are divided. Both science and religion are necessary because we can't live fully with only half of our psyche. For both aspects, the trouble resides in wrong perceptions and motivations.

Another point to consider is the focus of interest of each faction. Science is interested in space, in matter, while religion is interested in time. Religion is geared toward a future time when we will be dead or reincarnated. Science thinks about the here and religion thinks about the now up to then. They are different, but they should work together because they are made to marvelously complement each other. In a healthy individual, they are *not* separated. We have a default psyche that we attribute to the collective unconscious. This explains why a child deprived of love carries this pain from cradle to coffin. What is true for the physical and emotional is also true for the other levels. This is the inner compass I sometimes refer to. This is the little voice that knows how things should be.

Much of modern science, and thus the way we *perceive* the world, is a direct product of the reductionism of Grecian origin and of the Christian dogma.

This approach assumes that one must break a thing down into its smallest components in order to understand it. By doing so, unfortunately, we influence what we observe and take it out of its context. This limits us to the static physical nature of living beings and takes into account only the action of matter on matter. Feelings, thoughts, dreams, become mere consequences of chemical interactions and man is reduced to a mound of cells.

To explain further, our science attempts to assess causes and effects in a linear, sequential fashion, in order to measure, quantify, describe, and hopefully predict phenomena. This method produces extremely accurate information describing about half of observable phenomena. Because it is accepted that the visible aspect of the universe (mass) is less than 5% of reality, this means that we thoroughly understand about 2.5% of reality. We might say without exaggeration that we are ignorant.

Conventional medicine connected solely to this type of science studies illnesses and not as much health. It sees disease in a quasireligious, binary way (the fight of good against evil), and sees symptoms and diseases as the enemy, health care as warfare ("New drug in the arsenal against ...," "The war against cancer ..." etc.). This is a relic of a binary, masculine-oriented religion. Despite our laudable efforts, our science is not yet free of the masculine-polarity and binary penchant also seen in religions.

Medicine is particularly attached to the study of defective parts, while it should now also aim to understand the link between all the elements and how they influence one another. This is medical biocybernetics. But to change a mental pattern is not so easy.

The conventional medicine model is as follows: the patient presents a mix of symptoms to an expert. The doctor will prescribe drugs or maybe perform some procedure to remove the complaint with little participation from the patient. Disease is seen as something that just happens, with no connection to anything else in the patient's life. The patient is a victim, the doctor fights the "bad" nature that attacked the patient for no reason. Armed with painkillers, the doctor will fight nature, and the best way to do this is to trick the body into thinking that the situation has been settled. As I said previously, it is similar to shutting off the alarm while letting the burglar walk freely around in the house. In this view, symptoms must be suppressed, and if the treatment causes other symptoms, well, we will suppress those as well.

Also, the masculine polarity is interested in objects and their manipulation. Therefore, the medicine associated with it will add objects—drugs—or subtract them—surgery. It will focus only on them, and will have a whole battery of them. The feminine polarity will be more interested in the people, the individual in his uniqueness, his environment, and the system sustaining health in general.

The problem with the medical approach of the masculine polarity is that we are merely hiding the dust under the carpet and postponing the problems. Eventually, we will end up with an overmedicated patient, presenting multiple problems and a substandard quality of life, not to mention the impact on everyone's wallets. What we shut off on the physical level will eventually creep onto the emotional level, or one day we will be faced with a super virus that will not be controlled by any drug. Nature always strive for a return to balance, and chemical dams only force her to build something stronger than the dam. They don't hear, she says. So we will shout to regain balance!

As it is now, the price for medication is skyrocketing, taking precedence over other medical expenses. In agreement with the economic goals of a pancreas-driven world, pharmaceutical companies often act like children freed in a candy store. We end up with articles in the newspapers such as these:

"Last year, PhRMA admitted it was funding a campaign in Canada to oppose Internet pharmacies and lobby against Canadian drug price controls."[413]

"In one example cited in the suit, Medi-Cal paid $804.70 for a single bottle of the hypertension drug Atenolol. Providers, such as doctors, clinics and pharmacists paid $33.85 for the same amount of the drug made by Mylan laboratories. As a result, providers reimbursed by Medi-Cal for Atenolol pocketed $770.85."[414]

"The state initially sued Abbott Laboratories Inc. and Wyeth Pharmaceuticals (now 39 new companies have been added to those) in 2003, accusing them of reporting false prices that California then used to set reimbursement rates for Medi-Cal."[415]

In Canada:

"Between 1989 and 2000, 60% of the drugs were using active ingredients already available on the market and 76% were not better than the ones already on the market. Meanwhile, between 89 and 98% of the studies financed by drug companies were favorable to the drugs under study."

And similarly in the United States:

"The report found that from 1989 to 2000, only 35% of the new drugs approved by the Food and Drug Administration (FDA) contained new active ingredients, while the remaining 65% contained active ingredients that were already available in other products."[416]

"A 2001 report published by Families USA, examining the profits and spending of nine leading United States pharmaceutical drug companies, found that the average annual income of the highest paid executives was nearly $21 million."[417]

"Before its patent ran out, for example, the price of Schering-Plough's top-selling allergy pill, Claritin, was raised thirteen times over five years, for a cumulative increase of more than 50 percent—over four times the rate of general inflation. As a spokeswoman for one company explained, 'Price increases are not uncommon in the industry and this allows us to be able to invest in R&D.'" [418]

However, the general unsatisfactory results of conventional medicine with chronic diseases causes more and more people to turn their backs on this type of medicine and seek a more harmonious, less side-effect-prone approach. More than twenty-five years ago, when this movement began, some French medical doctors referred derisively to *medecine douce* (soft medicine) when discussing homeopathy, naturopathy, chiropracty, herbology, and the like. As if because it doesn't act as a "phallus erectus," it is therefore not effective. This reaction from some who deem themselves "scientific" has no link with a healthy Analytical brain, but more with an unruly mammalian one. They use labels to fight what is alien or threathening to their patterns. Labels are the weapons of the weak.

Religion followed a similar pattern. There was an all-knowing priest who would tell you what to do—or better yet, what to refrain from doing—in order to conform to eternal laws expressed faithfully by his church, the only repository of the truth, of which he, the priest, was the specialist and pure expression.

To last as long as the Egyptian civilization or the Chinese one did, it is necessary to respect the receptive feminine polarity, attracted to novelty, as well as the masculine one. Fear and greediness force a structure to hold onto a past that is no more, like an adolescent not wanting to part with the toys of his childhood. Any established structure protects itself and fights for permanence (linked to time) and hegemony (linked to space). Elements of change are not necessarily negative. When Jesus preached, "Love your enemies" (Matt. 5:44) and "If you love those who love you, what credit is that to you?" (Luke 6:32 NIV), he meant it as a warning against this rigid masculine polarity. Change, grow, and evolve, and then you can multiply. Or stay modular and die. In the evolution of humanity, these two polarities have played as well. So in fact, religion and science are the two sides of the same tool we use to understand and find our place in reality. They are two approaches to the universe.

When science tells us that: "improved spiritual health may be associated with improved behavioral and emotional health in such areas as diet, activity levels, communication skills, treatment compliance, reduced anxiety and depression, and improved mood states. These positive behavioral and emotional improvements in turn may be associated with heart disease reversal, reduced cancer mortality, enhanced immune function, and reductions in pain and other medical symptoms,"[419] we therefore cannot ignore that spirituality—or should I say a healthy Human brain?—is a key element to intellectual and physical health. To regain unity which is a necessary requirement to global health, we need to see a connection between both of these tools of the human psyche. To choose one over the other is not a solution. If we try to understand ourselves only through science, we nail our psyche to this physical earth and limit ourselves considerably. If we try to understand ourselves only through religion as it is now, we try to escape this world as we long for "home."

Science has done a full circle. It is thus normal that we look to cultures where religion and science walked hand in hand for a time, to find answers to our present dilemmas and evolve to a higher level.

Of course, the culture in which science and religion walked in this way must have, eventually, also suffered from the time when the masculine polarity prevailed.

We do not have enough information on Egyptian medicine; Hinduism is complex, although Ayurvedic medicine uses the same cybernetic system as does TCM; Zoroastrianism is almost dead; and the Muslim faith came too late, for the division between masculine and feminine polarities was already in full swing. We need something new based on something old and complete to add to these traditions. A system joining the scientific group while keeping its religious roots.

As we have seen throughout this book, there is such a system in which elements have been scientifically proven and accepted, or at least tolerated: traditional Chinese medicine uses it. It is both a science *and* a religion. Its scientific aspect explains and protects its religious aspect, and its religious aspects feeds its scientific aspect.

A system of health care that has endured many thousands of years, treating one-quarter of the world's population, proves through its longevity the veracity of its invisible directive system. Although many cultures have used this implicate order, only the Chinese have developed it to such an extent, and through using their Analytical brains. This allows us to understand it now.

Chinese medical concepts, though, are nothing like westerner medical concepts.

Since the beginning, TCM has used the method of inductive synthesis, recognizing the interrelatedness of phenomena. This is the feminine aspect—the religious aspect, one might say—our modern science lacks. The theory used by TCM as used by medical biocybernetics is a coherent, understandable, and usable science, with procedures and protocols based on centuries of observation of how nature works.

A science has no choice but to be rooted in its own system of beliefs. For our science, as we said earlier, it is the Greek-based philosophy of reductionism (separating into parts in order to understand the whole), as opposed to the unified theory used by Taoist philosophy.

The question asked in Chinese medicine is not, "What bug is causing this disease?" so much as "What weakness is causing this person to be susceptible to it?" It is the same question asked by medical homeopaths. This concept is not new even to Western medicine. Louis Pasteur, who popularized the theory that germs caused disease, admitted on his deathbed that his rival, Claude Bernard, was correct in arguing that the internal immunity of the host was a major factor in resisting disease. We can also find this assertion in both homeopathy and naturopathy. For this reason, I consider the four medicines seen through medical biocybernetics as "bio-Informative medicine." Bio-informative medicine is an adjective given to any corpus of knowledge or any therapy viewed through the angle of medical biocybernetics. As such, naturopathy, acupuncture, homeopathy, to name a few, will be used slightly differently in the frame of medical biocybernetics. They become bio-informative medicines. To me, though, only medical biocybernetics has the tools and the awareness to make a conscious and useful link between all of them, significantly enhancing their efficiency. Medicine restricted to the rational mind of the pancreas phase is solely concerned with objects. The feminine polarity does not act this way. It orients its interest toward system science and will use what is already in nature, namely, bio-informative medicines.

The goal of bio-informative medicine is to restore balance. Its guiding principle is to define what health is. Illness is seen as an imbalance between internal influences such as diet, exercise, rest, emotions, and thoughts; and external factors such as weather, trauma, or viruses. Health is not just the absence of symptoms. It is a state of resilience and strength of the being who can easily regain balance (homeostasis) on all levels: physical, emotional, rational, collective, and symbolic, regardless of what life throws at him.

Bio-informative medicines are more focused on increasing health to unseat disease and supporting the body's inherent qualities to regain health. They consider our bodies as designed to stay healthy and in balance, given the right conditions. Nevertheless, even TCM, which uses the autoregulated system of nature, has suffered from the overpowering masculine polarity. During the communist era in China, acupuncturists were forbidden to practice their art, which was judged irrelevant and a caste symbol. This brought the demise of a whole section of this medicine, as the feminine aspect wasn't understood anymore and was judged, maybe not without some reason, as too foggy. The rational mind is now destroying us because of its dissociation from the feminine polarity.

To give an example of the limits of the rational mind: the China Sea is infested with Nomura's jellyfish. These normally stay dormant, but become active when the proper conditions arise. The acidity and the lack of oxygen of these polluted waters seem to be their perfect habitat. In a beautiful TV documentary, they showed Chinese fishing boats bringing their fishing nets back on board. When they opened them on the deck, myriads of Nomura's jellyfish spilled out, but not a single fish. This is truly catastrophic.

The Analytical Brain, the Rational self, and the Personal Human

You see the sailors frantically running around. They want to kill these sea creatures and get them off their boats. They hurriedly slice them and throw them back into the sea. Unfortunately, although this kills them, it disseminates their spores and vastly multiplies their numbers.

In another documentary, when Chinese fishermen empty their nets, they are luckier. This time sea turtles are among their catch. Handling them as if they are things, they pierce the living turtles' eyes so to carry them around more easily. They have no qualms that this is an endangered species or the pain they inflict on the poor creatures.

There are many similar stories.[420]

Turtle meat and eggs are considered delicacies in China. One can find live sea turtles in tanks and bins in restaurants around Hainan Island, waiting for people to pick them out as their next meal. In Costa Rica, as soon as the turtles have laboriously laid their eggs, people poach these by thousands for resale. In the pictures I've seen, the people seem both happy and ignorant of the destruction they are causing. How can we keep on allowing this?

The other factor in this is that Nomura's jellyfish has only two enemies: a healthy sea—and sea turtles. The Analytical brain does not make the connection between the two facts. It just thinks: "free to poach and sell." Without a link to the feminine polarity, this mental state suffers from cretinism, however ingenious it thinks it is.

Our rationality is deeply colored by the state of the previous phases. Hence, it will find ways to explain our choices. I would like to give an example here. Dr. Fred Previc is a lecturer at Texas A&M University at San Antonio. For over twenty-five years, he did research for the United States Air Force. He wrote a book titled *The Dopaminergic Mind in Human Evolution and History* (Cambridge University Press, 2009), in which he tries to explain religion through the brain structure. His conclusion is that religion is solely a construct of the brain, and that at most we should consider it as the construct of the left hemisphere—thus masculine—which to him yields that atheists, agnostics, and spiritualists (which he curiously puts with the right hemisphere, thus feminine) as closest to the truth. His conclusion is ingenious but incomplete. He simply dismisses religious experience as a fabrication of the mind and as a result of brain function.

Further, Previc noted that "lesions (i.e., underactivation) of the right hemisphere are almost twice as likely as left-sided ones to produce complex visual hallucinations. ... Similarly, epileptiform activity (i.e., overactivation) in the left hemisphere is almost twice as likely to produce hallucinations and delusions." He thus agrees with me that an underperforming feminine polarity is an open door to illusions. And again: "Drugs that simulate [the dopamine] ... are hallucinogenic." Thus a functioning feminine polarity with an adequate amount of serotonin and noradrenaline is essential and the best protection against illusion, since the left hemisphere already dominates although it follows the right one by a few seconds.

Previc says: "The evolution of religion is linked to an expansion of dopaminergic systems in humans, brought about by changes in diet and other psychological influences." Maybe he was referring to the expansion of religions, not necessarily to their evolution.

Previc also proposes that the elevated situation of gods across religions is due to how the brain organizes space. The reality is that it relates to the fact that a two-dimensional image with a dot in the center, when projected as a three-dimensional figure, is similar to a mountain, with the center becoming the top of the mountain. Therefore our inner center corresponds to the physically most elevated point.

He writes: "The religious significance of upper space is further reflected in upward oriented behaviors and orientations during individual religious experience and practice. ... Upward eye shifts ... accompany meditation ... upward eye deviations promote generation of EEG alpha rhythm characteristic of the initial meditative state ... upward ocular deviations ... occur in mystical states induced by magnetic stimulation of the brain." This makes sense, as the medial prefrontal cortex and pineal gland are then central to the experience (heart-lung phases), and thus appear as upper from a geographic point of view compared to the visual field. The relationship of religion "and distant, upper space may partly account for the religious experiences and conversions frequently found in pilots and astronauts while flying high above the ground." I believe that pilots who work in

The Analytical Brain, the Rational self, and the Personal Human

space naturally tone both of their polarities. Their masculine polarity is stimulated as they cover a large expanse of space in little time. The sky is blue (lung phase), the sun shines (heart phase); you have here a "bath" of feminine polarity. It is thus not surprising that by stimulating these aspects, they have a tendency either to convert —or one would say, to regain a healthy balance.

Previc: "We have proposed that religiosity is primarily the construct of the left hemisphere and spirituality is primarily the construct of the right. This proposal is based on a substantial amount of cognitive evidence we have collected." I agree with this if he refers to empty religiosity.

He then goes on to associate spiritualists, atheists, and agnostics with the right hemisphere, while he classifies religious individuals as of the left hemisphere. He continues "as with schizophrenia in general, functional imaging studies point to a left-temporal predominance for delusions of a religious nature." Previc then discusses the well-known link between temporal lobe epilepsy and religiosity: "TLE [temporal lobe epilepsy] is associated with hyper religiosity and further argued the hyper religiosity was more likely to occur in left-sided TLE."[421] He continues:

> The final expansion of DA (dopamine) could have prompted the rise in abstract reasoning, human creativity in the form of art and music, and religious behavior. ... Both abstract reasoning and religious thought involve an emphasis on nonvisible (distant) space and time, and both are linked to the upper field. ... It might also seem strange that two ostensibly antagonistic processes—religious behavior and abstract (scientific) reasoning—may have co-evolved. ... both phenomena are concerned with abstract concepts and comprehensive frameworks with which to comprehend spatio-temporal events in the external environment.

His observation points the fact that both science and religion are now managed by the masculine polarity; the essence is missing in both. They are supposed to use both hemispheres.

Isis Code

Often, people who fall prey to sects match this description of a dominant left hemisphere and masculine polarity. Contrary to common lore, they do not have weak minds. Their Analytical brains can be quite active. They do have a weak Mammalian brain and an undeveloped or out of control Human brain, thus little discernment. The problem resides in the fact that religious realities linked to the right hemisphere, spirituality of a Celtic type for example, have systematically been persecuted for the last two thousand years by the prevailing masculine polarity and eradicated by the dominant churches. One has only to look at the history of Miko, the shaman women of Shintoism. With the rise of the masculine polarity they have fallen from the highest ranks of society (before 714) to the image of a prostitute in the Kamakura period of 1185–1333.[422]

To refrain from giving any spiritual direction to a child is not to protect him against lies; it is an open door to a lack of structure at the faith level. Better a limited religion then none. The valuable element for the child is to be allowed to ask questions and get patient, pertinent answers. He must understand that religions are expressed by mortal and limited humans. If he outgrows his religion, he will be mature enough to seek his own path.

Money and its associated power are an offshoot of the physical world, of the reproductive function (the kidney aspect of our system), and of the aspect of the system that controls it, the pancreas aspect. This aspect of the system, which is logical and rational as far as taking note of material elements, is nonetheless a cold, calculating tool, fed and controlled by the liver aspect, the emotional, subconscious side of humans. Nevertheless "it" thinks it is totally objective. Fittingly, the orbitofrontal associated to it manipulates objects, as found by research done in the laboratory of Dr. Lesley Fellows, a neuroscientist at the Montreal Neurological Institute and Hospital. The Mammalian brain, through its anterior cingulate cortex, refers more to a choice between actions.[423] The law should put forward universal rules of conduct because, as a result of this, the rational of one will not be the rational of the other.[424]

The time it takes to defend oneself is also unmanageable. All the papers to write, the sleepless nights, the emotional pain, the physical difficulty, the isolation; these all make one vulnerable and unable to defend oneself. If I were the aggressor, I would not lack energy, because then my masculine polarity would be directing the show. When challenged, the feminine polarity is not responsive. It retreats because it is receptive. If your feminine polarity dominates, then you are lost. Only the masculine one can act. If you are a woman with a limited masculine polarity, or one that you use only for the good of the feminine polarity in general, you are then powerless. The rationale of this man will not change, and he will reproduce to the infinite what he did to me—until he does it to a man with a predominant masculine polarity. For women who are alone, unless their masculine polarity is stronger than their feminine one, there is no hope. Indirectly, our justice system favors the aggressor. Hence the ancient culture of the knight who would defend "the widow and the orphan." That is, those of the feminine polarity who do not have a masculine one to protect them. Who protects the widow and the orphan nowadays?

Numbers: the Analytical Brain

In the Five Books of Moses, the pancreas phase corresponds to Numbers. This Book appears to be a collection of different literary genre with the rebellion story as its focal point. This is the story of earth-centered men and women: their census, tasks, lists, rebellion through dissatisfaction, crisis of leadership, life with neighbors, lies, not to mention preparation for war. It is the executive brain in action. The action unfolds in the shadow of Mount Sinai, which as we saw is the heart aspect, the idealistic self. The head count is made of those of "twenty years and upward," as if they were mere objects, which fits perfectly with the objectification and also the period of the pancreas phase (21-28 years old).

As we know, in the LIFE biosystem the Analytical brain has control over the Reptilian one, thus we find in Numbers many references to firstborns. The following renders perfectly this idea of control inherent to the psychology trait of this phase: "for all the firstborn are mine. When I struck down all the firstborn in Egypt, I set apart for myself every firstborn in Israel, from man or animal. They are to be mine. I am the LORD" (Num. 3:13, NIV). This God becomes possessive in the pancreas phase. Since this phase is controlled by the mammalian phase, an uncontrolled emotional phase will generate dissatisfaction, lies, and egoistic calculations. The People of Israel here become Israel, one individual and the object of God's wrath.

Isis Code

Repetition, a soothing element for the pancreas phase, can be found all over the text. This and other elements point to the profound therapeutic function of these sacred books. God's character changes, in harmony with the phase we enter upon. In fact, it is we who interpret him differently, as we move from one phase to the next. In the image of the insensitive pancreas phase and coldness of the Analytical brain, the God of Numbers gives an impression of unremitting severity, bordering on a lack of compassion. As example, let us take the death of Aaron, his supposedly beloved high priest:

> "Aaron will be gathered to his people. He will not enter the land I give the Israelites, because both of you rebelled against my command at the waters of Meribah. Get Aaron and his son Eleazar and take them up Mount Hor. Remove Aaron's garments and put them on his son Eleazar, for Aaron will be gathered to his people; he will die there." Moses did as the LORD commanded: They went up Mount Hor in the sight of the whole community. Moses removed Aaron's garments and put them on his son Eleazar. And Aaron died there on top of the mountain. Then Moses and Eleazar came down from the mountain (Num. 20:24–28, NIV).

Nevertheless, it is also here that we find an applied medicine against the ills of sin, which has a repercussion on the physical health, echoing the fact that the rational self has control over the physical one and that it is also the regulator and expressor of the whole system.

Indeed, Moses the healer is in action here more than anywhere else. His type of healing does not proceed from the Analytical brain, although the organization of the healing rituals puts this aspect to work. It nevertheless springs from the totality of the phases and aims to include the patient back in the holy community by giving the means to this end. A holy community is one in which the LIFE biosystem functions properly, with all phases harmoniously expressed.

Our conventional medicine has put aside this highly necessary aspect of the heart phase when it rejected the idea that illness might partly be the responsibility of the patient, the result of his "sins." These, of course, refer to a lack of hygiene of the physical, emotional, idealistic, rational, and/or social levels. Also, since all the levels of our being have an impact on the physical, it explains why we read in the Bible:

"I the LORD your God, am a jealous God, punishing the children for the sins of the fathers to the third and fourth generation of those who hate me" (Ex. 20:5, NIV).

"Maintaining love to thousands, and forgiving wickedness, rebellion and sin. Yet he does not leave the guilty unpunished; he punishes the children and their children for the sin of the fathers to the third and fourth generation" (Ex. 34:7, NIV).

"The LORD is slow to anger, abounding in love and forgiving sin and rebellion. Yet he does not leave the guilty unpunished; he punishes for the sin of the fathers to the third and fourth generation" (Num. 14:18, NIV).

The lack of hygiene we accept in our lives creates havoc in the system of our children to be, as this affects the epigenetic factors of our own system. We can observe this in Exodus, which is associated to the emotional phase owned by the masculine polarity. It is repeated twice as well as is present in Deuteronomy, which is associated with the social self.

"For I, the LORD your God, am a jealous God, punishing the children for the sin of the fathers to the third and fourth generation of those who hate me" (Deut.5:9, NIV).

"The third generation of children born to them may enter the assembly of the LORD" (Deut. 23:8, NIV).

In the emotional phase—Exodus—it is explained that sins occur to those who hate God; that is, those who separate themselves from the implicate order. In the same phase, the guilty is forgiven when he repents, but the mark is on him for three to four generation, as the epigenetic effects imprint the genes. There is silence about it in Leviticus, as it is implied that priests love the implicate order. Then, in the rational phase of Numbers, it is repeated again with the same text that is found in Deuteronomy, phase of the collective. It ties together the collective, emotional, and rational, as it is in the LIFE biosystem.

Isis Code

The beginning of manifestation (lung) controls the emotional (liver) which controls the rational (pancreas). Our expression in the world, our rationality that chooses what act to perform, is the child of the collective aspect and is colored by it. This is a vast subject that could be the matter of a voluminous book.

Iniquity concerns the heart stage of idealism. Transgression relates to the liver aspects and emotional phase, and sin to the kidney aspects and physical phase. Religion's first aim is to fight iniquity, as this aspect controls the collective aspect which then controls the emotional and feeds the rational aspect. It is also to say that you can't be without some beliefs, as we all have a symbolic self and this will be your "religion," feeding your rational self whatever those symbols are. In other words, one's scientific level depends on the idealistic one. Otherwise it is only the by-product of an analytical tool. You may have talent with it; it doesn't make you a great mind.

The first point established by Moses and Aaron is a link between sacred space and sacred time, providing a balance of the two polarities: space and time, masculine and feminine. This is the basics of health. For most imbalances, whatever therapy is aimed at the physical body will not hold if that aspect of the harmonious relationship of the two polarities is not taken into account first.

Moses is, to me, the first homeopath. Follows are three examples of this. I will let you judge for yourself. We know more about matter than did our ancestors, but we know far less about the energies of life or, if you prefer, information. These were not acts performed by credulous savages, but rituals taking into account the wholeness of human beings. It is through myths and rituals that human existence first acquired order and meaning, by modeling itself on the implicate order found in nature.

The first case is in Exodus 32:20. Moses returns from forty days of meditation, and he sees that while he was gone, all the structures he had put in place have vanished. Out of fear, Aaron allowed the people—here reduced to one entity named Israel—to make a golden calf to replace God. Israel was totally left to its own devices.

The Analytical Brain, the Rational self, and the Personal Human

Aaron could not impose his will or control the community while his idealistic aspect (Moses) was away. If there is vacancy in the Human brain, there is no more control over the Universal brain. Thus it is said that the people were out of control. This lack of control of the social self allows a free-for-all of the Mammalian brain, resulting in a depressive state. It generates a lack of control on the Analytical brain. Therefore, those people won't think properly as their emotions are uncontrolled. This eventually engenders fear, as the Reptilian brain receives no more control. According to the Bible, orgies ensued. When the idealistic aspect, personified here as Moses, comes back and sees the chaos generated by his absence, he has to take control by refusing the pattern that replaced him (diagram 16).

"And he took the calf that they had made and burned it in the fire; then he ground it to a powder, scattered it on the water and made the Israelites drink it" (Ex. 32:20, NIV).

Since gold is not soluble in water, I consider it would have the same effect as homeopathy.

In the materia medica from Dr. Michel Guermonprez,[425] we read that the homeopathic remedy aurum (gold) is given chiefly for heart rhythm problems (information-regulation brain) and for depression, which befits an out-of-control Mammalian brain. It is given for aggravated sulfur types, which is certainly what the biotype of these people would have been, given their absolute need of meat.

The aurum type is ameliorated by music, which explains why they were singing.

The second occurrence, in Numbers, relates to a woman suspected of infidelity by her spouse. It is mainly a ritual to dispel the husband's jealousy. Read aloud in Hebrew, it has the music of a mantra balanced in the movement of the two polarities, as if one made a circle to the right and then undid it and made a circle to the left (levogyre, dextrogyre): *"mei ha-marim ha-me'arrarim."* Language centers are on the left, and spiritual practices that rely on them offer a control over it, most of all when the music center on the right hemisphere is used. Many religions use reciting and—even better—singing prayers as an effective way to avert "temptations of the flesh." As a fakir plays music to subjugate a snake, by controlling and concentrating the mind on one point,

Isis Code

through vocalization the door is opened to the subconscious aspect of the feminine aspect. It mobilizes and fixates both the left hemisphere and pancreas aspects. Mantra practice—repeating short invocations over and over—is found also in the Hindu and Buddhist traditions and in most religions as prayers. The two polarities are at work and balance is restored. Scholars agree that the mantra "*mei ha-marim ha-me'arrarim*" could be translated as "the water of bitterness bringing the bane." Remember, the bitter taste is a stimulant to the heart aspect. This stimulated aspect has more control over the lung aspect, the Universal brain (diagram 20). And so the Bible says,

> Here the priest is to put the woman under this curse of the oath—"may the LORD cause your people to curse and denounce you when he causes your thigh to waste away and your abdomen to swell. May this water that brings a curse enter your body so that your abdomen swells and your thigh wastes away." Then the woman is to say, "Amen. So be it." The priest is to write these curses on a scroll and then wash them off into the bitter water. He shall have the woman drink the bitter water that brings a curse, and this water will enter her and cause bitter suffering. ... after that, he is to have the woman drink the water (Num. 5:21–24, 26, NIV).

Of course, if the woman is not guilty, her abdomen will not swell and she will have no fear in drinking the water. On the contrary, the bitters will stimulate her heart aspect and she will hold her head up high.

The third occurrence, in Numbers 21:3–9, relates Israel being short tempered after God delivered the Canaanites to it. We know that war affects adversely the lung aspect, mother of energies. The liver aspect then becomes uncontrolled, thus the short-tempered result. As a consequence of their overactive Mammalian brains, they spoke against Moses and God. We have a saying in French for someone who speaks ill of another: *avoir une langue de vipère*; to have a viper tongue. Curiously, here people die from viper bites, and the People come to see Moses. They acknowledge they have sinned and ask him for a solution to the curse. In homeopathy, the remedy used for such a condition—not the viper bite but the constant detrimental judging of others—is lachesis, which is made from ... snake venom.

Moses uses the same ill, a burning snake. This one is made of copper, a metal associated to Venus thus to the lung aspect. Whoever has been bitten by the viper and looks at the copper burning snake will not die.

Of course it is not given as a remedy for physical snake bites in general, but as the Bible is mainly a book of exorcism, it shows how to rebalance the wrongful information. The snakes are linked to kidney information, while copper and fire are linked to lung and heart energies.

Since the rational self is also related to nutrition, we find in the sacred text a discussion about two types of nourishment: that of the conventional digestive system and that of the second digestive system, which is concerned with information.

The first type is referred to here as *meat*. The second kind is *manna*, or the daily bread Jesus refers to. Now, which one precedes the other? Which one is the most important?

In Matthew 6:11 we find: "Give us this day our daily bread." Carl G. Jung tells us that St. Jerome translated a word from the Greek text that can be retraced only in this Gospel: "suprasubstantial." This can be translated as "subtle." So the bread Jesus is referring to is the one which feeds our individuality, not our personality. Now, which is the most important for eternal life, the bread for the personality or the one for the individuality? It is in fact information from the implicate order that brings us health and harmony, which feeds all aspects of the system.

If a phase is in difficulty, the pancreas aspect will have little energy and will be constantly solicited through need to restore balance to the system. Because the heart aspect precedes the pancreas aspect, if the Master of the Heart is not functioning, therefore if one does not receive his subtle daily bread, proper assimilation of the elements from the conventional digestive system will also fail. There will therefore be a waste of nutriments, waste accumulation, and although abundant, the material food will never be enough to satisfy this out-of-phase system. If the Master of the Heart is functional—which is allowed by a balanced masculine and feminine polarity, and by a healthy environment—the quality of the food prevails. Far less

Isis Code

food will be needed for nourishment, although the food will need to be of the utmost quality.[426]

The manna, gathered at sunrise, is the daily bread. When our lives are desacralized, which happens when we lose daily contact with our inner Isis and with the natural environment, the most divine elements of our lives pass unseen, unacknowledged. Our lives loses savor, since salt (base added to acid, masculine and feminine polarities working together) is not there to give flavor to things of this world, despite a great accumulation of these things. To those who have kept their sense of the ritual and of sacredness, the smallest element brings flavor to life, a joy that is the expression of the heart combined with the lung aspect. This is the daily bread. God always gives us our daily bread; we do not always eat it.

A series of experiments have shown that proper presentation enhance the assimilation of nutritive elements in food. In the same vein, Kurt Gray, a Canadian social psychologist, conducted a series of tests[427] to realize that "our social world impacts our physical world in a really fundamental way." Such things as good intentions, benevolence, and care positively enhance the reception of information. In one of his experiments, he found that a candy will taste sweeter if it is given to you by someone who chose it carefully and attached a loving note to it. The social aspect is the lung aspect of our system, which also refers to the natural environment. The Master of the Heart, part of the information-regulation function (heart) encompasses the five senses and dictates regulation of information. I name it the sense of synergy. This sense impacts how we *perceive* the world.

This idea of not having enough to eat was widespread in the '50s. Communists became major tenants of this notion as the principal cause of all human miseries. Although Simone de Beauvoir was not a communist, I will use her as a truly pancreas type of intellectual woman to exemplify this. She declared on November 13, 1959, in an interview censored by the CBC but broadcasted later after her death, that the source of all ills is that people don't have enough to eat. She admitted she had a profound disdain for bourgeoisies institutions, although she was born into a wealthy family that eventually lost its social status due to money problems. She was certainly bitter about that. She also added that the principle behind marriage was obscene, and that the only liberation for women was for them to go to work.

The Analytical Brain, the Rational self, and the Personal Human

As an educated woman, how could she not see the symbol behind marriage? This was understandable, though, since her parents were constantly fighting. Still, how could she not see that it is not marriage itself that is obscene, but the ignorance of those who submit blindly to their hormones without depth of feelings or an elevated ideal? Marriage was mainly to protect children and property. But the symbol behind marriage, the communion of two polarities, is the goal of evolution, so it shouldn't be vilified. But her perception dictated that as a pancreas type, she never wanted to be stuck in any situation. Her *perception* of the world made her believe what she believed. Her father would tell her she was a man in the body of a woman. He always wanted a son who would make him proud, so she became what Daddy wished for. At fourteen years old, the phase of the heart, she said that she didn't believe in a God anymore, because she didn't want to have to answer to anyone or anything. Also, she voiced that she could not find reasons to believe, neither in her mind nor in her heart. She quoted a philosopher, who affirmed that the synthesis implied in the existence of a God was both unrealizable and unthinkable. His first adjective—unrealizable—originates from the physical, thus kidney aspect and the second—unthinkable—from the analytical mental, thus the pancreas. Of course, from the point of view of his personality, he is right, for the personality cannot conceive a God if the person is totally stuck in the analytical phase. This was a perfect mirror to de Beauvoir's mind. Her Idealistic self didn't develop because she was stuck in the pancreas phase. This goes to show her deficiencies as well as those of the epoch she emerged from.

The world of the pancreas, limited to the masculine polarity is a prison, exceedingly limited indeed. The following short story illustrates this.

A boy named Tom accompanies his father to the cabin where lumberjacks pause for lunch. Tom sees in the distance an unusually tall and strong man. Extremely impressed Tom runs toward the giant, who is drying his sweaty forehead with a worn and dirty handkerchief. The boy pulls on the man's sleeve and asks: "Sir, I beg your pardon." Surprised, the man looks down, and then smiles as he sees the small boy admiringly looking up to him. "Yes?" he asks. "Sir, I beg your pardon, but why are you working so hard?" The lumberjack takes a deep breath and answers, "To earn money, son." The boy nods knowingly and asks, "Please tell, what you will do with the money?" The man answers, "I will buy a lot of food!" "Oh!" replies the boy. He follows the man inside and sits beside him. The lumberjack has his plate stacked with an enormous amount of food. The boy then asks: "And why do you need to eat so much?" The man's smile fades. Visibly irritated, he gestures for the boy to go away and says, "So I have enough energy to work!"

This is the repetitive prison of the personality: work, get money, pay for lodging and food, work, get money, pay for lodging and food, etc..... Without the individuality, it is sterile, senseless and without flavor.

The values of the rational self and of the Analytical brain we have endorsed did worse than objectify woman and nature. They rendered impossible this intimate connection we as humans must have with symbols and with nature. Only the symbols used for mass exploitation, such as money, politics, and patriarchal religions, were accepted. This way, the woman, normally the bearer of the manifested feminine polarity became doubly sequestered: first by becoming an object among many others, and then by seeing the window to invisible realities slammed shut, locked, and blackened. Crazy, hysterical, delusional, schizophrenic, emotionally unbalanced—she heard it all. She, who lives of the future, has been crucified to the present, the instantaneous, with only glimpses ahead of the merciless, lifeless, heartless economic world. All her dreams have been forced through the sieve of rationality. Out of this sifting, only the most crude, material, and lucrative were kept. The only manifestations accepted from her now are expressions of her masculine polarity.

This woman, which is also the inner aspect of man, is not exalted anymore, and hence Beauty silently withdraws more and more from this bleak world. All that is left of her now is price-tagged genital energy posing as beauty.

Harmony between head and heart, between the idealistic and rational selves is necessary in order to avoid a self-sufficient rationalism and an unethical economic system. Otherwise we limit our existence to its most rudimentary egoistic elements of immediate gratification, functionality, practicality, and economy, which are detrimental to others, to our Selves, and to the environment at large. Nevertheless,

The Analytical Brain, the Rational self, and the Personal Human

this period was necessary, and as a child matures and understands, we too have to develop and expand our Consciousness, which cannot be other than cellular.

Chapter 7
The Universal brain, the Social Self, and the Social Human

"*Before, we were strong. Now, everything is changed. We eat the white Man food and it softens us; we wear their heavy clothes and it weakens us. Before, summer and winter we would go daily to the river to bathe. This strengthened our skin. But the white colonials were shocked to see naked Indians so we had to stay away. Before, we were wearing aprons of bark and reeds. All winter we would work in the wind, naked arms, naked legs and we were not cold. Now, when the wind blows from the mountains it makes us cough. Yes we know, when you come, we die.*" **Chiparopai, an old Yuma woman at the beginning of the twentieth century, relating what happened with the colonials' arrival)**

*"From the pine tree
learn of the pine tree
and from the bamboo
of the bamboo"*
Basho

"From the fig tree learn its lesson ..." **Matt. 24:32 (ESV)**

"*The man who sat on the ground in his tipi to meditate on life and on its meaning had accepted a filiation common to all creatures and has recognized the unity of the universe; in this he was infusing to his being the essence of humanity. When the primitive man abandoned this type of development, he slowed down his evolution.*" **Chief Luther Standing Bear**, My People the Sioux **(1928)**

"*Anyone who has spent much time in the wilderness can easily recognize the parallels between it and the archetype of Sacred Space. Wilderness is*

Isis Code

difficult to get to and difficult to travel through. One passes a series of tests in order to exist within it. It is unlike the normal world in hundreds of ways. Above all, it pervades one with a kind of religiosity or mysticism—one of the most compelling things about nature is that it seems to implicitly suggest the existence of order and meaning."[428] (S.B. Bacon)

Phase of the Who Level

The Lung Aspect

- Isis
- Evening
- Green
- The universal, sensitive man
- The *who?* level
- Figuratively: rainbow
- Element: air
- Cell structure: membrane
- House analogy: the doors and the windows
- Bible: Deuteronomy (in the Biblical calendar, a new day begins after sunset)
- Collective unconscious (beginning of the cycle) + previous + supra consciousness (intuitive)
- Need: belief, openness
- Biocybernetic phase and function: from exchanges toward internalization
- Biocybernetic attached main phase: lung toward kidney (same or other level of the spire)
- Biocybernetic structure: part of the feminine polarity energies externalized in most women
- Women expressive, men receptive
- Level controlled by idealist level
- Level controls emotional phase
- Fed by analytical
- Brain aspect: right hemisphere and most particularly temporal lobe, right orbitofrontal including the ventral prefrontal, corpus callosum, anterior commissure, massa intermedia, resonance circuitry, pineal gland, caudate nucleus, fusiform gyrus, right frontal insula, and claustrum
- Sense: olfaction

The Universal brain, the Social Self, and the Social Human

- Expression phase: from 28–35 years, 63–70, 98–115. Masculine cycle: 32–40, 72–80, 112–120
- Main relational attitude: cooperation
- Psychological trait: love
- Emotion: when in difficulty or has no object to communicate with: depressed, sad, and anxious
- Type of attraction: fusion; fin'amor
- Function: exchanges
- Nutrition: information about oneself accessed through love
- Stimulation: exchanges, trust, spicy taste, hope, natural surroundings, care of all stages of life, travels, fitting in a community/ family, smoking and alcohol (detrimental in excess), politic (in the sense of care for the public interest in contrast to economic interest), economy (in the sense of responsible saving), charity, natural perfumes, essential oils, fusional love, rituals, altruism, country life, belief in a benevolent God, harmonious family life, humor.

"Love type": Fusional- fin'amor- selfless- eternal

This is the *who* phase, which actually is the first as well as the last phase, which through its completion will lead to the two possible tunnels I spoke of in the first chapter. The first one as "I am," implies the synergy of all developed and healthy phases working in synchrony as a biosynarchy—when Horus is finally mature enough to take his place in his kingship—or when the heart aspect has matured and the system uses it as main regulator . The second way will be experienced as a repetition, the individual staying on the same spire, or a regression, the individual going down a spire, a prey to consumerism since his identity is tied mostly to the physical world and its consequences. It resembles a snake and ladder game.

An old Irish fable tells of a fairy giving to a man named Conla an apple that never diminishes, no matter how much he eats from it. Conla, son of Conn of the hundred battles, lives on an island, the western terrestrial paradise. When Teigue, king of Munster, pursuing pillagers, arrives on this island, he meets Conla, who is holding the golden apple. Near him stands Veniusa, who is a daughter of Adam and a sister of Letiusa of Aliusa and of Eliusa, who both live in another paradise. Veniusa and Conla found happiness in their mutual contemplation without ever wanting or needing anything else.

Isis Code

On the same island, Teigue finds the tree with golden apples. On it, three birds are singing. One blue with a red head, the second red with a green head, the third green with a golden head. It is said that their melody is able to cure any illness.[429]

Taking into account the Celtic tradition of colors, the first bird which has a red head atop a blue body is the emotional self (red head), which controls and regulates the analytical /mental aspect (blue body). The second is the social self (green head), controlling the emotional one (red body), and the third is the idealistic self (golden head) in control of the social one (green body). The melody of these three heads/birds, expressing thus the implicate order of our LIFE biosystem, brings health. There are two paradises—one of the individuality, the other of the personality. The couple lives in the paradise of the individuality; therefore the golden apple of the heart aspect and idealistic self is eaten but always replenishes itself. They are in the western paradise, which is the lung aspect. We understand here that this island is individuality, with Conla as the heart aspect and Veniusa, a word play on Venus, the lung aspect. This apple does not make one "fall," but allows one to feed continuously and is a gift from the sun, the heart aspect, the Idealistic self.

At this age, twenty-eight to thirty-five years old, the individual should be able to know what he wants or not. If he had the chance to experience the raw power of nature, he has probably realized that there are forces beyond what one can understand. Even our different theories on evolution seem ridiculously limited. Take the example orchids give us.

From studies of their DNA, we've learned that orchids are one of the oldest family of angiosperms (flowering plants), arising between 74 to 84 million years ago. Of all flowering plants, 8% are orchids. The number of orchid species is about four times the number of mammal species and twice that of birds. They have developed such a highly specialized pollination process—to what aim, our evolution theories cannot say—that their chances of being pollinated are scarce. Added to this, their seeds lack the endosperm (nutritive element/pancreas), so they need to enter in symbiotic relationship with different fungi. Interactions between comet orchids and hawk moths are one example of this effect. It also explains why orchid flowers remain receptive for long periods of time. They are a perfect example of this Universal brain.

Our social evolutionary theory, which speculates that a woman chooses a man solely on the grounds of breeding purposes, has to be reevaluated. The example of the orchids should help us to better understand. Cooperation has value in itself and is one goal of evolution. This is not for a better survival rate or better reproduction, but for this expansion of Consciousness in diversity. Similarly, women will appreciate a man who expresses a depth of symbols. It is not coincidence that royalty, uniforms, spiritualists, and celebrity are attractive to women. They provide a possible expansion and sublimation.

If it were only for money, the teenager who plays the guitar and sings a song wouldn't be so popular.

Myths, fairy tales, and legends are not children tales. They are often echoes of the godly voice of the One. If they are inspired, they speak of this implicate order and of its archetypes. We have forgotten their meanings as our inner Isis has retreated far away to her lonely island, awaiting the return of Osiris.

Those two lovers have endorsed millions of names through time. Union of our personality and individuality in the fire of reciprocity, they are projected into the story, sometimes prosaic, sometimes magical, of millions of couples through the ages. Will you recognize yourself if I relate here a part of Siegfried and Brunhilde's story?

The Song of the Nibelungs[430]

...

Walking toward the field, it is then that to Siegfried's eyes as well as to those of the horse, Brunhilde appeared. She was lying down in the meadow, outstretched on the rocky ground, hands open to the sky. Slowly they approached her; she remained still. He leaned over the face of this beautiful woman ... came close to her mouth as to better feel the breath of life which he could feel was in her. Softly, he kissed her ...

Thus way was the miracle accomplished, this way ended Brunhilde's profound sleep.

And they remained long days and long nights, one close to the other, in the heart of the plain. The black stallion, guardian of the fire, was a constant companion. During these days and those nights, Brunhilde taught many wonderful, magical things to Siegfried. All these things she had known since her own birth, as if they were written in her flesh. She taught him the sky and its stars; the dance throughout the seasons of those ancient gods, the planets. She recounted how although they have retired from humans affairs, from afar they still hold vigil to the destiny of the living. She told him the art of the plants and the individual virtues each possesses to help or weigh down humans. She taught him the art of the runes ... for him to know the signs which are not yet written on our paths but come to us under veiled forms....

Then one morning ... she took Siegfried by the hand towards an unknown place. In a crater of a long extinguished volcano, the last rain had gathered, forming a large basin filled of water of an extraordinary limpidity. There, they abandoned their clothes and Brunhilde asked her companion to alternately look at the image of his own face and his own body reflected in the crystalline waters, then of her own face and body. She whispered: "When both images will become the same, when in this reflection you will not be able to distinguish who you are and who I am, then I will have taught you all I had to teach you ..." Siegfried does not understand. He looks but he still sees his self and hers.... A little later, seeing how pensive he had become, she goes on: "See, when we are both naked under the sun, your shadow is exactly similar to mine: it is because there is, somewhere hidden in your heart, a Siegfried which is absolutely similar to Brunhilde. ... And if I love you so, my Siegfried, it is because between your image and mine, I see no division." ...

In the night that followed, blood finally took over and, as the flesh requires, Brunhilde became Siegfried's wife. And it is not without distress that she mixed her body, replete of tenderness to Siegfried's tenderness. ...

Why the anxiety? She didn't know! She only knew that now, never would they be allowed to contemplate in the pure waters this image of themselves which would unite them for eternity. As infinite as their love was, now they would have to go their own way. And this path, similar to that of all other humans, would be heavy of ills, uncertainties and absences....

From that day, Brunhilde ceased to teach Siegfried the plants and the stars. Never did they go back to the crater and to its water full of secrets....

Then Siegfried started to dream of the boat waiting for him in the dunes, to dream of the sword which rested in the forest and of the helmet, instrument of the invisible....

Brunhilde herself guided him back to the sea, confiding him to the waves who had guided him to her. It is there, standing in the boat that Siegfried gave to Brunhilde, the golden ring of the Nibelungs.

That night, confronted by her solitude, Brunhilde searched in vain for the black stallion that had guarded the fire. In fact the horse had vanished the night she forsook her flesh to Siegfried's passion. In their fever, the lovers had forgotten about this creature of the shadow who had kept them in their light.[431] [My translation]

This is the hidden tragedy of our feminine polarity, of our individuality. Hers is the thirst for the absolute and a profound desire for unity, her only desire. His is the need of action in this world of manifestation and division. These two aspects live in us and take their place in our love lives. Sexual attraction really lasts only when it is echoed on all the other levels, a rare occurrence. A higher level of love must be experienced first to avoid the separation triggered by the material union. An image would be that we first kiss before we have intercourse. The highest must unite before the densest.

What happens when Brunhilde shows Siegfried everything she knows? He realizes that the Reptilian brain of the flesh highlighted the fact that they are two separate spaces and never will be fused here on earth. His Mammalian brain taught him that time separates us because our cycles are not always in harmony. The Human brain taught him that the personality is far from the world of gods but must be considered. The Analytical brain showed him that true love is indeed a rare thing. The Universal brain enlightened him that union and absolute respect of the other should spread to all life forms on earth, and that only at that level can the lover truly say "she is flesh of my flesh, soul of my soul."

The waters here are the energy of the Reptilian brain. They can be used as a mirror of the Human brain, as long as we don't give precedence to it. Otherwise the thickness of the personality hides from us the splendors of the individuality, and men cannot see the soul in women and women can't feel the spirit in men. This is as true for the individual as it is for the couple. Using other words, Leonardo da Vinci said that the body must not be a weight for the soul.

Freed from the bonds of time and space, legends, myths, and fairy tales speak to us individually with astonishing directness. When the tale is inspired, our idealistic self understands while our rational self claims: it makes no sense!

A majority of humans believe in a religion that has one God. Why is this so? The divine, or individual and thus indivisible, aspect is One. The feminine polarity is One. The right hemisphere works of the whole for the whole. This one God is thus fundamentally manifested through the feminine polarity and expressed by the masculine one.

Deuteronomy: The Universal brain

What does the Bible say about the last of the Five Books of Moses?

It is called Deuteronomy and it consists mainly of Moses addressing the assembled people on the Plains of Moab. It recasts Israel's mission and destiny. In the image of the lung controlling the liver, it reiterates the instruction of the Ten Commandments and shows what will happen if these are dismissed. In fact, if we look at the nature of those commandments, we see that six of them are not needed if we live alone and four are directed at our religious side. We can group them by twos and go counterclockwise, since they are stipulated *do not*, as in Egyptian tenets. If we bring those to five, three are so we can live harmoniously in the collectivity and two within ourselves. We start with the closer to God and end with our incarnation. We thus have the overarching synarchic idea: "I am the Lord thy God." Then the first commandment (the heart phase): "Thou shalt have none other gods before me./Thou shalt not make thee any graven image, or any likeness" (Deut. 5:7–8, KJV); then the liver phase: "Thou shalt not take the name of the LORD thy God in vain …/ Keep the Sabbath day to sanctify it" (Deut. 5:11–12, KJV); third, the kidney phase: "Honor thy father and thy mother … Thou shalt not kill" (Deut. 5:16–17, KJV); then the lung phase: "Neither shalt thou commit adultery./Neither shalt thou steal" (Deut. 5:17–18, KJV); and finally our incarnation as humans, the pancreas phase with the lies and coveting: "Neither shalt *thou bear false witness …/ Neither shalt thou covet … anything that is thy neighbor's*" (Deut. 5:20–21, KJV). This represents a complete cycle of the LIFE biosystem.

So, mainly the commandments are aimed at allowing us to live as humans in a community, to keep contact with our different selves, and to further our evolution.

Interestingly, Moses, who was not a man of words, finally finds his voice to address the community, fitting for the Universal brain. It is a sign that his previous phases were well developed and healthy. In the same vein, the lung aspect implies exchanges with the environment and between all of our aspects. The link with the physical body, its cells, and behavior is paramount. Knowledge that does not pass by the cells but is only intellectual is not true knowledge.

The monarchic setting here (Deut. 17:11–20, 12:20, and 18:16) is held to be the historical background of Deuteronomy. Gerald Gerbrandt, professor emeritus at the Canadian Mennonite University, has argued for a unified concept of pro-kingship in Deuteronomy. More precisely, he states that "the correct question with which to confront the Deuteronomist … is not whether he [God] was anti-kingship or pro-kingship. Rather, we need to ask what kind of kingship he saw as ideal for Israel, or what role kingship was expected to play for Israel."

We see here the association I previously made between the social aspect and the phase where Horus becomes king(the heart aspect). Gerbrandt adds that the function of the king is to lead the people in keeping a covenant. A true king should not be like those of other nations, leading their countries into battle and thus furthering division

Isis Code

among humanity. At the heart of this covenant was Israel's obligation to be totally loyal to Yahweh, thus to the truth that, as we saw, is contained in the LIFE biosystem.[432]

The Rev. Dr. Norman K. Gottwald, professor of Biblical studies emeritus at New York Theological Seminary, author of *The Tribes of Yahweh* (1979), has summarized Deuteronomy as: "the indivisible unity of One God, for One people, in One land, observing One cult."[433] But as the lung aspect opens on a new beginning, I would say that with Moses's death (kidney aspect) at the end of Deuteronomy, it is the closure of the old generation and its circumstances and the beginning of a new epoch and cycle. This is similarly found in other prominent writings, such as the *Iliad* or the *Epic of Gilgamesh*. This new cycle should be the arrival in the Promised Land, a land of milk and honey, which is universal, as is God. Indeed, if the five phases are lived under the light of the implicate order of the LIFE biosystem, and its inherent natural laws are followed, we will indeed have access to the Promised Land and we will thus become the Chosen People. Israel would then become the community of those who understood the implicate order and the unique nature of God. He is One, so we are One. With the closing of this Book, the Israelites have a choice: either be reduced to a self-serving tribe of the personality and masculine polarity, or open themselves to universality and understand the greatness of the God they have chosen to follow, in the image of the Kabbalah of their esoteric tradition, which influenced the world and bowed to the One.

In acupuncture, the points concerned with the regulation of information are positioned around the neck. These are linked to the Master of the Heart and to the heart aspect of the idealistic self. Fittingly, Exodus notes the labeling of the Children of Israel by God, and the term *hard neck* or *stiff neck* comes into play. This term is repeated in Deuteronomy. So here, God is describing people who have neither a mature idealistic self nor a functioning Master of the Heart.

As written in Deuteronomy 31:27: "For I know how rebellious and stiff-necked you are. If you have been rebellious against the LORD while I [Moses] am still alive and with you, how much more will you rebel after I die!" and in Exodus 32:9: "'I have seen these people,' the LORD said to Moses, 'and they are a stiff-necked people.'"[434]

"Circumcise your hearts, therefore, and do not be stiff-necked any longer" (Deut.10:16 NIV).

We see "stiff-necked" in Exodus, when the Mammalian brain dominates. God is saying that these people are dominated by their Mammalian brain, which is the foreskin over their heart aspects. In Deuteronomy, which corresponds to the lung phase, he again makes this observation. Because heart and lung work together, had the people changed, we would not see the adjective occurring again in Deuteronomy. This gives a good picture of the phase humans in general had reached back then.

To awaken and manifest our inner dreaming god is the aim of every human life. Nature, through her wide-open book, is the best educator to seek in order to achieve this goal. In my garden I can feel her heartbeats. Religions tell us that we can only find God through the inner self. It would be cruel then to make us live in a world that would be so much of the external—unless this happens because we give too much importance to the crust hiding the internal? In fact, we are similar to dolphins swimming in the ocean of the manifested world. We are air creatures, but we live in the sea of the manifested world. Every night, we come to the surface between this watery world and the other and breathe in. This allows us to remember and remain yet another day in this watery physical dimension. But this dimension is also the expression of the implicate order, although in a more rigid form.

Our physical aspect cannot follow us into the more subtle aspect of reality. Only our psyche can, as it also helps to keep our integrity and thus to remain alive, thanks to the regular breaths we take (figuratively) from that other dimension.[435] This is what we call the breathing of the soul. Nature survives like this too, through cycles. The Sabbath is an example of this cycle. It is a point when we, similar to dolphins, rise to breathe at the surface. So when I am one with the essence of nature, I am closer to the "other side" and to the source of all manifestation, which I can intuitively feel through the mirroring of this palpable implicate order. "Remember," nature seems to whisper to us.

Nature awakens in us the wholeness of this implicate order.

Isis Code

This phase of the social self (twenty-eight to thirty-five years old) is facilitated by the maturation of the frontal lobes and corpus callosum.

Much of the temporal lobe is an association and assimilation cortex. This part of the brain, involved in forethought, impulse control, judgment, organization, planning, and learning from mistakes, is mature. The area of the frontal cortex that matures last is the lateral prefrontal convexity. It is involved in higher executive function, such as the integration into a coherent structure of action of diversely sourced inputs (J. M. Fuster). The frontal lobes make half of the volume of each cerebral hemisphere and are the last to develop. The more differentiated granular areas of the lateral prefrontal cortex (Analytical brain) connect extensively with the parietal (Human brain) and the temporal lobes (Universal brain) and are, as qualified by Elkhonon Goldberg, of a heteromodal type.[436] This type which favors communication is expressed in the external world by our ability and our need to communicate with other humans and nature in general.

The third story I wish to share does not require any explanation and concerns this aspect. The Gwich'in are a First Nations/Alaska Native people who live in the northwestern part of North America, mostly above the Arctic Circle. Reminiscent to what the natural human must have been, the Gwich'in hunter has been known to dream of the animal he will hunt, which he can locate in the dream. When he goes to that place, the animal is there waiting for the hunter. Of course, this doesn't happen to just any hunter. He has to be a natural man, thus in synchrony, and must share an osmotic relationship with nature. I have personally experienced this on a few occasions. The spiritual life of those natural humans was not dissociated from their emotional, physical, social, and mental lives. As well, they were one with nature and considered that every individual was responsible for his own enlightenment and interpretation of experiences. Of course, natural men are now a rarity and native tribes too went through the phases of the masculine polarity, our hunter is an example to this effect.

This story was broadcasted on the CBC radio program *Ideas* on July 10, 2010.

I have shortened it.

The Legend of the Polar Bear

Not so long ago, a hunter went on an excursion to find some seal to eat for himself and share with his hungry tribe. He had a canoe of which he was extremely proud. Paddling close to the coast, he saw a white bear that was clumsily walking on thin ice. The hunter, laughing, yelled: "Hey! Bear! With my canoe I can go on the water faster than you can even swim!" Later, while he was hunting with his harpoon through a hole in the ice, he noticed the same bear, recognizable through his missing ear tip, looking at him from afar. "Ah! I have a weapon, I am stronger than you, miserable bear, and I can get my food easily while you go hungry!" He continued laughing and mocking the bear until the bear slowly turned around and withdrew.

A few days later, one of his sons finds a dead bear, still warm, with no sign of a fight around it. He gets his father who rushes to the site. The happy father starts to cut the bear open, when he is suddenly taken aback. In lieu of the heart lies a stone! The son looks at the heart, puzzled and speechless. Sadness suddenly overshadows this happy moment for the father. The red blood profusely stains the white fur and the snow. Remorse creeps into the father's heart and tightens its grip, for he just noticed that this bear has a missing ear tip. Sorry, he retells the story to his son, how his sarcastic, arrogant, and cruel attitude destroyed the bear.

The heart is the lungs of the spirit. Each beating receives information about the environment. It records if it is respected and wanted here. It listens to the breathing of the soul. This is its nest. The spirit of the bear was unwanted, so it left and the bear died.

It is as if we, humans, have internalized the lungs of the plant kingdom and are externalizing our heart function through speech, action and expression of subtle information.

Because plants bear their lungs externally, they are vulnerable to the quality of the environment. We bear our lungs internally. We are vulnerable not only to the quality of the physical environment but also of the internal environment. The ways we think, feel, and act, as well as the way others think, feel, and act toward us, through the Master of the Heart, have an effect on the way we breathe. Other animals can't consciously control their breathing. We can.

Isis Code

I have an absolute respect for nature and for the plant kingdom. Just ponder this fact. For fire to manifest itself on earth, plants had to exist and give off oxygen in the atmosphere. When the level reached 13%, then fire could exist. It is the same with Consciousness. There are necessary conditions. Fire and air work hand in hand.

We humans got to the top of the food chain because of our ability to use fire. This comes because of a certain affinity associated with our lung aspect. We now use electricity in the same way. More than any other animals, we are totally in affinity with electromagnetism and fire. But if we do not respect the plant kingdom, we are, literally, cutting off the branch on which we sit.

Breathing Exercises

Breathing exercises are therefore an invaluable tool to harmonize our external and internal selves, and by the same token our two polarities.

Meditation, yoga and similar relaxation techniques that incorporate slow, deep breathing have long been thought to aid blood pressure. In 2002, the Food and Drug Administration allowed the nonprescription sale of a medical device called RESPeRATE, to help lower blood pressure by pacing breathing[437]. In clinical trials[438] people who used the slow-breathing device for 15 minutes a day for two months saw their blood pressure drop 10 to 15 points. Why slow-breathing works "is still a bit of a black box," says Dr. William J. Elliott of Chicago's Rush University Medical Center, who headed some of that research and was surprised at the effect.[439]

In medical biocybernetics, we consider breathing a conscious action we can use to keep our bodies balanced and healthy, since the lung aspect is mother to all energies.

Breathing is the only bodily function we have to act on our autonomous nervous system. Breathing and oxygen are at the core of our existence on earth. Nevertheless, we pay little attention to them.

Breathing exercises, properly performed, are inexpensive, take little time, and have substantial positive health effects. For example they have been found to:

- positively influence the involuntary (sympathetic) nervous system, which regulates blood pressure, heart rate, digestion, and many other bodily functions;
- improve the venous return to the heart, which leads to improved stamina;
- improve the flow of lymph (improves immunity)
- benefit asthma sufferers;
- increase metabolism (therefore lowering weight);
- release endorphins (the body's own painkillers) into the system, which can help relieve headaches, sleeplessness, backaches, and other stress-related aches and pains;
- help clear and focus the mind;
- strengthen weak abdominal and intestinal muscles;
- benefit the appearance of the skin (in my personal observation, skin becomes softer and hydrated with daily half-hour alternate breathing); and
- fight and prevent neurodegenerative illnesses.

Step 1: Abdominal Breathing Technique

To practice deep breathing, you will need fifteen minutes of uninterrupted time and a timer. Obviously you want to be in an aerated area or better, outside. First, give yourself permission to perform this exercise, forgoing any inclination for your mind to multitask. Offer yourself this short period of time every day. Turn off your cell phone; make sure you can give your full attention to being, not doing. Loosen tight clothes and take off your shoes. Get yourself in a comfortable position. You may be sitting or lying down: it doesn't matter.

Set your timer for fifteen minutes, or less if you fear that fifteen minutes of sitting quietly might trigger anxious feelings. You can work your way up gradually. The first step is to have a relaxed body. Similar to a pilot preparing to fly and verifying all the instruments of his plane, check the relaxation level of the different body parts. Relax your hands, arms, shoulders. If you are sitting, sit straight with your head and back aligned (no slouching). Let go! Let your face droop. Too many of us hold tension in our jaws and foreheads.

If you are prone to headaches and wrinkles, relaxing your face can be most helpful. Put a smile in your head. You are now ready, espe-

Isis Code

cially if that made you take a deep breath, showing your diaphragm has released its tension.

If you haven't already, close your eyes. Concentrate on the movement of your breathing. Normally, when you breathe, your chest rises. Here, you want your stomach to rise so the air goes all the way to the bottom of the lungs. Feel your abdomen rising and falling. If your mind is sprinting or wandering, remind yourself that this is what you are doing *now* and that it is more important than anything else. You want to connect with your feminine polarity. Don't give up. Habits are difficult to change and your brain has to record this new activity as a daily one. Repetition here is the key to success.

Now, place one hand on your chest and the other on your abdomen. When you take a deep breath in, the hand on the abdomen should rise higher than the one on the chest. This insures that the diaphragm is pulling air into the base of the lungs.

After exhaling through the mouth, take a slow, deep breath in through your nose, imagining that you are sucking in all the air of the room for a count of four, or if you are outside that you are breathing in particles of light. Hold it for a count of six. Visualize these particles of light spreading everywhere inside you, in your head, arms, torso, and abdomen and all the way to your toes. Slowly exhale through your mouth for a count of eight. As all the air is released with relaxation, gently contract your abdominal muscles to thoroughly evacuate the remaining air from the lungs. You can hold here for a few seconds, muscles contracted, depending on how you feel. It is essential to remember that you deepen respirations not by inhaling more air but through completely exhaling it.

Try to breathe not faster than one breath every twenty seconds (three breaths per minute). At this rate your heart rate variability increases, which has a positive effect on cardiac health. There is an effect of entrainment between heart, brain, and lung. They become synchronized, working in unison, and are far more effective. If you've managed to fall asleep, it may be because you've been so used to functioning mainly on your sympathetic nervous system (fight or flight response) that changing mode for a few minutes gave you some much-needed rest. Don't feel guilty about it.

Breathing exercises such as this one should be done once or twice a day, or whenever you find your mind dwelling on upsetting thoughts or when you are experiencing pain. Once you feel comfortable with the above technique, you may want to follow to step two. In general, when you get used to it, your exhalation should be twice as long as your inhalation. Putting your hands on your chest and abdomen is only needed to help you train your breathing. Once you feel comfortable with your ability to breathe into the abdomen, this is no longer needed.

Step 2: Alternate Breathing Technique with the Elements

The nasal cycle synchronizes with brain function, due to selective activation of one half of the autonomic nervous system. In 1994, breathing through alternate nostrils showed effects on brain hemisphere symmetry on EEG topography. Breathing technique has a balancing effect on the functional activity of both hemispheres.[440] "These results suggest the possibility of a non-invasive approach in the treatment of states of psychopathology where lateralized cerebral dysfunction have been shown to occur."[441] David S. Shannahoff-Khalsa wrote in 2007 about the effect of this cycle and manipulation through forced nostril breathing on one side on the endogenous ultradian[442] rhythms of the autonomic and central nervous system.[443] A unique feature of the autonomic nervous system (ANS), although usually overlooked, is the lateralized ultradian rhythms, where one branch of the sympathetic nervous system (SNS) dominates one side of the body and one branch of the peripheral nervous system (PNS) dominates on the opposite side; and then the two systems switch dominance on the two sides. That is, when one nostril is blocked, the hemisphere on that side becomes predominant. In harmony with this, the electrical activity of the brain was found to be greater on the blocked side. Alternate breathing stimulates the two aspects. Step two technique is for period of breathing of fifteen minutes and up, once a day.

First follow the preparation to step one.

With the help of a finger, block the left nostril. Breathe in through the right nostril to a count of four, visualizing light-photons coming into your brain. As you hold for a count of four, visualize this light transforming into a roaring fire, burning all the impurities out of your

Isis Code

brain (hold for a count of four). Then the fire transforms into air in your lungs and cleanses them of all impurities (hold for another count of four). The air transforms into waves of water and cleanses your digestive system (hold for a count of four). The water transforms into powdery clay, the earth turns on itself, and cleanses all your gut and sexual organs (hold for a last count of four). You have breathed in four, held for sixteen and now will breathe out on eight.

Now blocking the right nostril, exhale while counting to eight until your lungs are emptied.

You start another cycle by breathing through this left nostril; verify that your body is totally relaxed.

Once you have practiced step two with ease and you wish to go to step 3, simply change to eight (inhale), thirty-two (holding four times eight), sixteen (exhale), which is the most valuable cycle for general health.

If you have health conditions, please ask your doctor whether these exercises are suitable for you.

Tied to breathing is speech. Singing allows us to control our breath and emotions and to stimulate the right hemisphere.

Choir Singing

Most of the connections between the areas responsible for movement and those that control hearing are on the left side of the brain in accord with the masculine polarity. The right ear connects to the left hemisphere; it is in affinity with short sounds, verbal, and organized sounds. The left ear is connected to the right hemisphere and manages music as well as longer and unorganized sounds. As researchers tell us, this is the ear privileged by men, and I see here a bridge that could be used for men to access their feminine polarity.

The remarkable movie "The King's Speech," starring Colin Firth, shows how the future King George VI was partially healed of his debilitating stammering by a therapy that encouraged him to sing the words he wished to speak.

In treatments for speech impediment, as patients learn to put their words to melodies, crucial connections form on the right sides of their brains. Brain imaging studies have shown that professional singers have overdeveloped this "singing center." Dr. Schlaug, a neurology professor at Beth Israel Deaconess Medical Center and Harvard Medical School in Boston, found that even after a *single* session, a stroke patient who had been unable to utter a word succeeded in saying "I am thirsty" by combining each syllable with the note of a melody.[444]

In this treatment, patients are also invited to beat each syllable with their hands. Dr. Schlaug says this additional element, combining movement and sound, makes the therapy even more effective. People who sing have been shown more likely to be happy. Daniel J. Leviton, who runs the Laboratory for Music Perception, Cognition and Expertise at McGill University, Montreal, and is a neuroscientist and author of a few books on the subject,[445] cites a number of studies that demonstrate that singing elevates the levels of neurotransmitters associated with pleasure, love and well-being. In one example,[446] people's levels of oxytocin, which, as we saw, is associated with pleasure love and bonding, were measured before and after voice lessons. The levels increased significantly for both amateur and professional singers. Other studies have shown that the levels of IgA, an antibody isotype helpful to the immune system, increased with choral singing.[447] As a powerful stimulant of the feminine polarity, music has also been shown to boost the levels of serotonin, especially when we listen to music we deem pleasant.

We should remember that before written language emerged, religious stories were passed on with the oral tradition of song. The Torah, Greek myths, the *Iliad* and the *Odyssey*, and even the Koran were all sung before they were put to paper. This was a way to prevent the masculine polarity's interference. With the writing of the texts, we forsook the most fundamental aspects of religion, which thereafter came under the control of the masculine polarity.

So when we sing, we tap into something that is not only one of the most ancient of human practices, we also stimulate the feminine polarity. According to Patricia Preston-Roberts, a board-certified music therapist in New York City, studies have associated singing with a lower heart rate, decreased blood pressure, and reduced stress. She uses song to help patients who suffer from a variety of psychological and physiological conditions.

"Some people who have been traumatized, often want to leave the physical body, and using the voice helps ground them to their bodies," Preston-Roberts says. "Singing also seems to block a lot of the neural pathways that pain travels through."[448]

As part of a three-year study examining how singing affects the health of those fifty-five and older, the Levine School of Music in Washington, DC, formed a senior singer's choir.[449] Those involved in the choir (as well as those involved in two of the arts groups involving writing and painting) showed significant health improvements compared to those in the control groups. Specifically, the arts groups reported an average of: thirty fewer doctor visits, fewer eyesight problems, less incidence of depression, less need for medication, fewer falls or other injuries. The average age of all the subjects was eighty.

These results were published in 2006. Even lead researcher Dr. Gene D. Cohen, who was director of the Center on Aging, Health, and the Humanities at George Washington University in Washington, DC, and who passed away in 2009, was surprised at how big of an effect the seniors' arts participation had on their health. He ends his report by writing the following:

> What is remarkable in this study is that after just a year into the study the cultural groups, in contrast to the control groups, were showing areas of actual stabilization and improvement apart from decline—despite an average age which is greater than life expectancy [eighty years old]. This pattern then continued throughout year two of the study. These results point to powerful positive intervention effects of these community-based art programs run by professional artists. They point to true health promotion and disease prevention effects. In that they also show stabilization and actual increase in community-based activities in general among those in the cultural programs, they reveal a positive impact on maintaining independence and on reducing dependency. This

latter point demonstrates that these community-based cultural programs for older adults appear to be reducing risk factors that drive the need for long-term care.

Breathing exercises as well as choir singing are particularly useful ways to harmonize all of our oscillatory systems, as well as enhance neurogenesis and our lung energy. Activation of the sympathetic division (masculine polarity) of the autonomic nervous system has been shown to inhibit the process of neurogenesis in primates.[450] Also, the harmonious exchanges with nature, and I include humans in it, should not be underestimated. As we have seen, the heart aspect controls the lung aspect. This implies that the quality of our personal, as well as global Master of the Heart, will profoundly influence our lung aspect. If we remember, Pherecydes of Syros in his writings has Zeus presenting Chtonie (the physical world) with a meshed veil sprinkled with the elements of his creation.[451] It is as if we are all individually part of this veil and influenced by it.

Ecopsychology

This has been intuitively felt by millions of individuals who now actively protect the natural environment. In 1992 Theodore Roszak wrote *The Voice of The Earth: An Exploration of Ecopsychology*. In it, he asserts that many of our personal, societal, and environmental predicaments stem from a disconnection with nature. The term *ecopsychology* can be attributed to Buddhist scholar and system theorist Joanna Macy. After voicing her concerns about the threat of nuclear war to a psychologist, she was labeled as suffering from neurosis, which she strongly rejected. With a group, she hypothesized that human's problems spring from nature's condition. She adds that the repression of the despair associated with this fact "produces a partial numbing of the psyche." Personally, I was surprised to observe, the few times I visited a club in France in which the participants live their lives in the nude, that these people seemed particularly sad. Ecopsychology's hypothesis is that since we are so interconnected with the web of life, we all feel the pain of nature — which includes humans— , whether or not we consciously are aware of it or acknowledge it. LIFE biosystem theory agrees. The late John E. Mack, professor of psychiatry at Harvard Medical School and

Isis Code

Pulitzer prize-winning biographer, urged his fellow professionals to consider that "when we hear expressions of distress about pollution or other forms of environmental destruction in dreams and other forms of communication, we not hear or interpret these simply as displacements from some other, inner source."[452]

Numerous ecopsychologists have since argued for a new definition of mental health that embraces our partnership with the natural world. Indeed, the lung aspect shows that our suffering is dictated by the general situation we as a group have generated. Those who are more in tune with their feminine polarity (for example, children, women, old people, evolved men, individuals from tribes that have not lost their connection with the natural world, spiritualists, leaders who have a functional Master of the Heart) are more sensitive or aware of it, but we are *all* affected physically and emotionally, some more than others. Psychologist Susan Linn says in *The Case for Make Believe*: "Time in green space is essential to children's mental and physical health."

Researcher and lead author Al Williams of the University of Nebraska-Lincoln studied the story books children are reading nowadays in schools and made this observation: "The natural environment and wild animals have all but disappeared in these books." Richard Louv, in his book *Last Child in the Woods: Saving our Children from Nature-Deficit Disorder*, states that: "This is the expression of a physical dissociation with the natural world." This implies a fortiori dissociation from the feminine polarity.

In his conceptualization of the field, on pages 319 and following of his book, Theodore Roszak presents us with eight principles of ecopsychology. They can be summarized and associated with the LIFE biosystem as such:

1. "The core of the mind is the ecological unconscious." Lung aspect.

2. "The contents of the ecological unconscious represent ... the living record of evolution." Collective unconscious of the lung aspect. Akashic record supported by the Master of the Heart.

3. "The goal of ecopsychology is to awaken the inherent sense of environmental reciprocity that lies within the ecological unconscious." Mirror quality of the lung aspect.

4. "The crucial stage of development is the life of the child." Each phase must be recognized, completed, and harmonized with the others.

5. "The ecological ego matures toward a sense of ethical responsibility with the planet." Function of heart-lung aspects.

6. "Ecopsychology needs to reevaluate certain 'masculine' character traits that lead us to dominate nature." The personality wishes to dominate all and be considered as the One, while the individuality wishes to merge into the One.

7. "Whatever contributes to small scale social forms and personal empowerment nourish the ecological ego." Lung aspect feeds on positive exchanges.

8. "There is a synergistic interplay between planetary and personal well-being. ... The needs of the planet are the needs of the person; the rights of the person are the rights of the planet."

Experienced by an individual, ecopsychology could be formulated as Steven Harper [not to be confused with Canada's prime minister, Stephen Harper] did in his book, *The Way of the Wilderness*. In it, he explains the healing aspect of reconnecting with nature and the new identity that can sprout from it.

> On a two-month canoe trip across the Northwest Territories of Canada, I was blessed with such an experience. Near the end of a long day of paddling the sun was low in the sky and my mind had ceased its normal chatter. I had the sensation of becoming my paddling and all that was around me. Stroke after stroke I was called to merge with my experience until "I" was no more. Only perception existed, a perception that was more complete, more whole than any I have known in a usual state of consciousness.[453]

Isis Code

This perception of the whole, as we saw previously, was also experienced by some astronauts. We find echoe of it in Jung's "collective unconscious" and "peak experience," which is similar to Freud's "oceanic" state, a preindividuation condition found in infancy, when the right hemisphere is expressed, and which could also still be found in some tribal communities. We can see it in some school of fish, flights of birds. "Let the little children come to me" refers to this ability for those who still live through their right hemisphere to receive from the unity that creation is. Their Masters of the Heart still function, unhindered. True, we are One.

Nevertheless, individualization was and is a necessary step to bring Consciousness to this physical realm. Reopening ourselves to the unity is a necessity to attaining a higher level of this Consciousness, to free ourselves from the illusion of death associated with space and matter and to reach human perfection through completude. It is also a sign of human maturity. It is the lung aspect of the beginning, but more evolved because we can now integrate Consciousness into space and time, instead of fleeing or being absorbed by it (diagram 12).

Brain Elements of the Universal Brain

Which elements in our brains are most associated with the lung energy?

1. The *anterior commissure*, which is a band of axonal fibers interconnecting the right and left temporal lobes and part of the neospinothalamic tract; a sensory pathway for pain, touch, temperature, itch, and crude touch. It is 18% larger in women (Allen and Gorski, 1992), and contains decussating fibers from the olfactory tract. Traveling up the brain stem, this tract moves dorsally. The neurons ultimately synapse with third-order neurons into the thalamus. From there, signals go respectively to the cingulate cortex, the primary somatosensory cortex, and insular cortex. A healthy thalamus (Reptilian brain) is thus a necessity for proper reception in the insular cortex. As with the corpus callosum, the anterior commissure is responsible for information transfer as well as inhibition within the limbic system, in accord with the LIFE

biosystem. It also allows the right and left female amygdala to communicate. This, coupled with the fact there are more neurons and they are more densely packed in the female amygdala, added to the gender difference in the hypothalamus, predispose women to be more emotionally and socially sensitive, perceptive, and expressive (Joseph, 1993). This would incite a woman to form and maintain attachments in a different manner from what we can observe in men.[454]

2. The *interoceptive cortex*, associated with the insular cortex, exists only in primates, is enormously enlarged in humans, and is phylogenetically the most recent. It appears that human pain is significantly reduced in the case of lesions to this precise region of the brain, generally known as the parieto-insular cortex. Eisenberger and colleagues also found that social rejection shared similar pathways with physical pain. This area, we won't be surprised, is involved in cardiorespiratory and visceral control, with the right anterior insula offering the ability to time one's own heart beat and blood pressure. The higher the gray matter volume, the more the accuracy and sensitivity in this sense of the inner body.

> The anterior insular cortex (AIC) is implicated in a wide range of conditions and behaviors, from bowel distension and orgasm, to cigarette craving and maternal love, to decision making and sudden insight. Its function in the representation of interoception offers one possible basis for its involvement in all subjective feelings. New findings suggest a fundamental role for the AIC (and the von Economo neurons it contains) in awareness, and thus it needs to be considered as a potential neural correlate of consciousness.[455]

This new view offers a substantive explanation for psychosomatic pain, as well as interpretation of information received by the Master of the Heart. The widely held hypothesis is that self-awareness is based on a mental image of the homeostatic condition of the body or, simply put, how one feels.[456] It is known that people with avoidant attentional orientation due to emotional damage do not do well on interoception. It seems that the insular cortex works with the mirror neurons, since it is said that the insula expresses feelings of empathy. In other words, affectivity (not the sensory components) will mirror what is happening to someone else as if it were a personal experience. It is true also that this aspect mirrors everything we see someone else do, as long as we have an idea of how it must feel. The Idealistic self working with the social one would explain this.

Functionally speaking, the insula is believed to process convergent information to produce an emotionally relevant context for sensory experience. More specifically, the anterior insula is related to olfactory (lung aspect), gustatory (pancreas-yin), vicero-autonomic (Master of the Heart) functions as well as to a wide range of emotions, while the posterior insula is related more to auditory-somesthetic-skeletomotor function (thus masculine polarity). Functional imaging studies have also implicated the insula in conscious desires, such as food cravings and drug cravings, these being seeked by an out-of-balance system. What is common to all of these emotional states is that they each change the body in some way and are associated with highly relevant subjective qualities. The insula is well positioned for the integration of information relating to bodily states into higher-order cognitive and emotional processes, as it is right beneath the meeting point of the temporal, parietal, and frontal lobes. It receives information from homeostatic afferent sensory pathways via the thalamus and sends output to a number of other mammalian-related structures, such as the amygdala, the ventral striatum, as well as to the orbitofrontal cortex, and motor cortices.

Using magnetic resonance imaging, a study has found that the right anterior insula was significantly thicker in people who meditate. Not surprisingly, there was as well a thickening of the middle prefrontal cortex (Lazar et al., 2005). The practice of looking inward and of reflection activates the insula and the middle prefrontal regions, especially on the right side of the brain. The LIFE biosystem agrees.

Another study, using voxel-based morphometry and MRI on experienced Vipassana meditators, was done to extend the findings of Lazar et al., which found increased gray matter concentrations in this and other areas of the brain in experienced meditators.[457]

The differential thickness in this region between groups is consistent with increased capacity for awareness of internal states by meditators, particularly awareness of breathing sensations. Years of practice and change in respiration rates were correlated with cortical thickness in the inferior occipitotemporal visual cortex and right anterior insula. Most of the regions identified in this study were found in the *right hemisphere*. Interestingly, despite the normal effects of aging on the prefrontal cortex, the average cortical thickness of the forty to fifty-year-old meditation participants was similar to the average thickness of the twenty to thirty-year-old meditators and controls, suggesting that regular practice of meditation slows the rate of neural degeneration there. [458].

3. A class of bipolar neurons known as *von Economo neurons* (VENs), or spindle neurons, are found in the fronto-insular cortex (FI) (more on the right than the left) and anterior cingulate cortex (ACC). Women have more gray matter in the insular cortices where we find those spindle cells. The cells' morphology and anatomical location suggest they are positioned to receive simultaneously different types of information. They may play a role in intuition, that "gut feeling" some people often experience. This is especially interesting since the FI is involved in processing the conscious monitoring of visceral activity, but also because the lung aspect is associated with the large intestine and receives the energy (information) from the pancreas aspect (digestive system), and is controlled by the heart (Master of the Heart—celiac plexus, vagus nerve). Of note, many of the sensory signals conveyed by the vagus nerve terminate in the insular cortex.[459] Also, those with irritable bowel syndrome have abnormal processing of visceral pain in this insular cortex related to dysfunctional inhibition of pain within the brain. Similarly of interest, in some case of autism "chronically inappropriate neuronal activity in the enteric nervous system and/or its vagal efferent" might have created the conditions for autism, since treatments that eliminate certain dietary antigens or improve intestinal health appear to bring improvement to the condition." [460]

We could say that the lung aspect is a type of unconscious Analytical brain. It is rational in the sense that it analyzes and reacts to invisible subconscious information. The insular cortex is especially active during social situations when we need to evaluate others, but also when we experience a whole range of feelings. Intuition allows us to overcome the uncertainty attached to these situations or to add unconscious information to what is being reported from the Analytical brain. Intuition is also renowned for quick decisions and for interpretation when the ACC is concerned. Phylogenetically, those spindle neurons are found only in great apes (orang-utans, gorillas, chimpanzees, and humans), with humans having the absolute highest percentage relative to neuronal quantity. Their clustered aspect in humans suggests that they might be more susceptible to dysfunction and degeneration, as found in autism and Alzheimer's disease.

4. The *resonance circuitry* is composed of the mirror neuron system, the superior temporal cortex, the insula cortex, and the middle prefrontal cortex. It enables us to be attuned, to resonate to each other. With the Master of the Heart and the structures that are associated with it (see chapter 5), it is the structure of the sixth sense of synergy. It allows emotional resonance. It has been shown that not only is this circuitry able to represent others' internal states, but it mediates the basic mechanisms of echoing, without which meaningful relationships would simply not exist (Iacoboni, Carr, Dubeau, Maziotta, and Lenzi, 2003). In this sense I can say to you the following: I am not asking of you to believe in what I believe. Only understand that I am telling you who I am. What is not your reality is, nevertheless, mine. Through your kind listening, you allow me to be and to say "I am." Don't try to wake me up. Only hold my hand and be with me in my dream. Let me tell you about it and allow you to penetrate in it. Then let me visit your dream. For we are all dreamers visiting a larger dream. Who is to say which dream is reality?

Justly, Carl G. Jung said that humans are enclosed in the psyche as everything we touch or that touches us is image.[461]

5. The *massa intermedia,* a structure that crosses the third ventricle between the two thalami, is present more frequently in females than in males, although it can be absent in the case of schizophrenia. Among subjects having one, this structure is on average 53.3% larger in women than in men. This, despite the fact that the men's brains are approximately 8% larger. A growing literature reports structural aberration and reduction of the thalamus, as well as of other midline structures in the brains such as the striatum (Mammalian brain) and superior temporal cortex (Universal brain), in patients with schizophrenia.[462] This suggests problems existed probably before the emotional phase, where physical deficits are eventually detected.

6. The *claustrum*, a thin sheet of gray matter attached to the underside of the neocortex in the center of the brain was previously considered either part of the insular cortex or basal ganglia. Established on more recent methods of scrutinizing the development of both the human and animal brain, including fMRI results, the hypothesis most accepted now is that it is a *seventh layer* of the cortex, in the insular region.

This proposition is labeled Filiminoff's hypothesis since he is considered the first to come to this conclusion. The neurons of the claustrum are required to take in different types of data (color, motion) across various modalities (visual, sound) merging them together so they are processed at the same time. It has even been shown that the same cells in it can process different types of information. Another study looked into its serotonergic innervation. Researchers Francis Crick and Christof Koch have compared the claustrum to the conductor of an orchestra.[463] The different parts of the brain (cortical subregions) must play in harmony; otherwise the end result will be a cacophony of sounds.

The largest bundle of connecting fibers, and one that has grown considerably in size when humans are compared to monkeys, is the *corpus callosum* (CC), which connects the two hemispheres and the four lobes. It is white and harder than the rest of the brain (callosum means hard) and has a central location in the brain. In vivo imaging has revealed that increases of the CC can be expected until the middle of the third decade. This makes it truly the last aspect of the brain to fully mature. The fastest increase is observed in the first decade of life.[464] In young children, transient projections between areas suggest additional cortical connectivity during early childhood (Carlson et al., 1988) with a subsequent pruning with maturation. The corpus callosum will nevertheless increase in size, with the splenium enlarging the most.[465]

"It [the corpus callosum] coactivates the unengaged and therefore underactive hemisphere, maintaining it ready to respond and distributing attentional capacity between the hemispheres."[466]

Houzel and Milleret (1999) state that: "the architecture of callosal axons ... suitable to promote the synchronous activation of multiple targets located across distant columns in the opposite hemispheres."

Marcel Kinsbourne ends by pointing that Engel, Konig, Kreiter, and Singer (1991) have found neurons on either side of the callosum that entrain in synchronous firing. This is in harmony with our oscillatory conception of brain structures.

Thanks to the corpus callosum, this increase in the connectivity between hemispheres has led to hemispheric specialization, as each hemisphere serves different functions. With it, it appears possible for the hemispheres to go their separate ways to a large extent, rather than to duplicate each other, as is seen in other mammals. Women, because they massively use their corpus callosum to access the left hemisphere, have a brain that is therefore symmetrically organized, maintaining a more bilateral representation of cognitive functions. "Women's brains appear to be more efficient than men's in the sense that an equal increase in volume produces a larger increase in processing capacity in women than in men."[467] Many studies have suggested that the difference in structure between women's and men's brain implies that women are better at computation while men are better at transferring information between distant areas of the brain. Women's brain also have more gray matter (computation), more convolution (more brain in less space), and have more volume in the structures linking the different brain modules. Now I believe these studies refer to the feminine polarity, which should be dominant in women. If men developed their feminine polarity, they would have access to a wider range of abilities. Men, by contrast, have more cerebrospinal fluid, more glial cells, more white matter (reception and transfer of information). It appears thus that evolution placed a priority of structures on both genders. Interestingly, in the rest of the body, it seems that the genders are reversed. The feminine body has an emphasis on the lymphatic system—white cells in the guise of fat—while the masculine body emphasizes muscle, the equivalent of computing power for the brain.

Considering that the lung aspect feeds the Reptilian brain and controls the Mammalian brain, it is not surprising that abnormalities of the corpus callosum have been found in cases of deficient affectivity (Raine et al., 2004).

Musicians who begin playing before the age of seven have a larger corpus callosum.[468] What happens in the case of an injured or underdeveloped corpus callosum? Transcallosal projections are necessary for the left hand to carry out verbal commands, and are required for the right hand to access spatial knowledge. Information received from the corpus callosum is essential to the development of the neocortex. This agrees with the feminine polarity feeding the masculine one. Generally, in women the posterior area of the corpus callosum, the splenium, is larger than in men. Age negatively affects the *anterior part*, the genu that inhibits task-irrelevant information, but does not influence the splenium section. However, a study has found that in the case of Alzheimer's disease, the white matter is selectively damaged in the splenium. Researchers observed that impairment of cognitive function and that extent of damage to this area of the CC were strongly correlated.[469] As well, deficits of callosal structures have been observed in a wide range of neurodevelopmental disorders as well as diseases: Alzheimer's disease, autism, attention deficit disorder, dementia, multiple sclerosis, and schizophrenia, to name a few.

We are not surprised to see the pineal gland tightly hugged between the *splenium* of the corpus callosum and the third ventricle.

7. The *olfactory bulb* is a structure in charge of the reception of olfactory sensory signals. A prominent brain component in most vertebrates, it is reduced in primates. In a study, early olfactory bulb ablation in rhesus monkeys caused spikes of aggressiveness, concomitant with lower levels of positive social behavior.[470] Our association of it with the Universal brain and its lung functions is therefore validated.

It is suspected that, because the olfactory bulb lies unprotected from the environment, pathogens can find a way to access brain structures during formation and cause illnesses such as schizophrenia (R. W. Doty, 1989). In the case of John Nash, discussed earlier, we know that his mother contracted scarlet fever and subsequently lost part of her hearing.

Patients suffering from schizophrenia manifest poor olfactory discrimination normally allowed by the right hemisphere (Yousen et al., 1997; Dade et al., 1998). Also, hallucinations indicate an inability for the right hemisphere to influence the left one. Added to this, in a majority of schizophrenic subjects, ventriculomegaly (expansion of the brains' ventricles) was revealed. Nonetheless, a 30% remission rate of this illness points to possible energetic and reversible causes in some cases.

In agreement with the LIFE system, we have seen that the atrial natriuretic peptide (ANP) secreted by the heart has an effect on this olfactory bulb (see chapter 2). Where active stem cells were found, adult neurogenesis will occur in the subventricular zone (SVZ), from which new neurons migrate to the olfactory bulb.[471] In other brain regions where neurogenesis exists, it appears that it is actively repressed by the local environment. We may thus say that neurogenesis depends on the health of the feminine polarity. The speed of maturation of these new neurons is experience dependent and varies between neurons. Cells born in the SVZ will migrate over a long distance to reach their target area, the olfactory bulb. This is in contrast to adult neurogenesis in the dentate gyrus (hippocampus).

Selection process in neurogenesis is critical. The survival rate of new neurons is quite low despite the large numbers of new neurons born to the olfactory bulb and dentate gyrus. In the dentate gyrus, it is estimated half of the new neurons will die within two weeks of birth. Various factors determine the rate of survival.[472] In contrast, in the olfactory bulb death appears to be later in the development process, when young neurons with extended dendrites are already covered with spines. The new neurons in the adult hippocampus are regulated by a range of factors, including stress (Gould et al., 1990), aging (Kuhn et al., 1996), environment (Kempermann et al., 1998), and activity (van Praag et al., 1999).

Stress, lack of sleep, and depression will reduce neurogenesis. Relaxation, breathing exercises, and other aerobic exercises performed joyously, will enhance positive emotions and stimulate the feminine polarity, thereby enhancing neurogenesis. Two meta-analyses have shown this. In 2003, Dr. Stanley Colcombe and Dr. Arthur Kramer analyzed the results of eighteen studies. The results clearly showed that aerobic exercises increase cognitive performance in healthy adults between the ages of fifty-five and eighty. Another study (Dr. Heyn et al., 2004) showed similar beneficial effects of fitness training for people over sixty-five who had cognitive impairment or dementia.[473]

In the first meta-analysis, Dr. Arthur Kramer, who is a psychology professor at the University of Illinois and director of the Beckman Institute for Advanced Science and Technology, suggested: "Doing ... aerobic exercise, 30 to 60 minutes per day 3 days per week, has been shown to have an impact in a variety of experiments. And you don't need to do something strenuous."[474]

Aerobic exercises are those designed to be low-intensity enough not to generate lactic acid, so that all carbohydrates are aerobically turned into energy. Thus, breathing exercises, swimming, gardening, walking in a continuous relaxed fashion, as well as low intensity dance, are the best choices to achieve this. We see here again that it is not the masculine polarity and its willpower that needs to be summoned, but the feminine one.

If your olfactory sense is deficient (lung, Universal brain), it appears you can't properly taste either (pancreas, Analytical brain). It seems this is true not only for the physical body, but also for the mind. The rational self of the Analytical brain reacts on the information it receives from the feminine polarity. When proper and complete information is not received, or is automatically dismissed, then the conclusions made by the rational self can only be false. An extreme example of this is paranoia. We encounter this illness in subjects who have suffered from early affective neglect. When they are adults, the analysis they make of a situation can be right, but the basis on which they elaborate their analysis is usually false.

In biology, a chemosignal, or pheromone, is a chemical secretion influencing the development or behavior of other members of the same species. This observation has led research to identify just how the body's chemical secretions can express emotional states and social cues. We ooze information on many levels.

We were all told that pheromones' function is to attract a member of the opposite sex. It now appears as though in humans, they might also convey information of a socio-emotional nature. A study by Denise Chen, assistant professor of psychology at Rice University, and published in the *Journal of Neuroscience*, analyzed how the brains of female volunteers handled and coded the smell of perspiration from men. Animals rely strongly on scent to communicate. Humans respond consciously to salient information, such as facial and vocal expressions.

Using functional magnetic resonance imaging, the research concluded that many parts of the human brain are involved in processing olfactory information. These include the right fusiform region, used for facial recognition; the right orbitofrontal cortex, which sends information to the insula; and the right hypothalamus, all associated with the feminine polarity or with women.

"With the exception of the hypothalamus, neither the orbitofrontal cortex nor the fusiform region is considered to be associated with sexual motivation and behavior," Chen wrote. She and fellow researchers concluded that the chemosensory information from natural human sweat is encoded holistically in the brain rather than specifically for its sexual message.[475] This provides an explanation of the neural mechanisms of human social chemosignals and opens the door to more subtle means of communication supported by the Master of the Heart.

Already, with honey bee queens, research has demonstrated that they produce a sophisticated array of chemical signals (pheromones) that influence not only the behavior but also the physiology and function of other members of the colony. For example, the queen mandibular pheromone (QMP) induces young workers to feed and groom her queen and leads bees into performing colony-related chores. In young worker bees, dopamine levels, levels of dopamine receptor gene expression, and cellular responses to this amine are all affected by QMP. These induced changes in their brains will modify the bee's behaviors.[476]

8. Information encoded in long-term memory for an extended period of time no longer needs the intervention of the hippocampus (Mammalian brain). Meanings of commonly used words, for example, will instead activate the frontal and *temporal cortex*. Activity in the temporal lobe elicits the memory of any particular fact, while activity in the frontal cortex indicates it reached awareness. The temporal lobe of the lung function is closely related to the Master of the Heart.

Thanks to the advent of functional MRI scans in the late 1980s, researchers have been able to study what areas of the brain are active when people engage in particular activities. What they found is surprising: higher intelligence scores were associated with less brain activation. Also, it was clearly demonstrated that the memory factor showed no correlations with brain activation.[477] This gives credence to the theory of Candace Pert, PhD. To her, peptides (protein is one or more polypeptides of more than fifty amino acids long) are the physical expression of emotion. In my view, allergies, which often are reactions to proteins considered alien by the organism, are provided here with an explanatory cause. In 1993, Pert appeared on Bill Moyers's landmark TV program *Healing and Mind*, where she clarified her theory of emotion to an intrigued audience. She gathered evidence that memory occurs at the point of synapse, where changes take place in the receptors. In her view, the *sensitivity of the receptors* is part of memory and pattern storage. This peptide network expands way beyond the hippocampus, to organs, tissue, skin, muscles, and endocrine glands. These all carry peptides receptors and can thus access and store emotional information. Therefore, emotional memory is stored in many places, not just the brain.[478] This strongly resembles an interface with the Master of the Heart. Also, as with the Master of the Heart, the autonomic nervous system is pivotal to this entire understanding. Its importance is much more refined than has been earlier deducted. Every peptide ever mapped by Pert can also be found in the autonomic nervous system. There is an emotional coding to the way our autonomic patterns are elaborated. Also, peptides are suspected of conferring magnetism to biological structures, another link to the Master of the Heart.[479]

Experiments have revealed that the hippocampus seems to be the gateway to expressing awareness of the whole emotional experience. Pert showed us that almost every variety of peptide receptor is also found in the hippocampus, which I mainly associate with the emotional self.

It appears memories, even subconscious ones, could be accessed through this peptide network, which is made of anything that has peptide receptors on it. Pert comments: "There are inhibitory chemicals and impulses that function to keep the emotion and information down. I think unexpressed emotions are literally lodged lower in the body."[480]

Temporal lobes also manage our states of awareness. Rhesus monkeys that have had their temporal lobes removed often demonstrate hypergenitality, in the form of homogenital and solitary genital behavior (Kluver, 1958). This makes sense in light of the lung aspect being associated with exchanges, whether these exchanges involve others or the environment.

Epileptic phenomena are more numerous for women than men. For example, women may have orgasms during TLE seizures, while men do not. Hypersexuality has been seen in people with lesions in their frontal and temporal lobes (Huws, 1991). Thus the temporal lobe exerts an inhibitory function over the Mammalian brain. This has been observed in association with limbic seizures (Andy, 1991; Persinger, 1994).

Studies of patients with brain lesions, as well as neuroimaging experiments, have revealed several regions that are crucial in social cognition, including the superior temporal sulcus, fusiform gyrus, amygdala, and prefrontal regions (Adolphs, 2001, 2003). The superior temporal cortex is significantly larger in women compared with men; 17.8% more volume in women. This is accounted mainly by one section of the superior temporal cortex, the planum temporale, which is almost 30% larger in women. In addition, the cortical volume fraction of the Broca area, associated to speech, in women is 20.4% larger than in men.[481]

9. In particular, the *ventral frontal cortex (VFC)*, consisting of the orbitofrontal cortex (OFC) of the feminine polarity and the straight gyrus (SG), is essential for normal human social behavior. The VFC contributes to social cognition in tasks involving facial recognition, attribution of intentions (Brunet et al., 2000), as well as perception of anger in others (Blair et al., 1999; Wicker et al., 2003). The ability to experience moral emotions, which appears to be linked to empathy, also involves the VFC (Moll et al., 2002). Damage to this area leaves intellectual functioning intact, but patients will exhibit an inability to feel empathy; a lack of basic social manners, strangely named "disinhibition"; and impairment

of reasoning about the mental states of others (Malloy et al., 1993; Bechara et al., 2000; Adolphs, 2001, 2003). It was observed that in patients with schizophrenia, the VFC gray matter volume is negatively correlated with social dysfunction, supporting the idea of it mediating social function (Chemerinski et al., 2002). It was observed that in nonhuman primates, the orbitofrontal neurons encrypt the rewards and pass that on to the dorsolateral prefrontal neurons of the Analytical brain, which then selects a response on the basis of this information (Wallis and Miller, 2003). The orbitofrontal cortex contains the secondary taste cortex, associated to the reward value of taste. This evidence thus demonstrates the orbitofrontal cortex as agent in decoding some primary triggers such as taste, in learning and reversing associations of visual and other stimuli to these, and plays an executive function in controlling and correcting reward-related and punishment-related behavior. It appears that, as implied by the LIFE biosystem, the right ventral prefrontal (orbitofrontal) cortex moderates social distress by controlling the anterior cingulate cortex.[482]

In premature babies' brains, the largest absolute volumetric reductions were found in the parieto-occipital and sensorimotor regions. However, deficits were also significant in the orbitofrontal and premotor regions in comparison to term infants (Peterson et al., 2003). Interestingly, the orbital region has a recognized function in word processing, as it has in some other domains, such as high-level vision (Kanwisher, 2010) and social cognition (Saxe and Kanwisher, 2003). This region is known to be quite sensitive to stress and mediates emotional responses (Rolls, 2002; Tranel, 2002). Lesions within this region may generate irritability, social impropriety, poor judgment, lack of drive, low tolerance of frustration, and rigidity (Rolls, 2002; Tranel, 2002). Those behaviors are often displayed by children born prematurely. Deficits in cerebral maturation of the orbitofrontal area could contribute to the delayed sensory integration of preterm infants (Rolls, 2002; Tranel, 2002). Interventions aimed at reducing stress in the preterm infant have been shown to improve frontal region cerebral white matter development (Als et al., 2004). The most striking difference in relative volumes was the large excess of CSF found in the four superior cerebral regions, this being a masculine feature.[483]

10. The FFA (fuisiform face area) is located on the ventral surface of the temporal lobe on the *fusiform gyrus*. It is in a slightly different area for each human and displays some lateralization, usually being larger in the right hemisphere. We thus assimilate it to the feminine polarity. Its function, outside of facial recognition, is to recognize the *unique*, in harmony with the function of the feminine polarity.

11. The *right hemisphere* is particularly involved in the processing of emotional data (Kinsbourne and Bemporad, 1984; Safer, 1981) as well as with the evaluation of information received by the subject from his own body (Luria, 1973). It regulates the sympathetic activity.[484] Right hemispheric lesions generate tachycardia, suggesting a reduction in parasympathetic cardiac innervation. Deficiency of the right hemisphere triggers shortcomings, such as problem in attention, left-side neglect, deficiencies in memory, orientation, thought ordering, general problem solving, reasoning, and social communication. Multitasking may prove difficult, if not impossible, as well as focusing on one topic during conversations. In the case of a right hemisphere stroke, the individual no longer acknowledges the left side of his or her body or space. We call this left-side neglect. These individuals, for example, will not eat food on the left side of their plate or read the words on the left side of a page. They will have problems with remembering information, such as street names and significant dates, and remembering personal information, such as birth date, age, or family names, as well as with learning new information easily. Daily irritants, such as something breaking, will be overwhelming. As for their social skills, these will be extremely limited, as these individuals will talk at the wrong time, be oblivious to nonverbal cues and body language, as well as to metaphors and humor.

12. As we have seen, the *pineal gland* is responsible for the production of pineal indolamine. The most researched and known is melatonin, made from serotonin and secreted in response to darkness. Other hormones secreted by the pineal are also

released at night, prior to REM sleep, and interact within the central nervous system. This gland contains the highest levels of serotonin (Sun et al., 2001), which plays a pivotal role in sleep, cardiovascular activity, respiratory activity, perception, memory, motor output, glucose regulation in the brain, as well as sensory and neuroendocrine function. Tryptophan, the essential amino acid converted into brain serotonin, is not synthesized by the human body. When deprived of it, brain chemistry and mood are altered. The cycle is as follows: zinc is needed for B6 production, which is requested to transform tryptophan into serotonin. Tryptophan is obtained by digestion in the gut and transported in the blood plasma to the brain, where it is converted to serotonin. Thus, the health of the gut, associated to the lung function, is fundamental to this process. Low levels of tryptophan have been documented in autistic children[485] and in sufferers of chronic pain caused by arthritis and lower back problems. It is thought that lower serotonin levels have an effect on lowering the pain tolerance threshold.

In *Evolving Brains*, John Morgan Allman explains that serotonin often "does not directly excite other neurons but instead modulates the responses of neurons to other neurotransmitters." The diffusion of serotonin has a strong influence on other neurotransmitters, although pyramidal neurons are directly excited by serotonin.

Tryptophan is particularly plentiful in chocolate, dried dates, eggs, fish, poultry, soy beans, Parmesan cheese, chickpeas, sesame seeds, sunflower seeds, pumpkin seeds, spirulina, and peanuts. As noted earlier, it is also richly present in colostrum and breast milk, but is lacking in formula milk, which will deprive the developing infant brain of nutrients essential for normal growth of the brain serotonin system. Since it is a precursor to melatonin, low serotonin implies less melatonin being produced by the pineal gland. This, among other things, will have a negative impact on sleep patterns. Also, in cases of failed mother-infant bonding, serotonin has been shown to be significantly reduced.[486]

Monkeys who suffered maternal deprivation were shown to have a deficit in serotonin transporters in the brain. Sarah

Hrdy, in her book *Mother Nature: A History of Mothers, Infants, and Natural Selection,* observed that "no wild monkey or ape mother has ever been observed to deliberately harm her own baby." How can humans harm their own offspring when no genes can be identified to account for this, or for the current epidemic of violence that has grown over this past generation (Prescott, 2001)? In adults, low serotonin or a deficient serotonin system are often associated with bulimia, depression, insomnia, impulsivity, and violent behaviors, particularly suicide. Findings have shown an association between childhood abuse and anomalous serotonin and cortisol function. Not surprisingly, with our observation of the general deficit of the feminine polarity, depression has become a worldwide leading health problem, a black tide engulfing entire lives.

Antidepressants have been prescribed for years in a bid to artificially boost serotonin levels by lowering its need through reuptake inhibition (SSRI). Is this a good idea? Virtually all brain tryptophan is converted into serotonin. In laboratory animals, dietary supplements can increase up to tenfold the serotonin concentration in the brain.

All SSRI drugs increase prolactin, often provoking hyper-prolactinemia. Following an extensive review of research, hyperprolactinemia is suspected as a factor in the loss of libido in men and women, erectile dysfunction, galactorrhea and amenorrhea, decreased bone density, a blockade of dopamine receptors, and even breast cancer.[487] The pineal gland, when acting on the pituitary gland, *suppresses* prolactin. In the case of a deficient feminine polarity, this does not happen. The mechanisms that direct equilibrium in dopamine, serotonin, prolactin, and melatonin are subtle and complex. Without a holistic approach we might do more harm than good.

For about eleven years now, Dr. Ann Blake Tracy (*Prozac: Panacea or Pandora?*) has persistently warned the public against artificially raising serotonin levels. Dr. Tracy has shown that an increase in serotonin produces rushes of insulin, dropping sugar levels, chemically inducing hypoglycemia (low blood sugar) in this way. Furthermore, it has been established that too much serotonin damages blood vessels, particularly in the lungs, and may also harm heart valves. This is because serotonin is a powerful vasoconstrictor.

The pineal gland and melatonin "exert a major influence in the control of brain electrical activity and have been shown to be involved in seizure and sleep mechanisms."[488]

With stress, an increase in adrenaline levels leads to a gradual increase in cortisol. Cortisol levels increase slowly but also decrease slowly. Cortisol remains in the system longer than adrenaline, continually exercising its effect. It increases blood pressure and blood sugar levels to aid in the flight or fight response. A chronic stressful state implies that adrenaline and cortisol are constantly secreted. This will lead to a higher level of fatty acids in the blood vessels (cortisol signals the metabolism of fatty acids to produce an emergency supply of energy). Some experts say that the high levels of cortisol affect neurotransmitters such as serotonin, lowering its level. Others argue that a high level of serotonin ends up creating adrenal exhaustion, where cortisol and adrenaline are eventually depleted, and the person experiences symptoms similar to those of chronic fatigue syndrome.

In 2000, a research team from Philadelphia's Jefferson Medical College, wondering about the long-term use of SSRI drugs, force fed rats for four days about 100 times the normally prescribed amounts of either the antidepressants Prozac and Zoloft or the obesity treatments Meridia and Redux.[489] Lead author Madhu Kalia, a professor of biochemistry, molecular pharmacology, and neurosurgery, disclosed that the nerve endings swelled and started to curl. Although the amount of drugs they gave the rats is not equal to what a human would normally ingest, she said: "You can't assume it will not have an effect on the structure of the brain." Similar effects have been seen in studies with the street drug Ecstasy. Also called MDMA, it was first synthesized and patented in 1914 by the German drug company Merck. It was formulated and prescribed as an appetite suppressant. In the 1970s, MDMA was given to psychotherapy patients because it helped them "open up" and express their feelings. This practice was stopped in 1986 when animal studies revealed that the drug caused brain damage. In those studies, baboons were treated with Ecstasy[490] and researchers used positron emission tomography (PET) to take brain scans of them. They found that this drug was so toxic to the brain, it damaged the axon terminals (nerve endings) of the serotonergic neurons. They retested the baboons seven years after discontinuing the drug. The damage was still present. Dr. George Ricaurte, a professor of neurology at Johns Hopkins University, also analyzed brain scans of people who had used Ecstasy. He documented the same damage of serotonergic neurons as observed with the baboons, especially in the *medial prefrontal cortex* of the Human brain and hippocampus of the Mammalian brain.

Melatonin has its peak at around two in the morning. During daytime, the daylight inhibits its release. Evidence proposes that melatonin not only controls cycles, but also has an immuno- regulating function, stimulates natural killer cell activity, allows cytokine expression, and inhibits immune cell apoptosis (cell death).

This is understood by the fact that one of the main targets of melatonin is the thymus, which is the main organ of the immune system. In an article published in the *Journal of Pineal Research*, Georges J. M. Maestroni explains that this is a two-way exchange. The thymus also influences production of melatonin in the pineal gland.[491] An intriguing observation is the protection from autoimmune diseases in natives of areas of West Africa (Greenwood, 1968), along with low calcification of their pineal glands. Indeed, Adeloye and Odeku (1967) working at a local hospital where an average of 2,000 skulls were studied every year, encountered fewer than ten cases of visible calcified pineal gland during a period of ten years. Among other possible disabilities, calcification of the pineal gland is shown to be closely related to a defective sense of direction (Bayliss et al, 1985). The pineal gland is a magneto sensitive organ, thus sensitive to electromagnetic waves from computer monitors, cell phones, microwave ovens, high voltage lines, etc. Electromagnetic fields suppress the activity of the pineal gland and reduce melatonin production.[492] EMF also affects serotonin.

But there is more to the pineal gland than melatonin and serotonin. "Recent observations documenting the ability of melatonin to stimulate electron transport and ATP production in the inner mitochondrial membrane also have relevance for melatonin as an agent that could alter the processes of aging."[493] It also should get the attention of chronic fatigue sufferers.

As well, researchers found that: "Much data has been gathered, indicating—in experimental conditions—a mutual relationship between the pineal gland and the thyroid. The confirmation of these relations in clinical studies in humans meets numerous difficulties, resulting—among others—from the fact that—nowadays—human beings, as well as animal species, used in experimental studies, have been living far away from their natural and original habitats."[494]

And: "It has been shown that melatonin suppresses the formation of cholesterol, reduces LDL accumulation in serum and modifies fatty acid composition of rat plasma and liver lipids. People with hypertension demonstrate lower melatonin levels vs. those with normal blood pressure. The administration of the hormone in question declines blood pressure to normal range."[495]

Also: "Integrin-mediated cell adhesions provide dynamic, bidirectional links between the extracellular matrix and the cytoskeleton. Besides having central roles in cell migration and morphogenesis, focal adhesions and related structures convey information across the cell membrane, to regulate extracellular-matrix assembly, cell proliferation, differentiation, and death."[496] This associates the extracellular matrix, an interface of the Master of the Heart, with the pineal gland. "The hormone may synchronize renal cell physiology with the photoperiod through cyclic cytoskeletal reorganization. Also, the participation of PKC (protein kinase) in the mechanism by which melatonin causes a cyclic increased water transport and microfilament reorganization is discussed."[497]

Placebo Effect and the Feminine Polarity

One evening last May, I was enumerating all my blessings to my mother. I told her that the only thing missing in the beautiful paradise I lived in were swallows, my favorite birds. The next morning, despite a lingering smoke from forest fires, I walked to my dock and sat there, motionless. The sun had already risen, so I closed my eyes. Suddenly, a familiar chirp caught my attention. Opening my eyes, I could see four swallows flying in great speed over the lake.

Observe your reaction to this text and you will know which aspect of the feminine or the masculine polarity dominates in you. If suspicion, derision, and disdain prevailed, then your masculine polarity overshadows any expression of your feminine one, as the fact I described truly happened. Therefore your feminine polarity should sense that this is a truthful statement.

Jesus's statement "Let the little children come to me" referred to the predominance of the feminine polarity, necessary to spirituality.

"'Go,'" said Jesus, 'your faith has healed you.' Immediately he received his sight and followed Jesus along the road" (Mark 10:52, NIV).

Are we surprised that the placebo effect—I prefer to call it healing by the Self—was not only known in those times, but was also respected?

The word "placebo" appears in the Roman Catholic liturgy from the thirteenth century: *"Placebo domino in regione vivorum"* ("I will please the Lord in the world of the Living").

The placebo effect, as well as spiritual cures (miracles), is instantaneous. Although we see that the patient does release endorphins, in the case of the placebo effect we have no explanation for the *immediate* improvement some patients feel. The field of the Master of the Heart, associated with the extracellular matrix, appears to be a strong contender for this effect. Only an aspect coming from the universal Master of the Heart can do this, as shown by instantaneous remissions of some physical illnesses through evangelical groups. It appears these don't last though. Working through or awakening the idealistic self of their fellow humans can alleviate some disorders.

We now know that the doctor/patient relationship is fundamental to the healing process.

I noted earlier in the heart chapter that research has shown that belief activates the middle prefrontal cortex. I trust that as seen similarly for other aspects of the brain, when information is accepted and not questioned, when it is in harmony with subconscious patterns, this aspect of the brain does not even need to be activated. It will be if one has to ask the question, "Do I believe in this?" Similarly, the anterior insular cortex is activated when a person needs to distance himself from a statement that he considers in opposition with his beliefs.

This implies that for the Master of the Heart to allow energies to flow, the feminine aspect has to be receptive. For women, they achieve this through receptivity to ideas as well as physical receptiveness. Men are receptive to the social aspect as well as to emotions. What they "believe in" is very much dictated by their social group, as well as by their subconscious and emotional patterns.

New research has shown that the Mammalian and Universal brains react most strongly to placebos. These are namely the areas of the anterior cingulate, left nucleus accumbens, right anterior insula, as well as the left dorsolateral prefrontal cortex of the Analytical brain, which is linked to awareness and short-term memory (and receives inputs from the ventrolateral PFC).[498] The lung aspect has control over the Mammalian brain. So, to say there is no rationality in the placebo effect is wrong, since the dorsolateral prefrontal cortex is also solicited. The same research, "Neurobiological Mechanisms of the Placebo Effect," does not stipulate if the study involves both genders.

Interestingly, a recent functional neuroimaging study (Kapogiannis et al., 2009) has shown that religious beliefs correlate with neural activity of the frontotemporal networks, which are also associated with embodied and conceptual representations of others' behaviors and emotions. In a similar vein, having confident religious beliefs was associated with changes in the neural activation of the dorsal (Analytical brain) and ventromedial prefrontal cortex (Human brain) during self-referential processing (Han et al., 2008). In particular, neural activity of the ventromedial prefrontal cortex in Christian participants showed that the subjects did not differentiate between the self and others in judgments of personality traits (Han et al., 2008). Others are self and self is others. Previous studies have suggested that religious beliefs may be associated with altered self-representations. In other words, religious beliefs change our identities and the aim of our personal selves, by changing the *orientation* and perception of our parabolic antenna of the idealistic self I referred to in the beginning of this work.

Two French doctors, Jean-Paul Giroud and Charles Hagège, authors of a medical guide, estimated that of the eight thousand remedies listed in the *Dictionnaire Vidal*, the reference prescription book for French medical doctors; at least half have no demonstrated pharmacological effect.[499]

A 2004 study published in the *British Medical Journal* found that in Israel 60% of doctors used placebos in their medical practice, most commonly to fend off requests for unjustified medications or to calm a patient. Legitimate doctors and pharmacists can open themselves up to charges of fraud by dispensing placebos, since sugar pills cost pennies for a bottle, but the price for a "real" medication has to be charged to avoid making the patient suspicious. A *British Medical Journal* editorial said, "That a patient gets pain relief from a placebo does not imply that the pain is not real or organic in origin ... the use of the placebo for 'diagnosis' of whether or not pain is real is misguided." Though not everyone responds to a placebo, neither does everyone respond to an active drug. The percentage of patients who reported relief after taking a placebo (39%) is similar to the percentage that reported relief after unknowingly taking 4 mg of morphine (36%).[500] A meta-analysis in 1998 found that half of the effectiveness of antidepressant medication is due to the placebo effect rather than the treatment itself.[501]

On Tuesday May 7, 2002, this was published in the *Washington Post* as a summary of all those meta-analysis on the subject: "The makers of Prozac had to run five trials to obtain two that were positive, and the makers of Paxil and Zoloft had to run even more."[502]

As I said previously, through their general social acceptance, conventional drugs carry an even stronger placebo effect than all the fads, especially on men. This is not because "men are more rational."

Indeed, there are two kinds of placebo. One is for the subject oriented toward the feminine polarity, so the patient is in affinity with the practitioner (the individuality and qualitative count); and one is for the subject oriented toward the masculine polarity, for whom the most salient feature of a therapy is that society at large accepts this drug as effective (the numbers and linear aspect count). Great changes in society come from people in touch with their feminine polarity, as the right hemisphere is in charge of novelty. Also, men in general, when faced with another man who represents a contradiction to their mental patterns—especially if their wives agree with the other man—will often hold on to their position, indeed a singularly reptilian behavior.

Really, we *don't own our thoughts. We fish for those in affinity with us from the ocean of thoughts.*

"The way cognition is distributed throughout the cortex is graduated and continuous, not modular and encapsulated ... this pattern applies particularly to the heteromodal [right hemisphere] association cortex probably less to the modality specific association cortex [left hemisphere] and least to the primary projection cortex [Reptilian and Mammalian brain], which remains strongly modular."[503]

Once we study ancient Egyptian culture and its myths, we come to realize that social and moral values arise naturally in a life coherent with genuine knowledge. This is in affinity with and the expression of the implicate order, and is thus congruent with the natural world and its cycles. The interconnectedness of all nature implies that the sociality of the lung aspect is its direct consequence and an expression of a fundamental and universal unity. The new discoveries of coherence in living systems as seen in the new biophysics as well as our intuitive insight validate this.

"From this perspective, culture is the creation of meaning and knowledge in partnership with nature, in which every social being participates. The coherent society is the society of natural beings living in harmony with nature's creative process."[504]

"The growing weight of evidence indicates that animals are rarely solitary; that they are almost necessarily members of loosely integrated racial and inter-racial communities, in part woven together by environmental factors, and in part by mutual attraction between the individual members of different communities, no one of which can be affected without changing all the rest, at least to some slight extent."[505]

As humans we have designed religions to serve this reality of universality. Nevertheless, religion and spirituality are not interchangeable. It is possible for someone to be religious but not spiritual. Religion is quantitative, spirituality is qualitative. You can add an *s* to religion but you can't say spiritualities. Well, you can, but then it is only religion under another name. Spirituality is the esoteric aspect, associated to the feminine polarity; religion is the exoteric aspect, associated to the masculine polarity. Spirituality is universal and eternal; religions are not, even though the leaders of dominant religions were universal and spiritualist in their teaching. Religion is an ensemble of practices subsequently accepted by a group of people as identified in a certain time and space and accepted by others later on. It is linked to a certain time and space, to certain events in history, and often to someone who in the past represented the essence of spirituality. He was a model to follow, a lamp in the dark. He would indicate the path and the direction. He would not judge you on your abilities but on your sincere desire of perfection. It should be the common language of the heart of a certain group and their way to link themselves to what is best in humanity. It is an extended family. In our societies, it was the fabric. For many groups religion is not the gatherer anymore, so we need to find alternatives, since spirituality is fundamental for humans' sanity. Comprehension of the esoteric aspect is only given to those with a functioning feminine polarity.

For some, baseball or hockey is a type of religion. "This hockey player has been admitted to the Hall of Fame." Sports are often the only religion of those who do not have the ability to develop their spirituality. Sports are easy. Like religion, you are part of a group; like religion, your feelings are triggered. Some religions have you clap your hands and don't ask much more than that. You have nothing else to do other than recognize the group and say "I love Jesus." You don't have to make any effort; he already did that for you!

The destruction that can occur at the end of a game can be compared to the violence of a religious group that believes it represents "the best," "the only true" religion. It is the masculine polarity expressing its wish for predominance. Sports use the masculine aspect of religion. There is no coincidence that the prize you win in a sport is often a cup or a bottle—a grail—which are feminine symbols. Sports should put the emphasis on team playing and consider the other team as friends who are trying to make us better at the sport both teams enjoy. Not as adversaries who put their ball in our net, a country fighting and winning over another one!

For example, Jesus did not say to his disciples to avoid the Temple. On the contrary, he defended it. Divisions arose among his followers because some were not politically engaged in the Jewish cause, and were therefore dubbed traitors. In Jesus's time, the religious landscape was made of a multiplicity of sects that fought over rituals and interpretations of texts. Jesus was there to show that what was essential was the link to God and not all these battles. He was pointing to the essential. The essential is within, not without. The *without* only serves as an expression of the *within*.

Let us recapitulate. What have we seen thus far? As explained through the Taoist interpretation of the universe and as confirmed by science, everything manifested subsists through the two polarities at play. Everything does, from cells to planets. Those polarities bind different elements together, enabling the expression of life on this earth. The masculine polarity is fed by the feminine one. No feminine polarity? Then no lasting structure on this earth. Generally, take the envelope away from the seed and it will not grow. But similarly, take the male DNA away from the seed, it generates something with no structure. Both feminine and masculine polarities are essential. What we are starting to understand, and what we have shown with the LIFE biosystem, is that in humans, this polarization exists not only on the physical level but also on the emotional, mental, and spiritual ones.

In any given civilization, at all levels, these two polarities are constantly at play, setting off rhythms and therefore cycles. As day gives way to night, in every living process the masculine protects and melts into the feminine and the feminine feeds and merges into the masculine. The feminine polarity is receptive in comparison to the expressive masculine one and vice versa. This can be compared to the process of breathing. To inhale requires the use of a different function than to exhale. Together they represent a complete cycle. If you don't breathe in, you can't breathe out.

The Universal brain, the Social Self, and the Social Human

When a structure looks for permanence, it will inevitably defend itself against the forces of evolution, which it deems as an aggressor. It will use its masculine polarity to fight for hegemony, impose its will (laws) and wage wars, and always try to expand, since this is what the masculine polarity does. But as nothing is permanent on earth, it will eventually implode, destroyed by inner or outer forces unless it can nurture new seeds of itself through its feminine polarity. What can't regenerate itself degenerates. Our brains function this way.

As we have seen in the first chapter, *spiritual* came to Middle English from Old French—*spirituel*—which in turn came from the Latin word *spiritualis*, from *spiritus,* which means of breathing. This implies the lung function in our biocybernetic system. Thus, to be a genuine spiritualist or to truly understand religion, you need a functioning feminine polarity. The seal of any great religion should therefore be evaluated through the management this religion has of the feminine polarity. The essence of a religion, its spirituality, should be understood in the language of the right hemisphere, in a holistic and universal point of view, not as it is done most of the time in a segmented, left-hemisphere-oriented, linear way of thinking.

Spirituality implies ability. To be spiritual is a grace. In the Koran this is expressed by God saying that only those who deserve it can be spiritual. Our brain, in its propensity toward wholeness, is engineered to believe linked to this sixth sense of synergy, the expression of the Master of the Heart. Nonetheless, as I said in the introduction, some people are born, or become, blind or deaf or mute or, in the case of synergy, insensitive. This sixth sense gives us the taste of life in what it has of beautiful, of harmonious, and of universal. It makes the other senses more active. It is the holistic conclusion of the interaction of the other senses. Those synergistically challenged are not able to discern truth as an echo of Consciousness. They can easily rationalize and understand something from a point A to a linear point B, but they don't "know." They can be brilliant with a stellar IQ, but they are always outsiders. They can penetrate with their mind and with their actions, but they are always still on the surface of reality. In religions they remain on the surface of a dogma and can't penetrate the universality of it. Therefore they need to defend it, although in fact they defend themselves.

Take this example. When I was about ten years old, sometimes alone or with my sister, I would stop by the church before going home. I would pray for my grandfather and my grandmother. I loved to kneel down in front of the wooden statue of the Virgin. It felt as if she were smiling tenderly at me. In my nighttime prayers I would ask her to cover me with her light blue cape and keep me safe. I felt a close connection to Mary. I did not really know she was considered Jesus's mother. She was just the lady who calmed me at night, a kind of surrogate mother who would be there when I was scared and oh so lonely. I could feel her presence. Sometimes this presence continued in my dreams and I would dream of her. She was the lady with the diaphanous face who came twice into my room and smiled at me just before I fell asleep. I saw only the pure face made of light streaming with love for me, and she was so beautiful that both time I gasped. With my rosary wrapped around my small hands, her face seemed only at a few inches from mine. I didn't even question her presence. I did not think, "This is Mary" or "There is someone in my room." I was in rapture with the purity of her face, with the beauty. When I would realize I was not breathing anymore, the vision would vanish and I would breathe deeply. Maybe I fell asleep while praying and this was a dream. Well, I still wish for more dreams such as those.

So one day when I was alone, I stopped by the church and went inside. There was no one around. I knelt in front of "my" statue and prayed for a while. When I got up, I noticed the Bible which was open near the altar, and felt irresistibly drawn to it. My heart was full of love and reverence. Climbing the few steps to the altar, I wondered if I would be tall enough to read in it. My heart was pounding. It was a huge book with pages bordered in gold. I whispered "Thank you," because for some reason the sight of it made me joyful. I got close enough to see the opened page, but I did not have time to read. I heard someone yell, "You! Get down from there, immediately!"

I was stunned. Looking down from the top of the stairs, I saw the white-haired priest gesturing for me to come to him. He looked *very* angry. I couldn't understand what was wrong, so I went to him. He said, "This is the Bible, it is sacred. You have no right reading it!" I was a few inches from him, looking up to him. There was white foam on the right corner of his mouth. "You will go to hell! Get out of here!"

I ran out feeling numb, not understanding what I had done wrong. I asked my mother if she knew the priest and she replied that he was a holy man. I therefore never told my parents what had happened, because they probably would have sided with him. At the time I could not understand how he could see any wrong in what I had done, given the love in my heart. I felt, somehow, I had every right to read the Bible and that he should have known or felt that.

Another day, sometime later, I again went to the church. This time, the doors were locked, which was unusual. I tried to look inside but couldn't see anything. Not wanting to miss my prayer, I just knelt on the ground near the side entrance. A few minutes later the door swung open and the vicar came out. I was ready to run away, but he said: "God bless you!" I must have looked puzzled because he added, *"I saw you pray."* He asked me what I was praying for. He seemed touched by my answer, for he replied: "You will go to heaven, my child." He laid his hand on my head and whispered something I couldn't hear. I shivered. I realized then that priests were mortal men and could be rigid and heartless or sensitive and open. Interestingly enough, a few years later I asked my mother what had happened to the vicar. It seems he and the priest didn't get along and the vicar was assigned to another, lesser church.

As I said, religions evolved during the different cycles humanity sailed through. Thus, most religions started with a mother goddess concept or an association with water. The Celts, as an example closer to my origins, had a mother goddess of the name of Danu (which eventually gave its name to the river Danube). As we know, water is associated to the kidney phase, the feminine aspect of the masculine polarity. In the same tradition similar to the Indian Brahmins came the druids, and their vast knowledge of plants. Trees became sacred symbols and it is not by accident that Osiris's casket was hidden in a sacred tamarind tree. This corresponds to the liver phase. The following heart phase is associated with the importance of kingship; we get to that phase if the wars associated with the liver phase have not already destroyed civilization. This era of kingship precedes an era when material knowledge dominates in its masculine form of analysis. At this time, the knowledge associated with the esoteric aspect of religion—spirituality—allowed by the heart phase should dominate. This in turn would shape the exoteric aspect of religion.

Isis Code

Globally, the phase we are living through at this time is of the Analytic brain. Because our feminine polarity is deficient, we have dismissed both the esoterism and exoterism of our profound identity and are left only with great technologies to manipulate matter. These are only empty tools. Of Thor and his hammer, only the hammer is left, now totally useless.

Of course, with our weak feminine polarity, only a minority can express the LIFE biosystem in a truly healthy manner, although it would benefit everyone. This majority with an ailing feminine polarity is solicited to take decisions which will impact the life of future generations. Nevertheless, look at the singer Madonna's popularity when she became interested in the Kabbalah teachings. The need for spirituality is always there, as long as we still are humans. It only needs to be awakened.

Finally, with the lung aspect, we come to the understanding that by definition, a true religion can only be universal in its structure. The essence of religion becomes the crucial aspect. How people live this religion will depend on their geographic situation and culture, but the basic immortal elements will be the same. We then see what the ancient civilizations, including the Celts, showed us: there is no such separate thing as religion. There is only a way of living and being, which is the fundamental aspect of any culture. Spirituality becomes a separate element of our lives when we have lost touch with the integrity of our being. That is why Mohammed said that a person should pray each second of the day. It is a state of being, not something you "do." Our present world and its disintegration can vouch for that.

Since the heart is linked to emotion analysis, it is normal that a community (lung) forms around a charismatic guide who always represents the heart aspect. Of course the religion is only as good as those forming it. The guide, as it is often the case, might give a teaching that is not fully understood by his or her followers because they are still stuck in a previous phase. In fact, they will adhere only to what makes them vibrate. At least they have a direction, food for thought, and a model imprinted in each of their cells, making the path easier for them to follow. Also, it is essential to live in a community that shares the same ideals, since disciplines are then easier to follow. It is easier to live a certain way if many are doing it.

This is like birds forming a *V* in the sky. They travel together in this formation because it is easier than flying alone. In the manifested world, quantity counts.

The true guide, who has by essence a universal and eternal vision of life, recognizes other guides and will often start his teaching as an extension, a reactualization and reminiscence of lost details of what is already accepted in an actual religion. He will try to fully express *the source, the implicate order.* True guides don't fight each other, since their inner compasses make them adhere to the same implicate order: Ma'at.

Schisms are of the masculine polarity and of the personality. They are not associated with spirituality, which is, by definition, universal and in affinity with the feminine polarity. Imposition of rituals and beliefs is similar. No great guide has come along and said that everything that was thought and taught before was wrong. Usually the criticisms deal with the mortal and masculine aspects of a religion: what men did with the teachings, how they used and interpreted them in their daily lives, how their intellect twisted certain aspects of it to satisfy mortal desires of the personality. In fact, problems do not reside in religion per se, but in the phase from which the religion sprouted or is interpreted. In every healthy religion, we should find elements of the general matrix, of spirituality.

Let us say we are in a phase where the masculine polarity dominates. Then the practical religion should put the emphasis on the feminine polarity: belief in the oneness of creation/creator, alms giving, fasting, prayers and meditation. I would add an element that was not needed until the modern era, but was at the base of any religion before the patriarchal era: respect of nature, breathing, universalism. Nonetheless, the core of religion must express all of the phases, including knowledge seeking, education, and building of social structures associated with this religion, as well as protection of nature. Following the rituals linked to these elements should vouch for a balanced biocybernetic system and thus allow physical and psychological health.

If we take the Muslim faith as an example, we find (1) pilgrimage (kidney), (2) prayer (liver), (3) belief in God (heart), (4) fasting (pancreas), and (5) almsgiving (lung). These are the five pillars of Islam.

Isis Code

Islam means those who surrender to God. They are also the five pillars of any religion. The aspect associated with the lung phase—almsgiving—is therefore one aspect in the general protection of the creation, which is a main component of the lung aspect. In the Koran, "Do not kill the she-camel" (Koran 7:72) refers to these stages. As Mohammed said to his companions, "There is a time for the she-camel to drink and there is a time for you to drink." The LIFE biosystem shows us that the she-camel is the whole creation, including all of the natural kingdoms, and is directly associated with the feminine polarity. Even if the she-camel does not drink at the same time men do (the masculine polarity), Mohammed asks the companions not to kill the she-camel. Otherwise, what befell the Thamud tribe will befall them. That is, self-destruction will occur. In this story of the she-camel, the manifestation of the greatness of God, his creation, is eventually destroyed, cut in pieces in the fashion of the Analytical brain, killed by those who had an interest in seeing it disappear.

This is the era we went through. The masculine polarity and its values dominated us even to the point of killing everything associated with the feminine polarity in its lung aspect (respect of cycles in life, of life in general, community, natural world, beliefs, virtues, etc.). Even the fruits of this feminine polarity were and are being destroyed. In the story of the she-camel, when the people try to kill the prophet Saalih, everything is destroyed. This only indicates that when we will try to destroy our heart aspect, we simply destroy ourselves. In the image from the Gospel we referred to earlier: "He who falls on this stone will be broken to pieces, but he on whom it falls will be crushed" (Matt. 21:44, NIV). The interpretations some have made of the Koran's camel story have truly appalled me, and indicate the level of spiritual immaturity of some, as well as their profoundly disturbed LIFE biosystem.[506] I strongly believe that we should see the common, universal elements in each religion and see how every religion shares these same ideas. All the rest are details that will not enhance our lives. How the texts were poorly translated is due to the immature feminine polarity of the translators. How many texts were cut because they did not suit the masculine polarity, or how others might have been added, we will never truly know. But starting with the basic aspects, the skeleton is what this world needs now. Therefore, we should look at the implicate order behind the text. That is one of the very few ways to reach the true meaning.

As another example, the ancient Egyptian religion put tremendous importance on not lying. Now that we know lying can cause the demise of the whole system, we see how important truth is. Ma'at, which I associate to the implicate order, represents the attitude with which justice was applied, rather than the detailed rules. For what good are the rules if there is no comprehension of the essential? Listed in the *Papyrus of Ani*, the principle of Ma'at is represented as forty-two negative declarations to Rekhti-merti-fent-Ma'at. This is therefore a counter clockwise movement. Interestingly enough, these negative declarations are in harmony with how the brain controls executive actions: through inhibition. Inhibitory control and working memory act as the primary and rudimentary executive functions, which set the stage for more complex executive functions, such as problem solving, to develop.[507] Inhibitory control and working memory are among the earliest executive functions to appear in the period when the right hemisphere is still in charge. With maturity they develop, and the DLPC, OFC, and maybe MPC will become in charge of task inhibition.

42 Confessions (Papyrus of Ani)[508]

I have not committed sin.
I have not committed robbery with violence.
I have not stolen.
I have not slain men and women.
I have not stolen grain.
I have not purloined offerings.
I have not stolen the property of the god.
I have not uttered lies.
I have not carried away food.
I have not uttered curses.
I have not committed adultery; I have not lain with men.
I have made none to weep.
I have not eaten the heart.
I have not attacked any man.
I am not a man of deceit.
I have not stolen cultivated land.
I have not been an eavesdropper.
I have not slandered.
I have not been angry without just cause.

I have not debauched the wife of any man.
I have not debauched the feminine aspect of Man.
I have not polluted myself.
I have terrorized none.
I have not transgressed [the Law].
I have not been wroth.
I have not shut my ears to the words of truth.
I have not blasphemed.
I am not a man of violence.
I am not a stirrer up of strife (or a disturber of the peace).
I have not acted (or judged) with undue haste.
I have not pried into matters.
I have not multiplied my words in speaking.
I have wronged none, I have done no evil.
I have not worked witchcraft against the King (or blasphemed against the King).
I have never stopped [the flow of] water.
I have never raised my voice (spoken arrogantly, or in anger).
I have not cursed God.
I have not acted with evil rage.
I have not stolen the bread of the gods.
I have not carried away the khenfu cakes from the Spirits of the dead.
I have not snatched away the bread of the child, nor treated with contempt the god of my city.
I have not slain the cattle belonging to the god.

It is easy to see that later, Moses simplified these into the Ten Commandments that, basically, express the same structure.

In Egypt, adopting Ma'at as a principle and as a system allowed Egyptians to embrace the diversity of complex needs associated with a state composed of people from varied backgrounds and often conflicting goals. It is the same with any multicultural society. For harmony to exist, the system, the organization leading this society, must encompass and take into consideration all the cultures therein. The LIFE biosystem is such a system. If the right order expressed in the concept of Ma'at is not followed, then we end up with the concept of Isfet: chaos (Reptilian brain), lies (Analytical brain), and violence (Mammalian brain). As we have seen, this is the masculine polarity separated from the feminine one. Therefore, when we allow this to happen, the feminine polarity cannot be expressed.

This is exactly the world we live in right now. Everything is lies and manipulation. Acts against Ma'at, the implicate order, generate chaos, diseases, and wars. This is proven by history, in any country. It is the basic meaning of the word *sin*. In French, the same meaning comes from the word *péché*, which is from the Latin *peccatum*, which is linked to "error." It appears that in the New Testament, the Greek word *hamartia* (τ μαρτία) can be translated as "sin," although it signifies a wrong outcome and does not necessarily imply a moral or ethical breach. It is more on the lines of a result of karma, as of the Hindu tradition, or even to the Muslim one where it is said that "God" is responsible for the people not seeing clearly, as well as the Jewish tradition in which "God" made the pharaoh not listen. In fact, lies bring obscurity on the general system, and the grace of the feminine polarity cannot operate anymore. If we refer to the Celtic tradition, which was more associated to a language, an attitude, and a social tripartite structure than to a specific geographic area, we encounter such philosophers as Pelagius (ca. AD 354 – ca. AD 420), for whom the Christian concept of sin was totally alien. To him, confession of sin was not obligatory. If the need arose through a feeling of guilt, a person was to confide in his soul-friend. In the early Christian Celtic society, the soul-friend was also a companion, and was sometimes of the other gender. Such couples would wander and bring spiritual education to the people. This was frowned upon by the established Catholic Church which, understandably, wanted to exercise total control over its members.

"The soul-friend [anam chara] who acted as a spiritual guide and counselor—not confessor—to young monks and converts was part of the Druidic practice."[509]

Heretics were Christians who refused to adopt the Catholic faith in its integrality. The meaning of heretic is "to choose." In a period that saw the rise of the masculine polarity and its Analytical brain, the Catholic Church's control was seen as a necessity.

The centers of the many heretic movements that arose in the twelfth century could be found in northern Italy and Languedoc. Later, some blossomed also in Flanders and Belgium. At the time, the first Humiliati of northern Italy were earning their simple living as weavers. Their wives, who mostly belonged to the noble families of Milan, also formed a community under Clara Blassoni. So many others eventually joined them that it became necessary to open a second convent. In France, the Cathar movement, with some similar elements of dogma, stood as living testimony to the inadequacies of the Church. As with all those other groups, which often lived in monasteries, they questioned the foundations of existing authority. They wished to live a Christian life but felt the present dogma of the Catholic Church didn't allow that. For most, they mistrusted some basic doctrinal notions, such as: original sin, the identity of Christ, the new sacrament of marriage, the link between the devout and the rest of creation, the cult of the saints, the presence of Christ in the Holy Host, the pejorative view on women, the ecclesial hierarchy, the validity of sacraments performed by a clergy accused of clerical concubinage (nicolaicism) and of the buying and selling of ecclesiastical offices and pardons (simony), interpretation of the texts, and more. Also, people wished to receive a clear preaching, in a language they could understand.

This revival of evangelical fervor rose in conjunction with many factors. In France, the north was impoverished and overpopulated, which rendered the south extremely attractive. There was a general disintegration of social structure and dissatisfaction with the Church. At the beginning of the eleventh century, the Capetian kingdom was composed of a multitude of independent principalities free of royal dominance. The princes who ruled these principalities were issued from the aristocratic Carolingian families, but by 950 the local counts took power into their own hands, self- allocating the right to judge, until then an exclusive royal prerogative. Then, those who were in charge of defending the ecclesial goods, even those not associated to a seigniory, assumed the same power. These new lords, to defend their lands, gathered around them hordes of horsemen later called knights. Starting in 1030, knights also became part of the nobility. At the same time, for protection, the people grouped around castles. This is how the first fortified villages came into existence, between 950 and 1050. With this feudal age, consecutive to the dislocation of the Carolingian empire, dukes and castle owners, all powerful on their own lands, often seized the ecclesial patrimony, distributing without scruple the episcopal charges to whomever they pleased, often to close family members and friends. The fact that these individuals had no aptitude in religious matters was not considered a problem. Thus, princes installed their families in monasteries. The lords had no qualms in collecting the majority, if not all, of the parochial revenues. Some even placed their released serfs in charge of religious instruction. This created an osmotic situation between the aristocratic and clerical strata of society. In a genuine desire on the part of some people to reform these practices, abbeys such as Cluny came into existence. William III, Aquitaine duke in 910, created this abbey, which he immediately and directly placed under the pope's protection. With it, the monastic ascetics were reintroduced under the following of the rule of St. Benedict. This monastery became in charge of reforming the other ones.

Eventually, the Albigensian Crusades, initiated by Louis IX against the Cathars, led to a political as well as religious domination of northern France over the southern part.

During the same period, St. John Chrysostom (Homilies 46:1) said, "to put a heretic to death is to introduce upon earth a crime beyond atonement." There is no disputing, then, that leading Christians knew full well the teachings of their religion, and it was not that dissenters should be murdered. It is horrible to think that groups devoted to serving their fellow humans and to live chastely and humbly would be put through such torture. Those, the best people of Occitan, were either murdered or had to flee to Italy. Hunger, absolute destitution, fear of being captured, anguish and solitude, this became their lot. Exiling themselves, the faydit chevaliers—knights who lost everything because they either were Cathars or protected Cathars—were ruined and destroyed because of the war and because, in the name of God, the Inquisition confiscated their lands. We have to remember that all those who were found guilty lost their belongings to the Church, which distributed them back to the accusers.

In the same vein, in 1209 the Catholic Church triggered a vast military operation that was more political than religious. As a reward, the Crusaders who fought against the Cathars in the Albigensian Crusade not only would go to Heaven and have all their debts erased, but could appropriate the lands and possessions of the Cathars or of any who were found to be favorable to the Cathar faith. When we know Cathars were from many of the noble families of south of France, we understand what happened. After starting at Béziers (twenty thousand killed) the crusade lasted for more than twenty years. Many fled, and whole communities reorganized themselves in Italy, mainly in Cremone, Piacenza, Pavia, Genoa, and other cities.

Isis Code

What happened to those who stayed? One example summarizes the lot: In 1211 the town and castle of Lavaur, in the Midi-Pyrénées, was taken by Simon de Montfort. The chatelaine, Dame Guiraude of Laurac was thrown alive into a well and stoned to death while her brother, Aimeric-de Montréal, and all his soldiers were hanged and their throats slit. Four hundred men and women were burned in one of the most horrible pyres of the Crusade.

What were their beliefs that made them so impossible to bear alive? For these families who believed in elements that had been only recently abandoned by the Church, there was the deliberate choice of a woman, a grandmother, a matriarch, who would guide the religious opinions of her children and grandchildren (Michel Roquebert). Seen through our LIFE biosystem, this makes sense. The religious aspect is the result of the coordination of the social, idealistic, and emotional selves. In these, women are expressive on the social and emotional aspect. It is women who guide the spirituality of their families. Examples of young girls educated or growing up in a Cathar house under the tutelage of an aunt who was a "Perfect" are numerous, if we judge the abundant witness statements made to the Inquisition. It is the women who allowed the implantation and survival of the Cathar faith. In the hearts of their family, they could imagine themselves ending their days in religion in one of the Cathar castrum, surrounded by the affection of their grandchildren and the respect of everyone. These religious people drew a logical and coherent system from Scriptures, which, in principle, ignored gender inequality. In the Cathar faith there was a spiritual aspiration that saw man and woman united as brother and sister in a love freed from the chains of incarnation, walking together hand in hand toward the divine light of knowledge. This is hinted in the Muslim faith, which says that women should consider men as their brothers. It is also implicit in the Christian faith when Jesus says to live with your wife as you would with a sister.

The symbolism of ichthus, the fish, attributed to the Christian faith, recalls the myth of Osiris, where the male organ was not recovered because a fish ate it. So the true function of Christianity is to bring back the element of Osiris that Isis can't recover, in this way bringing back our Osiris aspect. Curiously, of the two fish of the beginning, associated to the miracle of the two fish and five loaves of bread, only one fish remained. In this lone fish we can read five Greek letters. Is this mere coincidence? In fact, we could say that these heretic groups represented the second neglected fish of the feminine polarity. With reason, the people from Occitan considered that the teaching from the heretics and the one from the Catholic Church complemented each other perfectly.

The Church did not offer women means to consecrate their lives, other than by being taken away from their families. The heretic faith did. When Constantine, in the fourth century, ensured the Christian triumph, bishops happily agreed with Lactantius, whose main concern was to present Christianity in a form that would be palatable to philosophical pagans. The first council of Nicaea, AD 325, reads: "Concerning those who have given themselves the name of Cathars, and who from time to time come over publicly to the catholic and apostolic church, this holy and great synod decrees that they may remain among the clergy after receiving an imposition of hands." The wind turned and the Cathars were suddenly seen as despicable heretics. People could now be saved for eternal life by killing them as cruelly as possible, when before they would have been sent to hell with no possibility of parole. Here is a rationale of the masculine polarity at its worst.

Of course the masculine polarity also had its grip on some of the Cathars. Their vision often appears as dualistic, which is a mark from the Reptilian and Mammalian brains' governance. Since pain and ugliness existed, this world had thus been created by the devil and, consequently, everything manifested in it was from the devil. Anyhow, this was the understanding we can gather from the forced statements given in front of the Inquisition. Was this a basic tenet of their faith? They were right in the fact that this level of manifestation could be considered an aspect of hell, as it is the most retracted and condensed, in the image of the kidney energy of the physical self. But as being the most receptive, with the right vision and work, it could also be transformed into a paradise.

This hell aspect was accurately described by Dante Alighieri (ca. 1265–1321). Moreover, the art of the troubadours of the Cathar country of Languedoc was the inspiration for Dante's famous *Divine Comedy*. The view he offers us in it is the one the resurrected Savior would have had when descending through the levels of purgatory to get to the devil in his abode. We see him, this Satan, in the frigid misery of his ice palace, a giant weeping irrepressibly, partly frozen in the ice at the center of the world, which to Dante was the lowest level of hell. The middle, as we have seen, is where the pancreas regulates from on earth when the masculine polarity dominates. Because this place of desolation is immersed in a thick fog where one can't see, Dante has to be walked through it holding the hand of a guide [his soul?]. The various levels and types of sin simply express the material, manifested world. The devil at its center is shown as the sovereign of the material existence, in the image of Seth, the center of our material life. God is nowhere to be seen because he is raised on high, a long way away, in a spiritual realm quite different from all this.

The Cathar cosmogony is illustrated here. Man has the choice to rise to God's abode, a function of the feminine polarity, but earthly attractions, associated to the masculine one, weigh him down. The difference here with the LIFE biosystem is that for me, "God" is not a stranger to this lower realm. He is in it also. It is our vision and perception that is limited and binary. Our vision is imperfect. In the total scheme of eternity and infinity, everything has its place. This world is perfect, but we nevertheless need to understand that polarity necessarily begets pain. If this ice hell can be traced back to Gnosticism, that would show the direct link of Gnosticism with the Zoroastrian religion and mythology, which dates to approximately 600 BCE. In the Gathas, seventeen hymns attributed to Zoroaster himself and which are the oldest writings of Zoroastrianism, the devil is not defined. At most we could build "the devil" from putting together two or three words, *angra* or *aka* and *mainyu*. In the only instance in these hymns where *angra* and *mainyu* appear together, the concept spoken of is that of a mainyu—mind or mentality—that is angra—destructive, inhibitive. Hence, a problem with the Analytical brain is expressed, not the physical world in itself. *Aka*, seen in Yasna 30.3, is the word for evil and is not used in conjunction with angra or mainyu. So we have angra mainyu, *destructive mind or thought,* which is the antithesis of spenta mainyu, the bounteous mind. Ahura Mazda is said to have used this spenta mainyu, this psyche, to create the universe. It agrees with Carl G. Jung, who said that everything is psyche; with Einstein, who said that everything is energy, and with the LIFE biosystem, which adds the reality of the Master of the Heart's field. Our biosystem also pinpoints the logical brain as the problematic element of our incarnation, as it is our conscious inhibitive brain (destructive mind or thought). It also shows that when the system fails, we are not living in the "truth." Zoroaster does not define evil otherwise. Scholars accept that the Avestan angra mainyu "seems to have been an original conception of Zoroaster's."[510]

The symbol for Ahura Mazda, the winged disk farahavar, made its first appearance on royal seals. It is thus associated with the Egyptian deity Horus, again in agreement with the LIFE biosystem.

So to return to the Cathar faith, our physical world is not a creation of the devil. But as for it being a world of separation (devil = division), yes.

For some Cathars, they could not reconcile the pain of this world with the perfection of their God. Their God, since he is all love, could not be associated to or accept this pain. But pain is the result of a separation, a division from harmony and cohesion. It is therefore inherent to change and evolution in a polarized world, since to move from one structure to another implies a restructuring that in itself implies separation, thus pain, very much the chattel of the social brain.

Cathars were Universalists in that they believed in the ultimate salvation of all humanity.

Not surprisingly, some Catholics disliked Dante as much as they hated the Cathars. Dante thus failed in his political ambitions through the opposition of Pope Boniface VIII. And the Cathars lost their lives.

What the Cathars believed in can be summarized by analyzing their coins. Cathar coins showed an Ankh-like Tau cross, with the Persian Yehud Rosette, and an equinoctial cross. The latter refers to the days and nights being of the same length. The goal was for the two polarities to be balanced.

The Cathar Perfects, known as *Parfaits* in French, always travelled in pairs, just as the Celts did, and just as Jesus instructed the apostles to do, most probably following Essene practice. Perfects were successful healers and doctors, using herbs and stones, like the Essenes a thousand years before and the druids more recently. Ordinary devouts of the Cathar faith had to find their own paths to purity, and rules were not enforced for them. Only through experience would they find their own identity.

Although the Cathars thought this world was a creation of the devil, they also had to know that reproduction was necessary. Otherwise, what would happen with the imperfect souls? They needed a body in order to find salvation. They kept three Lents in the year, were vegetarian, refrained from consuming fat, and fasted on Mondays, Wednesdays, and Fridays, at which time they had bread and water. They included wine in the humble repast at Easter, and fish on some other occasions, mainly during the weekend, as they followed the "apostle fast". Eating this way did not amplify the personality and opened the door to their individuality, since the pancreas is the regulator and the ladder from personality to individuality. Most of all, Perfects refused to kill.

It is essential to be clear that for the Greek philosopher Plotinus and those who reflected on the human path to perfection, the intellectual (noetic) processes of the more evolved levels are intuitive rather than logical, and are associated with the Human brain more than with the rational one. Plotinus treats this as a minor step in passing from one state of the masculine polarity to the other of the feminine one, from the Analytical brain to the heart-lung aspect. But, as he experienced, stopping the inner chatter of the rational self might prove difficult.[511]

To reach the level at which illumination might occur, Plotinus explains that the soul must not reflect upon material forms or anything manifested, even less contemplate them as *other*, but must *become* them and experience a mirroring contemplation. When he speaks of the soul, he refers to this aspect in us that wants to be part of, and conscious of, the whole, elevate itself, and commune with all other creatures and God himself. This is also the feeling one has when "falling" in love. Obviously this is chiefly possible through the feminine polarity, not the masculine one. This we have seen in the legend of *Nibelungs*. In this intuitive flow, the mind is unselfconscious and loses its awareness (as we may experience when we are totally absorbed in an action). However, this state cannot be sustained for long, for the human mind seeks the awareness given by the

Analytical brain, so we fall back to a masculine level as soon as we wish to analyze the event. In the same vein, as Carl G. Jung pointed out, when a revelation is given, a numinous effect appears, and just as the mind wants to analyze it, [thus nails it to space and time], it disappears.

The Low (or Late) Middle Ages encloses a period that spanned from the twelfth to the fourteenth centuries. It begins with the golden age of France's culture with the birth of Gothic art and courtly love. It is a time of a second medieval rebirth, following the one of the ninth century—the age of the Carolingian—and preceding the end of the Middle Ages in the fifteenth century. The Crusades helped the propagation of antique knowledge in all of Occident, and allowed contact with copies, translations, and commentaries from the ancient authors of the south Mediterranean regions.

From the time of ancient Rome to the Middle Ages, poets enumerated the "five lines of love." These are: gazing, contemplating, or mirroring (lung-heart); communicating, speaking (liver-lung); caressing (heart-liver); kissing (pancreas-yin); coitus (kidney). By linking these together, we have the LIFE biosystem and a pentagram. Fin'amor would refrain from kissing and coitus where separation inevitably occurs in an attempt to raise the energies to sublimation. Courtly love almost became an institution in the twelfth and thirteenth centuries. This was a consequence of the search for knowledge, considered a *trésor*, a treasure. Many written work used this adjective/noun in their titles. In the middle of the thirteenth century, although some poets still wrote in the Provençal language, the langue d'Oc, the race of the troubadours, had almost vanished. The change in France's domination had created a change in morals. Selfishness and pettiness replaced the noble generosity that was part of the lord's responsibility and obligation. Sadness replaced the joyous court, where the dame of the castle used to receive the compliments of the troubadours.

Dante's achievement was to have created unity from disparate elements that were rolling in the human mind for the last five thousand years. The Cathar ideas were part of this unity.[512]

Dante brings us to fin'amor. Five qualities were necessary for fin'amor to exist.

Isis Code

1. generosity, given by the heart

2. hardiness, a gift of the liver

3. courtesy, a lung aspect of the chivalry

4. humility of the kidney, since pride cannot help the lover in his relationship

5. *domney*, a term that does not exist anymore. The language d'Oc even had a verb associated with it: *dosnoyer*. It was the art of relating to the ladies following a code. It is associated to the pancreas aspect of the system. It was a type of platonic love similar to the exchanges between Cathars (Perfect) of the south of France. Generally, courtly love and fin'amor were secret and between members of the nobility. This represented an enlightened Analytical brain aspect.

This code of love was born in the centuries when the scepters of some kings were iron rods with which they hit their subjects. The popes in turn denounced the kings with the weapon of excommunication. The powerful lords whipped their vassals and husbands beat their wives and children. This "rule of Solomon"[513] eventually fell, but regarding children it sadly remained in use until the beginning of the last century and had insidious effects on our civilisation. As an example, we can read what Henry IV wished for his son, the young dauphin and future Louis XIII (born 1601). This text was addressed to the dauphin's governess, Mrs. de Montglat:

> *Madame, Je me playns de ce que vous ne m'avez pas mandé que vous avyez foueté mon fils; car je veux et vous commande que vous le foueter toutes les fois qu'il sera opyniatre, ou fera quelque chose de mal, sachant bien par moy-mesme qu'yl n'y a ryen au monde qui luy face plus de profyt que cella, ce que je reconnoy par espérience m'avoyre profité; car estant de son age, j'ay esté fort foueté; c'est pourquoy je veux que vous le fassiez; ce que vous lui ferez éntendre. Adieu, Madame de Montglat ...*

Translation : Madame, I resent that you have not whipped my son ; because I want and order you that you whip him every time he shows to be stubborn, or does something bad, knowing full well myself that there is nothing in the world that could benefit him more than this, which I recognize by my own experience; since when I was his age I was whipped a lot; this is why I want you to do it; and tell him that I specifically asked this of you ...

In this, the king directly orders her to beat his son. He argues, probably intuitively knowing she would not accept easily, that this was salutary and necessary. We have to remember that not so long ago in Christianity, husbands could beat their wives "for the salvation of their souls." It is also seen in Muslim culture, although at one point the Prophet Muhammad criticizes men for beating their wives excessively.[514] When I studied the Koran and used the filter of the implicate order, a few verses did not fit in the ambiance of the rest. One of them is 4:34.

"Men have authority over women because God has made the one superior to the other [elsewhere it is said that one is not superior to the other except through his piety], and because they spend their wealth to maintain them. Good women are obedient. They guard their unseen parts because God has guarded them. As for those from whom you fear disobedience, admonish them, and send them to beds apart and beat them."

For what I read, this is the verse used by some Muslim to excuse beating their wives. Again, the masculine polarity divides. Instead of taking the whole of the Koran to understand the principles behind it, and to adapt it to the phase we are now going through, men took one line that suited them. We could answer with another verse, 16:123: "If you punish, let your punishment be commensurate with the wrong that has been done you. But it shall be best for you to endure your wrongs with patience." It is also said in 15:89: "We will surely punish the schismatic, who have broken up their scriptures into separate parts."[515] In the phase we are going through, we should understand that a man or a woman beating a child, a woman, or anybody damages the collective feminine polarity. It is a sign of degeneration. Only for the purpose of immediate self-defense can such acts be excused.

Isis Code

Even though the term courtly love appears only in one Provençal poem (as *cortez amors* in a late twelfth century lyric by Peire d'Alvernhe), it is closely related to the term fin'amor (refined love), which does appear frequently in Provençal and French, as well as in German, where it is translated as *hohe Minne*. Eleanor of Aquitaine brought the ideals of fin'amor from Aquitaine, first to the court of France when she was married to Louis VII, and then to England, when she married King Henry II. Her daughter Marie, Countess of Champagne, brought the elements of courtly behavior to the Count of Champagne's court. Courtly love found its expression in the lyric poems written by troubadours, such as William IX, Duke of Aquitaine (1071–1126), one of the first troubadour poets and also grandfather to Eleanor of Aquitaine.

According to Gustave E. von Grunebaum, several notions from Arabic literature of the ninth and tenth centuries uphold ideas such as "love for love's sake" and "exaltation of the beloved lady." Avicenna or Ibn Sina, the Persian psychologist and philosopher, developed the idea of the ennobling power of love in the early eleventh century through his treatise *Risala fi'l-'ashq* (*Treatise on Love*). The final element of courtly love, the concept of "love as desire never to be fulfilled," was at times implicit in Arabic poetry and further developed as a doctrine in European literature, in which all five elements of courtly love were present.

In the twelfth century we find the first singers of the fin'amor, singing courtly love to their ladies. They take the name *trouveurs* (those who found divine inspiration through the subtle love), or *troubadours* in the language of Oc of the south of France and *trouvères* in the language of Oïl of northern France. We can find them performing in all the illustrious courts and active cultural centers. Among the best known are the courts of William IX, Duke of Aquitaine, of Eleanor of Aquitaine, of Marie of Champagne, of the counts of Toulouse.

William IX (1071–1126) himself is regarded as the first troubadour and the first poet in medieval vernacular writing. His granddaughter, Eleanor of Aquitaine (born 1122), was one of the greatest patrons of this art of the *trouveurs*. She gathered around her the most famous poets, not to mention Chrétien de Troyes, known today for his Arthurian chronicles.

The Universal brain, the Social Self, and the Social Human

It was also a time of epic poems (*chansons de geste*), monasteries, Cathars and other heretics, and the rediscovery of the Orient. Medieval poets (*trouveurs*) created new poetic forms, new images, and could play with existing ones, driven as they were by the quest for the absolute, as well as by a desire to enjoy the present moment. The ecstasy found its realization in the discovery and love dialogue that followed, and its justification in the search for love. Courtly love is not associated with religion, thus fin'amor had to be found individually, as revealed by true love, in which the *trouveur* was transcended and transmuted by the feelings he harbored for the other. In this, he was not interpreting a god but a loved one, in whom he saw the Self reflected and therefore his Self. In this experience, he loses consciousness of his rational self. In the reality of this love, he feels thankful to the other and discovers a sensual and divine communion that inhabits his soul. It allows his individuality to vibrate, and through his personality to express concretely and sacredly all the subtleties of human love. The shared love and ecstasy, through his poems, finds an echo in the hearts of others. The poet becomes a universal flame and his songs reach far from the space and time he lives in, all the way to us, and comes to ignite and sacralize our lives.

King Arthur, it appears, was a Celtic high king who was transformed by Chretien de Troyes as a legendary figure that embodied all the themes evoked by the fin'amor tradition. This, added to the popularity of Geoffrey of Monmouth's twelfth century *Historia Regum Britanniae* (*History of the Kings of Britain*), awakened the archetypes associated to this legend in whoever read the books, putting them in contact, subconsciously, with their feminine polarity.

Chapter 8
I, Horus, the Androgynous King

The Synarchic Brain, the Conscious Self, and the Complete Human (Perfect)

"The universe begins to look more like a great thought than a great machine." **Sir James Jeans (English physicist and mathematician)**

"The order of things that I ... deal with ... is that in which all materials having a 'historical' and 'scientific' value are the ones that matter the least; conversely, all the mythical, legendary, and epic elements denied historical truth ... acquire here a superior validity and become the source for a more real and certain knowledge ... From the perspective of 'science' what matters in a myth is whatever historical elements may be extracted from it. From the perspective that I adopt, what matters in history are all the mythological elements it has to offer." **Evola**[516]

"L'ordre matériel est un emblème, un hiéroglyphe du monde spirituel." *("The material nature is only a symbol, a hieroglyph of the spiritual one.")* **Pierre-Simon Ballanche**

"Ars imitator naturam, natura Deum." *("Art imitates nature and nature God.")* **Plotin**

"If this outer carcass of man seems marvellously created, consider that it is nothing compared to the soul which has informed it. In fact, whatever man might be, he always incarnates something divine." **Leonardo da Vinci**

"Individuality is only possible if it unfolds from wholeness." **David Bohm**

Isis Code

"And the LORD God said, "The man has now become like one of us, knowing good and evil. He must not be allowed to reach out his hand and take also of the tree of life and eat, and live forever." **Genesis 3:22 (NIV)**

"We should by this time know what human beings are for. They have evolved as cooperative creatures, and their further biosocial evolution quite clearly lies with the further development of their cooperative capacities. This is what human beings are for. And it is upon the solid foundations of the development of their cooperative capacities that all their other capacities may be developed. We need, then, to recognize that the rearing and education of children must be designed to enable them to realize their cooperative capacities to the optimum. And by "cooperative capacities" we mean the ability to love." **"What Ought We to Do?" Ashley Montagu**

The Perfect Human: I Am

As evening approached, the disciples came [to Jesus] and said, "This is a remote place, and it's already getting late. Send the crowds away, so they can go to the villages and buy themselves some food." Jesus replied, "They do not need to go away. You give them something to eat." "We have here only five loaves of bread and two fish," they answered. "Bring them here to me," he said. And he directed the people to sit down on the grass. Taking the five loaves and the two fish and looking up to heaven, he gave thanks and broke the loaves. Then he gave them to the disciples, and the disciples gave them to the people. They all ate and were satisfied, and the disciples picked up twelve basketfuls of broken pieces that were left over. The number of those who ate was about five thousand men, besides women and children. (Matt. 14:15–21 NIV)

The five loaves correspond to the five organs and their associated structures and functions, and the two fish represents the two polarities. When we put them all together, we can feed a thousand times what they individually stand for. Curiously, the Christian sign is a lonely fish, in the image of the thousands of years of masculine dominance.

Between the "who" phase and a new cycle of the "what" phase, we find the "I am," which is the culmination of all the developed phases working harmoniously and synergistically—a biosynarchy. The

I, Horus, the Androgynous King

feminine polarity is necessary to this expression of the individuality. When it is written in Genesis that it is not good for "man" to be alone, it refers to the masculine polarity and means that the manifested world cannot exist without the feminine polarity. The two of them working together in harmony is the only way to make this world "good." Good is an adjective frequently given to God. As is written in Luke 18:19 (NIV): "'Why do you call me good?' Jesus answered. 'No one is good—except God alone." Hence, with the expression of the feminine polarity reduced, this world becomes godless and is destined to vanish, similar to so many lost civilizations.

I used to smile at the native tradition, in which men thought they had to go at sunrise and acknowledge the sun otherwise this one would not rise. I thought it was pretentious, but now I see that there was an element of truth in this ritual. Indeed we have the power to limit the manifestation of the feminine polarity. The feminine polarity is also necessary to access a higher phase in the spire of evolution toward a greatest Consciousness. Without the polarities working together, humans remain slaves to their instincts, although they might believe they are living their lives. They are merely monkeys in suits. In truth, they are an empty shell.

In order to evolve, both personality and individuality are necessary. If one has not learned the lessons of the limited personality and has not sought maturation, then the next cycle of life is but a repetition. If, on the contrary, evolution's bell has vibrated in the whole body, the seed of the being is enriched and the person may attain a higher spire in his evolving spiral (diagram 12). The hereafter will be different. For those who have a memory of this reality (reincarnation), this is the never-ending life. This is when we reach and eat of the tree of life which stands in the center of the garden of Eden.

"Lest they eat of it and become like us." (Genesis 3:22)

In the same way, religion should mainly be concerned with the feminine polarity. If this had happened before, religion would not have had significant expansion in space, which is by essence the result of a masculine action. Now things are different because of the phases we went through; expansion of the masculine polarity is not needed anymore for the Church. We could therefore instigate a renewal on a solid base. The feminine polarity gives permanence

Isis Code

to a religion. The masculine polarity allowed it to spread. Jesus was aware of that when he said to Peter, "And I tell you that you are Peter, and on this rock I will build my church" (Matt. 16:18 NIV). This refers also to the tree of knowledge of good and bad. Curiously, the Hebrew name Seth, son of Adam, and the Egyptian name Set are philologically identical, and both children are born as third child.[517] As we have seen earlier, the name Seth means among other things, "pillar of stability." Thus it is not illogical to equate Peter and Seth. Peter therefore represents the personality of the Church.

The presence of John besides Jesus irritates Peter. These are the two possible centers of our lives represented in the LIFE biosystem as heart—the Human brain associated to the individuality, represented by John—and pancreas—the Analytical brain associated to our personality, represented by Peter.[518] This is the cross Jesus said we have to bear if we wish to follow him. "Jesus said to him: "If I want him to remain until I come, what is that to you? You follow me!'" (John 21:22 ESV). John the beloved represents the feminine polarity of Jesus's teaching; he was there at the beginning and will be there at the end.

We would not be surprised at Easter to hear the music of a great composer titled *The Passion of St. John,* but we might raise an eyebrow at hearing the title, *The Passion of St. Peter.* There is no coincidence in the fact that in the paintings of the great masters, such as of Leonardo da Vinci, John is painted as if he were a woman. In the bestselling novel *The Da Vinci Code,* some characters in the book claim that the figure is not John but Mary Magdalene. Da Vinci used physical truth to express a more profound one. I would not be surprised, either, provided in the future our technology allows us to see the details of the past, to discover that the apostles Leonardo painted in *The Last* Supper closely resemble the actual apostles— and that he was one of them.

Leonardo da Vinci and the Vitruvian Man

The profound truth Leonardo appears to express is that there were two churches. One, of the feminine polarity, was hidden, and one, of the masculine, dominated. If you remember the persecution by the Inquisition, you understand that for hundreds of years, art in its

I, Horus, the Androgynous King

many forms was the chosen form of expression of the full-fledged but hidden heretics. Thus, we will not be astonished to find similarities between Dante and Leonardo da Vinci. Both saw nature as a product of the divine intellect (in the sense of noetic, psyche, and associated to the heart phase) and art as its imitation. This is why I name a creation *art* only if it expresses a healthy LIFE biosystem. Otherwise it is a lie or a confabulation of the rational mind, and often the expression of a disturbed LIFE biosystem.

In Leonardo's *The Last Supper*, it is clear that Peter represents the masculine polarity of the Catholic Church while John represents the feminine aspect of the universal, eternal one. To give an example of this, in da Vinci's painting, the movement of Peter's hand toward John's neck seems quite obvious, in the knowledge that the masculine polarity wants to "cut off the head" of the feminine polarity. But as we know, Mary is crushing the head of the snake, not the reverse. Also, it appears that John is holding a knife and Peter is holding his wrist or is it the reverse? Hard to say, since Judas is in the forefront of the painting. John will kill Jesus? Peter will kill John? No. What could this mean?

The feminine polarity accepts "what is written." What is written is the path humanity must tread upon to get to its ultimate evolved state. This implies the sacrifice of Jesus, the predominance of Peter who, three times, disowns Jesus, and of Judas being allowed to perform a paid act of treason. Jesus himself said that his death was "written." Where? In the implicate order. The rational mind, here represented by Peter, always wishes to find a way out of it, if his immediate interest is not served by the foreseen events. In that sense, the feminine polarity of the group—John—accepts what is about to happen while the masculine aspect of the group—Peter—doesn't. So yes, accepting the course of event without a fight is, to Peter, similar to killing Jesus. In fact, Leonardo explains that Judas is only a tool, a mere puppet, in the role that was preordained. By our choices we all choose to fulfill a role in the great scheme of evolution, although we might not be conscious of this.

The different esoteric groups linked to Christianity promoted the feminine polarity. They were hidden but had a vital social function. They kept the essence of the faith but were persecuted. Why were they persecuted? The masculine polarity eventually considered

Isis Code

the feminine one as dangerous. Although at the beginning those representing the masculine polarity ignored the feminine one representatives as being, at the most insignificant, this changed when an increasing influence on the population was observed; that could not be accepted. The masculine polarity wants total control. The feminine polarity is mobile and open to elements of other structures, while the masculine one is conservative, fundamentalist, modular, and exclusive.

Leonardo da Vinci knew, or intuited, that the evolved human had to integrate this feminine polarity. This he showed particularly well in his vastly popular drawing called the Vitruvian Man. He viewed this drawing as a cosmography of the microcosm. If we assimilate the biocybernetic system to it, we find the left leg as being the lung, the right one the liver. This is in harmony with the most ancient Egyptian sculptures, in which the right foot represented will and the left one love. I was extremely interested when I visited the Louvre to see how the Egyptian civilization changed, from the first statues with a left foot forward to the later sculptures with their right foot forward. The left arm is associated to the kidney aspect (note that in some cultures, the left hand should not be used as it is linked to impurity and it was later wrongfully assimilated to women) and the right arm is the pancreas and personal will. The head assimilated here to the heart aspect is the pivot of our subtle existence (illustration 7).

As depicted in the drawing, the man is positioned in a square, a symbol of linear thinking and of the physical world. His center or point of balance is the genital organs. If, on the other hand, he is positioned in a circle, a symbol of circular thinking, the navel, the source of life, becomes the center. As we have seen, this corresponds to the mouth of the Master of the Heart, prominent in statues of the Buddha. Are there other elements in this drawing pointing toward this direction? Yes.

In the circle, the Vitruvian Man has the peculiar characteristic of arms raised higher than can be useful for a strict study of proportions. The finger touching the outside of the circle and the outside of the square is the middle one, linked to the Master of the Heart. Coincidence? Although the middle finger is often the longest one, why not use the index finger for such a purpose? Look at the effort made to bend the index finger. Also, the position of the feet is unusual. The left

I, Horus, the Androgynous King

foot, that of love and lung is perpendicular to the right one. In its static position, the weight of the body is therefore on the right leg. If the figure takes a step, though, the left foot will have to be first. In ancient Greek statuary, the two polarities were expressed by one aspect of the body in action (expressive) while the other was bearing an object and relaxed (thus receptive), as seen in Polycletus bronze sculpture. The perfect body was not only one of proportions but also of dynamism. It represented the standard of physical beauty because in it the two polarities were harmoniously expressed. Within the square of the masculine polarity, the head is the limit of the space in which the man lives. In the circle however, he has an environment not allowed by the square.

We could go further and draw a Seal of Solomon in the circle and have the same symbolism, although not a teaching as to the effect of the evolution of men from the masculine polarity aspect toward the reintegration of the feminine. Indeed, all the pharaohs of Egypt were seen as reincarnations of the victorious Horus and thus had to integrate their feminine polarity, the mark of a true king.

We know that Leonardo da Vinci studied the harmonic laws of nature, which were expressed to him through the sensible forms. Like Plato before him, he thought that Beauty was the splendor of Truth. He was well aware of the vast symbolic world already recognized by St. Paul in I Cor. 10:11 (NIV): "These things happened to them as examples and were written down as warnings for us." Everything would come to the ancients in a symbolic form, thus through their idealistic selves. To them, the world itself was a symbol, an image of the reality. Hence, in his paintings, Leonardo da Vinci used Dionysus, as an archetype. Dionysus, god of the elements of nature, androgyne, and placed in caves that were painted red, the color of the feminine principle for the tradition, white being that of the masculine one[519]. Red also is Adam, the universal Man (in Hebrew, Adam means red). Closer to us and to Leonardo da Vinci, in Geoffrey of Monmouth's *History of the Kings of Britain*, the red dragon is a prophecy of the coming of king Arthur, a Celtic type of Horus. Otherwise we find it in the story of Merlin, who to me is none other than the spiritual aspect, the individuality of king Arthur:

The tale is sourced from the Historia Brittonum first told by Nennius and then by Geoffrey of Monmouth. King Vortigern (early to mid

Isis Code

400s?) tries to build a castle at Dinas Emrys (north-west Wales). Every night the building materials for the castle walls and foundations would disappear through unseen forces. Vortigern consults his wise men, who advise him to find a boy devoid of natural father, to sacrifice him and sprinkle his blood where he wants the fortress to be erected. Vortigern finds such a boy (who later becomes Merlin)[520]. On hearing that he is to be put to death to solve the demolishing of the walls, the boy dismisses the knowledge of the advisors. He tells the king of two serpents (dragons) which for centuries have been fighting and remained hidden there. Vortigern immediately excavates the hill, freeing the dragons. They continue their fight. Depending on who tells the story, when and its interpretation, either the white or the red dragon wins. The boy tells Vortigern that the white dragon symbolizes the Saxons and that the red dragon symbolizes the people of Vortigern[521]. But the dragons might have a wider significance. In tradition red and white signify the two principles, feminine and masculine. The masculine polarity is white and the feminine polarity is red[522]. Symbolically, the serpents are those of Egyptian and Chinese traditions which because they are symbolic and archetypal have wings (dragons). Hence the family crest of some families had a red dragon. Later, at the time of the adoption of coats of arms dragons were considered as evil[523]. This was construed from Revelation. But again, this red dragon of Revelation is the great prostitute, the feminine aspect of the masculine polarity as it is not put in comparison with the white one. Some legends say that King Arthur (the human individuality- the red dragon) was mortally wounded but that he is to return.

The Celtic tradition, as that of the Pythagorean or of the Cathar one, are expressions of the feminine polarity to come.

Dionysus (whom the Romans called Bacchus), sung and introduced into Greece by Orpheus, bears the two principles of the ancient dragons, expressed in all traditions and by exemple manifested in the Roman gods Libera and Liber. It is the expression of the two polarities developed in humanity, when the individuality will eventually take its rightful throne and will allow the return and reign of the heart aspect, of king Arthur or under other names Jesus, Buddha, Mithra, Mahomet, Osiris ect....In this sense, Leonardo echoes the Zohar: "All form in which we cannot find both the male and female principle is not a complete and superior form."

During the time of Leonardo, every artist, from the sculptor to the paper manufacturer, including the wool maker in Milan, had a mysterious watermark in their art, expressing their knowledge of an esoteric world. Autodidact, Leonardo had studied the Flower of Life's form and its mathematical properties, as well as the platonic solids, and he must have abundantly used the golden ratio in his artwork. Not surprisingly, the oldest example of the Flower of Life can be found in the Temple of Osiris at Abydos, Egypt. Another interesting point is that humanists in the time of Leonardo were fully aware that Bacchus was a solar divinity. The Platonicien of Florence viewed Bacchus as the sovereign priest, twice born. The hagiography of symbols used in visual arts also associated some specific animals to symbolize Bacchus: the rabbit, the antelope, and the leopard. How fascinating that these are generously scattered in the tapestries of the *Lady and the Unicorn*.

The Sixth Sense of Synergy: À Mon Seul Désir —The Lady and the Unicorn

From the humble weavers fleeing the Inquisition to Italy, to the elaborate tapestries such as the *Lady and the Unicorn*, we can certainly follow a heretic thread. Why is the unicorn in the title? Why not the lion? Obviously, the unicorn and the Lady are the main characters. It appears that the presumed author, Jean d'Ypres, Master of Anne de Bretagne, was well aware of the symbolic language, as can be seen in other works of his. Where would this knowledge come from?

Among the artists working in Paris from about 1450 to 1485, the mysterious Coëtivy Master was one of the most important. Whereas his contemporaries worked exclusively as manuscript illuminators, he painted in a variety of formats, including wooden panels, stained glass and ... tapestries. He worked mainly for members of the royal family and of the court. He painted books of many different types, ranging from devotional manuscripts such as books of hours to works such as Dante's *Divine Comedy*, Augustine's *City of God*, and at least five copies of Boethius's *Consolation of Philosophy*. Although in the fashion of heretics he is known under different names, and his origin is disputed (Ypres is in Belgium but the man is believed to be from Amiens because of technical and stylistic similarities to artists active in the north of France), we know him also under the names of

Isis Code

Colin d'Amiens and Nicolas d'Ypres. His name also derives from that of his patrons, Olivier de Coëtivy and Marie de Valois, daughter of the French king Charles VII, who, as we have seen, was the king put on the throne of France by Joan of Arc. It appears that the Coëtivy Master was the father of Jean d'Ypres, to whom the tapestries of the Lady and the Unicorn are attributed.[524] Is it really surprising that his son would have gathered so much symbolic knowledge?

Jean d'Ypres also produced the *Ship of Temptations* allegory, in which the five senses are caricatured as five female follies. The equivalence of mind with male and senses with female essentially goes back to Philo of Alexandria and even further back. It is clearly a gender stereotype of long history that comes to a resolution through a change of context. The female in question is the physical world, the feminine aspect of the masculine polarity. Yes, the senses can imprison us, but they can also take us to a higher knowledge in the same way they triggered the development of the cortex as we know it. To me, whether we speak of Leonardo da Vinci or of the Cluny tapestries, the message is similar: we must spiritualize our matter, personality, masculine polarity and incarnate our Consciousness, individuality, and feminine polarity. This marriage of personality with individuality is a unity made visible. It is the manifestation of Solomon's seal (diagram 24) as well as a balanced LIFE biosystem. It will inevitably manifest Consciousness on earth.

The lung phase will awake in us the awareness of this sense of synergy with its knowledge that we are all intimately interconnected with the universe in which we have our existence. This might be finally the enlightenment the sages of old told us about, the sacred marriage. There, solitude does not exist anymore, completude appears and, with it, perfection and its fruit; happiness.

In the six Unicorn tapestries, the intention and message are as such.

When visiting the museum of Cluny in Paris, I was taken by the sacredness surrounding these tapestries. Only silence could properly echo their beauty. One could sit and take in the wonderful tapestries in the same way you listen to a symphony at a concert hall.

As for the order of the tapestries, we can follow a few. The first, from the bottom of the body to the top, would give the touch (the hands lay lower than the head—heart), then the taste (mouth—pancreas), then the smell (nose—lung), then the hearing (ears—kidney), and then finally the sight (eyes—liver). This is the same order as the one followed in the Middle Age as seen in the *Bestiaire d'amour* of Richard de Fournivcal or *Le livre du Trésor* of Brunetto Latini. This is also in harmony with the LIFE biosystem. Although the succession is correct, the beginning of *manifestation* or incarnation is the genetic aspect associated to the kidney, hence the Reptilian brain, physical self, and hearing sense. À *mon seul désir* is then well associated to a sixth sense in affinity with the heart and attachment, as a prolongation and refinement of the senses associated with the lung and heart, that is, smell and touch. To express these truly, we thus have to understand that the sixth sense is the beginning and the end, as Mary, Queen of Scots' motto implies: *"En ma fin git mon commencement"*—in my end is my beginning. This is the circular thinking of the feminine polarity, the ouroboros of old.

In the nature-loving context of these tapestries, where dozens of plants and animals are free and so precisely depicted, one is compelled to view the Lady and the message of the tapestries as an hymn to love through exchanges with the environment and sublimation of the senses. The style, millefleurs, was a motif used back then in prayer carpets of south Persia. In the sixth panel, the Lady comes out from the tent, from her inner sanctum, which bears the colors associated with monarchy. This tent is scattered with golden flames that have been used before to signify sainthood[525] in the same way the jasmine flower signifies chastity. In a few paintings from Fra Angelico (1395–1435) as well as others, we see such flames adorning the head of angels (illustration 8). The link to the evolved heart aspect is undeniable.

The hairdo of the Lady and of her servant also imitate these flames. The Lady, both in dress and appearance, resembles Botticelli's *Primavera* more than she does a cloistered nun. There, dressed in a gown similar to the one she wears in the smell (lung aspect) tapestry but of another color—blue for lung and red for À *mon seul désir*—she takes from a treasure chest a heavy jeweled chain. In those times, as we recall, treasure was another word for knowledge. Is it a necklace? If yes, it is similar to the chain she wears at the waist. Thus, the above and below have been made the same. In acupuncture the

Isis Code

meridian of the neck of information-regulation as well as the one at the waist are associated to the feminine polarity.

Some consider the tapestries a wedding gift, but only one family's coat of arms is represented, so the owner was not married. There are so many symbols involved in this work of art that a thick book could be devoted to its analysis and interpretation. I consider every detail meaningful. I will summarize some of those. The panel associated to touch, thus to the Heart and Idealistic self, shows the Lady dressed as a queen. Her gown lined with ermine fur is a symbol not only of kingship but also of purity, perfectly fitting for the heart function. Her headdress, similar to a crown, is in the shape of a flame. Again a symbol of kingship. She is in contact with her feminine polarity as she holds the horn of the unicorn, protruding from the third eye site. As we saw before, a horn could represent inner divinity. The axis kidney (horn) and heart are thus working harmoniously in her. The unicorn is active compared to the lion, as it is standing while the lion is sitting and not holding the banner. The lion has a quasihuman face and looks straight at the viewer. In fact, lions are symbols of monarchy and the heart aspect. It is the masculine aspect of the feminine polarity. It is therefore natural that it would have a human face. It represents the viewer who, to vibrate to this tapestry, must have a functioning feminine polarity. Only in the *À mon seul désir* is the lion really expressive, to signify that only when the feminine polarity is totally expressed can the masculine one be.

The ascetic interpretation of the tapestries has the senses terminating in an act of renunciation of the physical world. As the fin'amor was not a religious movement, these tapestries are in fact the epitome of this philosophy. They all the more represent the masculine and feminine polarities in a mystical marriage through the sublimation of the senses and the acquisition of a particular knowledge, represented by the necklace in the adorned chest. In acupuncture, the points associated with information-regulation (heart phase) are situated around the neck. It is spirituality at its best.

After Napoleon started guiding France, he reintroduced the monarchic institution (heart) and its continuity (kidney), religion and its rituals (heart-lung), arts and sciences (pancreas), and established the civil code (liver). To him, enlightened governance was impossible otherwise. What is universal is timeless. It cannot become anachronistic or archaic, unlike what belongs to a world submitted to the masculine polarity. Those with a nonfunctioning feminine polarity cannot see the use of it. They see everything through their segmented, archaic masculine polarity.

This movement of integration of the feminine polarity can be seen through art expressed by sensitive painters. In the pre-Raphaelite group, we can see this emphasis. One of the best expression I found is in the painting *The Accolade* by Edmund Blair Leighton, in 1901. The pre-Raphaelite group took the imitation of nature for its central theme, and its members were fascinated by the medieval culture in which they saw spirituality and creativity. In this painting, the woman, who obviously is a queen, represents the feminine polarity while the knight identifies with the masculine one. She is the heart aspect. He is dressed in red, so has integrated his feminine polarity with the receptivity of the kidney aspect. She dubs him (a ritual which marks the transformation from personality to individuality as this person receives a new name), in the image of the feminine polarity, individuality received by the personality. The three onlookers, a man with a venerable beard, a monk, and a young man, respectively represent the balanced pancreas aspect (wisdom), lung aspect (religion), and liver aspect (youth). It is an inspired piece of art (illustration 9).

Knowledge and the Feminine Polarity

We are on the threshold of a new kind of revolution, never experienced before. We can individually and globally accept or refuse this shift: from the explicit, analytical, subconsciously controlled, verbal, analytical left hemisphere to embrace the implicit, synthetic, integrative, Conscious, silent, bodily based, sensitive right hemisphere. Only then will we be whole and will the masculine polarity be healed. Already we feel drawn to this. For what is the ultimate and only desire of the masculine polarity? Union through penetration. And what is the only ultimate desire of the feminine polarity? Union through embrace.

Women thus penetrate emotionally and socially, while men penetrate physically, mentally (patterns), and, if they succeed in developing their feminine polarity, symbolically.

We received a great deal from nature. The physical aspect (kidney) gave us the physical structure. The emotional aspect (liver) gave us our modalities (behavior). The idealistic/symbolic aspect (heart) gave us the faculties. The rational aspect (pancreas) gave us the application tools. Finally the collective aspect (lung) gave us the virtues. To become a perfect human involves weaving all of these elements together, working toward a common goal that will become more than the sum of the parts—a biosynarchy. This is our personal six tableaux.

For this shift to happen, we finally have to realize that there are, in fact, three types of knowledge. There is the knowledge of the analytical, rational mind. It is focused on one point and on phenomena; it is linear. Because it is analytical, it is fragmented. This is the knowledge associated with our masculine polarity. It is tributary of the quality of our physiology and of our schools, universities, and the availability of pertinent information. Our physical and emotional bodies have a strong impact on it. When there is mention in the Bible of people who are deaf (physical mind) and blind (emotional mind), it is implied that their masculine polarity is not functioning. The masculine polarity is not functioning if the feminine polarity is not. They are codependent.

In the Koran, there is an emphasis on the fact that God prevents evildoers from understanding. God here acts through their personal LIFE biosystem. It implies reincarnation, for how can a baby be an evildoer? Such people, the evildoers, are limited to the material world. When—through natural evolution— they will finally realize that this world is a prison in which they are choking, they will ask for air, and for change, and they will receive it.

The second type of knowledge is that of the idealistic self. A cellular, intuitive knowledge, it has a holistic approach. It is open to the environment at large through the lung, the Universal brain. It is inhibited by our masculine polarity, depends on personal biological dispositions, and is influenced by the encounters we make and the awakening of our inner compass by elements of our lives. It is attached to the feminine polarity. It is wavelike, oscillatory.

The third and last type of knowledge is the perfect synergy of these two. This is the Knowledge we should strive for.

The binary knowledge of the rational mind, especially in our society, censors the knowledge from the feminine polarity in the same way our left hemisphere has control over the right one. This explains why "mysteries" were taught only to individuals who had gone through initiation, often at great risk (see chapter 7). Indeed, only when we are in physical danger does the right hemisphere regain its dominance, as when we were children. It is primordial to understand that through suffering we will end up reconnecting with our feminine polarity. For people in whom the masculine polarity dominates, this is often a necessity to access this second birth. "There comes a point where the mind takes a higher plane of knowledge but can never prove how it got there. ... All great discoveries have involved such a leap" (Albert Einstein).

Einstein was referring to the idealistic mind. The feminine polarity wishes to rise.

The leading voice of intellectual culture in the United States in the nineteenth century, Ralph Waldo Emerson, influenced such men as Nietzsche and William James, Emerson's godson. Oliver Wendell Holmes considered the speech Emerson gave in 1837 and entitled "The American Scholar" as being America's "intellectual Declaration of Independence." Emerson wrote,

> There is one mind common to all individual men. Of the works of this mind history is the record. Man is explicable by nothing less than all his history. All the facts of history pre-exist as laws ... The creation of a thousand forests is in one acorn, and Egypt, Greece, Rome, Gaul, Britain, America, lie folded already in the first man. Epoch after epoch, camp, kingdom, empire, republic, democracy, are merely the application of this manifold spirit to the manifold world.

Could America rediscover this innate knowledge brilliantly expressed here by Emerson? Toward the end of his essay, he asserts that: "every history should be written in a wisdom which divined the range of our affinities and looked at facts as symbols. I am ashamed to see what a shallow village tale our so-called History is."

Isis Code

He understood that an innate humanity, common to all of mankind, operates throughout the ages in the shaping of events. Did he feel the implicate order? It appears as such.

"History is for human self-knowledge ... the only clue to what man can do is what man has done.

The value of history, then, is that it teaches us what man has done and thus what man is."

In *The Science of Synthesis,* Debora Hammond wrote,

> Bertalanffy described the symbolic dimension of culture as an emergent property unique to human society that could not be reduced to biological drives, suggesting that "symbolic universes" are the most important part of the individual's behavioral system. While it is questionable whether humans are truly rational creatures, they are certainly symbol–creating and symbol-dominated. Humans are thus creatures of two worlds, as biological organisms living in a universe of symbols.[526]

Thus we find our two centers: one for the physiological life—the pancreas for the personality and its rational self; and the second—the heart for our individuality, feminine polarity, and human self.

Symbols are like quanta, they find their definition through the eyes of the observer. Hence the color red can mean blood and danger, or the love of Christ, or the devil, the need to stop, sex, vitality, or the feminine. It depends on who sees the symbol and its surrounding context. As well, Egyptian ideograms, by their disposition and level of interpretation, could signify many different things, depending on the level of knowledge of the reader and how the wheels of his biosystem turned. This explains why dictionaries of dreams don't work.

The population might interpret symbols on one level and priests another way. When a person's two polarities are functioning, he acquires depth and meaning from his environment. He literally lives in another world. Consequently, dreams and everything that occurs in his life takes on significance, order, and meaning.

I, Horus, the Androgynous King

In the same way mathematicians have a symbolic language they use to do their computations, every manifestation is symbolic because it is the expression, a fractal aspect of the implicate order and relates to it. Appearances, manifestations are only an unfolding in time and space for a brief moment of a formula.

The observer has different choices. He can reduce what he sees around himself to crude particles of the physical self detached from the whole. He can add emotion and mainly feel these elements as friend or foe (emotional self). He can view these only as able to serve or not his image of himself (idealistic self). He might also analyze these quickly through his rational self and ponder about their utility and material worth. He can also empathize or not with these elements (social self), and finally incorporate what he sees around himself in his life experiences or in the big picture. All depends on the observer.

In the Dagpo Kagyu knowledge base, which encompasses all the branches of the Kagyu school of Tibetan Buddhism, as well as in the Mahamudra, a set of advanced Buddhist meditation methods, we can find the same seal. The supreme attainment of Buddhahood, of perfection, lies in five *paths* to the ten bhumis (stages of bodhisattvahood). The five paths are as follows:

1. Accumulation (kidney)
2. Unification (liver)
3. Seeing (heart)
4. Meditation (pancreas)
5. No-More-Meditation (lung)

The goal, or Buddhahood, is considered to be achieved at the end of the No-More-Meditation phase, which to us is represented by the advent of Horus, the androgynous king.

As we have seen, there was a time when science and religion were not divided. Priests were doctors or doctors were priests. Human society was a consequence of the religious ideology, not the reverse. We forget this when we read and interpret sacred texts. In ancient Egypt, long before Moses, and for four thousand years, the same basic principles guided the people. As with any civilization, ideas became more complex with time, similar to different colors of paint.

Isis Code

They were often dressed in emotional hues and diverted by politics and businesses in order to gain of money and power. Like the sun rising on one side and setting on the other, religions rose with simplicity and died in the complexity of the phenomenological world.

For a new civilization to be born, we need the seed as well as compost, the decay of the last one. Add some water and light and suddenly, in the silence of death, a sprout unfolds in the sun, tender, feeble of body, but strong in energies, as if it has condensed all the past lessons. The difficult environment seems only to stimulate these vital forces. A particular ability to use reserves and to open itself to surrounding energies, added to a desire to communicate new ones, are at the heart of this new seedling. Then the plant grows and becomes stronger, sprouting roots everywhere, and eventually fights with its environment to gain total control. With time it becomes rigid, until it starts cracking down; and the environment, through its suppleness, erodes it and feeds from it. At the end only a mound remains with a seedling at the center, feeble, tender, unrolling its head toward the sun. Two principles are constantly at play: the receptive feminine one and the expressive masculine one.

The Egyptian and Babylonian seeds were eaten by the Jews through Moses and Aaron and given to the Jewish faith. Since humanity was going through a lunar period (kidney), that aspect survived. The Jewish seed was eaten by the Greeks through Jesus' teachings and gave us Christianity. The solar aspect of this religion was recognized by Paul, and he somewhat transformed Jesus, now representing an archetype, into a Roman Mithraic god and/or Sol Invictus. That resulted in the Christian exoteric tradition. The Jewish/Christian seed was eaten by the Ebionites sect and the Muslim faith was born. The Muslim faith influenced the Christian one and allowed the Renaissance of the twelfth century and the one that spanned from the fourteenth to seventeenth centuries. From the Greeks to the Romans and colored by the offshoot of Muslim faith (science, medicine) we are now children of those cultures, and our religion and science bear the DNA of these, whether we accept it or not.

Basically, those three faiths—Judaic, Christian, and Muslim—are different aspects of one, all different dressing up of the implicate order, of the LIFE biosystem. The Koran names the people who follow these teachings People of the Book. It is only men who divide

I, Horus, the Androgynous King

because of the predominance of their out of control masculine polarity and Analytical brain. The same can be said for Taoism, Buddhism, and Confucianism. One could say their DNA is the same. They sprouted from the same origin, the same brain. They differ only by their function in the human psyche and the phase they concentrate on (diagram 3). Moses didn't ignore the preexisting Jewish culture or Egyptian and Babylonian science and religion. He underlined certain aspects of them, which at the time had been left aside by the priests. Jesus didn't try to expel the poisonous elements in the Jewish faith, but emphasized what had been forgotten or put aside. As told in Matt. 13:24–30, he let the wheat and the weeds grow together so that one day— when the time of Horus finally comes— we could differentiate easily between the two. Mohammed didn't push away the Jewish and Christian faiths or the tribal culture of Arabs, but added to them what had been put away or ignored, and tried to expel from it the corroded elements. The rituals of many Christian military orders bear a striking resemblance to the Mithraic rituals so favored by Roman legionnaires. All adapted the rituals to a certain time, space, and people, eventually creating a thriving community in which the members would recognize and support one another. But the revelation of these religions is not entirely manifested yet. It will be when our heart aspect matures. The paintings and statues of the Roman Mithra wearing a Phrygian cap indicates that this will happen in the era of the Aquarius. Mithra will slay the bull, which we understand is the kidney aspect of humanity. Thereafter Mithra will eat the bull and thus incorporate it into himself.

In our modern world, the masculine aspect of the way people lived distanced itself from the feminine aspect, resulting in two elements that seem to oppose each other: science and religion. Religion itself took an antifeminine stance and became male oriented, even though it should have expressed the feminine polarity. In all of these cultures the same elements heralded the birth, growth, and death of religion. There was all over the world a preeminence of masculine values over feminine ones. It was also natural that science, as it is known in the west, eventually "won" over religion.

This goes also for our economic system. I was discussing this with a young financial advisor. When I asked him if he saw cycles in the ups and downs of the market, he produced a chart for me. I

Isis Code

was surprised—although it was not really surprising. Without even wanting to, bankers themselves have a natural rhythm in mind.

I will end with this thought of Carl G. Jung:

> That the Christ is the Self of humans, that was already expressed, implicitly in the Gospels, but we have never clearly draw the conclusion that Christ=Self. This is a new sense given to the Incarnation, it is another stage of it, a new phase in the manifestation of Christ.... The Self is a living person and has always been. This is a concept shared as well by the hindu philosophy, buddhism, taoism, a certain branch of Islam and christianism. My psychology is integrated also in this august assembly."[527]

And I would like to humbly add to this list the *LIFE biosystem*.

Chapter 9

In Isis's Footsteps: Applications

"Indeed, the attempt to live according to the notion that the fragments are really separate is, in essence, what has led to the growing series of extremely urgent crises that is confronting us today."[528] David Bohm

"The Self, but I collect It through diverse experiences and I recompose It." Carl Gustav Jung, 1954

"As soon as daybreak, say into yourself: today I will meet an indiscrete, an insolent, a liar, a jealous and an egoist." Marcus Aurelius (121–180)

All of the traits mentioned here by Marcus Aurelius are a direct result of an imbalanced LIFE biosystem. The indiscreet is of the Mammalian brain, the insolent of the Human brain, the liar of the Analytical brain, the jealous of the Universal brain, and the egoist of the Reptilian brain. These traits are all expressions of deficiencies of the different phases. Only through the understanding and application of the implicate order, of the LIFE biosystem, can we rectify these in generations to come.

As individuals are unique, I have refrained from falling into the binary trap of good versus bad, right versus wrong. Everyone can draw their personal conclusions, devise their own hygienic rituals, and apply the system to their personal lives in harmony with their level of maturation and who they are. Possible applications of this system are infinite. Nevertheless, the LIFE biosystem, confirmed by brain research, has taught us many general, simple principles that are applicable to all. I have highlighted here a few of them.

Isis Code

In the face of physical reality, we *are unequal*. Our bodies are different, and our genetic and epigenetic makeup is as individual as our fingerprints. These will influence our medium, the body, in its ability to translate our core identity and our perception of the world. Prenatal awareness is therefore suggested for all, as is promotion of the feminine polarity.

The brain, like the rest of our body, matures in phases that are different in length for boys and girls. These phases are shorter for girls (around seven years) than for boys (eight years), and this is confirmed by observation, brain research, and TCM. In the same way pregnancy cannot be shortened, phases cannot be jumped over or abridged without detrimental effects on later ones. Education in harmony with these and respect of life cycles are suggested.

The different phases of maturation are associated to different structures, highlighting the differences between genders. For the same action, men and women will use different brain structures. These can be divided into feminine and masculine polarities. We generally qualify as *woman* the person who preferably expresses her feminine polarity, while a *man* would express his masculine one on the physical level. These are not restricted to the genital aspects of a body. Homosexuality is a biased term, as human's sexuality is not limited to the physical aspect, but incorporates the emotional, symbolic (archetypal), analytical, social (environmental), and synarchic (spiritual) aspects as well. Relations between two humans are always a question of difference in potential on any of those levels. At this point of our evolution, we are masculinized because our rational phase dominates all the others; we are passing through the phase of the rational self. Differences in potential are therefore less obvious and exchanges are less rewarding. Also, this masculine polarity, because it is not communicating properly with the feminine one, is ill. Rectification of the concept of what makes a *man* and a *woman* is suggested, as well as promotion of the feminine polarity in every domain.

Humans are a part of and an extension of nature. If nature is ill, humans will also be ill, and these illnesses will be manifested on any of the five levels of human expression. Nurturing as well as a close link with and respect of nature are essential for the optimal development of the human child. Therefore, the protection of natural habitats,

In Isis's Footsteps: Applications

education toward respect of nature, protection of the integrity of the different elements of nature (water, air, soil, light) that feed us, as well accessibility to healthy natural habitats are essential for all.

Individual optimal development is founded on healthy exchanges between feminine and masculine polarities. If these are not properly expressed through the structures built by Man, whether these are religions, governments, education systems, cities, or families, the individual has little means to succeed in properly developing his own polarities and thus will fail to reach his potential. Analysis, auditing the different manmade structures through the understanding of the two polarities is suggested. Our actual structures express mostly the masculine one. The result of this is that the general human LIFE biosystem presents an undernourished and poorly protected feminine polarity. Side effect of this can be seen through the sky-rocketing numbers of depressed individuals, neurodegenerative and cardiovascular diseases, obesity, cancers, social problems as well as environmental ones.

Our rationality is not rational when it is separated from the feminine polarity, in the same way someone with a challenged right hemisphere will fail to see that he has a problem. Our world is masculinized, and this masculine is diminished and diseased since it cannot properly function without a healthy feminine polarity. Indeed, the feminine nourishes the masculine and the masculine protects the feminine. At the present time, though, we are regressing and going toward a general collapse that nobody will be able to circumvent (diagram 29).

The only way to stop the downhill trend is not through economic restructuring—which would be as effective as a Band-Aid on a gangrenous leg—but a profound collective reflection on our core values, added to the promotion of the feminine polarity at all levels of society. Our current governments, born from the competitive masculine point of view, are ill adapted to conduct this reflection. This movement should be spearheaded by a worldwide association of people devoted to the protection and promotion of the feminine polarity in all countries. It is urgent that this be set in place. The association's function would include, but not be limited to, voicing its suggestions and warning of bad practices concerning the feminine polarity in countries, promotion of the respect of nature, and most

Isis Code

importantly, promoting the reintegration of the sacred point of view, thus of dignity, in the population's daily lives.

Religions, traditions, civilizations, art—which are universal and eternal—are all expressions of the same LIFE biosystem. Common points can be found and cherished in order to put a stop to competition between religions, as well as to correct attitudes and actions that are the result of an inflated and ill masculine polarity.

The prefrontal cortex expresses difficulties of other parts of the brain in the same way our rational mind expresses patterns associated with those deficiencies. Our present consumerist society is but an epiphenomenon resulting from the disequilibrium in our LIFE biosystem.

In the image of the brain structures associated to the masculine polarity, our vision of the world and of ourselves is now modular and fragmented. To reconnect all of our different aspects is essential to further our evolution. Reintroduction, enhancement, and understanding of the feminine polarity are necessary and not only an option.

Environment, Urbanism

"Merely changing the shape or conformation of a protein can alter its biological properties."[529]

The whole sector of the agro-industry is presently living a crisis. The earth is exhausted. More fertilizer, more pesticides are not the solution. GMOs are not the solution. The masculine polarity, totally inflated and disconnected from the natural world, is bursting at the seams. The modular values, solely directed toward profit and conveyed at this time by this industry, are far from those of the feminine polarity. Whether or not we are aware of this, we are all dearly paying for this. We often think we are saving money, when in fact we are wasting our environment and our health. We can see this through the many chronic illnesses that plague the general population. What money we "save" on the environment, we pay double through the health sector.

In Isis's Footsteps: Applications

There is no reason our industrial economies should be in debt. Something went wrong somewhere; our sole interests in sectors associated to the masculine polarity.

It is time to stop saying, concerning those who try to raise awareness, that they are using "scare tactics" and high time for us to use common sense.

Despite the relentless mockery of those who say they will not be fooled and thus would not buy organic products, many of us have eaten organic produce for years and still do, even when we didn't have the means to purchase them. We know that conventional agricultural processes deplete the soil of essential nutrients, and eventually those nutrients will not be in the products we eat. Selenium is one example of this. Although it is still controversial to say that organic produce is more nutritious than conventional produce,[530] we can say that it tastes better when produced properly. We also wish to support local small farmers, this being one way of mending the social fabric. I particularly like the "adopt a farmer" approach. Like Claude Béland, the past president of Caisses Populaires Desjardins in Québec, I believe in cooperatives to address some of our actual problems. Of course the philosophy is different. The size has to be adjusted for every voice to count, and for communication and interaction to be possible. People will have to choose between making money and being only a number, or living a social, human life.

As for choosing organic produce, we also know that an accumulation of herbicides, pesticides, and heavy metals is not harmless, since this ends up in the fat tissue of our bodies. Our nervous system is made of this tissue, including our brain. The axons of many neurons are covered in a fatty substance that is called myelin. It has several functions, one being to increase the rate at which nerve impulses travel along the axon. It can be as fast as 120 meters/second. Painstaking research has been done by so many dedicated scientists about the effect of pesticides, herbicides, and the like on the mammal organism. We don't hear much about them. The website on fluoridation[531] has a lot of information on some of these, and I put as an endnote the following text in lieu of summary.[532] The text that is of interest to us can be found in appendix. This was more than twenty years ago and not much has changed since then. Under "advance informed agreement," we should have had a label stating that the food we

Isis Code

consume is composed of living modified organisms (LMO), which to me is the same as GMO. Why would countries been warned that they are importing LMOs, but not their citizens? The precautionary protocol has been put in place but does not address the issue of chemicals already in use. Is it because the findings jeopardize some monetary interests? Was Franklin D. Roosevelt right? Is corporatism dangerous to the individual's well-being?

> The profound concern generated by the 1991 statement of a respected group of scientists (see appendix) led to an international effort to identify chemicals that disrupt hormonal systems. What a part of the public learned in 1991 was that no chemical had been tested for its hormonal-disrupting potential. The regulatory communities back then were committed to testing every chemical in use (more than sixty-five thousand). In order to achieve this goal, several years were spent on developing the protocols for testing and prioritizing chemicals for these tests. The unresolved problem is that even if we can eventually proceed with these testings, how about the combined use of these chemicals?

Also, vested interest groups have opposed the integration of new knowledge of sustainable farming by putting their ideas into law. However, soil bacteria and fungi do not care about laws. In the meantime, contrary to rumors, to get the same yields, conventional farmers have to use more plant food, more herbicides, and insecticides on their crops. Wouldn't it be wiser to learn to work with the soil and let it produce many of the plant food ingredients? Wouldn't this pollute the ground water and air less? The soil would be alive, which is certainly better if we want food that is alive. We could appreciably reduce the need for one of the most expensive soil nutrients we are currently using.[533] Many serious farmers have worked on the subject, but it is as if we don't hear them. This is another case of inflated masculine polarity in the agribusiness, or more exactly in this case the "agri-abusiveness".

Prince Charles had been warning about these practices for a long time and has been labeled *mad*. Everything was done to ridicule him, with resounding success, so his voice would not be as strong against GMOs and damaging farming practices. They succeeded. In Canada, some from the agribusiness, maybe thinking themselves more astute than their European counterparts, decided not to feed beef to beef cattle, but to still "enhance" the feed intended to herbivores with dead animal parts.

The same intellectual pattern is used by the biotech industry. For some of them, a gene is a gene, whether it comes from an elephant, a man, a scorpion, or a tomato. The problem resides in the fact that this type of science is analytical and can't grasp the whole picture.

At the root, we have science and its limitations; we have laws that should prevent us from poisoning our neighbors, and ourselves; and we have a business pushing to make more money. Is voluntary action of the industry, executed for and by the industry, an acceptable risk? On the one hand we ask them to be profitable and on the other hand we ask them to be ethical. This, unfortunately, is a conflict. Then we have the people, uninformed, with blind faith in their government to keep watch over too many details. This can't work.

To have a company, which has a vested interest in saying so, affirm that GE (genetically engineered) bacteria will not survive in the gut and then have research show us the contrary is quite scary. A British study found that genetically engineered DNA does transfer into bacteria in the human digestive system. According to the FSA gene uptake study[534] entitled "Evaluating the Risks Associated with Using GMOs in Human Foods," researchers fed a hamburger and a milkshake, both of which contained genetically engineered soy, to study participants. One of the problems associated to the transfer is that effectiveness of antibiotics may be reduced, since the bacteria in the human body could develop immunity to the antibiotics that serve as marker genes in biotech crops.[535] Researchers also found an herbicide-resistance gene from a Roundup Ready variety of engineered soy in bacteria in the small intestines of three out of the seven study participants. Studies with mice demonstrated that naked bacteriophage DNA is taken into mammalian intestinal cells, and even appears in systemic tissues (Schubbert, 1997).

If companies who are patenting and profiting from the promotion and selling of engineered products could assure us that they can clean up their mess afterward, that they can reverse all harm done,

Isis Code

and that they will pay for this, we could feel more secure. But they can't.

The ones who will pay in the end are us the people: the children, the mothers, the fathers, the grandparents. Every catastrophe, we eventually pay for emotionally, physically, and financially. Unlike car manufacturers who are responsible for their products, companies that patent seeds and drugs seem almost untouchable.

Pushing for organic production is not an emotional issue, contrary to what detractors claim but a health concern. One has only to study present organic production to see that it has nothing to do with what our forefathers did. Scientific knowledge has ushered us into an agricultural revolution because we know which plants work together. We know the exact chemical balance and microorganisms needed for an utmost performance. And most of all, we know that the soil is a living organism. Sure, when I decided one morning to grow a few vegetables, my lack of knowledge meant a very small crop. (But oh, so good!) I do not know what to grow where and when. When biotech activists discuss organic culture, we are mostly given a contradictory and emotional picture of the pro-organic group: hippies ignorant of technical and scientific discoveries, at the whim of a fickle nature, living in fear and superstition, dreamers, relics of a feudal system where basic foodstuff was enormously expensive and readily available only to the ruling class. They give this image that organic is somewhat unclean and unsafe. This reminds me of a teacher saying to my then eight-year-old son: "Can you imagine if we would grow only organic produce, we would be infested with bugs of all sorts!" Or this kid's exclamation when my son admitted in class that he ate organic fruits: "Shame on you!" Some say organic food tastes the same, but after having eaten conventional food for twenty-five years and organic food for the same length of time, I can say that they don't taste the same! Some say organic food is much more expensive. They have never compared organic products, say the President's Choice or Organics, with conventional products. In Alberta as well as in Quebec, I have found some organic products that were the same price, and sometimes even less expensive. Of course, the more people eat organic, the less expensive it will be, as with anything else.

In Isis's Footsteps: Applications

With the mad cow scandal, the price of organically fed and grown cattle in western Canada was evaluated. The result was that it would cost only a few pennies more to eat organic beef. But the people would have to ask for it globally. All of us are the one who change things. We don't have kings to go to and complain to anymore. And anyway, kings now would not be allowed to say what they think. We have to act in our daily lives and express ourselves through the choices we make. Every little act counts immensely. Maybe these foods are more expensive upfront, but they are an investment for your health and for your children's future. Isn't it worth it?

The GMO approach could be accepted in specific cases to engineer plants by taking genes from other plants, but not from other kingdoms. Engineer plants to eradicate the use of pesticides, not to multiply the use of it. And always use the precautionary principle. Knowledge is a sword with two edges. Coupled with greediness, it always destroys.

In Ethiopia, four hundred people have signed a petition saying they received no compensation after thy were expelled from their land, which had been taken over by Karuturi Global Ltd., the world's largest exporter of cut roses, with 250,000 acres under rose cultivation. Karuturi is also under the magnifying glass of New York-based Human Rights Watch, which in a new report highlights the forced dislodgment of thousands of indigenous people in the Gambella region of Ethiopia, where this company is a key operator. Bangalore-based Sai Ramakrishnan Karuturi has denied any wrongdoing. In a documentary I heard him proudly state that with the quantity of land he will cultivate; he will become an influential player in the market and will have the power to play with the prices, thus garnering respect from other businessmen. Can we say that this dream of his is the result of a balanced LIFE biosystem? Unfortunately for us, he is not the only one thinking this way. It appears that many people whose only interest is in making money may suffer from early affective deprivation; that is, a lack of bonding during their childhood. The LIFE biosystem explains it as an out-of-control Mammalian brain, engendering an inflammation of the Analytical brain. This aspect, being the natural regulator, therefore takes over the entire place, rather than controlling the one-fifth normally dedicated to it. If we look at industrialized countries, this is what we see. This aspect is like a cancer spreading its tentacles over everything, sapping the vital energies of the entire world, environment and humans alike.

The evicted Ethiopian farmers say their families have farmed and grazed their animals on their land for generations. One farmer, who spoke on condition of anonymity, said: "We are for development of our country, but we cannot develop our country when land is in the hands of the government. ... You can work on your land, and all of a sudden, they push you out of your land."[536]

The winter 2009 issue of *Food First News* reports that the World Summit on Food Security in Rome issued a declaration that the world is now hungrier than ever before. Significantly, this is not the result of food shortage, with worldwide food production one and a half times more than is needed to feed every man, woman, and child on the planet.

"The root cause of this insecurity is the food system itself, which is controlled by a handful of global monopolies. In fact, the crisis comes at a time of record global profits for the world's agro-food corporations. Archer Daniel Midland, Cargill, Monsanto, General Foods, and Wal-Mart all posted profit increases in 2008 of 20% to 86%. For Mosaic, a fertilizer subsidy of Cargill, profits increased by a stunning 1200%."[537] The World Food Summit did nothing to confront the market-hunger crisis.

Consider that there are two kinds of wealth. The one linked to the masculine polarity—quantity of money, food, power, sex, houses, powerful cars, and planes—allows for our physical survival and drags our point of view toward objectification. Everything is an object we can add, sell, and buy. Then there is the wealth connected to the feminine polarity—the quality of our relationships with ourselves, with others and the environment, the quality of our environment (air, earth, food, water, sun). Everything becomes a subject. Some countries are rich in one aspect and extremely poor in the other. A balance should be sought, with the feminine polarity, which generates everything else, always protected and enhanced. The wealthier/healthier countries often are not those we think are.

Recently, the US Department of Agriculture reported that one in seven Americans cannot afford enough food throughout the year.

Why, in the most productive farming country in the world, are there so many hungry people? Not because of food shortages.

With more than 1 billion people in poor countries unsure of how they will afford their next meal, many chronically food-insecure countries are selling their land, as Raphael Grojnowski reports in the same issue of *Food First News*. Sudan, Ethiopia, and Cambodia, for example, have already sold nearly 40 million hectares of their *best* agricultural land to foreign investors, mainly from the Middle East, China, and South Korea. This could be qualified as a classic land grab, like those from the colonial past. This time, though, these are not perpetrated by kings, but by businesses walking, it seems, hand in hand with governments.

Spurred by the global food-price crisis in the volatile world food market, wealthy but food-deficient countries are buying up vast tracts of land, especially in Africa. There they expect to grow food and fuel long distance. Since these countries promise new technologies and employment to some of the world's most neglected areas, many poor governments are rushing to attract these new investments, seeing them as a providential solution to their problems.

Regrettably, these land deals are negotiated in secrecy and are having devastating effects on local farmers and their families. To make room for the new foreign monster farms, these farmers are being dispossessed of the land they cared for. In their place, vast monoculture plantations are being established, using industrial farming techniques that have been shown to generate extremely damaging environmental effects, such as chemical contamination of scarce water supplies.[538]

While many organizations are relentlessly drawing attention to this devastating practice, the UN and other agencies seem slow to act. UN agencies and the World Bank announced plans to draft a code of conduct for such "foreign land acquisitions." The problem is that those guidelines follow a nonbinding and voluntary code[539].

The secret behind China's success was its farm policy of the 1990s. Investing with small farmers became the engine of growth. At this time, the Chinese government wisely does not wish to move toward the mega farms prominent in so many other countries, because a plot of land is a form of social security for the 850 million rural residents. Sadly, 25% of these farmers have now exiled themselves to cities, pushed by the decline of the soil that has resulted from conventional farming techniques and its consequence: industrial pollution. Dr. Han Jun, an expert on rural policy at the Development Research Center of the State Council, a member of the Chinese government who has helped to draw up the cabinet's rural policies for over seven years, said that more than *twice* as much nitrogen fertilizer now has to be used on the average hectare of Chinese farmland, compared to before. "That became the engine of growth for the country," he said. Now it is becoming its demise. High-rises are springing up everywhere, although 70% of the apartments are unoccupied since the migrators do not have the means to pay for them.[540] Also, the Chinese are buying vast tracts of lands in other countries, including Brazil, so they can grow genetically modified soy.

In Saudi Arabia, Turki Faisal Al Rasheed gave a speech on food security and agriculture at the Jeddah Economic Forum of 2010. He said that there is a food shortage, though it appears distribution and market games (some countries hung on to their rice crops instead of distributing them, thus raising its price) are the real culprits. Commenting on the reluctance many people have about the genetically modified seeds, he said: "We don't have choices. People were against this when there was abundance of food. Now when there is a shortage you have to live with it. ... We must use the highly genetically modified seeds. We must heavily depend on the new technologies of the most optimum water management."[541]

Lim Li Ching of the Third World Network said: "The same failed World Bank policies on agriculture do nothing to address the root problems of food insecurity, including the collapse of many agricultural sectors due to structural adjustment, declining terms of trade, asymmetrical trade rules and financialisation of food commodities."[542]

Like the 2008 World Development Report states, the World Bank's plan[543] for the future contains a mix of approaches, but the overall thrust is to back "concerted efforts to integrate smallholder farmers into growing agricultural markets and supply chains." This market-focused approach downplays the role of government and involves significant support for agribusiness and foreign investment.[544] This is not progress.

For its part, BlackRock Global Allocation Fund has added to its range of resources with the launch of a global agriculture fund. This new Luxembourg-domiciled BGF World Agriculture fund is managed by Richard Davis and Desmond Cheung from the firm's $35 billion natural resources team. [545]

I realize that, because the environment is in such a miserable state, this chapter sounds more like a ranting whistle-blowing discourse than anything else. It is not my preference to denounce. The people involved are only following a pattern, and I would rather be *for* something than *against*. But we are ill, and just because most of us are does not make it normal or acceptable. So let's just say that there are ways to live harmoniously with the environment.

The native who rose early and offered his life to creation understood something. Everything we have we took from somewhere, so we don't own anything. We should be grateful, and every day we should give back a little through the goodness of our hearts and the respect of our actions to calm the acidity of this nature we still abuse. Our hand does not ripen the fruit. The only one who deserves something from nature is the one who cares for it, tends it, feeds it, respects it, and loves it. This implies, among other things, organic growing, limitation of land grabbing, respect for diversity, connecting pockets of virgin nature so that the wildlife can roam freely, and urban plans that respect the elements of nature. At this point, we generally steal from nature without ever giving back. If conventional cultures are so good at not needing water and growing things where nothing grows, then let them use all the dead spaces natural farmers can't use. Don't let them lease fertile lands that they will ruin. We poach and hope not to be seen doing it. We use and abuse "our" natural resources to the limit, and sometimes totally outside legality. This legality only implies ownership, and therefore not necessarily caring for the equilibrium and health of nature. Governments shamefully dodge their own environmental laws and vision, as seen here in Canada, a country that should be spearheading environmental protection. Money talks and humans run to its service, shamefully kneeling in front of it, their identity totally attached to it. And they will argue it is all about creating jobs for the people.

Isis Code

What can we do in our cities to enhance our health? This could be urban art, where water, fire, air, plants and sun are valued and reintegrated into the general plan. Giving prizes to the best city gardens is a fantastic idea. Cities and boroughs should hire designers in the field. I have seen such creations in an outdoor shopping center. How inspiring it was to be close to a fire in the freezing winter, to feel its warmth! How beautiful and inspiring in summer, as seen in Paris, to stroll through small parks where mothers and infants, old people can walk and sit for hours. Where lovers kiss on a bench under venerable trees and friends share a good book in the sun? Why only create parks for sports, which are associated to the masculine polarity, and nothing for the feminine one? When I lived in France, I loved to stroll every weekend through the larger parks, where I could find a nice coffeehouse and have a piece of fresh baguette bread and an espresso. However poor you are, as I was, this helps you keep hope and feeds your dreams. You are not isolated.

And how about fountains, such as those found in the central places in the south of France, where children gather during hot weather? Why not use the tops of some buildings for private sunbathing enclosures, as well as for gardens and beehives, as it is done in some places?

Why not a few natural beaches in summer where the water is good? Instead of amorphous suburbs, why not create a community through summer marketplaces and promote initiatives that enhance the communal fabric? Why not truly welcome the new residents, instead of having only a tax invoiced to them? Why not put in common what the inhabitants of a community do instead of staying anonymous? Why not invite people of the borough to open businesses in a place people can walk to, so that the poor, old, and young have a pleasant destination to walk to? For this we need a general plan of the borough, where people participate, not just as an extension of the city but as a self-reliant model.

The biocybernetic system finds its place even through the structure of the cities. Changes and improvement to urban environments would be easier to manage and readily applicable with this model. Dr. DeBavelaere has taken the example of the city of Paris to explain that a wise use of the LIFE biosystem can protect against ineffective investments, and moreover can enhance the life of its inhabitants. In other words, if one opens a certain type of business in the right area

In Isis's Footsteps: Applications

of the city, he has more chance of it flourishing. Also, when elements of the community are placed at the right place, strolling through the city becomes a pleasure and one is less tired of walking. A more thorough analysis could be done with the history of the greatest cities and of their landmarks. Here is only a superficial one.

Paris and the LIFE Biosystem

Paris is renowned for the pleasure one feels in strolling its many areas. The locale of the city itself, in the midst of the Parisian bowl, is already a factor of stability and equilibrium, as it corresponds to the central function of nutrition and analysis (pancreas). This city will feed the world, be a center of commerce as well as an intellectual center. This central position thus attracts people associated with those things. Perusing the Paris map and taking into account its arrondissements as well as its landmarks, we cannot fail to observe that this city does well as a projection of the LIFE biosystem. The regulator is positioned in the middle, the information/regulation elements to the north, the natural environment all over but mostly to the east, the defensive elements to the west and the passage from life to death as well as death to life (birthing) in the south (diagram 18). The snail design of the arrondissements allows for expansion, so elements had to move on their axes in accord with the expansion of the city. So let us look at Paris as if it were our brain.

We find in the south of the city the Reptilian brain:

The Catacombs of Paris, dubbed the municipal ossuary, were created at the end of the eighteenth century. Saint Innocents Cemetery, close to Saint-Eustache Church in the district of Les Halles, was used for nearly ten centuries and became the origin of infection for all the inhabitants of the district. The Council of State, on November 9, 1785, ordered the removal and the evacuation of the cemetery. The Catacombs now hold the remains of approximately six million Parisians, transferred there between the end of the eighteenth century to the middle of the nineteenth century, following the progressive closing of different cemeteries for health reasons.

Bones were removed from the Innocents Cemetery after the blessing of the place on April 7, 1786, and continued until 1788, always at night and according to a ceremony. Priests sang the burial service as carts were filled with bones and covered with black veils. Thereafter, this place was used, until 1814, to collect the bones of all the cemeteries of Paris. The Montparnasse cemetery is also situated in the south.

We also find the private maternity hospital, the Maternité Sainte-Félicité in the fifteenth arrondissement, which is one of the top ten Parisian hospitals, and where there are the most numerous births per year (3,200).

Between the Reptilian brain of the south and the Mammalian brain of the west we find the Bois de Boulogne, which is renowned not for the park itself but more for the nightly sexual encounters there. We also find there the Parc des Princes, dedicated to sports.

To the east we find:

The Jardin des Plantes, considered the botanical garden of Paris, is in the fifth arrondissement. An ancient royal garden, this park has developed during the centuries and now contains the gardens, the Museum of Natural History, and the Ménagerie (mini-zoo). The Parc Zoologique de Paris, the zoo, is located in the twelfth arrondissement, which is near the Bois de Vincennes. If you were to climb up the sixty-meters-high artificial rock within the zoo, known as the Grand Rocher, you would get an expansive, magnificent view of Paris, as well as see goats, sheep, and vultures and as other species. We also find the Paris Institute of Technology for Life in the fifth arrondissement.

To the north we find:

EPRA (Echanges et Productions Radiophoniques). This is a network of 160 partner radio stations that produce and receive programs. Here we should also find religious monuments as well as castles. These instead are found in the center of Paris. The Louvre, the Sainte-Chappelle, the Conciergerie, the Tuileries, the Luxembourg Palace, even Notre-Dame de Paris basilica are all in the center of Paris. The heart aspect is really the center of Paris.

The Moulin Rouge, with its accent on marvels is a landmark found to the north, in the feminine polarity. The red district is situated there as well, but would probably be more successful closer to the Reptilian brain district. More to the south, between the Reptilian and Mammalian brains, we find the cabaret Crazy Horse—which is more associated with sexy bodies with an orientation toward the genital, than visual entertainment—is well positioned in the west.

As for communications, we find here three train stations and the highest density of underground railways. The north-south axis is also associated with religion, so we find spreading from south to north: the Panthéon, initially dedicated to St. Geneviève, guardian of Paris, plundered during the French Revolution, and transformed to a mausoleum for distinguished French citizens. We also find here the Sacred Heart Basilica where the kings of France are entombed.

In the center:

With the castles and main churches, we also have Les Halles, the central market of Paris, located in the first arrondissement. The main universities are also here. This agrees with the pancreas function.

Finally, to the west:

Not surprisingly, we find the Eifel tower and La Défense (the Defense), a prime business district, as well as the Invalides, the French Army Museum, and the tomb of Napoleon Bonaparte, all symbols of masculine power. Avenue de la Grande Armée and Champs-Élysées are found in this area. At the limit of the defense aspect of the Mammalian brain and bordering the symbolic aspect of the heart and the Human brain, the Arc de Triomphe stands with its eternal fire. It was designed in 1806 by the architect Jean Chalgrin, and was a gift from Napoleon to his new young bride, Archduchess Marie-Louise. Why not have such visions for our cities?

Sustainable Government: Elected Synarchy or Biosynarchy, the Only Possible True Democracy

"Do you know the only thing that gives me pleasure? It's to see my dividends coming in." John D. Rockefeller[546]

It is impossible to ask of a government dedicated to economic priorities, thus essentially of the masculine polarity, to decently manage what is of the jurisdiction of the feminine polarity without, consciously or not, trying to make money out of it. We have designed our government as an androgynous structure, but it cannot be so. We have designed parties that fight to be elected by any astute means. Once in power, they undo what has been done by their predecessors and hurriedly put in place elements that will serve them in the future. This is costly to the general population and does not help progress in any area. Once in power, these politicians keep their partisan outlooks. They do not necessarily work for the good of the country, despite what they claim, but certainly for the profit to their political parties. If it is political, it is not humanistic. This is unacceptable. I remember when Brian Mulroney, former prime minister of Canada, left power, he was proud to say in front of his conservative buddies that the coffers of the party, empty when he came into power, now were full. The partisan is never far from the surface.

So how can we fix this? Some have suggested proportional elections, which mean that the parties will share power based on the numbers of votes they receive. Then we end up with many captains all pulling in different directions. The LIFE biosystem points to a different solution. The biosynarchy would suggest five parties, two for the masculine polarity, headed by a prime minister; two for the feminine polarity, headed by a great mother; and one, in charge of the economy, who communicates with both and must answer to both. Instead of having a horizontal stratum, we have a spectrum, joined in the middle by the two centers: the heart aspect and the pancreas aspect. This would mean five parties, each assigned to a particular function and working for the good of the group, but most of all for the good of the country and the world. What we would elect is not a party but the representative of those parties, who will fill one of the seven (5 + 2) main functions. This would be done while keeping in mind the mandate that the masculine polarity protects the feminine one, and the feminine one feeds the masculine one. Parties can still exist and be associated to the different functions, but when someone is elected he becomes part of the system. He is not of his party anymore and has to sever all links with it. His "buddies" cannot ask him for favors. The parties don't make the structure. The structure is in place and the elected members are there to fulfill a function. People vote for the best person to fulfill those functions. We call this a biosynarchy because the members work together. Laws will always be there and must be improved upon, but we are now faced with a provincial government or a state suing a federal one and the general population paying for this? There is a name for this: an autoimmune disease. We could adapt our present system to express this elected biosynarchy.

Persuaded by intellectuals and merchants that the institutions of monarchy and religion were the culprits of all human miseries, instead of being the result of individuals suffering from an inflated masculine polarity, we got rid of kings and queens as well as of religions. The problem is, these were supposed to be, in their mandate, expressions of the feminine polarity and awakeners of our Self. They are archetypes, fundamental to our equilibrium. Religions (spirituality) are associated with the heart energies and its associated brain structures, and royalty to the heart aspect and its associated brain structures. Spirituality should have worked hand in hand with monarchy. Their rituals were attached to the quality of informative energies. In the royal tradition, for instance, we find archetypes supported by symbols of purification (lung aspect) and solar energies (heart) necessary to the proper performance of the feminine polarity. This we can see with the use of the fleur-de-lis, the bees, the lion, eagle, unicorn, as well as many others. The Sun King of Versailles was, although imperfectly, an expression of this.

To explain how a government would function through a healthy LIFE biosystem, I will take the example of the Canadian government, as this is the one I know best. The structure I describe, though, can be applied to the American government, or any government for that matter.

I remember that morning, because we Canadians made it special. It was the nomination of our new governor general, representing our head of state, the Queen. Many will smile or cringe at this: the Queen. For my American brothers and sisters, this is part of the function the American president carries, although not specifically expressed. It is implied. How could we come to a stage when we don't even know what the universal and eternal aspects of these refer to?

I remember watching Michaelle Jean, tears rolling down her cheeks when she heard the song *Ne tuons pas la beauté du monde* (Don't

Isis Code

kill the world's beauty). Did she know why she was crying? Like a prophecy, this song stayed as a backdrop of this nomination. To me, that is what such a nomination should be, to safeguard Beauty. But a governor general, as the Queen in our desacralized and truncated society, has no power, no voice, and no truly recognized function. They are mere puppets used to distract attention and uphold a semblance of clear conscience. This shows the state of our general feminine polarity: ridiculed, gagged, and abused. How do we change this?

We could first become conscious of what beauty is or is not. Plato said Beauty is the splendor of Truth. Truth is the child of our love and wisdom. So look around you. This is the state of our love, of our knowledge. Do you recognize yourself in it?

Jesus reportedly said: "The truth will set you free" (John 8:32 NIV). Thus, Beauty is necessary to the freedom Jesus refers to. Of course Jesus was speaking of the ultimate truth, the one above phenomenon, the one of the implicate order generating the Beauty Plato mentioned.

Truth is the manifestation of the level of our love and of our wisdom, of the relation between our masculine and our feminine polarities.

At a governmental level, I could say that the function of the governor general is more of the Beauty than of the economy. The prime minister is more of the economy than of Beauty. This is why this type of government could become the best if both understood their functions. It is the only one that could truly express the two polarities. The governor general's office expresses the feminine polarity, the prime minister's office expresses the masculine one. Their synarchic work in society gives the truth of a nation that can be measured through its GNP as well as through the quality of life of its people. Think of the prime minister becoming emotional during a song. We would feel ill at ease because it is not his function. The function of a president is tricky because he has to manifest both polarities at the same time. This is impossible to ask of a human being. You may change a president, put all the faults on his shoulders and kick him out of office. The truth is that the machinery behind him is at fault because the feminine polarity has no place in it.

In Isis's Footsteps: Applications

In April 2002, the then president of the United States, George W. Bush, declared that human dignity was conditional to four elements: that people could raise their children in peace, that they could own property, that they could have freedom of religion, and education.

This should be extended to all countries.

As for United States foreign diplomacy, I remember his father, George H. W. Bush, saying to an audience of Japanese people when he was president: "on a level playing field, we can out produce, outcompete, and outsell anybody, anywhere in the world." No mention of time here. This is a nice case of overinflated masculine polarity. Can the government in America allow for the free expression of the two polarities? No. The problem is that the system is so archaic, if someone totally honest and dedicated were elected president, he could hardly change anything and would only be swallowed by the machinery. A change in the point of view has to occur first.[547]

The office of the governor general is responsible for the feminine aspect of society and should be allowed to speak up in matters of education, communication, environment, and all humanistic interests. It is too easy to say, when a governor general wishes to give his opinion, that it is a political concern and therefore out of his reach. Everything can be viewed and transformed into a political issue (read financial).

I have heard strong voices calling for the removal of this function of the governor general. The reason for this is that they see the position as obsolete. What is archaic is the monarchy. Not the one with kings and queens, but the one that expresses exactly that: governance by one. Whether we call it a party, a president, or a king changes nothing.

In Canada, we have the structures already in place to govern in the most democratic way, not only through masculine structures, but also feminine ones. This governance would not be only through the right and the left (which presently are no more left or right but all around the center), but also up and down. In the image of the two hemispheres of the brain, each has a particular, distinctive, and complementary function. The governor general function cannot be reduced to a part of a minister as it appears to be now. It should be

Isis Code

as valued as that of the prime minister's office. Since it represents a feminine function, it is suited only for someone who has shown a healthy and developed feminine polarity. The feminine function expresses itself through the domains associated to communications, emotions, arts, and spirituality, while the masculine function expresses itself through the intellectual and physical domains. The masculine is responsible for structures, for the plumbing. The feminine is responsible for the water running through it. The feminine polarity interests itself in the fabric of society, while the masculine interests itself in its economic aspect. The senate—an expression of the feminine aspect—should have its members nominated by the governor general, not by a blatantly partisan prime minister. The prime minister shouldn't choose who will be governor general, as this should be a vocation. In fact, we could have our own in-residence king or queen, a Great Mother or Great Father.

A present-day prime minister or president expresses first the ideas of the party that promoted him to power. He has a debt to this party. It is impossible to ask a human being to put his party in the back of his head and put country first. His party would soon dismiss him. But this is possible for a governor general. The governor general and prime minister are the first couple, symbolically speaking. The way the prime minister protects the governor general and the governor general stands up in a moral way, making the prime minister proud and strong, will affect all sectors of society. The governor general has a visceral, intellectual, and symbolic link to the country in all its tradition, history, environment, people, and regions. The governor general function is to empathize with the physical country and its people, becoming it. The prime minister does not have the same link. He is a minister. Sure he might have an emotional, economic, and social link to the people and country, but he can detach himself from those. This is not the case for the person who represents a feminine function.

The governor general should be chosen by the biosynarchy governing the country, mainly by the governor in place, by the prime minister in place, by the representative of the people in the five main functions. The time in office should be longer than what it is. The governor general should remain as long as the synarchic group deems it appropriate, as this function is linked to time while the prime minister's office is more interested in space. For the kidney-heart

In Isis's Footsteps: Applications

axis to function, a hereditary position should be considered. Not hereditary in the genealogical sense, although this could also apply, but in the sense that the person filling the function has a strong voice in the choice of his or her successor.[548]

It is well known that political parties have ties with businesses. Which makes sense since their chief interest at this time lies with the economic aspect. For them, if the economy goes well, all goes well, but this is a myth. Even if the economic aspect is important, we all know from personal experience that if our health, emotional lives, or the environment are in difficulty, the economy counts for little. What did Napoleon and King George III have in common? Both vehemently opposed businesses entering the doors of government.

Solutions will only be brought forward with the restoration of the feminine functions and structures on an individual as well as collective level. Only then will the masculine polarity be healed. The feminine polarity is the only one able to control, harmonize with, and properly nourish the LIFE biosystem. But how can I say that our current form of government does not succeed in expressing both polarities? It is easy.

Associated to the Master of the Heart, the feminine polarity plays the role of information and communication agent. It makes the connection between the down-to-earth facts of the environment and the subtle world of thoughts and feelings (spatial reference). With a healthy feminine polarity we should find the entire above well represented and respected in our daily life. Do we?

We have also witnessed that the feminine polarity is related to the Analytical brain in its nutritional element, and to the social aspect in its purifying aspects. Associated to the economy, the feminine polarity will prefer quality over quantity. Associated to the social, its purpose is to maintain the purity of the environment, whether it is physical or psychological. Without purity, as every computer owner knows, information cannot circulate properly. The feminine polarity also manages and controls exchanges between outer and inner, past and future, visible and invisible. By communicating with the past, the inner, and the invisible, it is guardian of traditions. By communicating with the future, outer, and visible, it supports maternity and primary education. By communicating with living nature and its rhythms, it respects all forms of life and its cycles. It allows the integration of the system into the environment, as well as allows its proper functioning. This implies that the structure responsible for the feminine polarity should be a moral authority (quality of information). Finally, it is responsible for the manifestation of beauty, harmony, and evolution in general. With a blooming social feminine polarity, we should find these elements clearly in view and respected in our society.[549]

Do we?

After the wedding of Prince William to Kate Middleton, I heard Oprah Winfrey say on national television: "We all know fairy tales don't exist." Truly, seen through the Reptilian, Mammalian, and Analytical brains they don't, but through the Human one they certainly do. Otherwise, why were we so many of us around the globe riveted to our TV screens on that early morning? That fairy tales are more and more difficult to express I agree, but don't exist?

The ability we have of living the fairy tale within us allows, little by little, its manifestation in the external world. The prince, which is our masculine polarity, works hard to find his princess. And when he does, nothing can stop him from being with her, and they do live happily ever after.

It is our personal selves that reduce everything to this finite dimension. We should not say to children that fairy tales don't exist. That is like putting a lid of lead on their hearts.

Considering William and Kate, here is a young couple who generously gave us the opportunity to project this aspect of us on them. How long will it last? That is actually none of our business. They showed us their souls and opened up their hearts to us. Of course, now the lives of their personalities are different. This is their life, not ours. Their wedding created a marvelous ambiance that, for a few hours, made us forget about the greyness of this world. They showed considerable courage and generosity in sharing what is most private, rare, and essential in a human life—their secret garden. I wish them well, and so should we all. If I may object to one thing, I would have given them a year away so they could be all to each other, in the same way the Bible says that for a year the husband should be to his wife and her to him without social interfering. This is prenatal awareness. Their commitment would have been cemented, and on these foundations they could have built a more resilient feminine polarity for the benefit of their future family and also of their whole country.

On a global level, countries all have functions that fit more in one of the five phases. Reflecting on this, we could save a lot of human and environmental energy if we could orient ourselves toward this. For example, why push a country to plant a profitable crop because *at this time* there is a demand for it, when this crop doesn't fare well in that area and jeopardizes ancient , but less profitable cultures? So we could repeat what was disastrously done in Spain, where olive trees, hundreds of years old, were unceremoniously destroyed to make place for greenhouses so they could grow tomatoes and cucumbers faster than other European farmers? And employ underpaid Puerto Rican labor? If the forte of this country is not of the pancreas phase, we could help enhance its true strength. Of course all aspects are in every country, but because of environmental, cultural, historical and economic factors, some aspects would benefit from being enhanced.

As an example, lung-type countries of the world could put emphasis on this function, and as a global world we should not go against it. This aspect should be protected as it concerns everyone. The environment is under a country's jurisdiction, it is the country's responsibility, but no one owns the environment. Now, who "owns" the government owns the environment. That is not acceptable. The World Resources Institute pointed to the fact that all the frontier forests are in jeopardy. The three main countries possessing the most remaining frontier forest are Russia, Canada, and Brazil. They can thus be considered as having an important lung function for the world. Their forests should be protected from clear-cutting. And the protection of the forests should be considered more important than the money some individuals can make from them. The greater good should be assessed.

Do I think we should have an international association overseeing and protecting the feminine function of the world? An association composed of great mothers and great fathers of the world? Indeed, and it is almost too late. This cannot be a gathering of people in charge of the masculine aspect of their country, but of those in

Isis Code

charge of the feminine aspect. There are none? Create the function then. Economy has nothing to do with it, thus the political should be excluded from it, in the same way it is said that princesses are not allowed to take political stance. I did like Princess Diana's answer to those who had no interest in seeing her address problems linked to the feminine polarity: "I am not a political figure. I am a humanitarian." In this, some royal families have a strong draw toward the common good. This could usefully be exploited for the common health of the feminine function.

Prenatal Awareness

In 1982, I was part of a group of people working at drafting the lines of a new association. Its goal was to emphasize, through education, the observation that education and care given to parents before and during pregnancy would lead to a healthier community. We all had been motivated to create this association for different reasons, but the first inspiration came from Omraam Mikhaël Aïvanhov, the French philosopher I referred to earlier, who dedicated his life to explaining how this is a necessary step toward the betterment of humanity and to stop its downhill slide. In Switzerland I met with the ex-president of the AGIEM (General Association of the French kindergarten teachers) who was particularly keen on being part of such an association to help future parents. She told us how she had observed, for the forty years she worked with small children, how these upon entering kindergarten already had a personality that no later intervention seemed able to profoundly modify. Her belief, as ours, was that conception and pregnancy were the only periods when we could, if not influence this personality, at least enhance the qualities of the physical body, which would be the tool of this being for his or her entire life.

During my years in Paris, I worked on these ideas and met with professionals who actively researched that domain. The question on everyone's minds was: Does the mother and her environment really has that much influence on the fetus's development? The answer from research was a resounding yes! I also helped for three years in writing a book on the subject with the naturopath I referred to earlier. With others from the group, I was invited to express my ideas on

In Isis's Footsteps: Applications

radio and in associations while I was teaching relaxation techniques, nutrition, and reflexology to pregnant women.

I came to realize that the subject is emotionally charged, and that the people interested in it have to first exorcize the demons of their own in utero and infancy period. Someone with no legs who has never danced can hardly show you how to dance. The teaching stays a bundle of intellectual notions, it is not alive. For this exorcism, instead of psychoanalysis, I found that a holistic knowledge and life could help. We come into this world with a cry, not with a smile. We exit this world often with a smile more than a cry. Separation seems therefore to happen more at the beginning than at the end. Luckily for me, after I had studied the subject for at least seven years, life gave me the chance to carry and give birth to two wonderful boys. My ideas on pregnancy became more profound.

What I learned from my experiences only reinforced my understanding that we do not come to this earth as a blank page. I also understood that we go through different cycles, periods in which our expression on earth concentrates on different domains. This is how I was able to divide the growth of a child to adult into five different stages. Of course, I am not the only one who came to this conclusion. In every country where attention was given to prenatal care, education became the backbone of that civilization. Interestingly enough, the country interested in those questions would also see a blossoming in all domains. The LIFE biosystem is perfectly positioned to deal with prenatal awareness.

To psychoanalysis, the LIFE biosystem answers with psychosynthesis endowed of the biocybernetic system: psychocybernetics and sociocybernetics. Freud would say of the term psychosynthesis: "As we analyze ... the great unity which we call his ego fits into itself all the instinctual impulses which before had been split off and held apart from it. The psycho-synthesis is thus achieved in analytic treatment without our intervention, automatically and inevitably."[550]

Once the analytical mind is conscious that synthesis is necessary to mental health and to health in general, it stops hindering cohesion, starts promoting it and everything falls naturally in its rightful place.

Isis Code

Legend has it that the day Horus will reign in our hearts over Seth, Osiris will reappear, crowned with fire and streaming with love, and that in his eternal embrace with a revered Isis, they will bring peace and joy to the world.

On the tortuous roads of this quest, I continuously sought Her Presence. Then one day, sitting near a pure spring, surrounded by the green smell of mosses, I saw Her there, for a brief moment smiling back to me.

It compelled me to write this book in the hope that one day, looking in your mirror you too might also witness Her smiling back to you. For in that instant, Isis and Osiris in you will be reunited, happiness will flow in you and the world will be a better place.

Epilogue

Here are a few facts, as per the US National Institute of Mental Health (NIMH):

> In any given year, approximately 18.8 million American adults[551] aged 18 and older will suffer from a depressive disorder. These figures translate to 12.4 million women and 6.4 million men.
>
> Of this amount, half can be considered as suffering from a *major* depressive disorder.
>
> Major depressive disorder is now the *leading* cause of disability in the U.S. and in established market economies, worldwide.

Now though, depressive disorders are appearing earlier in life. Fifty years ago, the average age of onset for depression was 29 years old — which marks the beginning of the social self phase—whereas recent statistics indicate it at just 14.5 years, in today's society—which marks the beginning of the Idealistic self phase of teenagers. Depressive disorders often co-occur with anxiety disorders and substance abuse.

As we have seen earlier, depression is a symptom of a profoundly deficient feminine polarity. It is not a solely personal problem, but a societal one that we collectively and individually dearly pay for, on every level (diagram 29). To the physical distress of the poor, we have now added emotional distress. Although silent, the pain of depression is quite real and profound. To this war we are losing since we are not fighting back, we are now sending our teenagers.

Isis Code

This spells disaster and underlines the need to urgently act to rectify this situation. But how?

Our civilization started its march a long time ago. With the Reptilian brain, which was controlled by food availability and territory (pancreas aspect), humans have seen the utility of controlling space and their environment (lung aspect) and started devising ways to permanently secure for themselves ownership of this environment.

With the Mammalian brain, larger groups formed since to impose hegemony numbers count (lung aspect). They created rules and laws to protect their own group against the others. A more powerful group could control a larger territory; the more violent or wicked group could dominate all the others.

With the Human brain, ideology created unification within those groups which became even more powerful as sacrifice is limitless. The group with the highest ideology won over the other groups, God was with them. Identity showed itself to be a powerful tool of cohesion as well as of division. Religions, kingships and different types of government evolved from this phase, replacing the tribes and clans. This phase is controlled through the fear of demise (kidney phase).

With the Analytical brain, we see the rise of mercantilism and businesses while the kings and queens are decapitated of their powers. Kings, religions and governments had won over groups with no leaders or no ideals, but lost in the hands of businesses. We see the rise of consortiums and agribusinesses as well as business despotism which now rules over the leaders of all countries. Since this stage is controlled by the laws established in the Mammalian phase, actors here tried for many years and by any means to circumvent and manipulate these laws. And they are succeeding. This is the phase our civilization is in now, although the majority of the population has been blocked to a maturity level of the Mammalian brain. This was caused by a general absence of identity (outing of religions and kings and establishment of financial globalization) and a consequential out of control consumerism, expression of an "hippopotamesque" masculine polarity.

The phase we are entering upon is that of the Universal brain. Since there is no more regulation on it due to the deficiency of the general

Idealistic self, the general system is spinning out of control. This can be observed through many symptomatic events but notably through this high level of depression throughout the world. In India alone this level reaches 36%. We can see this social phase active as groups from the population try to be heard as well as by the rise of environmental consciousness. This phase supersedes the previous ones. Here, businesses will fail. The environment, through natural catastrophes, general destruction of it and human physical and mental illnesses will win over them. They will be ruined. Since the Idealistic Self is the only one which can control this phase, our only hope is for a new ideology, a new way of life shared by a vast number. An ideology which echoes in everyone because it is the expression of the implicate order, of the model on which we and nature are all built.

In other words, our only hope for survival is through the awakening and strengthening of our general feminine polarity, of Isis.

Yes, Beauty can save the world.

Today, on this beautiful spring morning, I rose early. Sliding open the white shutter patio doors leading onto a lanai, I am greeted by a landscape of ocean, palm trees, and rolling fields. Stepping out, I breathe in the air charged with the wonderful fragrance of the blooming plumeria trees, I hear the breaking waves, and the singing of the birds of Hawaii. Looking afar to the ocean, I wonder if this morning again I will be welcomed by the display of my friends, the spinner dolphins, frolicking in the rising sun.

Sitting in a rattan chair, I happily await, admiring the architecture of this place. It is a two-story building and reminds me of a cloister. Indeed, people here seem quieter, more respectful of nature, more civilized in a sense. The sun appears over the rooftop. This is my last day here. Leaving this place always feels like dying a little. I swallow my sadness and invite in thankfulness.

I wish I could live here and be pregnant again; then this child I would bear, in all his cells, would have the imprint of this beautiful nature, of its sounds, smells, sights, and warmth. Together with my love and the *aloha spirit* of this island lost in the middle of the Pacific, this child would indeed be blessed and growing up would vibrate to Isis more

Isis Code

easily. He would desire to promote peace, he would become a true knight or, if a girl, a true lady. I do wish for every mother-to-be the opportunity to stand in such surroundings.

At least while I was here, all the cells newly born to my body received the imprint of the visible and invisible qualities of this place. These new cells bear this peace, this aloha spirit, these palm and plumeria trees, this ocean, these birds, and these dolphins I now see leaping in the distance.

Because of what I lived here, Isis is a little stronger. For there is only one Isis, made up of all the elements of creation, made up of all the flowers, of all these invisible stars above us, made up of all the people with their dreams and their hope. I am part of her, and so are you.

In our anonymous daily lives we can all strengthen the general feminine polarity by protecting nature and childhood, by expressing love and beauty, and by creating sacredness. And also, we really need this new woman and this new man, this new breed of knights and ladies, sitting around the round table of the world, working together, in harmony, as one.

Epilogue

Taijitu = Implicate order
= LIFE biosystem
= Fractal structure and function

Sierpinski triangle = Fractal form

One Force

Divides into two polarities
(feminine and masculine)

Each exists as a fractal

- Masculine aspect of feminine polarity
- Masculine polarity
- Feminine polarity
- Feminine aspect of masculine polarity

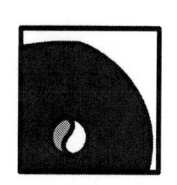

The fractal repeats indefinitely:
One Force divides into two polarities

diagram #1

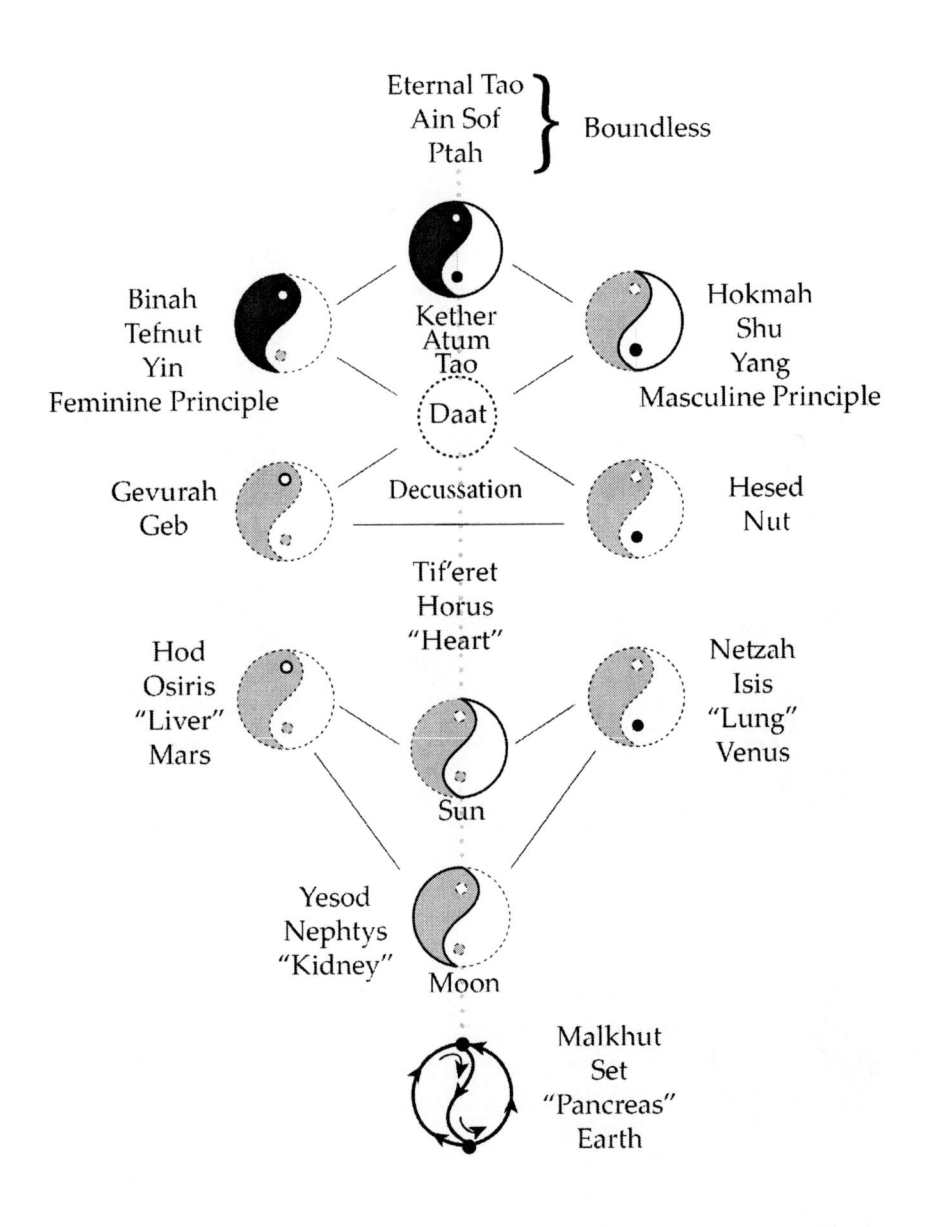

Egyptian Tradition, Tree of Life (Kabbalah), Tao and LIFE biosystem

diagram #2

Epilogue

4. Christianity
"Heart"
Holy Day: Sunday
Planet: Sun
Heart Aspect: Hochmah
Mental world-symbolic
aspect

5. Science, Confucianism,
Greek Philosophy,
Pragmatism
"Pancreas"
Holy Day: None or all
Mental World (Analysis)
Humanity is here but with
a global maturity of #3

2nd Cycle: must
have assimilated
all previous ones
to get there:
Islam "Lung"
Holy Day:
Friday (in
French, <u>ven</u>dredi)
Planet: Venus

3. Roman Empire
"Liver"
Planet: Mars
War (Lower Aspect)
and Discoveries
Emotional World

1. 1st Cycle:
Egyptian,
Sumerian,
Babylonian, Celtic,
animist traditions,
as well as
Zoroastrianism,
Taoism and
Hinduism.

2. Judaism
"Kidney"
Holy Day: Saturday
Planet: Saturn and
Moon (lower aspect)
Physical world
Higher Aspect: Binah

"People of the Book" in LIFE Biosystem

diagram #3

Isis Code

Levels of Incarnation:
Symbolic
Collective
Analytical
Emotional
Physical

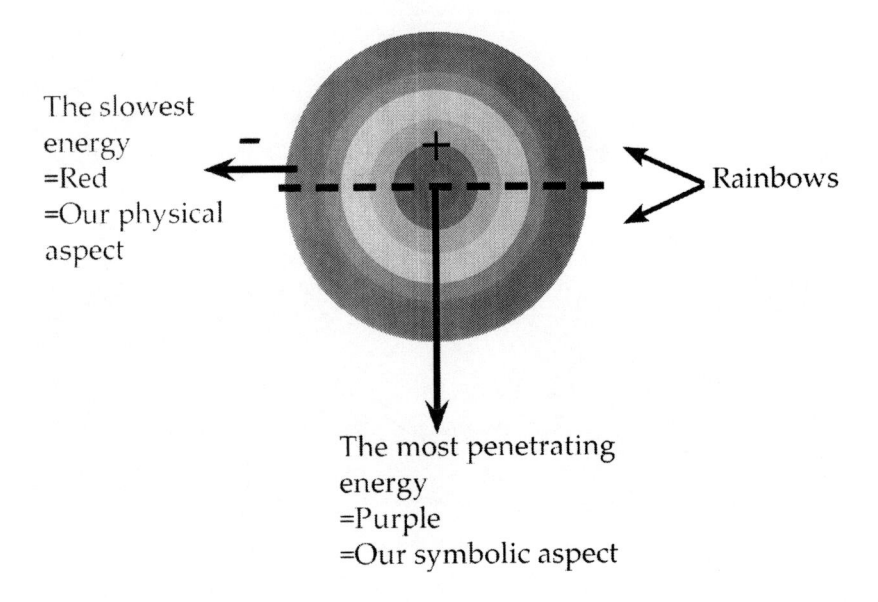

The slowest energy
=Red
=Our physical aspect

Rainbows

The most penetrating energy
=Purple
=Our symbolic aspect

Our symbolic aspect is central. Our physical aspect is the most receptive to subtle information as well as the most peripheral.

diagram #4

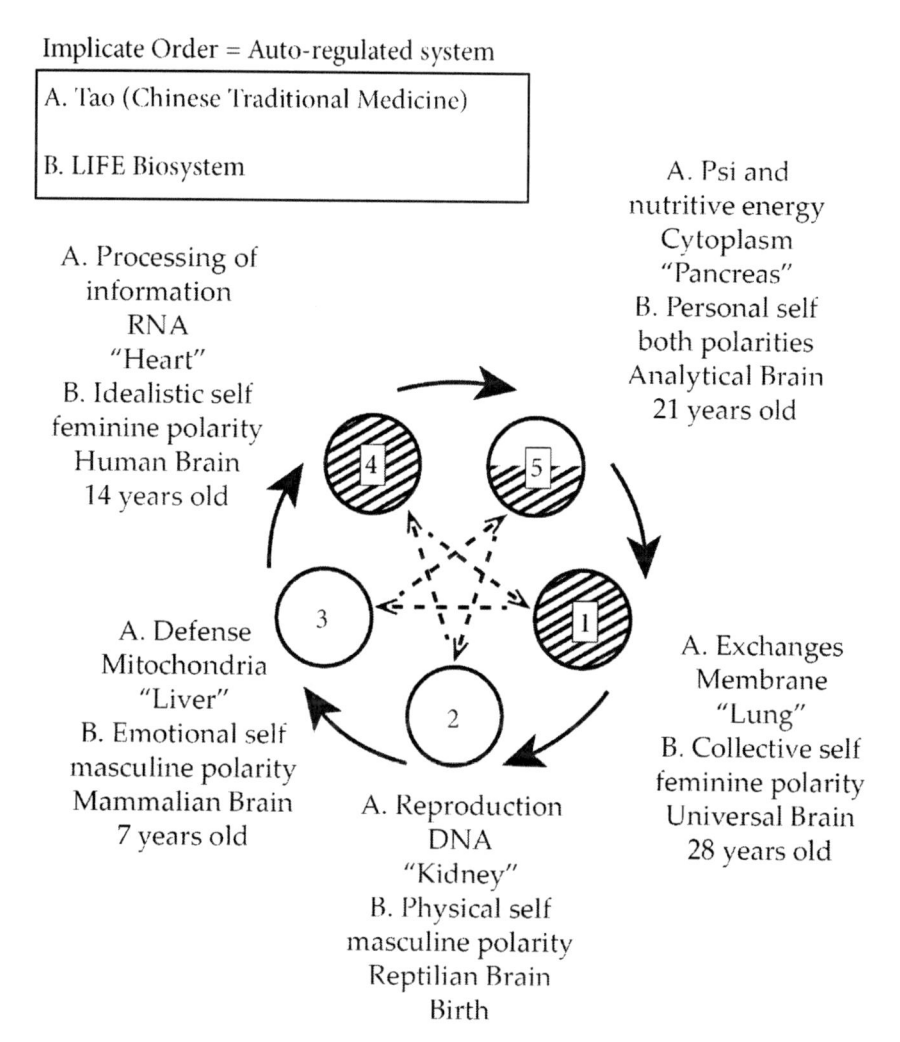

Epilogue

Implicate Order = Auto-regulated system

A. Tao (Chinese Traditional Medicine)

B. LIFE Biosystem

A. Processing of
information
RNA
"Heart"
B. Idealistic self
feminine polarity
Human Brain
14 years old

A. Psi and
nutritive energy
Cytoplasm
"Pancreas"
B. Personal self
both polarities
Analytical Brain
21 years old

A. Defense
Mitochondria
"Liver"
B. Emotional self
masculine polarity
Mammalian Brain
7 years old

A. Exchanges
Membrane
"Lung"
B. Collective self
feminine polarity
Universal Brain
28 years old

A. Reproduction
DNA
"Kidney"
B. Physical self
masculine polarity
Reptilian Brain
Birth

2+3+ △5 = Expressive nature,
Masculine polarity (as per
Medical Biocybernetic)

1+4+ ▽5 = Receptive nature,
Feminine polarity (as per
Medical Biocybernetic)

generation ⟶
inhibition ---→

diagram #5

575

Isis Code

Wave Function (oscillations) and LIFE biosystem

Electromagnetic Wave with LIFE Biosystem Elements

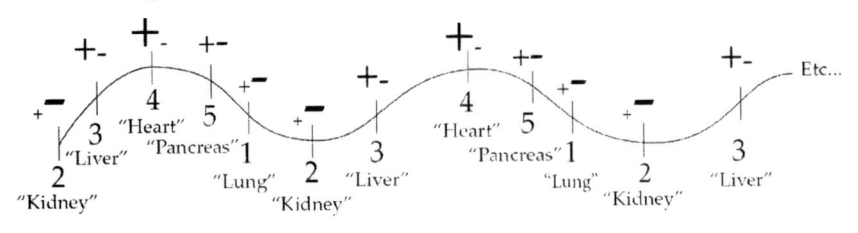

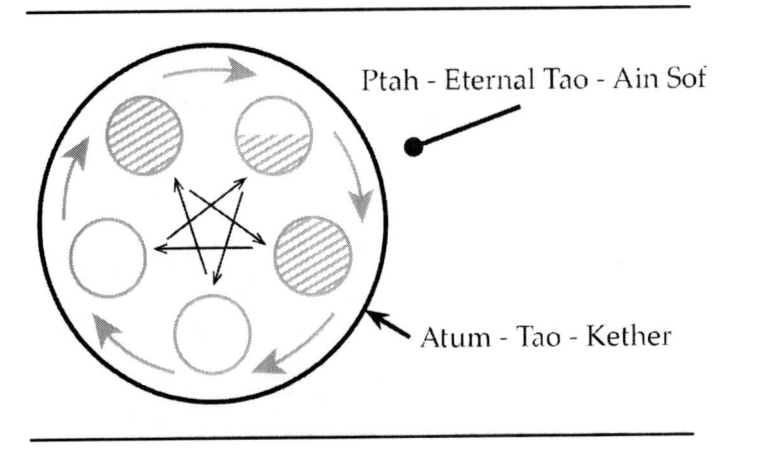

Ptah - Eternal Tao - Ain Sof

Atum - Tao - Kether

Hygieia Pentagram

Pythagorean Pentagram from the Pentemychos
or
Pentagram from Agrippa's Book

Different Representations of the Same System

diagram #6

Epilogue

"Heart"
=Center of the feminine polarity

"Pancreas"
=Center of the masculine polarity

"Liver"

"Lung"

"Kidney"

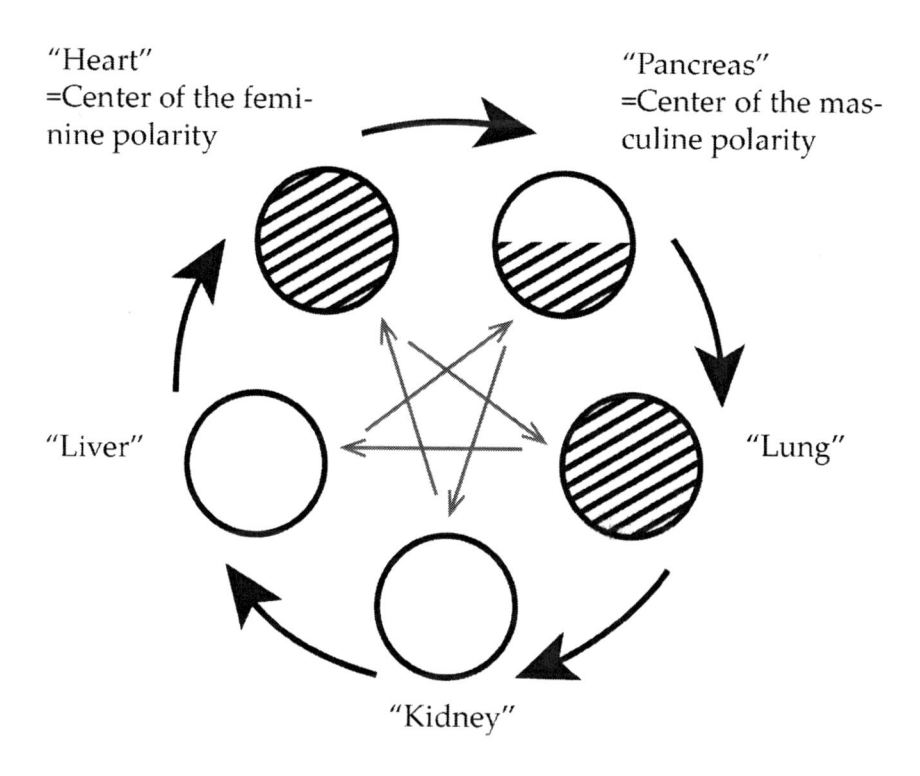

Femine polarity: "Lung" + "Heart" + "Pancreas" Yin, Master of The Heart and associated structures

Masculine polarity: "Liver" + "Kidney" + "Pancreas" Yang + Triple Warmer and associated structures

Feminine and Masculine Polarities

diagram #7

Isis Code

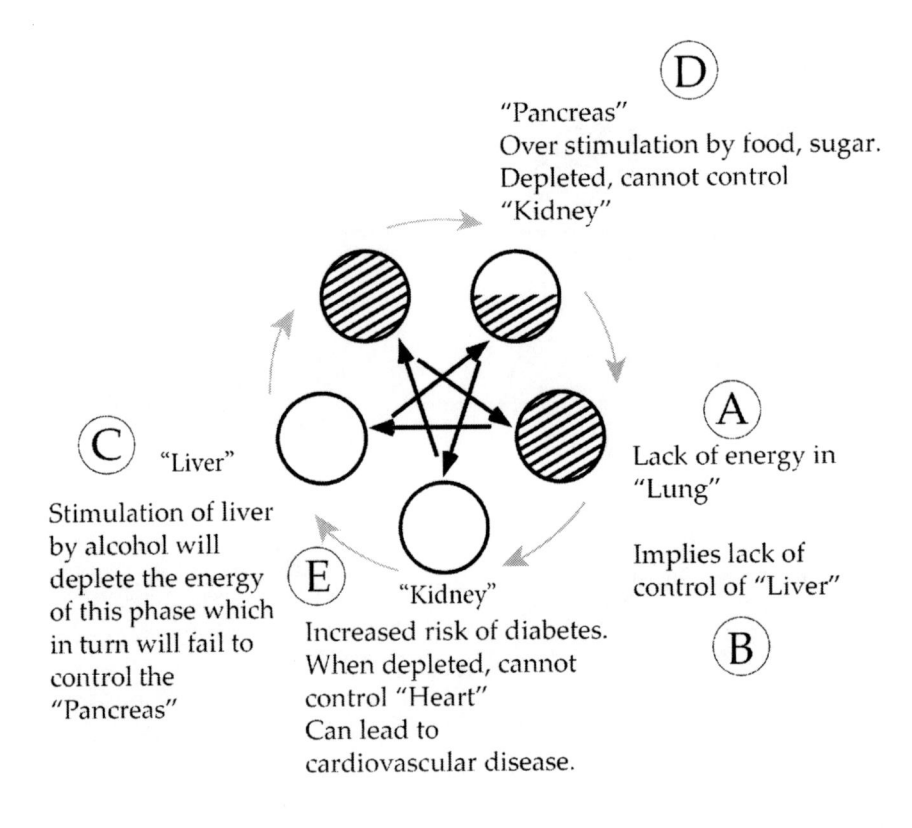

D
"Pancreas"
Over stimulation by food, sugar.
Depleted, cannot control
"Kidney"

C "Liver"
Stimulation of liver
by alcohol will
deplete the energy
of this phase which
in turn will fail to
control the
"Pancreas"

E "Kidney"
Increased risk of diabetes.
When depleted, cannot
control "Heart"
Can lead to
cardiovascular disease.

A
Lack of energy in
"Lung"

Implies lack of
control of "Liver"

B

Biopsychosocial (BPS) is found in the Medical Biocybernetic Model

*Feinberg School of Medicine test on 4,681 people"
What is true for the individual is also true for society

diagram #8

Epilogue

"Heart"
Center of Self

"Pancreas"
Center of self

"Liver"

"Lung"

"Kidney"

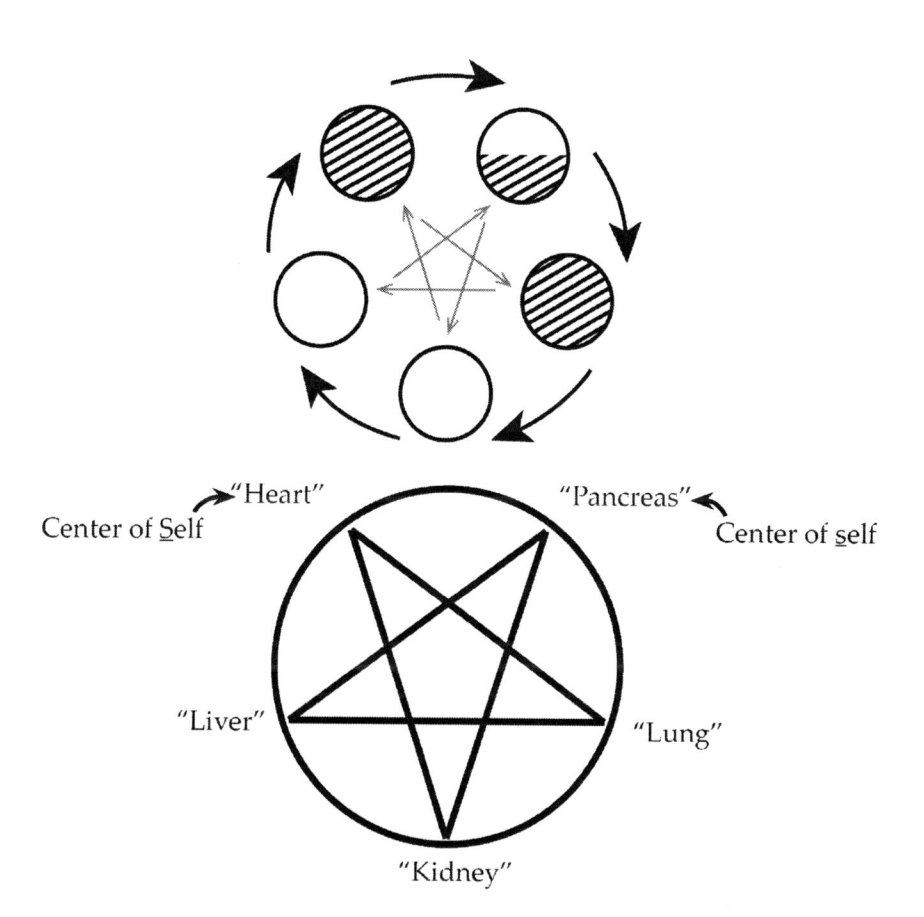

When both polarities work in harmony, the two centers are functional and the expression of the self is constructive and healthy = Dantian functionality and Perfect Human

diagram #9

Isis Code

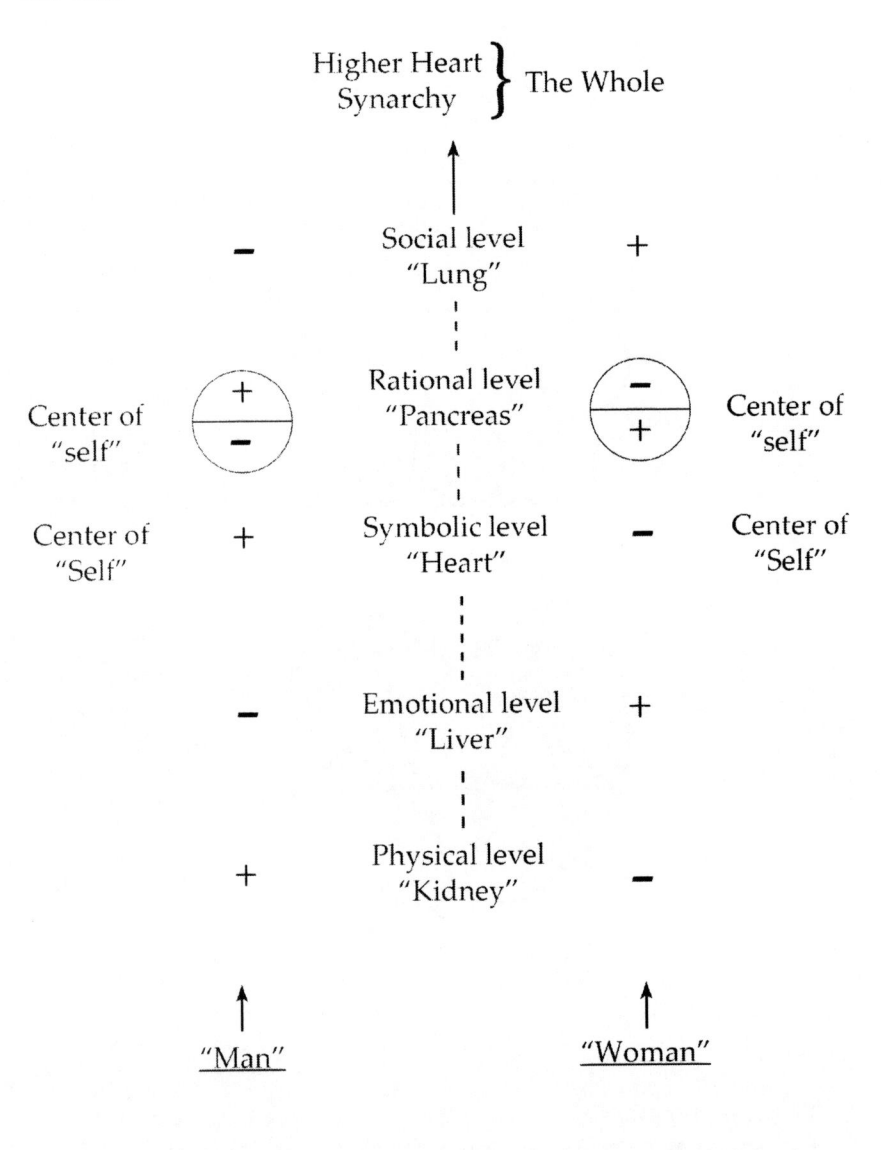

```
                    Higher Heart  ⎫
                    Synarchy      ⎬  The Whole
                                  ⎭
                           ↑
                           │
              −      Social level      +
                      "Lung"
                         ┊
                ⎛  +  ⎞                ⎛  −  ⎞
  Center of    ⎜─────⎟  Rational level ⎜─────⎟  Center of
   "self"      ⎝  −  ⎠    "Pancreas"   ⎝  +  ⎠   "self"
                                  
                         ┊
  Center of      +    Symbolic level    −    Center of
   "Self"              "Heart"                "Self"
                         ┊
                         ┊
              −      Emotional level    +
                      "Liver"
                         ┊
                         ┊
              +      Physical level     −
                      "Kidney"
                         
                 ↑                     ↑
               "Man"                "Woman"
```

+ = Expressive of energy (output)
− = Receptive of energy (input)

Man and Woman: Receptivity, Expressiveness

diagram #10

Epilogue

Human
(14 - 21 y. old)
Parietal Lobe

Analytical
(21 - 28 y. old)
Frontal Lobe

Mammalian
(7 - 14 y. old)
Occipital Lobe

Reptilian
(0 - 7 y. old)
Cerebellum

Universal
(28 - 34 y. old)
Temporal Lobe

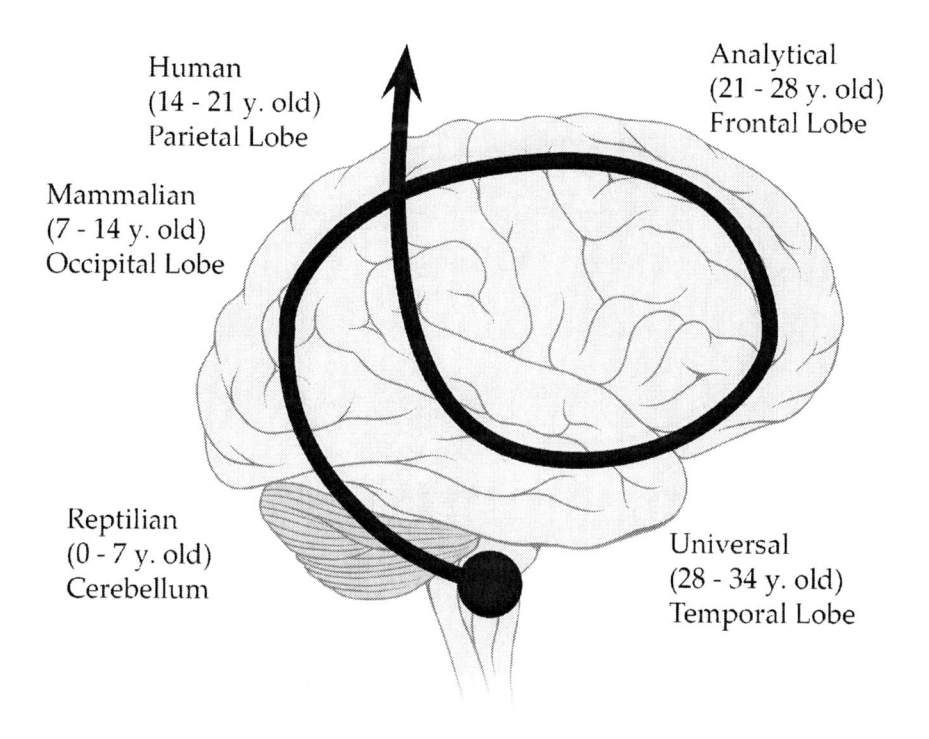

The last aspect to mature is the corpus callosum

Maturation and Evolution of the Human Brain

diagram #11

Isis Code

Synarchic brain	"I am"	Conscious Self	Complete human	Master of the Heart
Reptilian brain	"What?" "Where?"	Physical self	Instinctual Human	Kidney
Mammalian brain	"When?"	Emotional self	Emotional Human	Liver
Human brain	"Why?"	Idealistic self	Symbolic Human	Heart
Analytical brain	"How?"	Personal self	Rational Human	Pancreas
Universal brain	"Who?"	Social self	Social Human	Lung

Human
Idealistic self

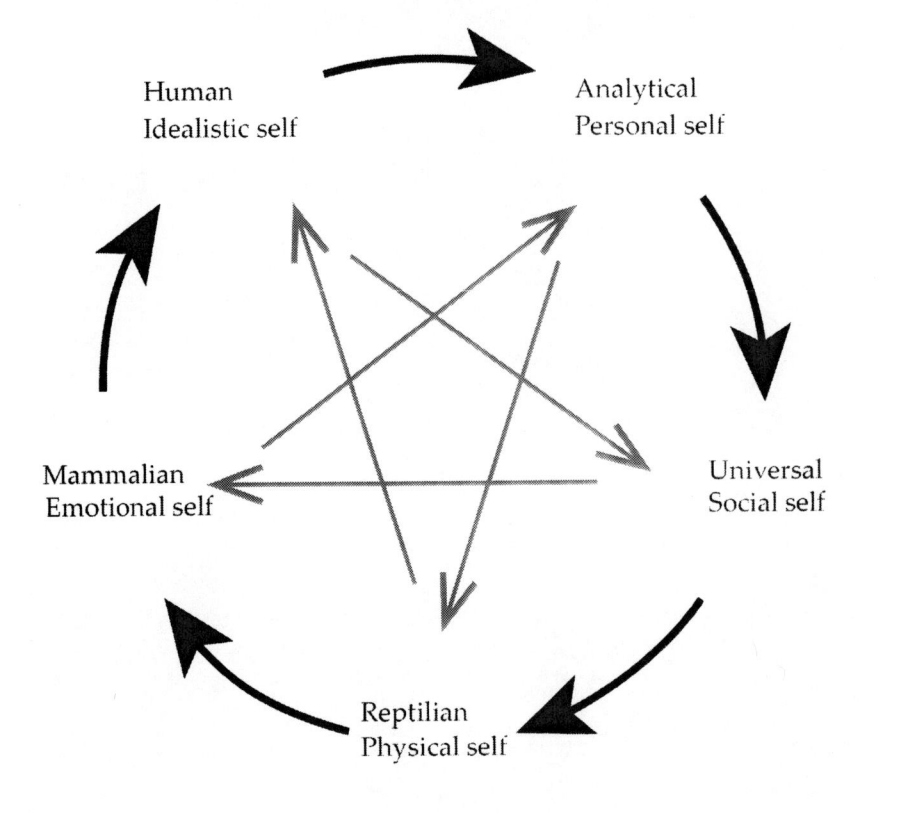

Analytical
Personal self

Mammalian
Emotional self

Universal
Social self

Reptilian
Physical self

Healthy and Mature LIFE Biosystem

diagram #12

Epilogue

Human brain
Upright Position
Up/Down
"Why?"

Analytical brain
"How?"

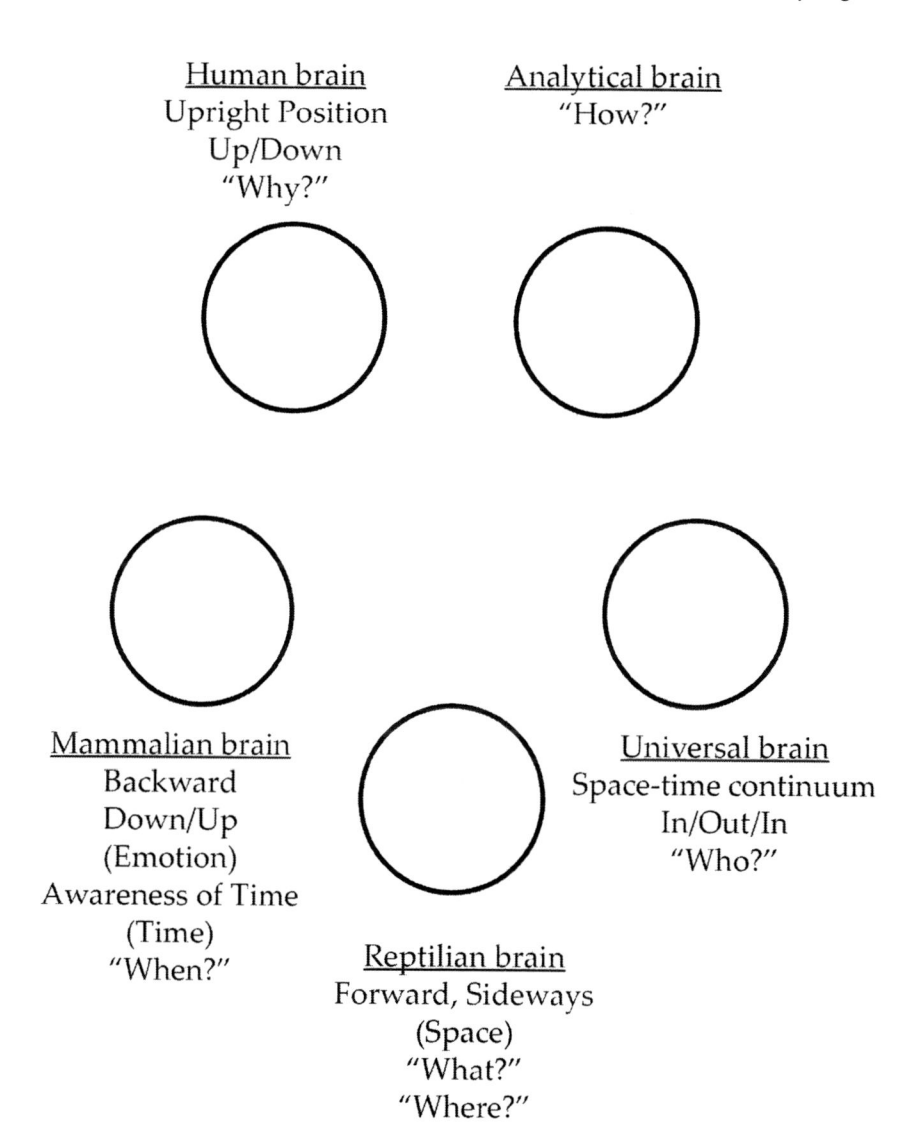

Mammalian brain
Backward
Down/Up
(Emotion)
Awareness of Time
(Time)
"When?"

Universal brain
Space-time continuum
In/Out/In
"Who?"

Reptilian brain
Forward, Sideways
(Space)
"What?"
"Where?"

All the phases expressed in harmony= "I am"

Diagram #13

Isis Code

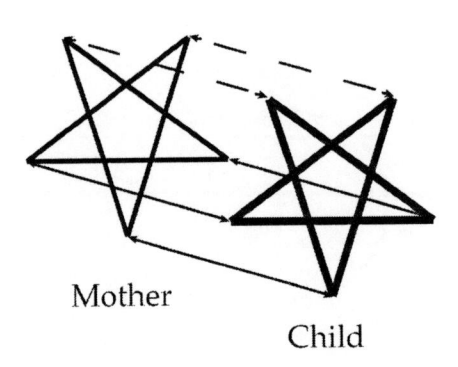

During pregnancy, her feminine and masculine polarities influence the flowing of energy in the child's LIFE biosystem.

Mother

Child

The child's LIFE biosystem mirrors that of the mother's and synchronizes to it. It will slowly become independant but always linked through the Master of the Heart.

Mother and Child

diagram #14

Epilogue

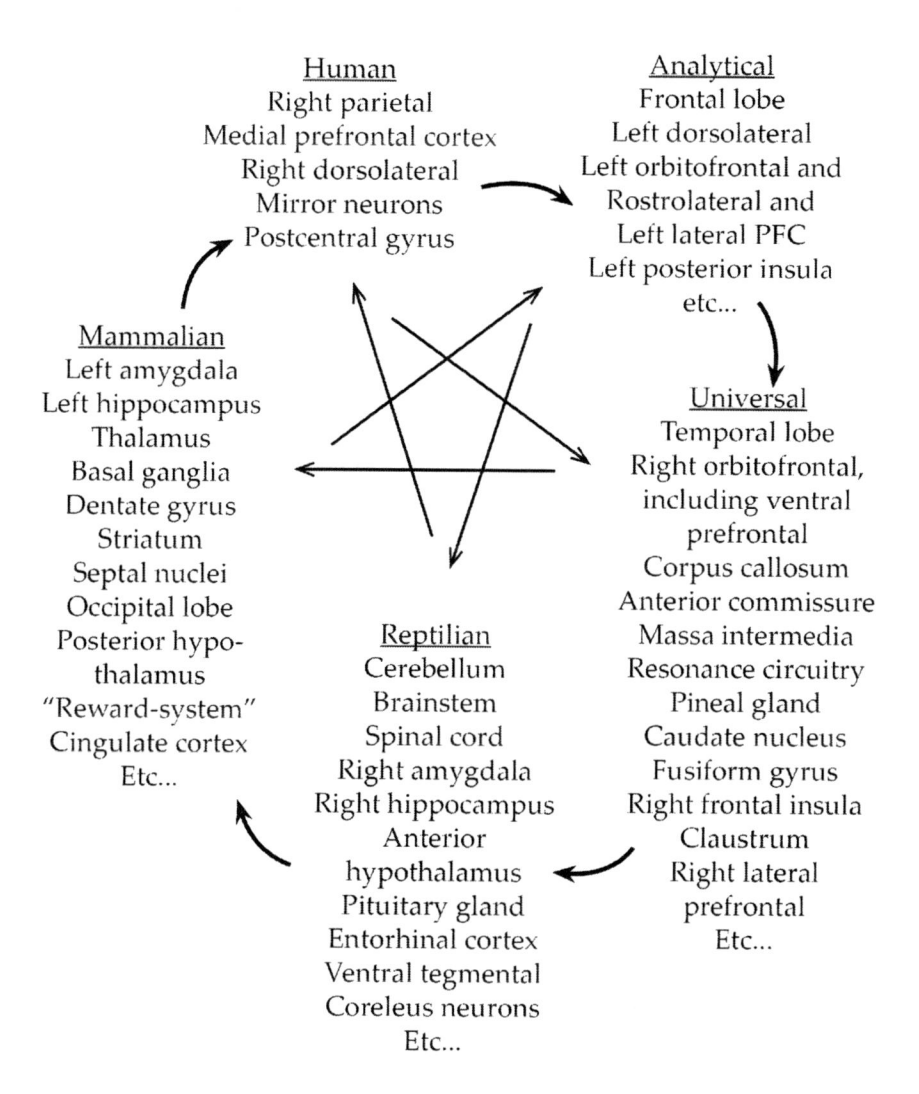

<u>Human</u>
Right parietal
Medial prefrontal cortex
Right dorsolateral
Mirror neurons
Postcentral gyrus

<u>Analytical</u>
Frontal lobe
Left dorsolateral
Left orbitofrontal and
Rostrolateral and
Left lateral PFC
Left posterior insula
etc...

<u>Mammalian</u>
Left amygdala
Left hippocampus
Thalamus
Basal ganglia
Dentate gyrus
Striatum
Septal nuclei
Occipital lobe
Posterior hypo-
thalamus
"Reward-system"
Cingulate cortex
Etc...

<u>Reptilian</u>
Cerebellum
Brainstem
Spinal cord
Right amygdala
Right hippocampus
Anterior
hypothalamus
Pituitary gland
Entorhinal cortex
Ventral tegmental
Coreleus neurons
Etc...

<u>Universal</u>
Temporal lobe
Right orbitofrontal,
including ventral
prefrontal
Corpus callosum
Anterior commissure
Massa intermedia
Resonance circuitry
Pineal gland
Caudate nucleus
Fusiform gyrus
Right frontal insula
Claustrum
Right lateral
prefrontal
Etc...

Different brain aspects as per LIFE biosystem and inhibitions of different aspects of the brain

Masculine polarity = Reptilian + Mammalian + Part of Analytical

Feminine polarity = Human + Universal + Part of Analytical

diagram #15

Isis Code

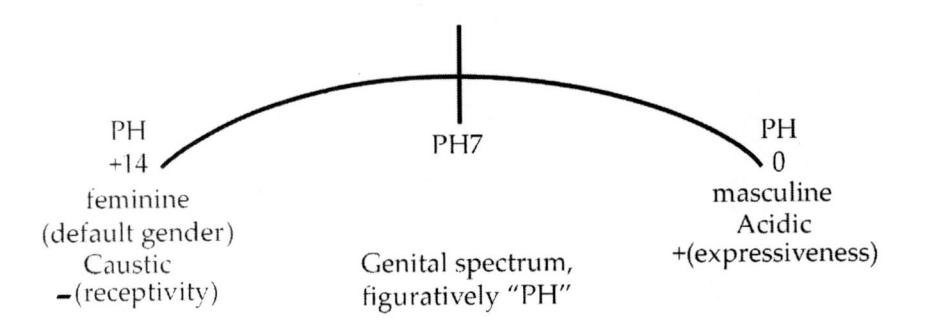

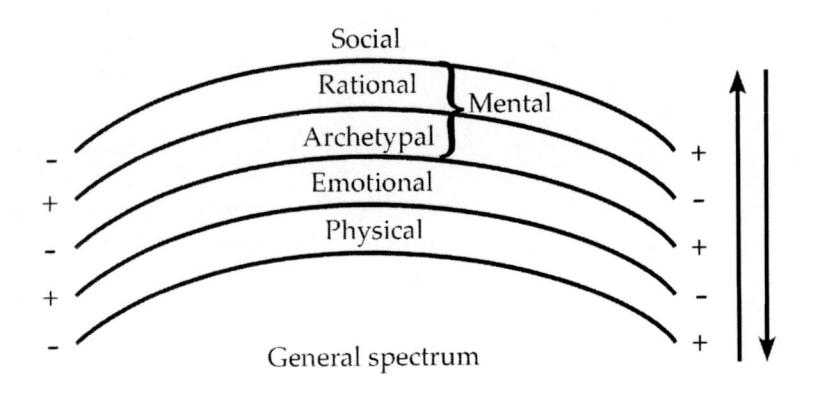

PH
+14
feminine
(default gender)
Caustic
−(receptivity)

PH7

Genital spectrum,
figuratively "PH"

PH
0
masculine
Acidic
+(expressiveness)

Social
Rational
Archetypal } Mental
Emotional
Physical

General spectrum

diagram #16

Epilogue

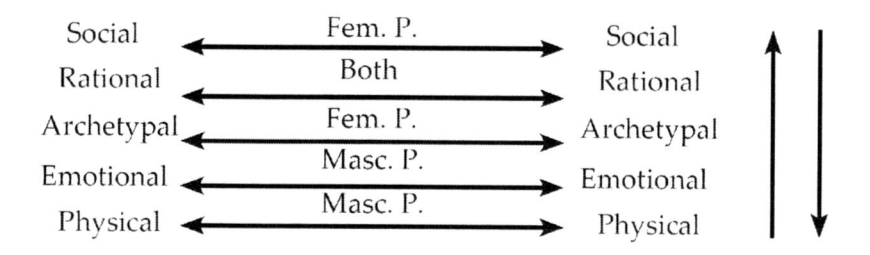

Social	← Fem. P. →	Social
Rational	← Both →	Rational
Archetypal	← Fem. P. →	Archetypal
Emotional	← Masc. P. →	Emotional
Physical	← Masc. P. →	Physical

Two people meeting, their masculine and feminine polarities vibrate together. There is no dynamism between "mirrors". The difference in potential explains attraction and repulsion.

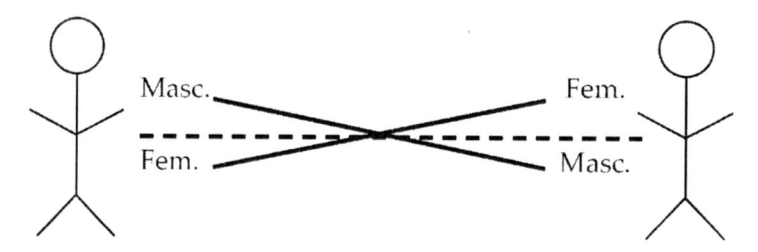

Two masculine polarities interact with two feminine polarities.

diagram #17

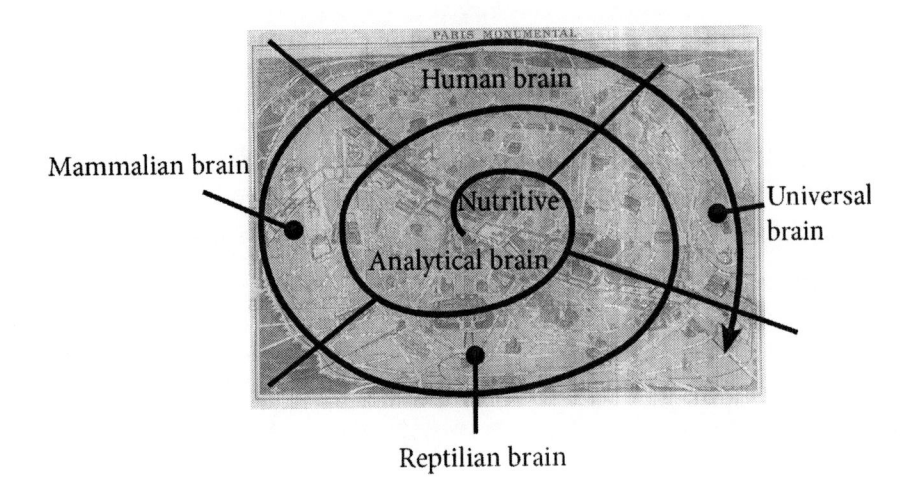

Isis Code

Diagram with concentric ovals over a map labeled "PARIS MONUMENTAL", showing: Human brain, Mammalian brain, Nutritive, Analytical brain, Universal brain, Reptilian brain

1- Maternities (major), catacombs, jail, cemeteries (Reptilian brain)
2- Sports, defense, sexuality, exhibition, war museum, Eifel Tower (Mammalian brain)
3- EPRA (Radios), Train Stations, Communications (Human brain)
4- Food Market (Les Halles), Main Universities (Analytical brain)
5- Natural parks - Zoo de Vincennes (Universal brain)

Applied Medical Biocybernetic: Paris & the LIFE biosystem

diagram #18

Epilogue

"LIVER" (sprouting) Generated by "KIDNEY" (seed, water)	"Let the water under the sky be gathered to one place, and let dry ground appear." And it was so. [10] God called the dry ground "land," and the gathered waters he called "seas." And God saw that it was good. [11] Then God said, "Let the **land produce vegetation**: seed-bearing plants and trees on the land that bear fruit with seed in it, according to their various kinds." And it was so. [12] **The land produced** vegetation: plants bearing seed according to their kinds and trees bearing fruit with seed in it according to their kinds. And there was evening, and there was morning—the third day.
"HEART" (light, information) The most "yang"	[14] And God said, "Let there be **lights in the vault** of the sky to separate the day from the night, and let them **serve as signs** to **mark sacred times**, and days and years, [15] and let them be **lights in the vault of the sky to give light on the earth**." And it was so. [16] God made two great lights—the greater light to govern the day and the lesser light to govern the night. He also made the stars. [17] God set them in the vault of the sky to give light on the earth, [18] to govern the day and the night, and **to separate light from darkness**. And God saw that it was good. [19] And there was evening, and there was morning—the fourth day.
"LUNG" (air, collectivity)	[20] And God said, "Let the **water teem with living creatures**, and let **birds** fly above the earth across the vault of the sky." [21] So God created the great creatures of the sea and **every living thing with which the water teems** and that moves about in it, **according to their kinds**, and every winged bird according to its kind. And God saw that it was good. [22] God blessed them and said, "Be fruitful and increase in number and **fill the water in the seas, and let the birds increase** on the earth." [23] And there was evening, and there was morning—the fifth day.

diagram #19.1

Isis Code

"PANCRE-AS" The regulator and most expressed bearer of the implicate order, holds the two polarities, nutrition phase	[24] And God said, "Let the land produce living creatures according to their kinds: the livestock, the creatures that move along the ground, and the wild animals, each **according to its kind**." And it was so. [25] God made the wild animals **according to their kinds**, the livestock **according to their kinds**, and all the creatures that move along the ground **according to their kinds**. And God saw that it was good. [26] Then God said, "Let us make **mankind in our image**, in our likeness, so that they may **rule over** the fish in the sea and the birds in the sky, over the livestock and all the wild animals, and over all the creatures that move along the ground." [27] So God created mankind in his own image, in the image of God he created them; **male and female he created them.** [29] Then God said, "**I give you every seed-bearing plant on the face of the whole earth and every tree that has fruit with seed in it. They will be yours for food.** [30] And to all the beasts of the earth and all the birds in the sky and all the creatures that move along the ground—everything that has the breath of life in it—I give every green plant for food." And it was so.

diagram #19.2

Epilogue

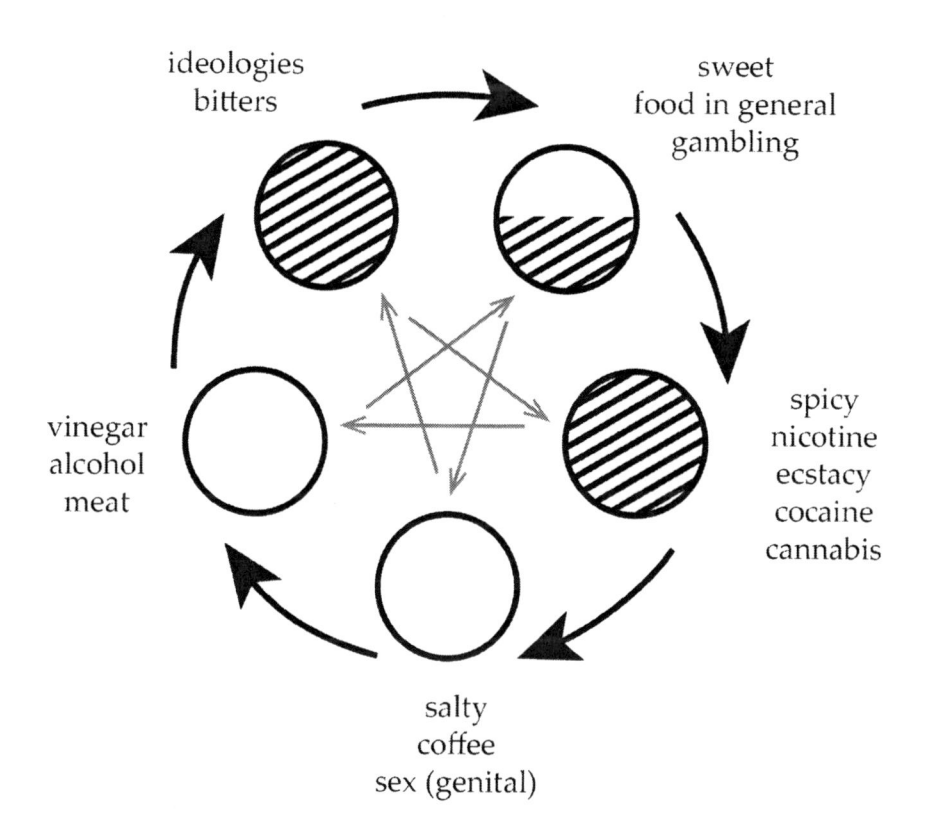

ideologies
bitters

sweet
food in general
gambling

spicy
nicotine
ecstacy
cocaine
cannabis

salty
coffee
sex (genital)

vinegar
alcohol
meat

Stimulation of the Different Phases
Through Taste and Main Addictions.

diagram #20

Isis Code

Archetypal Symbolism = Masculine aspect of the feminine polarity. Women project this aspect on men (Tif'eret, the Sun) ≠ Masculine polarity nor men.

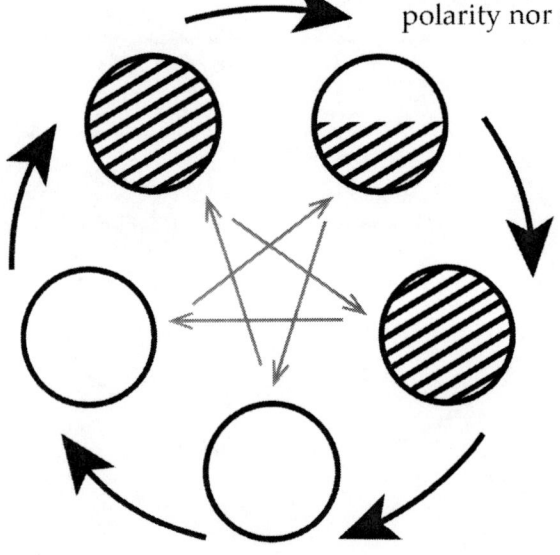

Genital = Feminine aspect of the masculine polarity. Men project this aspect on women (Yesod, the moon) ≠ Feminine polarity nor women

Because women identify with how they are considered, men have masculinized women.

diagram #21

Epilogue

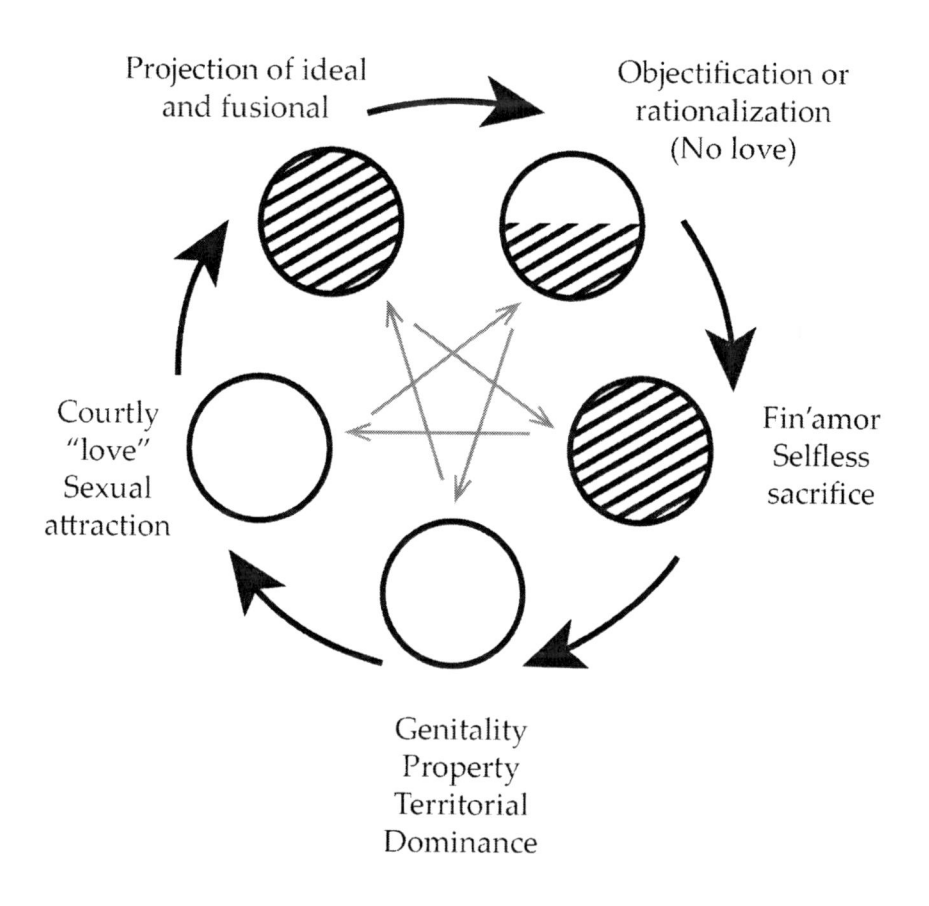

Projection of ideal and fusional

Objectification or rationalization (No love)

Fin'amor Selfless sacrifice

Genitality
Property
Territorial
Dominance

Courtly "love" Sexual attraction

Types of Love and Love Cycle

diagram #22

Isis Code

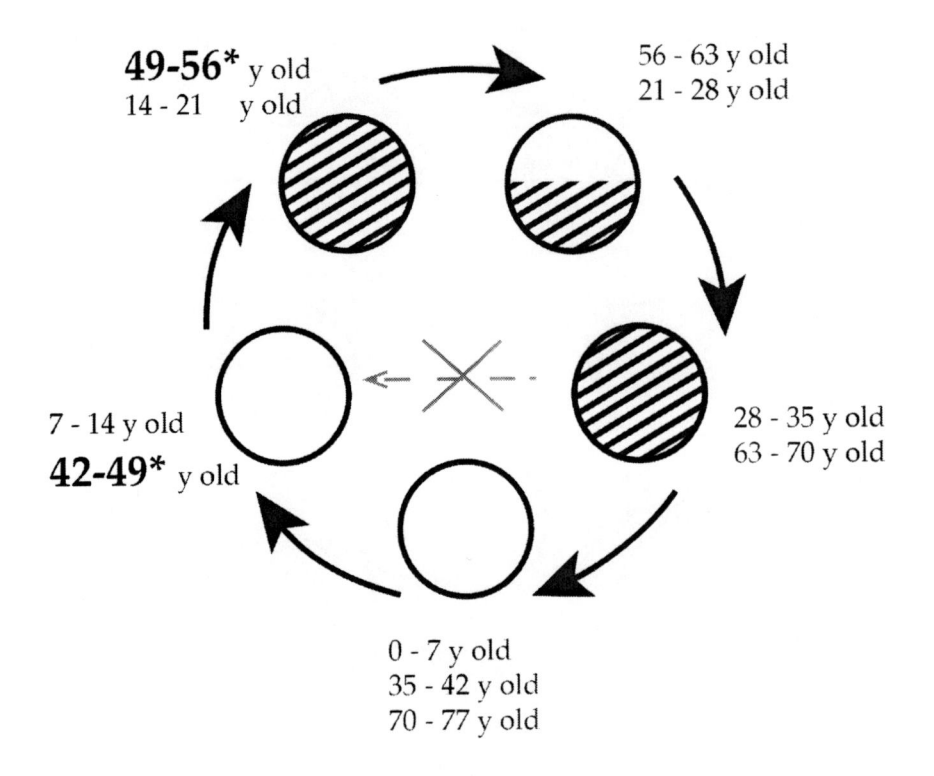

49-56* y old
14 - 21 y old

56 - 63 y old
21 - 28 y old

7 - 14 y old
42-49* y old

28 - 35 y old
63 - 70 y old

0 - 7 y old
35 - 42 y old
70 - 77 y old

*Midlife depression in women; an identity problem as well as a social problem with the feminine polarity

Ages and Phases

diagram #23

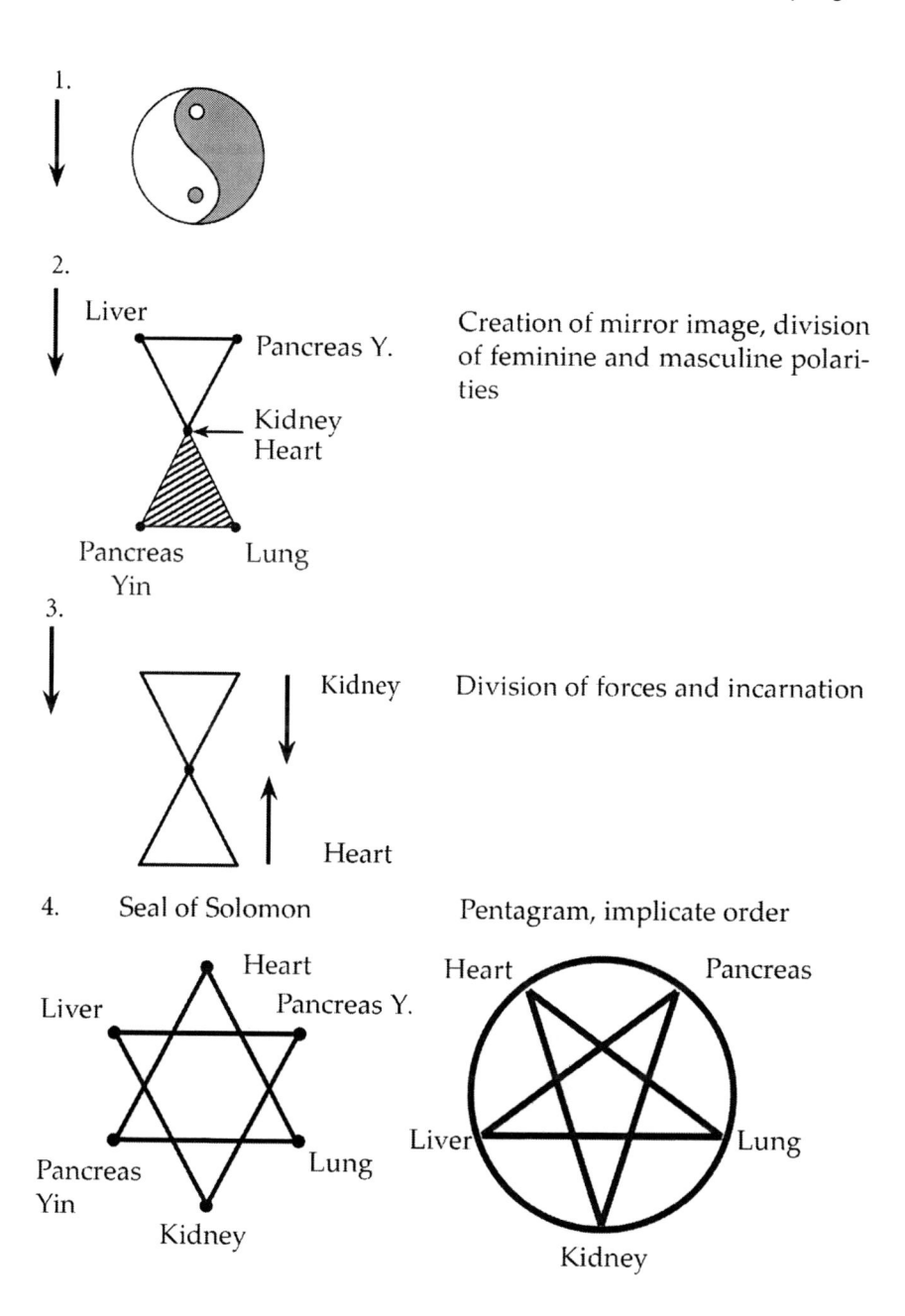

Epilogue

1. ↓

2. ↓   Liver   Pancreas Y.   Creation of mirror image, division of feminine and masculine polarities
Kidney
Heart
Pancreas Yin   Lung

3. ↓   Kidney   Division of forces and incarnation
Heart

4. Seal of Solomon   Pentagram, implicate order

Heart   Pancreas Y.   Heart   Pancreas
Liver   Liver   Lung
Pancreas Yin   Lung
Kidney   Kidney

Seal of Solomon

diagram #24

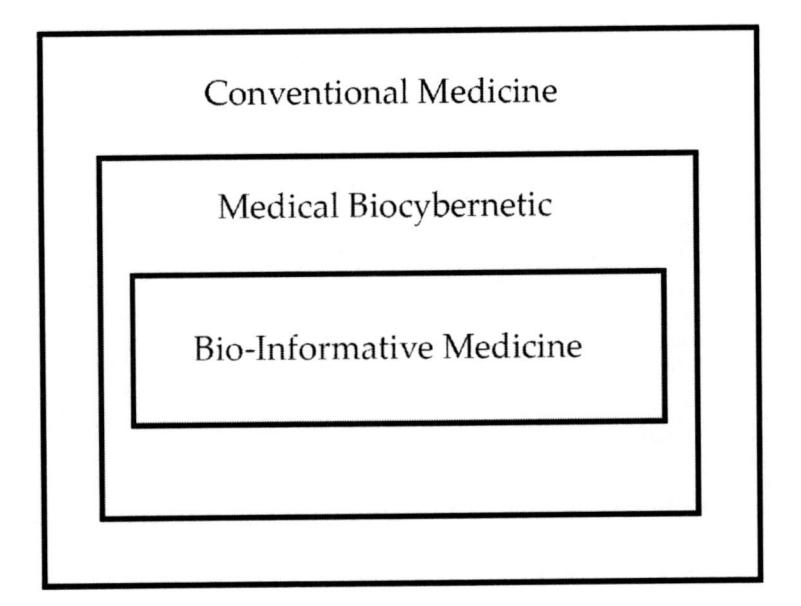

Isis Code

```
┌─────────────────────────────────────┐
│        Conventional Medicine        │
│   ┌─────────────────────────────┐   │
│   │     Medical Biocybernetic   │   │
│   │  ┌───────────────────────┐  │   │
│   │  │  Bio-Informative Medicine │   │
│   │  └───────────────────────┘  │   │
│   └─────────────────────────────┘   │
└─────────────────────────────────────┘
```

Ultimate Conventional Medicine

Diagram #25

Epilogue

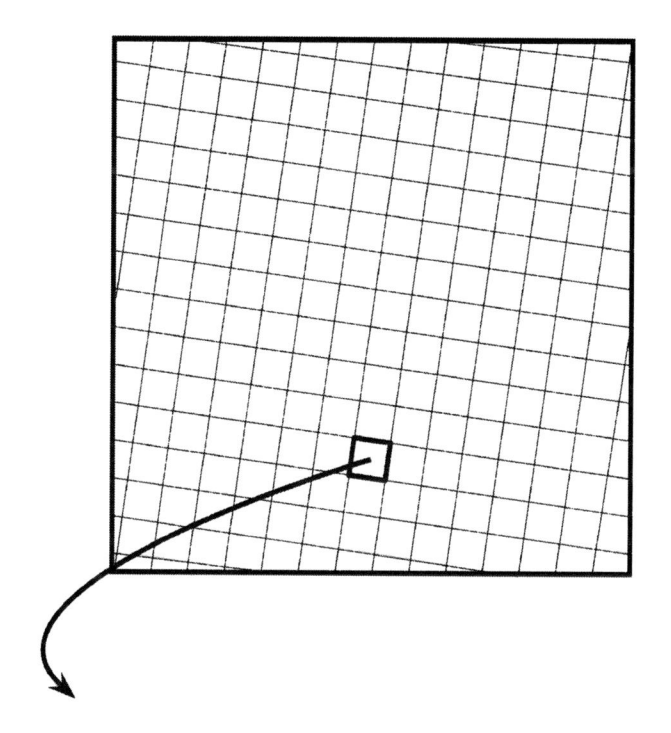

Smallest aspect of Master of the Heart
Similar to the cell from a beehive
Interfaces with Triple Warmer

Master of the Heart
also known as Akasha of Hindu Tradition

diagram #26

Isis Code

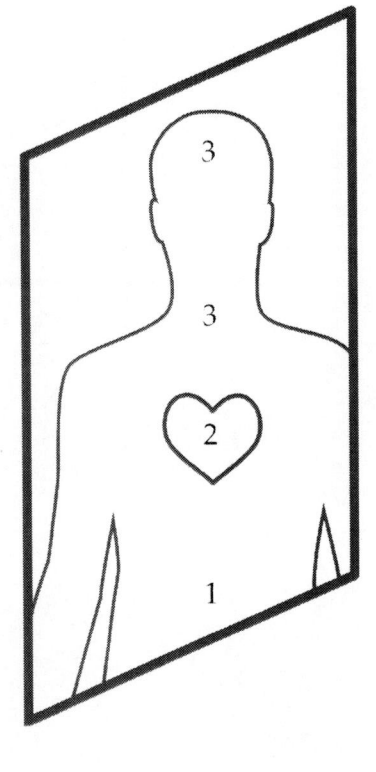

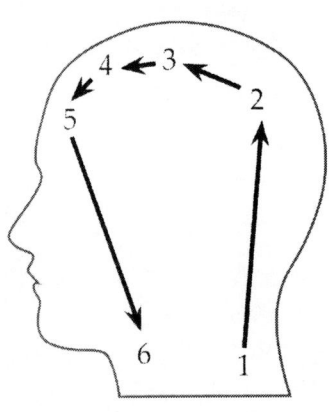

1. Absorption of energy through the skin, the celiac plexus, and the eyes.

2. Assimilation of energy through the cardiac plexus, the blood, heart, and lungs.

3. Regulation of energy through the pineal gland, as well as the carotid and celiac plexuses.

Elimination of energies by speech, thoughts, feelings, facial expressions, and general behavior.

1. Inhibitive or stimulating information received from the environment through the organs dedicated to energy absorption.

2. Superior cervical sympathetic ganglion

3. Pineal gland

4. Hypothalamus

5. Hypophysis

6. Endocrine glands

Master of the Heart, schematized physiology, as per medical biocybernetic

diagram #27

Epilogue

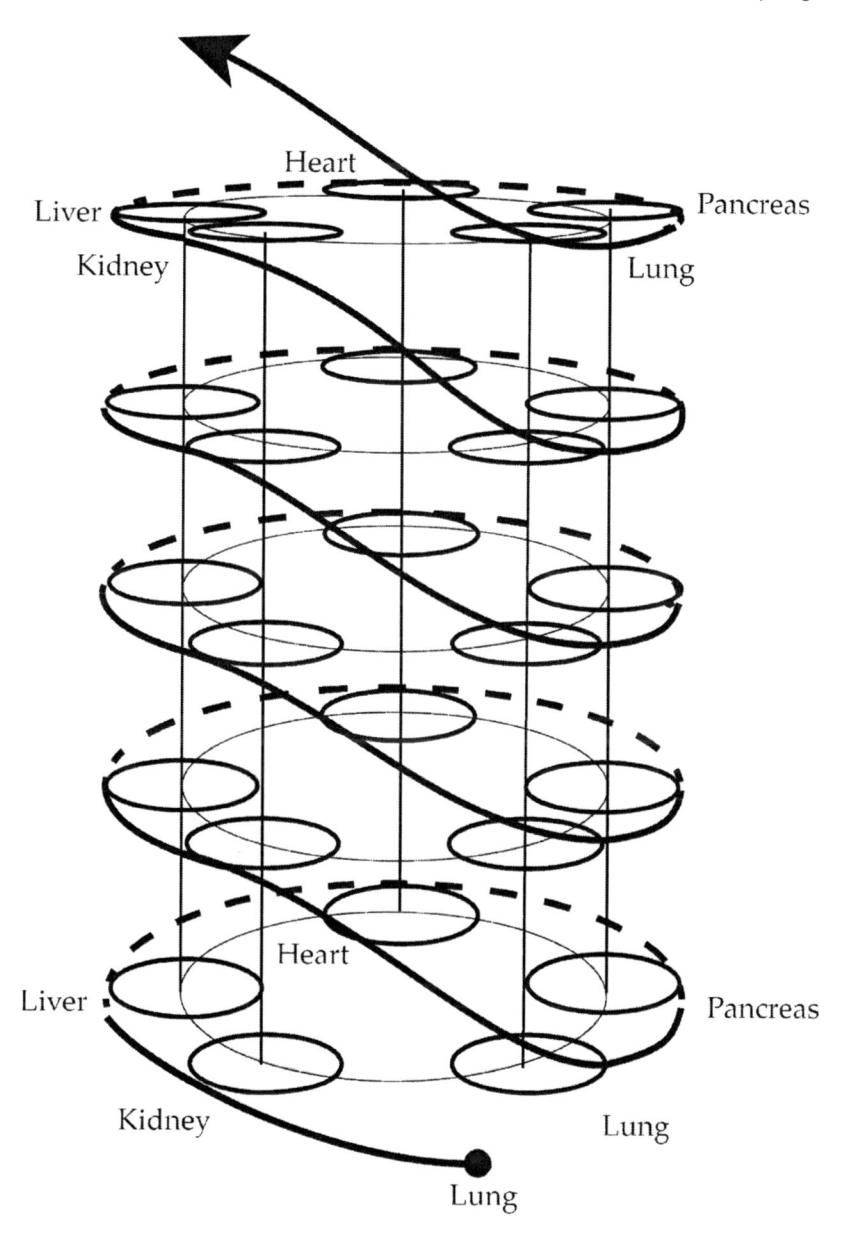

Spiral of Evolution

diagram #28

Isis Code

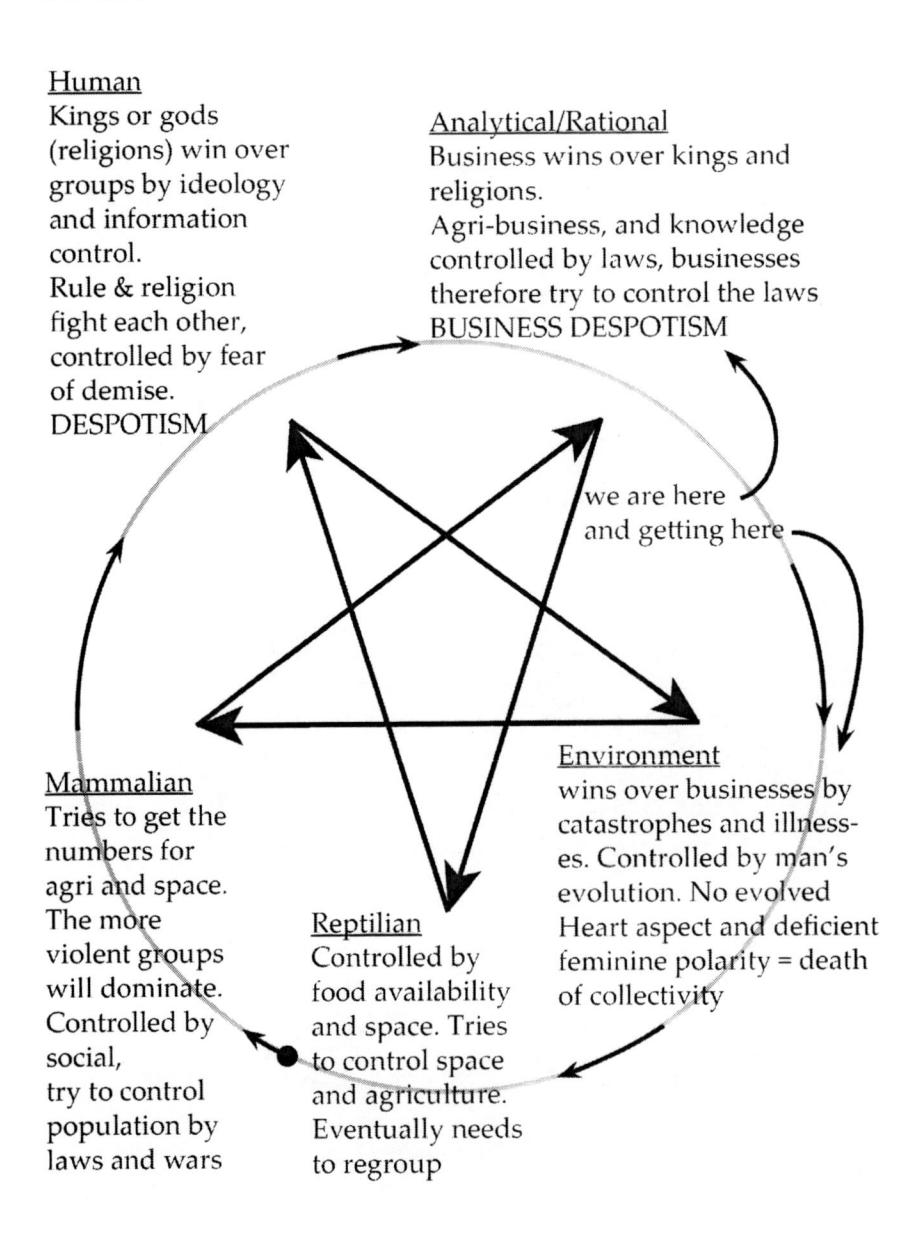

Human
Kings or gods (religions) win over groups by ideology and information control.
Rule & religion fight each other, controlled by fear of demise.
DESPOTISM

Analytical/Rational
Business wins over kings and religions.
Agri-business, and knowledge controlled by laws, businesses therefore try to control the laws
BUSINESS DESPOTISM

we are here
and getting here

Mammalian
Tries to get the numbers for agri and space. The more violent groups will dominate. Controlled by social, try to control population by laws and wars

Reptilian
Controlled by food availability and space. Tries to control space and agriculture. Eventually needs to regroup

Environment
wins over businesses by catastrophes and illnesses. Controlled by man's evolution. No evolved Heart aspect and deficient feminine polarity = death of collectivity

The feminine polarity allows proper functioning of masculine polarity.

Phases in society when feminine polarity is deficient.

diagram #29

Appendix

In July 1991 a multidisciplinary group of experts researching endocrine disruption reached consensus and published "The Wingspread Conference Statement."[552] This statement sent shock waves through the international scientific and regulatory communities. The scientists stated that the "cancer paradigm is insufficient" when assessing chemicals. That certain "chemicals can cause severe health effects other than cancer" that are of "a profound and insidious nature." The scientists stated,

We are certain of the following:

- A large number of man-made chemicals that have been released into the environment, as well as a few natural ones, have the potential to disrupt the endocrine system of animals, including humans. Among these are the persistent, bioaccumulative, organohalogen compounds that include some pesticides (fungicides, herbicides, and insecticides) and industrial chemicals, other synthetic products, and some metals.
- Many wildlife populations are already affected by these compounds. The impacts include thyroid dysfunction in birds and fish; decreased fertility in birds, fish, shellfish, and mammals; decreased hatching success in birds, fish, and turtles; gross birth deformities in birds, fish, and turtles; metabolic abnormalities in birds, fish, and mammals; behavioral abnormalities in birds; demasculinization and feminization of male fish, birds, and mammals; defeminization and masculinization of female fish and birds; and compromised immune systems in birds and mammals.
- The patterns of effects vary among species and among compounds. Four general points can nonetheless be made:

(1) The chemicals of concern may have entirely different effects on the embryo, fetus, or perinatal organism than on the adult;

(2) The effects are most often manifested in offspring, not in the exposed parent;

(3) The timing of exposure in the developing organism is crucial in determining its character and future potential; and

(4) Although critical exposure occurs during embryonic development, obvious manifestations may not occur until maturity.

(5) Pesticides in the organism hijack some hormones receptors thus jeopardize the action of natural hormones.

- Laboratory studies corroborate the abnormal sexual development observed in the field and provide biological mechanisms to explain the observation in wildlife.

- Humans have been affected by compounds of this nature too. The effects of DES (diethylstilbestrol), a synthetic therapeutic agent, like many of the compounds mentioned above, are estrogenic. Daughters born to mothers who took DES now suffer increased rates of vaginal clear cell adenocarcinoma, various genital tract abnormalities, abnormal pregnancies, and some changes in immune responses. Both sons and daughters exposed in utero experience congenital anomalies of their reproductive system and reduced fertility. The effects seen in utero DES-exposed humans parallel those found in contaminated wildlife and laboratory animals, suggesting that humans may be at risk to the same environmental hazards as wildlife.[553]

Pesticides and herbicides also leave residues in the air, soil, and water. Children are real sponges for these, and we can link these pesticides and herbicides to increases in childhood cancers. I always thought Canada was somewhat protected and better than the United States on this level. Research has shown that the levels of pesticides in children are twice as high here! DDT, which was banned more than thirty-five years ago, still shows in 99% of the population. Our imports from countries that use those chemicals after they were banned here can explain this. This is one more reason to eat locally grown food. In children six to eleven years old, the level of chlorpyrifos for corn, wheat, soy, and cotton culture has shown up in 93% of the samples and are four times higher than the admissible level. In a study, 287 chemical products were found in the umbilical cord. Of these, 180 are proven carcinogens, 217 are neurotoxic, and 208 are teratogen (congenital malformations). Dr. Eric Dewailly of the National Institute of Public Health (Quebec) made three thousand samples ranging over ten years. Almost anything in the mother goes to the embryo. We have no clue how these different chemicals interact in the developing organism.

Despite some efforts, a lot still has to be done to protect our children and the environment. More disturbing to me is the fact that many manufacturers of fluorinated pesticides are also pharmaceutical companies: Bayer, DuPont, Novartis (which even sponsors a website against breast cancer), Bell, Solvay, Zeneca, to name a few. We already know that companies specializing in genetic engineering are often pesticide and herbicide producers. We have to be realistic; they are in it to sell more, not less. There are obvious conflicts of interests here, given the results of so much research. But they need to please the shareholders: us.

With the information I've gathered, I do not wish to eat genetically engineered food, but the government does not give me choice. In Canada, 85% of the population asked for the labeling of genetically engineered products. When it was supposed to be discussed in Parliament, everyone had more urgent matters to deal with, the spotlight went elsewhere, and the "against labeling" bill rapidly passed, unnoticed.

In my opinion, it is totally undemocratic that mothers are not allowed to know what they are feeding their kids. It shouldn't even be questioned. This was very painful to me then, as it is now, I am sure, for many other mothers.

Organic food is good for the planet and good for us. Some say that the crop yield is much lower with organic produce. That's like saying an athlete runs faster with drugs and a whip, so let's beat him and give him drugs. But what is the end result of conventional agriculture in fifteen years? We can see it now.

There is a difference between genetically manipulated and genetically engineered. Of course, everything on earth has been genetically manipulated by the environment, weather, circumstances, human selection, etc. But this was done in harmony with the phases of the implicate order and its biocybernetic system. What is not acceptable is when you mix the genes from different kingdoms, playing dice with them, and when a company or its affiliates that sells chemicals (pesticides and herbicides) also sells genetically engineered seeds and medicinal drugs. Take, for example, the development of plants that are resistant to the popular herbicide marketed as Roundup. You can spray the weed killer everywhere and not worry about damaging your crop. But do you think that some of those genes in the Roundup-resistant plant will not end up in weeds, which will then be resistant to herbicides? We were told that this was impossible—but it happened.

In 1999, Roundup used in genetically engineered (GE) soy was classified as carcinogenic by the Environmental Protection Agency in the United States and considered mutagenic. Also, the mood of the one thousand attendees of the EcoFarm Conference turned somber in January 2011 when the announcement was made that the Obama administration had decided to deregulate Monsanto's genetically modified (GM) alfalfa.

"This is a sad day for the future of sustainable agriculture," pioneering organic farmer Larry Jacobs, president of Jacobs Farm/Del Cabo, summarized in the press release published by the Ecological Farming Association, which organizes the annual conference in Pacific Grove, California. Alfalfa is the fourth most important crop grown on American soil, behind corn, soybeans, and wheat, alfalfa hay being a staple diet of dairy and beef cattle. According to Michael Pollan, an author and advocate of sustainable food systems: "93 percent of alfalfa hay is grown without any herbicide at all," which means that the GM alfalfa seed developed by Monsanto in order to resist its Roundup herbicide "is a bad solution to a problem that doesn't exist."[554]

Appendix

One of the reasons it is a lousy idea is that alfalfa pollen is exceptionally volatile, migrating easily for miles courtesy of insects and breezes. In other words, GM alfalfa is a prime candidate for the harmful transgenic contamination of conventional crops through cross-pollination—a phenomenon that biologists simply call gene flow. The Supreme Court accepted that much in a ruling handed down in June 2010, in the case brought by the Center for Food Safety and other plaintiffs against the USDA after a premature deregulation of GM alfalfa.[555]

The catastrophic chain of events that could result from the planting of GM alfalfa includes the fact that organic farmers rely on organic alfalfa to feed their animals. In the absence of GMO labeling in the United States and Canada, the organic label was the only non-GMO guarantee offered to consumers. In fact, 83% of organic consumers say they purchase organic food specifically to avoid GMO, according to the Stonyfield Farm Case Study (2009).

Contamination of fields by GMO products heralds the end of organic crops. Not only does this jeopardize the organic practices and labeling of organic farms, putting their very survival at risk, but it goes against the liberty of choice we the people should have. In democratic countries such as the United States and Canada, this should not happen. We also clearly know now that some GMO plants cause their own problems.[556] It is, as shown in diagram 29, a consequence of businesses trying to take control of the legal sphere.

Genetically modified canola has escaped from the farms and is thriving in the wild. Some of the highest densities are found along a trucking route into Canada. "That's where the most intense canola production is and it's also the road that goes to the canola processing plants across the border," said ecologist Cynthia Sagers of the University of Arkansas, referring to a canola processing plant in Altona, Manitoba. The GM canola, which was engineered to withstand herbicides that kill weeds, is the only thing growing on that route. "In some places along the road where department of transportation had sprayed for weeds, the canola was blooming brilliantly," Sagers said. Of the 288 canola plants the researchers tested, 231 were transgenic or genetically modified. Perhaps most significant is that two of the plants had amalgamations of herbicide resistance that had not been developed commercially. "That suggests to us there is breeding going on, either in the field or in these roadside populations, to create new combinations of traits," continued Sagers.

She added that the findings raised questions about whether the escaped GM canola might pass on manmade genes to wild species like field mustard, which is an agricultural pest. "It is conceivably a very large problem." [557]

"Biology doesn't know any borders," said Rene Van Acker at the University of Guelph, Canada, who has done extensive research on the extent and behavior of escaped GM crops in Manitoba. Van Acker said the study, like others done in Canada, raises red flags over plants used to grow pharmaceutical drugs and industrial oils in GM plants. Such crops would have to be "confined" and kept out of the food system, said Van Acker, "and that starts to worry me."[558]

Research has also shown that once you spread more than fifty pounds of nitrogen on an acre of land at one time, you significantly reduce the soil's ability to produce nitrogen. Fertilizers are substantially influenced by the activity of fungi and bacteria in a soil. The more dormant these life forms are, the easier it is to use the same program year after year and do fairly well, provided there is regular rain or irrigation.[559] However, it has been shown that over time, this type of schedule causes more disease problems and destroys soil structure, especially if manmade NPK fertilizers (nitrogen, phosphorus, and potassium) are used.

Let's now talk about another clever human action: feeding animal proteins to herbivores. It resulted in transmissible spongiform encephalopathy (TSE), popularized under the term mad cow disease. For many people in the agro-industry, a protein is a protein. When I was working as an assistant for an agricultural TV show, I listened one day to people saying that we would not talk about how many cows or sheep a farmer had, but about the quantity of protein he could sell. They also argued that there was no possible harm in feeding protein from dead animals to an herbivore, since protein is excellent for health. Amino acids enhance health, right? So why not find the cheapest amino acids? Economically—and this is the first and often only factor in the agro-industry—it made sense to feed the leftovers of dead animals to other animals. We ended up feeding parts humans didn't want to herbivores. This included giving blood to veal calves instead of milk. Then mad cow disease hit the front page, people got scared, and the laws turned around. Proteins are made by the genes in cells and affect every aspect of cell behavior. Are GMOs recognizable proteins for the organism? Cancers can develop when proteins are not working properly.

For example, it has been known since Louis Pasteur that living matter molecules are homochiral, which means that in a living cell, proteins are only levogyre (or leftward coiled) and sugars only dextrogyre (or rightward coiled). With sugar we know that the body recognizes only the dextrorotary one. L-Glucose, made in the laboratory, is hardly available in nature. Although it tastes the same as the natural glucose, living organisms can't use it as a source of energy. Also, long-term reactions to its use are unknown.

Why wouldn't it be the same with proteins, when we know that the slightest change of shape affects a protein's function? Within inert matter, molecules are heterochiral, which means that they have an equal probability to be dextrogyre or levogyre. So, homochirality is a specificity of living beings.

Plainly: when genes from other kingdoms are introduced into an organism, creating a transgenic organism (commonly called a genetically modified or genetically engineered organism), the results for the organism and its environment are almost always unpredictable. The intended result may or may not be achieved in any given case, but the one almost sure thing is that unintended results—called nontarget effects—will also be achieved. How did we come to allow individuals, and their for-profit companies, control what touches everyone? Legislation from an international association with no ties whatsoever with commerce (which is not the case for individual governments) should be in charge of allowing—or not allowing—what might affect the whole community. As such, water, air, sun, agriculture, animals, earth, subtle energies, and seeds should be protected. No companies, groups, associations, or governments should be allowed, in any country, to act in a way that might have negative repercussions on all. Nobody owns the elements. Most importantly, this international group should put forward what enhances the health of the feminine polarity and be a whistle blower to the elements working against it.

When common sense tells you that feeding meat byproducts to an herbivore can't be sensible, those who wish this to be a good thing find linear ways to circumvent common sense and explain "scientifically" that they are right. They raise the monetary issue and accuse you of not wanting to help feed the poor. They transform it into a moral or ethical question. That is not science. People who follow their intuition in this matter are called credulous, hysterical, and the like.

Prion diseases of animal and man, bovine spongiform encephalopathy being one of those, are neurological diseases with amyloidal deposition of the respective proteins. Until now, all known pathogens (viruses, bacteria, etc.) contained nucleic acids. This was necessary for reproduction. Contrary to a central dogma of modern biology, the prion (PrPC), a protein structure, can and does self reproduce. In the prion's case, it has been observed that the healthy form had 42% of alpha-helix content and no beta sheet, while the infected prion (PrPSc) has a 30% alpha-helix and a whopping 43% of beta sheet. This evidently leads to amyloid formation. Are there human diseases featuring amyloid formation? Yes: type 2 diabetes, Alzheimer's disease, Parkinson's disease, Huntington's disease, Creutzfeldt-Jacob disease, and congestive heart failure.

Seeds extracted from cells with the prion state can convert the normal form of the protein, PrPC, into the infectious form, PrPSc. Several molecules of a particular protein high in glutamine and asparagine form a highly structured amyloid fiber when they convert into the prion state. A few factors are important to consider: the prion variant responsible for mad cow disease has the remarkable ability to bypass the species barrier to transmission. Current knowledge leads us to assume that unapparent stages of prion infection suggest a nonexistent species transfer. In fact, this unapparent infection precedes overt disease.

The link here with genetically engineered products is that both of them concern proteins.

The DNA makes the RNA which makes the protein. Before science accepted that proteins could reproduce themselves, they couldn't forewarn that this could be possible since it went against their dogma. If something of this sort happens with genetically engineered proteins

Appendix

and leads to calamities, what will the companies now profiting from the technology tell us? That they didn't know? That the scientific community is at fault?

Remember the thalidomide scandal? Babies were born with malformations because their mothers took a drug, thalidomide, while pregnant to relieve morning sickness and as a sleeping aid. It was made available in nearly fifty countries under at least forty different names. Estimates range from ten thousand to twenty thousand babies born with birth defects. Two-thirds survived past their first year of life. Canada was the last country to stop the sales of the drug, in early 1962, well after it had been proven unsafe. In the United States, pharmacologist and MD Frances Oldham Kelsey refused Food and Drug Administration (FDA) approval for an application from the Richardson-Merrell company to market thalidomide, saying further studies were needed. Free samples were nonetheless distributed by the pharmaceutical sector to doctors. It is still unknown if the disabilities of the survivors might be passed on to their own children, though the World Health Organization has cited evidence that this could be the case. This tragedy might have been avoided had the proper tests been done and the results not been faked. The other main side effects of the drug were fatigue, constipation, peripheral neuropathy, vein thrombosis, pulmonary edema, and refractory hypotension. Forty-six countries were duped? And we are told that this is the only case when a company sought monetary gain instead of the truth? At least shareholders were happy when the drug sold successfully. This is an invisible, blind, dangerous, self-interested power; truly a loose cannon.

All the amino acids of living beings are made of left-handed isomers. We can't assimilate right-handed ones. The toxicity and teratogen effect of the thalidomide drugs was due to the fact that half of its molecules are right-handed isomers. Such a small detail ruined the life of thousands of people who trusted their doctors, who only wanted to avoid morning sickness and wished for a better night's sleep. The problem is also that now, to get a new patent, some drug companies will simply patent a different isomer under another name.

If we play with genes that form proteins (made of amino acids), how can we be sure that there will not be any teratogen or undesired effect? How can anyone responsibly assure us that all of the host reactions can be foreseen? They presume that a protein is a protein, a gene is a gene, and that they can play with them any way they want. They say nature has done this for thousands of years. Not if you take into account the LIFE biosystem. Different kingdoms are of different phases. Even the Bible emphasizes "each in their kind" in Genesis. It is not for the pleasure of adding empty words to a text.

For people to argue that the precautionary principle is backward and would stop progress is irresponsible and dangerous. Sure they will get the approval of groups with a vested interest. Politically, it would be labeled as an "intelligent" move.

In 2003, the Swedish government authorized an ex gratia compensation of SEK 250,000 for each of the 150 individuals that were subjected to thalidomide, sold in that country as Neurosedyn. The people of Sweden paid for the lies of a drug company. How is it that the people and the environment in general pay for the harm done by for-profit companies? We can see this everywhere and at all levels, from petrol companies leaking their oil into the oceans, to money-market brigands threatening global financial institutions. They take from the pockets of small earners so they can still make a profit at the end of the year for their CEOs and shareholders. It is pure madness. Meanwhile, John and Jane pay taxes after taxes. In their small apartment, they watch TV and learn all this and have to cut off the heat because it is too expensive. Work is precarious, and the Friday beer is not enough anymore to make you feel optimistic. Anyway, a few weeks ago they closed the pub where they used to gather with friends.

Some members of the British Parliament argued that Americans have been fed genetically engineered foodstuff for at least the last fifteen years, without their consent, and that no deaths can be directly linked to this. How could a link be found if no one knew genetically engineered products were fed to the general population? What about the increase in allergies[560] and asthma? This implies immune reactions, including type I hypersensitivity, in which a person's body is hypersensitive and develops IgE antibodies to typical proteins. Can anyone prove that GE foodstuffs and pesticides are not responsible in part for the increase in cancer and neurologic disorders? Was this population used as guinea pigs?

Appendix

Allergy is a common disorder and more than 50 million Americans suffer from it. Allergies are the sixth leading cause of chronic disease in the United States, costing the health care system $18 billion annually. (By the way, death linked to medicinal drugs is ranked fourth). When I researched the hypothetic causes for various allergies, I read about the protein problem, but also that some believe it is a result of having fewer parasites in the gut, parasites that would release a chemical to neutralize our immune system, preventing the host from killing the parasites. This suggests those who are for human hosting numerous parasites believe we are somewhat devoid of them, since the level of allergy sufferers is so high. Why would anyone think we are "cleaner"? With all the people in our society going to restaurants and fast-food chains, with the traditional annual cleansing forgotten for the last forty-five years, how could this be so?

I have seen many patients, including myself, much healthier after getting rid of any parasites. I have not seen patients better because they have parasites. It is as if we are back in the eighteenth century, when dirt was considered protective against illnesses. People would wash and change clothes only once a year. We have learned the necessity of physical hygiene. We must not regress; on the contrary, we have now to apply hygiene at the emotional, mental and social levels…

Notes, References, and Selected Bibliography

[1] As an example, we have cars that allow us to travel greater distances in a shorter time, providing us with less loss of time and less physical strain. One result of this, besides pollution, is that physicians now exhort us to walk more. We now know walking is necessary to keep bodies and minds healthy. We are engineered for it. Unfortunately, we have designed cities to facilitate transportation, not pedestrian happiness and well-being.

[2] Franklin D. Roosevelt: "Address at Madison Square Garden, New York City," October 31, 1936. Online by Gerhard Peters and John T. Woolley, the American Presidency Project. http://www.presidency.ucsb.edu/ws/?pid=15219.

[3] See chapter 5

[4] See chapter 8

[5] http://www.mentalhealthhope.com/Alarming%20Statistics.pdf

[6] Carl G. Jung, *Correspondance 1950-1954*, Albin Michel, p. 34, translated from French.

[7] Claire Dunne, *Carl Jung: Wounded Healer of the Soul: An Illustrated Biography* (New York: Continuum International Publishing Group Ltd, 2001), 3.

Chapter 1

[8] David Bohm, Coherence and the Implicate Order (London: Routledge & Kegan Paul, 1980).

[9] Alceste Bonanos et al., 2006.

[10] Ian Stewart, *Why Beauty is Truth: The History of Symmetry* (New York: Basic Books, 2007), 242.

[11] Ervin Laszlo, *Science and the Akashic Field: An Integral Theory of Everything* (New York: Inner Traditions, 2007).

[12] Personally, I had this experience not through meditation but through a dream. As I didn't know at the time what it was, I found it very disturbing. At the *top*, there was *nothing*.

[13] Jose Beltran Jimenez and Antonio L. Maroto, "Cosmological Electromagnetic Fields and Dark Energy," *Journal of Cosmology and Astroparticle Physics* 2009, (March 2009) doi:10.1088/1475-7516/2009/03/016

[14] David S. Walonick, PhD, "A Holographic View of Reality," 1993, http://www.statpac.org/walonick/reality.htm

[15] Papus, La Cabbale, Tradition Secrète de l'Occident, p.155

[16] Jung, *Correspondance 1950-1954*, p. 36.

[17] David Walonick, "A Holographic View of Reality," 1993. This is the best summary of fractal holographic elements and the implicate order I could find. http://www.statpac.org/walonick/reality.htm.

[18] "If a line segment intersects two straight lines forming two interior angles on the same side that sum to less than two right angles, then the two lines, if extended indefinitely, meet on that side on which the angles sum to less than two right angles." "Parallel Postulate," last modified June 19, 2012, http://en.wikipedia.org/wiki/Parallel_postulate.

[19] Mark David Glucina, and Kozo Mayumi, "Connecting Thermodynamics and Economics, Well-lit Roads and Burned Bridges," *New York Academy of Sciences* 1185, (January 2010): 11-29, doi: 10.1111/j.1749-6632.2009.05166.

[20] Mae-Wan Ho, Bioelectrodynamics Laboratory, Open University Walton Hall, Milton Keynes, UK (2004).

[21] C.A. Meier, ed., *Atom and Archetype: The Pauli/Jung Letters, 1932-1958* (Princeton, NJ: Princeton University Press, 2001), 14.

[22] Werner Heisenberg, *Physics and Beyond*, 1971.

[23] Jung, *Correspondance 1950-1954*, p. 51.

Isis Code

[24] http://www.reportlinker.com/p0360446/Schizophrenia-Market-Forecast.html.
[25] "Socioeconomic Status, Structural and Functional Measures of Social Support, and Mortality," The British Whitehall II Cohort Study, 1985-2009 Am J Epidemiol 24 April 2012: kwr46lvl-kwr46l. Silvia Stringhini, Lisa Berkman, Aline Dugravot, Jane E. Ferrie, Michael Marmot, Mika Kivimaki, and Archana Singh-Manoux: Correspondence to Dr. Silvia Stringhini, Institute of Social and Preventive Medicine, Lausanne University Hospital, Lausanne, Switzerland.
[26] David S. Walonick, PhD, "A Holographic View of Reality" 1993, http://www.statpac.org/walonick/reality.htm.
[27] Paul Pietsch, *Shufflebrain: The Quest for the Hologramic Mind*. (Boston: Houghton Mifflin, 1981).
[28] György Buzsaki, *Rhythms of the Brain* (New York: Oxford University Press, USA, 2006), 371.
[29] Elkhonon Goldberg, PhD, *The New Executive Brain*, Oxford, p.274
[30] Ibid.
[31] C. A. Meier, ed., *Atom and Archetype: The Pauli/Jung Letters 1932-1958* (Princeton: Princeton University Press, 2001), xli.
[32] Meier, *Atom and Archetype: The Pauli/Jung letters 1932-1958*.
[33] See chapter 5
[34] The nervous system is an oscillatory system.
[35] a Greek theological, philosophical, and scientific term usually translated into English as 'nature' and 'physical world'.
[36] Meier, *Atom and Archetype: The Pauli/Jung letters 1932-1958*, 176.
[37] "If two systems are in thermal equilibrium with a third system, they are also in thermal equilibrium with each other." Wikipedia.
[38] Wikipedia, definition Cybernetics
[39] Bohm, *Coherence and the Implicate Order*.
[40] For a good summary, see "A Holographic View of Reality," David S. Walonick, PhD, http://www.statpac.org/walonick/reality.htm.
[41] Entrainment: causing a gradual phase shift so that the oscillation becomes synchronized with the entraining rhythm or signal. (1974)
[42] R. I. **Kitney**, "An Analysis of the Nonlinear Behaviour of the Human Thermal Vasomotor Control System."
[43] György Buzsaki, *Rhythms of the Brain*, (New York: Oxford University Press), 115.
[44] Ibid.
[45] Allan DeBavelaere, MD, N.D. and Nicole DeBavelaere, HN, "The Biocybernetical Revolution Applying an Elementary Biocybernetic System to Medicine and Society: Complexity, Democracy, Sustainability."
[46] This can be seen in patients who have lost the use of one brain hemisphere. The remaining one develops and takes up the functions left without an expresser.
[47] X. Gu, and N. C. Spitzer, "Distinct Aspects of Neuronal Differentiation Encoded by Frequency of Spontaneous Ca^{2+} Transients," *Nature*, 1995, 784-787. R. D. Fields, F. Eshete, B. Stevens, and K. Itoh, "Action Potential-Dependent Regulation of Gene Expression: Temporal Specificity in Ca^{2+}, cAMP-Responsive Element Binding Proteins, and Mitogenactivated Protein Kinase Signaling," *Journal of Neuroscience* 17, (1997) 7252-7266.
[48] Gordon Globus, "Quantum Consciousness Is Cybernetic."
[49] *Agricultures Magazine*, Purdue University (Spring 2008).
[50] http://users.rcn.com/jkimball.ma.ultranet/BiologyPages/P/Photoperiodism.html.
[51] A. Sehgal, A. Ousley, Z. Yang, Y. Chen, P. Schotland, Howard Hughes, "What Makes the Circadian Clock Tick: Genes That Keep Time?"
[52] A. DeBavelaere, - Introduction à la Médecine Holistique, de l'Homme à la Société - Westphal pub.
[53] Articles evidencing the existence of energy meridians, Compiled from the Internet by Fred Gallo, PhD, http://www.eftuniverse.com/index.php?option=com_content&view=article&id=2479.
[54] "The endoplasmic reticulum (ER) has emerged as a major site of cellular homeostasis regulation, particularly in the unfolded protein response, which is being found to play a major role in cancer and in many other diseases. Here, we address ER-mediated signaling and regulations in the context of environmental challenges in cancer, such as hypoxia, angiogenesis, and chemotherapeutic resistance, and we discuss how ER-resident molecular machines become deregulated and involved in cancer-related pathology. Further exploration of how the ER senses, signals, and adapts to stress

Notes, References, and Selected Bibliography

may redefine and deepen our understanding of its functions in cancer pathobiology. [*Cancer Res* 67, no.22 (2007):10631–4]

[55] Kaoru Sakatani, "Concept of Mind and Brain in Traditional Chinese Medicine," *Data Science Journal*, 6, Supplement (April 2007).

[56] Elkhonon Goldberg, *The New Executive Brain Frontal Lobes in a Complex World*, Oxford p.210

[57] M. R. DiMatteo, K. B. Haskard, and S. L. Williams, "Health Beliefs, Disease Severity, and Patient Adherence: A Meta-Analysis," *Medical Care*, 45, (2007): 521–528.

Chapter 2

[58] By the year 2020, the World Health Organization (WHO) estimates that depression will be the number two cause of "lost years of healthy life" worldwide.

[59] World Health Organization http://www.who.int/mental_health/management/depression/definition/en/.

[60] "Global Depression Statistics" published on ScienceDaily online, July 25, 2011, quoted by http://www.depression-statistics.com/.

[61] Homeopathy, a therapy that addresses the feminine polarity, has been ostracized since its beginning. Can hundreds of thousands patients and medical doctors be so wrong in the observation of its efficiency? The reality is that experienced homeopathic medical doctors have achieved results.

[62] Roy C. Ziegelstein and Brett D. Thombs, "The Brain and the Heart: The Twain Meet," *European Heart Journal* 26, (2006):2607–2608 doi:10.1093/eurheartj/ehi576.

[63] Jung *Correspondance 1950–1954*, p. 97.

[64] See the section discussing bonobos in chapter 3

[65] Burr and Northrop, 1935.

[66] Karl H. Pribram "Holonomic Brain Theory and Motor Gestalts: Recent Experimental Results Professor Emeritus, Stanford University, James P. and Anna King University Professor and Eminent Scholar, Commonwealth of Virginia Abstract of the keynote lecture at the 10th Scientific Convention of the Society for Gestalt Theory and its Applications (GTA) Vienna/Austria. March 1997 http://gestalttheory.net/conv/prib.html

[67] Laszlo, *Science and the Akashic Field*.

[68] http://www.sciencemag.org/content/286/5439/548.abstract.

[69] An interesting fact here is that the neocortex of a rodent for example is smooth while one of a primate or human has groves. All human brains have the same pattern of gyri and sulci therefore these are not driven solely as a result of a need to save cranial space.

[70] Roger Lewin, "Is Your Brain Really Necessary?" *Science* 210 (December 12, 1980) and *Alternative Science News*, (September 9, 2002). [http://www.alternativescience.com/no_brainer.htm. Also http://www.enidreed.com/serv01.htm.

[71] John Nolte, *The Human Brain: An Introduction to Its Functional Anatomy*, (Philadelphia: Mosby Publishing, 2002).

[72] Suzanne L. Tyas, David A. Snowdon, Mark F. Desrosiers, Kathryn P. Riley, and William R. Markesbery, "Healthy Ageing in the Nun Study: Definition and Neuropathologic Correlates," *Oxford Journals Medicine, Age and Ageing* 36, no. 6 (2007): 650–655.

[73] Norman Doidge, "The Brain That Changes Itself," 20–24.

[74] Dr. Mark A. Smith. is a cognitive neuroscientist. For more information, see his website: http://markashtonsmith.info/

[75] Norman Doidge, MD, *The Brain that Changes Itself* (New York: Penguin), 55.

[76] "Signs of PCB toxicity vary with the species affected, from no clinical signs to signs comparable to DDT toxicity. The toxicity of PCBs had been known since before its first commercial production through research done by producing companies themselves in the 1930s; however, these conclusions were dismissed as negligible. The toxicity of PCBs to animals was first noticed in the 1970s, when emaciated seabird corpses with very high PCB body burdens washed up on beaches. Since seabirds may die far out at sea and still wash ashore, the true sources of the PCBs were unknown. Where they were found was not a reliable indicator of where they had died. At the time of necropsy, petechial and ecchymotic hemorrhages will be seen on the heart (which usually stops during systole) and the myocardium will appear whitish in color. The lungs appear congested and darkened in color and in some cases there may be blood-tinged exudate in the bronchioles. The liver may be affected with fatty degeneration, focal necrosis or cirrhosis. As of 2004, there were at least 165 organochlorides

Isis Code

approved worldwide for use as pharmaceutical drugs, including the natural antibiotic vancomycin, the antihistamine loratadine (Claritin), the antidepressant sertraline (Zoloft), the anti-epileptic lamotrigine (Lamictal), and the inhalation anesthetic isoflurane. Diagnosis is based on clinical signs and necropsy examination. The differential diagnosis for organochlorine toxicity is tetanus and strychnine poisoning. Definitive diagnosis can be made by laboratory analysis of organochlorine residues in brain or fat tissue. The brain is the best tissue for organochlorine determination as the brain is the lethal site of action, but an approximate level can be obtained from fat analysis. *The chief target organ of pesticides is the brain because pesticides seek out lipids and the brain is highly lipid with a high density of acetylcholinesterase which is the target enzyme of pesticides.* They inhibit the enzyme acetylcholinesterase. This enzyme controls the metabolism of our neurotransmitter acetylcholine.

[77] "Acetylcholine is the major neurotransmitter in the brain and what Dr. Sherry Rogers calls our primary happy hormone. It is the basic chemical that makes the brain, all the nerves and the muscles work. Pesticides disrupt this hormone. They also inhibit the conversion of tryptophan into serotonin, which is our temporary happy hormone. Organophosphate pesticides are taken directly into our nervous system (the brain, spinal cord, and long nerves) and then transformed into chlorpyrifos oxon, which is actually 3000 times more potent than the original compound.

[78] "Then what is frequently done is to mix organophosphate pesticides with organochloride pesticides and then they inhibit the other happy hormones norepinephrine and dopamine as well as serotonin. Neurotransmitters are chemicals in the brain that are responsible for numerous functions within the mind and body, especially our mood states, the nervous system and cognitive functioning. When neurotransmitters are not functioning properly then a variety of conditions develop. Pesticide exposure is believed to be linked to serious conditions like Alzheimer's, Parkinson's, autism, clinical depression, anxiety disorders, hyperactivity and attention deficit. The reason Dr. Sherry Rogers calls these neurotransmitters our happy hormones, because they control how we feel. When they are not produced in adequate numbers or functioning properly then feelings of sadness, depression, and hopelessness develop. It should be very clear at this point as to why we have an epidemic of depression in our society." Cynthia Perkins, http://www.holistichelp.net/pesticides.html. References: Sherry Rogers, "Depression Cured at Last," (1997). Janice Wittenberg Strubbe, "The Rebellious Body," (1996). Pamela Gibson Reed, "Multiple Chemical Sensitivity," (2000). Al. Sear, "How I Keep the Critters Out" (2011).

[79] Alister Doyle **Oslo**, "Pesticides May Damage Brain," *Reuters*, March 31, 2009.

[80] Elkhonon Goldberg, *The New Executive Brain: Front Lobes in a Complex World*, 48.

[81] Kaoru Sakatani, "Concept of Mind and Brain in Traditional Chinese Medicine."

[82] W. J. Freeman, *Societies of Brains. A Study in the Neuroscience of Love and Hate*, (Lawrence Erlbaum Associates, Hove, 1995).

[83] PhysOrg. "Manipulating the Brain Network Could Improve IQ," PinkElephant, June 28, 2009. "I'd be careful trying to boost the brain's connectivity willy-nilly. Some nasty side-effects may emerge if it isn't done just right: like grand mal seizures, for example ... There's probably a good reason why most modern brains aren't as optimal as they could theoretically be. There's got to be a cost to genius; some get lucky and don't have to pay the piper; others may end up in institutions or dead ... or maybe something more prosaic, such as failing to procreate ..." http://phys.org/news163824097.html.

[84] R. Rosenman, *Integr Physiol Behav Sci.* (1993): 28(1)

[85] Katkin Hantes and Reed, 1984.

[86] Paul Pearsall, Gary E. Schwartz, and Linda G. Russek, "Organ Transplants and Cellular Memories," *Nexus Magazine*, (April May 2005).

[87] Barbara B. Walker and Curt A. Sandman, "Visual Evoked Potentials Change as Heart Rate and Carotid Pressure Change," *Psychophysiology* 19, no. 5 (September 1982) 520–527.

[88] Dulce Zamora, "Death from a Broken Heart," November 14, 2003, Medicinenet.com.

[89] L. M. Macrea, M. R. Tramer, and B. Walder, "Spontaneous Subarachnoid Hemorrhage and Serious Cardiopulmonary Dysfunction- A Systematic Review," *Resuscitation* 65 (2005):139–148.

[90] Paul A. Wilson, Giorgio Lagna, Atsushi Suzuki, and Ali Hemmati-Brivanlou, "Concentration-Dependent Patterning of the Xenopus Ectoderm by BMP4 and Its Signal Transducer Smad1. Development," *The Company of Biologists Limited* 124, (1997): 3177–3184. Mesoderm cells under the ectoderm secrete BMP-4 inhibitory signals, and as a result cause the overlying cells of the ectoderm to

Notes, References, and Selected Bibliography

develop into neural cells. The cells in the ectoderm that circumvent these neural cells do not receive the BMP-4 inhibitor signals. As a result, BMP-4 induces these cells to develop into skin cells.

[91] "The Heart of Neurofeedback," brainandhealthblog.com, http://www.brainandhealthblog.com/2005/12/29/the-heart-of-neurofeedback.

[92] http://www.cardio-coherence.com/Fr/Studies.htm.

[93] Ibid.

[94] C. B. Clayman, ed., *The American Medical Association Encyclopedia of Medicine*, (New York: Random House, 1989),142-143.

[95] Rémi Quirion¹, Michel Dalpé,¹ André De Lean¹, J. Gutkowska¹, J.; Marc Cantin,¹ and Jacques Genest,¹ "Atrial Natriuretic Factor (ANF) Binding Sites in Brain and Related Structures."

[96] J. S. Floras, "Sympathoinhibitory Effects of Atrial Natriuretic Factor in Normal Humans," *PubMed, US National Library of Medicine National Institutes of Health*, 81 no. 6 (June 1990):1860-73.

[97] J. Andrew Armour, "Neurocardiology— Anatomical and Functional Principles," *Can J Cardiol*. 13, no. 3 (March 1997): 277-84. "Intrinsic Cardiac Neurons Involved in Cardiac Regulation Possess Alpha 1-, Alpha 2-, Beta 1- and Beta 2-Adrenoceptors," Department of Physiology and

[98] Biophysics, Faculty of Medicine, Dalhousie University, Halifax, Nova Scotia. jarmour@is.dal.ca.

[99] Takeshi Kubota, Kazuyoshi Hirota, Hitoshi Yoshida, Satoshi Takahashi, Noriyuki Anzawa, Hirofumi Ohkawa, Tetsuya Kushikata and Akitomo Matsuki, "Original Investigation Effects of Sedatives on Noradrenaline Release from the Medial Prefrontal Cortex in Rats," *Psychopharmacology (Berl)*, 146, no. 3 October 1999): 335-8. DOI: 10.1007/s002130051125.

[100] Biosynarchy: bio meaning life and synarchy being a system of which the elements work together toward a common goal. In terms of a political system, I speak solely of an elected synarchy in opposition with its previous use— a nonelected, elite group governing a country or the world— and solely through the context that the different elements governing are part and true expression of the implicate order or LIFE biosystem.

[101] B. C. Goodwin, *How the Leopard Changed Its Spots: The Evolution of Complexity*, (London: Weidenfeld and Nicolson, 1994).

[102] Le Robert, *Dictionnaire historique de la langue française*.

[103] Willis Harman, and Howard Rheingold, *Higher Creativity: Liberating the Unconscious for Breakthrough Insights*, (New York: Jeremy P. Tarcher, Inc., 1984) 5.

[104] Dr. Blomquist's *Integrated Theory of Intelligence*, http://www.supraconsciousnessnetwork.org.

[105] Koltzoff and Schroeder, "Artificial control of sex in the progeny of mammalians," *Nature*, 329 (March 1933).

[106] Mark J. Plotkin *Medicine Quest: In Search of Nature's Healing Secrets* (New York: Viking Press, 2000).

[107] Koltzoff and Schroeder, "Artificial control of sex in the progeny of mammalians," *Nature*, no. 329 (March 1933).

[108] S. A. Ishijima, M. Okuno, H. Odagiri, T. Mohri, H. Mohri, "Separation of X- and Y-chromosome-bearing Murine Sperm by Free-flow Electrophoresis: Evaluation of Separation Using PCR," *Zoolog Sci* no. 3 (June 9, 1992): 601-6.

[109] Plamen Ch Ivanova, Qiani D. Y. Mab, and Ronny P. Bartschba, "Maternal-Fetal Heartbeat Phase Synchronization."

[110] "The heart is the most powerful generator of electromagnetic energy in the human body, producing the largest rhythmic electromagnetic field of any of the body's organs. The heart's electrical field is about 60 times greater in amplitude than the electrical activity generated by the brain. This field, measured in the form of an electrocardiogram (ECG), can be detected anywhere on the surface of the body. Furthermore, the magnetic field produced by the heart is more than 5,000 times greater in strength than the field generated by the brain, and can be detected a number of feet away from the body, in all directions, using SQUID-based magnetometers. Prompted by our findings that the cardiac field is modulated by different emotional states (described in the previous section), we performed several studies to investigate the possibility that the electromagnetic field generated by the heart may transmit information that can be received by others." "Science of the Heart: Exploring the Role of the Heart in Human Performance An Overview of Research Conducted by the Institute of HeartMath."

[111] Before the first day: higher heart— the Spirit of God was hovering over the waters. First day: *lung*— separation of Light from Darkness. Second day: *kidney*— expanse between the two waters and sky. Third day: *liver*— water under the sky gathered, dry ground appeared. Fourth day: *heart*— the

Isis Code

beginning of seasons thanks to light. Fifth day: *pancreas* – every moving creature and he told them be fruitful. Sixth day: *lung* – God created man in his own image; in the image of God he created him; male and female he created them. Seventh day: kidney – God rested: See schema 2.

Chapter 3

[112] An example of this is Previc's picture of a The Stargazer Rat. Hyper-dopaminergic activity— dopamine is found more in the left hemisphere than right— moves a rat's field of vision up. http://neuropolitics.org/defaultfeb08.asp.

[113] Carl G. Jung, *Correspondance 1950-1954*, Albin Michel, p. 21, translated from French.

[114] Buzsaki, *Rhythms of the Brain*.

[115] Daphne Maurer and Catherine J. Mondloch, Catherine, "Neonatal Synesthesia," www.psych.mcmaster.ca/maurerlab/Publications/Maurer_NeonatalSynesthesia.pdf.

[116] Bin Gao, Won-Il Jeong, and Zhigang Tian, "Liver: An organ with Predominant Innate Immunity," Section on Liver Biology, Laboratory of Physiologic Studies, National Institute on Alcohol Abuse and Alcoholism, National Institutes of Health, Bethesda, MD.

[117] M. Lee, J. L. Saver, K. H. Chang, et al., "Low Glomerular Filtration Rate and Risk of Stroke: Meta-Analysis," *BMJ*, (2010), DOI:10.1136/bmj.c4249.

[118] Mayumi Kobayashi1, Nobuhito Hirawa, Keisuke Yatsu, Yusuke Kobayashi1, Yuichiro Yamamoto, Sanae Saka, Daisaku Andoh, Yoshiyuki Toya, Gen Yasuda, and Satoshi Umemura, "Relationship between silent brain infarction and chronic kidney disease."

[119] Norman Doidge, *The Brain that Changes Itself* (New York: Penguin), 226.

[120] Elkhonon Goldberg. PhD, *The New Executive Brain* (New York: Oxford University Press), 173.

[121] Grossman et al., "Social Cognition and Affective Neuroscience," 2007. Nakato et al., "Human Brain Mapping," 2009.

[122] R. Lemlich, "Subjective Acceleration of Time with Aging," *Percept Mot Skills*, 41, no.1 (August 1975): 235-8.

[123] "Une heure sur terre," CBC broadcast, 2/12/2011.

[124] M.L. Laudenslager et al., "Total Cortisol, Free Cortisol, and Growth Hormone Associated with Brief Social Separation Experiences in Young Macaques," *Dev Psychobiol*, 28, no. 4 (May 1995): 199-211. P. Rosenfeld et al., "Maternal Regulation of the Adrenocortical Response in Preweanling Rats," *Physiol Behav* 50, no. 4 (Oct 1991): 661-71.

[125] Linda Folden Palmer, "Stress in Infancy."

[126] James W. Prescott, PhD "The Origins of Love & Violence: An Overview," March 28, 2002, http://ttfuture.org/bonding/love_violence.

[127] C. D. Williams, "Kwashiorkor: A Nutritional Disease of Children Associated with a Maize Diet," *Lancet* 226, (1935): 1151-2.

[128] Wikipedia.

[129] M. C. Latham, "The Dermatosis of Kwashiorkor in Young Children," *Seminars in Dermatology* 4 (1991): 270-2.

[130] Wikipedia, baby hatch

[131] Joseph Rhawn, "Environmental Influences on Neural Plasticity, the Limbic System, Emotional Development & Attachment," *Child Psychiatry and Human Development*, 29 (1999):187-203.

[132] Joseph Rhawn Joseph, *Neuropsychology, Neuropsychiatry, and Behavioral Neurology* (Plenum Press, 1990), 124.

[133] R. A. Spitz, "The Psychogenic Diseases in Infancy— An Attempt at their Etiologic Classification," *Psychoanalytic Study of the Child*, 6 (1951): 255-275.

[134] Ashley Montagu, PhD, SD, LittD, was born in 1905 and became an internationally renowned anthropologist who chose to direct his numerous published studies on the significance of the maternal-infant relationship for the human condition to the general public. Ashley Montagu has taught and lectured at Harvard, Princeton, (where he was chair, department of anthropology), University of California, New York University, and many other institutions of higher learning. He has published over sixty books, among them: *Life Before Birth; Touching: The Human Significance of the Skin; On Being Human; The Nature of Human Aggression; The Natural Superiority of Women; Growing Young; Living and Loving; The Peace of The World; Man's Most Dangerous Myth: The Fallacy of Race; Human Evolution; The Elephant Man;* and *Anthropology and Human Nature*. He is coauthor with Floyd Matson of *The Human Connection* and *The Dehumanization of Man*. He is the writer and director of the film *One World or None*,

Notes, References, and Selected Bibliography

described as one of the best documentaries ever made. ("Ashley Montagu" James W. Prescott, PhD, Institute of Humanistic Science).

[135] American Humanist Association.

[136] Richard Ferber, *Solve Your Child's Sleep Problems* (Fireside, 1985-2006).

[137] http://web.archive.org/web/20070108073244/www.violence.de/tv/rockabye.html.

[138] http://www.violence.de/tv/rockabye.html.

[139] Mary Neal, "Vestibular Stimulation and Development Behavior in the Small Premature Infant," *Nursing Research Report*, 1067, 3:1-

[140] "Perspectives On Violence: James W Prescott, PhD," http://birthpsychology.com/free-article/perspectives-violence-james-w-prescott.

[141] M. K. Floeter and W. T. Greenough, "Cerebellar plasticity: Modification of Purkinje Cell Structure by Differential Rearing Monkeys," *Science* 206 (1979): 227-229. M. Coleman, "Platelet Serotonin in Disturbed Monkeys and Children. Clinical Proceed of the Children's Hospital," *Science* 27, no. 7 (1971):187-194. G. K. Bryan and A. H. Riesen, "Deprived Somatosensory-Motor Experience in Stumptailed Monkey Neocortex: Dendritic Spine Density and Dendritic Branching of Layer IIIB Pyramidal Cells," *The Journal of Comparative Neurology*, 286, (19891): 208-217.

[142] E. R. de Kloet et al., "Brain- Corticosteroid Hormone Dialogue: Slow and Persistent," *Cell Mol Neurobiol* 16, no. 3 (June 1996): 345-56.

[143] M. R. Gunnar et al., "Stress Reactivity and Attachment Security," *Dev Psychobiol*, 29, no. 3 (Apr 1996): 191-204. M. R. Gunnar, "Quality of Care and Buffering of Neuroendocrine Stress Reactions: Potential Effects on the Developing Human Brain," *Prev Med* 27, no. 2 (Mar-Apr 1998): 208-11.

[144] Christiane Fielder, "The Sexual Paradox," 366.

[145] Allan N. Schore, "Attachment Trauma and the Developing Brain," 110.

[146] Mark Stibich, "Top 10 Causes of Death for Ages 15- 24," *About.com Guide*, June 15, 2007.

[147] Dardo Tomasi,[1] Linda Chang,[2] Elisabeth C. Caparelli,[1] and Thomas Ernst,[2] "Sex Differences in Sensory Gating of the Thalamus During auditory interference of visual attention tasks."[1]

[148] Michael Levin, *Feminism and Freedom*, (Transaction Publishers, 1988) 210.

[149] Neuroscience 2000 Conference, November 7, 2000.

[150] Norman Doidge, *The Brain that Changes Itself* (New York: Penguin) 81.

[151] I also believe they have an effect on our global oscillatory system, influencing future cells metabolism.

[152] Ibid.

[153] Stefan H. Bracha, Tyler C. Ralston, Jennifer M. Matsukawa, "Does 'Fight or Flight' Need Updating?"

[154] Laudenslager et al.: "Total Cortisol, Free Cortisol, and Growth Hormone Associated with Brief Social Separation Experiences in Young Macaques."

[155] Terry M. Levy, *Handbook of Attachment Interventions*.

[156] Indeed, theory and research suggest that error monitoring is associated with both cognitive and emotional processes (C.S. Carter, Botvinick, and Cohen, 1999; Cameron S. Carter, Braver, Barch, Botvinick et al., 1998; William Joseph Gehring, 1993; W.J. Gehring and Fencsik, 1999; Kiehl, Liddle, and Hopfinger, 2000). ERP (event-related potential) studies have shown that error monitoring is associated with a response-locked negative component, termed the error-related negativity (ERN) (Falkenstein, Hohnsbein, and Hoorman, 1998; William Joseph Gehring, 1993; W.J. Gehring, Goss, Coles, Meyer et al., 1993). This one is maximal over the midline central and frontal brain sites. Psychopathy has also been associated with perseveration (see review by Newman, 1998), and response inhibition abnormalities (Kiehl, Smith, Hare, and Liddle, 2000; Lapierre, Braun, and Hodgins, 1995).

[157] In MRI studies harm avoidance was correlated with reduced grey matter volume in the orbito-frontal, occipital and parietal regions all associated with the feminine polarity.

[158] D. Barch, "Cognitive and Affective Neuroscience of Psychopathology, Paralimbic Dysfunction Hypothesis of Psychopathy: Cognitive Neuroscience Perspective."

[159] http://www.mspca.org/programs/cruelty-prevention/animal-cruelty-information/cruelty-to-animals-and-other-crimes.pdf.

[160-153]These notes have been taken from many websites that posted them without indicating their sources, notably a thread by "Cherri." See also http://dogsrcute.webstarts.com/.

[161] A. Walsh, *The Science of Love*, (Buffalo, NY: Prometheus Books, 1991).

[162] A. P. Weil, "Thoughts about Early Pathology," *Journal of the American Psychoanalytic Association*, 33 (1985): 335-352.

Isis Code

[163] Justin M. Carré,[1] Patrick M. Fisher,[2] Stephen B. Manuck,[3] and Ahmad R. Hariri,[1,4] "Interaction between Trait Anxiety and Trait Anger Predict Amygdala Reactivity to Angry Facial Expressions in Men but Not Women," *Soc Cogn Affect Neurosci* (2010): doi: 10.1093/scan/nsq101. First published online December 22, 2010.

[164] Jessica L. Wood, Dwayne Heitmiller, Nancy C. Andreasen, and Peg Nopoulos, "Morphology of the Ventral Frontal Cortex: Relationship to Femininity and Social Cognition," *Oxford Journals, Social Cognitive and Affective Neuroscience* (December 22, 2010). 10.1093/scan/nsq101.

[165] Fausto-Sterling, *Sexing the Body: Gender Politics and the Construction of Sexuality* (2000).

[166] "Symmetry of Homosexual Brain Resembles That of Opposite Sex Swedish Study Finds," *Science Daily* (June 18, 2008). Proceedings of the National Academy of Science (June 17, 2008). http://www.sciencedaily.com/releases/2008/06/080617151845.htm
Ivanka Savic and Per Lindström, "PET and MRI Show Differences in Cerebral Asymmetry and Functional Connectivity between Homo- and Heterosexual Subjects," Proceedings of the National Academy of Sciences, 2008, DOI: 10.1073/pnas.0801566105.

[167] "It is not uncommon for a male rape victim to blame himself for the rape, believing that he in some way gave permission to the rapist (Brochman, 1991). Male rape victims suffer a similar fear that female rape victims face—that people will believe the myth that they may have enjoyed being raped. Some men may believe they were not raped or that they gave consent because they became sexually aroused, had an erection, or ejaculated during the sexual assault. These are normal, involuntary physiological reactions. It does not mean that the victim wanted to be raped or sexually assaulted, or that the survivor enjoyed the traumatic experience. Sexual arousal does not necessarily mean there was consent. According to Groth, some assailants may try to get their victim to ejaculate because for the rapist, it symbolizes their complete sexual control over their victim's body. Since ejaculation is not always within conscious control but rather an involuntary physiological reaction, rapists frequently succeed at getting their male victims to ejaculate. As Groth and Burgess have found in their research, this aspect of the attack is extremely stressful and confusing to the victim. In misidentifying ejaculation with orgasm, the victim may be bewildered by his physiological response during the sexual assault and, therefore, may be discouraged from reporting the assault for fear his sexuality may become suspect (Groth & Burgess, 1980). Another major concern facing male rape victims is society's belief that men should be able to protect themselves and, therefore, it is somehow their fault that they were raped. The experience of a rape may affect gay and heterosexual men differently. Most rape counselors point out that gay men have difficulties in their sexual and emotional relationships with other men and think that the assault occurred because they are gay, whereas straight men often begin to question their sexual identity and are more disturbed by the sexual aspect of the assault than the violence involved (Brochman, 1991)." National Center for Victims of Crime, www.ncvc.org/ncvc/main.aspx?dbName=DocumentViewer&DocumentID=32361.

[168] A. DeBavelaere, N. DeBavelaere, *Einstein, Jung and the Master of the Heart*, Westphal Pub.

[169] Jung, *Correspondance 1950–1954*, p. 123. "Je suis tout à fait de votre avis quand vous dites que l'être humain ne vit totalement que dans sa relation à Dieu, qui lui fait face et le determine."

[170] David Jones, MD, *Textbook of functional Medicine* (Institute for Functional Medicine, 2005).

[171] The frog and the Cow, Aesop's Fables

[172] deprotonates weak acid, (although strong acid deprotonate, or give their H+ more rapidly)

[173] Luke 17:19; Matthew 9:22; Mark 10:52; Luke 18:42; Mark 5:34; Luke 7:50; Luke 8:48.

[174] Amina Hamed, Ahmad Al Obaidi, "Expired Breath Condensate Hydrogen Peroxide Concentration and pH for Screening Cough Variant Asthma Among Chronic Cough," *Annals of Thoracic Medicine*, 2, no. 1 (2007): 18–22. http://www.thoracicmedicine.org/text.asp?2007/2/1/18/30357.

[175] "The Seal of Solomon was a magical signet ring said to have been possessed by King Solomon, which variously gave him the power to command demons, genies (or jinni), or to speak with animals." Wikipedia.

[176] N. Sandstrom, J. Kaufman, S. A. Huettel, "Males and Females Use Different Distal Cues in a Virtual Environment Navigation Task," *Brain Research: Cognitive Brain Research*, 6, (1998):351–360.

[177] Saucier et al., "Are Sex differences in Navigation Caused by Sexually Dimorphic Strategies or by Differences in the Ability to Use the Strategies?" *Behavioral Neuroscience*, 116, (2002):403–410.

[178] Georg Grön, Arthur Wunderlich, Manfred Spitzer, Reinhard Tomczak, and Matthias Riepe, "Brain Activation During Human Navigation: Gender-Different Neural Networks as Substrate of Performance," *Nature Neuroscience* 3, no. 4 (April 2000): 404–408.

Notes, References, and Selected Bibliography

[179] www.newyorker.com/online/2007/07/30/slideshow_070730_parker?viewall=true#showHeader

[180] "The world produces enough food to feed everyone. World agriculture produces 17 percent more calories per person today than it did 30 years ago, despite a 70 percent population increase. This is enough to provide everyone in the world with at least 2,720 kilocalories (kcal) per person per day" (worldhunger.org).

[181] http://www.waterinfo.org/resources/water-facts.

[182] Al Sears, MD, "How the Feds Set Frankenstein Free on the Farm Terrorizing Farmers, Making Us Sick," http://www.alsearsmd.com/pdf/Frankenstein-On-The-Farm.pdf.

[183] "U.S. Spent $140 Million of Haiti Earthquake Aid on Controversial Food Exports in 2010." "In 1986, Congress also approved the Bumpers Amendment to the Foreign Assistance Act, which bars the government from helping farmers abroad increase the yields of crops that could compete with staple American exports. Critics like Jean-Baptiste say it's the reason current USAID programming in Haiti focuses on developing export goods like mangoes, cacao and coffee, but largely ignores staples like rice and corn. One important USAID program in Haiti does include some rice and corn growers. The $127 million Watershed Initiative for National Natural Environmental Resources hopes to boost Haitian agriculture through training, better seeds and the use of new fertilizers. USAID claims to have helped 9,700 farmers increase their output since the program began in 2009. Many Haitian farmers rejected the program, however, after they discovered that 475 tons of seeds were hybrids donated by Monsanto, the world's largest developer of genetically modified seeds. Unlike traditional crops, hybrids do not produce new seeds that can be collected and planted the following growing season, meaning farmers in Haiti would need to begin purchasing the seeds from Monsanto or another company once donations stopped. http://www.asafeworldforwomen.org/environment/env-latin-america/1895-us-aid-to-haiti.html.

[184] Chirag J. Patel, Jayanta Bhattacharya, Atul J. Butte, "An Environment-Wide Association Study (EWAS) on Type 2 Diabetes Mellitus," *PLoS ONE*, published online May 20, 2010.

[185] Jeremy Rifkin, *Beyond Beef* (Plume 1993).

[186] Nemeroff C. H. Mayber, S. Krahl, J. McNamara, A. Frazer, T. Henry, M. George, D. Charney, S. Brannan, "VNS Therapy in Treatment-Resistant Depression, *Neuropsychopharmacology* 31, no. 7 (2006)1345-55.

[187] E. Goldberg, R. M. Bilder, J. E. Hughes, S. P. Antin, and S. Mattis, "A Reticulo-Frontal Disconnection Syndrome," *Cortex* 25 (1989): 687-95.

[188] See table online at http://ajp.psychiatryonline.org/article.aspx?articleID=174101.

[189] Ziad Kronfol and Daniel G. Remick, "Cytokines and the Brain: Implications for Clinical Psychiatry," *Am J Psychiatry* 157 (May 2000): 683-694.

[190] This inflated R&D amount seems to become part of the expenses in the search for a cure. Thus it seems we pay twice for the remedies. Once as a donation and then when the government buys the medicine.

[191] US Environmental Protection Agency New Chemical Registration Candidates. See the table at http://www.epa.gov/opprd001/workplan/newchem.html.

[192] This analogy between the auditory and the kidney makes sense since sounds are linked to the lowest frequencies of the electromagnetic spectrum. Humans ears can detect sounds situated between 20 Hz and 20 kHz. This corresponds to a range between ELF (extremely low frequency) and VLF (very low frequency). In the same way, the electromagnetic spectrum ranges continuously from the ELF to the gamma rays. Below this, NASA reports symptoms linked to 18 Hz infrasound. These can be as diverse as watery eyes (NASA found that the eyeballs vibrate at 18 Hz), shortness of breath, anxiety, panic attack, pressure in the chest. "An experiment conducted on May 31, 2003, by a team of UK researchers, exposing some 700 people to infrasound of 17 Hz hidden in musical parts have resulted in an average of a quarter of listeners reporting anxiety, extreme sorrow, revulsion and fear, chills down the spine and pressure in the chest. Although infrasounds were physiologically received by all, only those 20 to 25% were *conscious* of its effects." These symptoms have been also reported from people suffering from allergies or food sensitivities. (See 'infrasound' on Wikipedia.) Infrasound can be found in natural phenomena of earthquake, flood, fire, and the aurora borealis, and is used by certain animals, notably elephants and whales, as a defense system. Jet pilots in mission are bathed in infrasonic waves. This is a problem since it can influence many physiological and psychological systems: "Pilot damaging effects include decrements in vision, speech, intelligence, orientation, equilibrium, ability to accurately discern situations, and make reasonable decisions." Fan, ventilators, cars, air conditioning can all be emitters of infrasound. The

Isis Code

symptoms are similar as those experienced in the presence of certain types of "ghosts." This fact leads sceptics to conclude that all supposed ghost sightings can be similarly explained. This would reassure them, but I strongly doubt this is the case. The infrasound wave can travel kilometers, and an aftermath of earthquakes can shatters windows kilometers away.

[193] See chapter on the analytical brain aspect.

[194] Neurophysiologist E. Roy John (2005) p.145.

[195] Lena, S. Parrot, O. Deschaux, S. Muffat-Joly, V. Sauvinet, B. Renaud, M. F. Suaud-Chagny, and C. Gottesmann, "Variations in Extracellular Levels of Dopamine, Noradrenaline, Glutamate, and Aspartate across the Sleep-Wake Cycle in the Medial Prefrontal Cortex and Nucleus Accumbens of Freely Moving Rats," *Journal of Neuroscience Research* 81 (2005): 891–899.

[196] Thomas J. Gould, "Addiction and Cognition," *ScienceDaily* (December 1, 2009).

[197] Catharine A. Winstanley, Quincey LaPlant, David E. H. Theobald, Thomas A. Green, Ryan K. Bachtell, Linda I. Perrotti, Ralph J. DiLeone, Scott J. Russo, William J. Garth, David W. Self, and Eric J. Nestler, "TosB Induction in Orbitofrontal Cortex Mediates Tolerance to Cocaine-Induced Cognitive Dysfunction," Departments of Psychiatry and Basic Neuroscience and Charles River Laboratories CSS, The University of Texas Southwestern Medical Center.

[198] "Biologists have surmised that transporter proteins of this type, which sit in the cell membrane, carry molecules through the otherwise impermeable membrane by shifting between at least three distinct structural states, controlled by ion gradients. In the first state, there is an outward-facing cavity. A compound will enter this cavity and attach to a binding site whereupon the protein will move to a second state with the cargo locked inside. The third state is formed when the protein opens up a cavity on the inward-facing side to release the compound into the cell. The switch between outward and inward-facing sides works rather like a 'kissing gate' in which the cavity is either on one side or the other but there is never a direct channel through the protein. However, until now, scientists had never observed the structural details of these three states in a single protein and theories about how the mechanism worked in detail were based on stitching together their observations from different transporters." "All Three Structures of Single Transporter Protein Revealed," *ScienceDaily*, (April 22, 2010). T. Shimamura, S. Weyand, O. Beckstein, N. G. Rutherford, J. M. Hadden, D. Sharples, M. S. P. Sansom, S. Iwata, P. J. F. Henderson, A. D. Cameron, "Molecular Basis of Alternating Access Membrane Transport by the Sodium-Hydantoin Transporter Mhp1," *Science* 328, no. 5977 (2010): 470 DOI: 10.1126/science.1186303.

[199] C. Gianoulakis, "Influence of the Endogenous Opioid System on High Alcohol Consumption and Genetic Predisposition to Alcoholism," *J Psychiatry Neurosci* 26, no. 4 (September 2001): 304–18.

[200] Thomas J. Gould, "Addiction Science and Clinical Practice," December 2010

[201] F. A. Bambico, N. Katz, G. Debonnel, G. Gobbi, "Cannabinoids Elicit Antidepressant-Like Behavior and Activate Serotonergic Neurons through the Medial Prefrontal Cortex." (Low doses of a specific cannabinoid agonist enhance serotonin activity in the medial prefrontal cortex, but high doses were ineffective and decreased serotonin.) *The Journal of Neuroscience* 27, no. 43 (October 24, 2007): 11700–11711, doi: 10.1523/JNEUROSCI.1636-07.2007.

[202] Georg Grön, Arthur P. Wunderlich, Manfred Spitzer, Reinhard Tomczak, and Matthias W. Riepe, "Brain Activation During Human Navigation: Gender-Different Neural Networks as Substrate of Performance," *Nature Neuroscience* 3 (2000): 404–408, doi:10.1038/73980.

[203] T. Hafting, M. Fyhn, S. Molden, M. Moser, E. Moser E, "Microstructure of a Spatial Map in the Entorhinal Cortex," *Nature* 436, no. 7052 (2005): 801–6. doi:10.1038/nature03721. PMID 15965463.

[204] Nicole C. Berchtold, David H. Cribbs, Paul D. Coleman, Joseph Rogers, Elizabeth Head, Ronald Kim, Tom Beach, Carol Miller, Juan Troncoso, John Q. Trojanowski, H. Ronald Zielke, Carl W. Cotman, and Stephen F. Heinemann, ed., "Gene Expression Changes in the Course of Normal Brain Aging are Sexually Dimorphic," The Salk Institute for Biological Studies, La Jolla, CA, August 12, 2008, http://www.pnas.org/content/105/40/15605

[205] From Marie Louise Von Franz, Number and Time, p. 33)

[206] Ancient Egypt, Royal Ontario Museum, http://www.rom.on.ca/programs/activities/egypt/learn/bakaakh.php.

[207] *A Mithraic Ritual*, (London, 1907) 102.

[208] Laszlo, *Science and the Akashic Field*, 31.

Notes, References, and Selected Bibliography

Chapter 4

[209] Bing Liu, Jun Li, Chunshui Yu, Yonghui Li, Yong Liu, Ming Song, Ming Fan, Kuncheng Li, Tianzi Jiang, "Haplotypes of Catechol-O-methyltransferase Modulate Intelligence-Related Brain White Matter Integrity," NeuroImage 50 (2010): 243-249.

[210] Encyclopedia Britannica article.

[211] Melissa Frederikse, Angela Lu, Elizabeth Aylward, Patrick Barta, and Godfrey Pearlson, "Sex Differences in the Inferior Parietal Lobule. Cerebral Cortex," *Cereb. Cortex* 9, no. 8 (1999): 896-901, doi:10.1093/cercor/9.8.896.

[212] "Science of The Heart: Exploring the Role of the Heart in Human Performance An Overview of Research," conducted by the Institute of HeartMath.

[213] Curiously, in another Jewish wedding, the rabbi informed the couple that it was a sign they were now part of the history of Jewish people and as such must never forget what happened to this community in the past.

[214] Michel Foucault, *The History of Sexuality, Vol. 3: The Care of the Self*, (New York: Random House, 1986).

[215] Oxford Classical Dictionary, David M. Halperin, s.v. "homosexuality," 720-723.

[216] *Psychophysiology*, (September-October 2001).

[217] Robin W. Simon, Anne E. Barrett, "Nonmarital Romantic Relationships and Mental Health in Early Adulthood: Does the Association Differ for Women and Men?" *Journal of Health and Social Behavior*, 51, no. 2 (June 2010): 168-182.

[218] http://www.cardio-coherence.com/Fr/Studies.htm

[219] Cognition, Brain, Behavior 11 (2007) 635-646A. Fisher 643

[220] "Brain Areas Critical To Human Time Sense Identified," *Daily University Science News*, http://www.unisci.com/stories/20011/0227013.htm.

[221] Intelligence 38 (2010) 293-303

[222] Ann M. Graybiel, "The Basal Ganglia," Magazine R509.

[223] Elkhonon Goldberg, *The New Executive Brain*, (New York: Oxford University Press) 211.

[224] Alessandro Treves, Ayumu Tashiro, Menno E. Witter, and Edvard Moser, "What Is the Mammalian Dentate Gyrus Good For?"

[225] F. Ferino, A. M. Thierry, and J. Glowinski, "Anatomical and Electrophysiological Evidence for a Direct Projection from Ammon's Horn to the Medial Prefrontal Cortex in the Rat," *Experimental Brain Research* 65, no. 2, 421-426, doi: 10.1007/BF00236315.

[226] Indeed, it was suggested, following MRI scans, that estrogen might have a stimulating effect on neuron proliferation (Tanapat et al., 1999), dendritic spine increases (Gould et al., 1990) and synaptogenesis (Woolley et al., 1996) in the hippocampus. Estrogen has also been reported to similarly induce myelination in the rat brain (Prayer et al., 1997).

[227] Michio Suzuki, Hirofumi Hagino, Shigeru Nohara, Shi-Yu Zhou, Yasuhiro Kawasaki, Tsutomu Takahashi, Mie Matsui, Hikaru Seto, Taketoshi Ono, and Masayoshi Kurachi, "Male-Specific Volume Expansion of the Human Hippocampus During Adolescence."

[228] Jean A. Frazier, Director of the Child and Adolescent Neuropsychiatric Research Program at Cambridge Health Alliance, Harvard Medical School, Boston, Massachusetts.

[229] Claire-Dominique Walker, Sophie Deschamps, Karine Proulx, Mai Tu, Camilla Salzman, Barbara Woodside, Sonia Lupien, Nicole Gallo-Payet, and Denis Richard.

[230] Gilberto Paz-Filho, Ma-Li Wong, and Julio Licinio, "Circadian Rhythms of the HPA Axis and Stress."

[231] Mark Hyman, MD, "Systems Biology, Toxins, Obesity, and Functional Medicine," 13th International Symposium of the Institute for Functional Medicine, http://www.ultrawellnesscenter.com.

[232] T. L. Lassiter, S. Brimijoin, "Rats Gain Excess Weight after Developmental Exposure to the Organophosphorothionate Pesticide, Chlorpyrifos," *Neurotoxicology and Teratology*, 30, no. 2 (March-April 2008):125-30.

[233] Rene J. Huster, Carsten Wolters, Andreas Wollbrink, Elisabeth Schweiger, Werner Wittling, Christo Pantev, Markus Junghofer, "Effects of Anterior Cingulate Fissurization on Cognitive Control During Stroop Interference, 'Human Brain Mapping 30, no. 4, (June 20, 2008):1279-1289.

[234] Murat Yücel, Geoffrey W. Stuart, Paul Maruff, Dennis Velakoulis, Simon F. Crowe, Greg Savage, and Christos Pantelis, "Hemispheric and Gender-Related Differences in the Gross Morphology of the Anterior Cingulate/Paracingulate Cortex in Normal Volunteers: An MRI Morphometric Study."

Isis Code

[235] M. Remondes and E. M. Schuman, "Role for a Cortical Input to Hippocampal Area CA1 in the Consolidation of Long-Term Memory," *Nature* 431, no. 7009 (2004): 699-703.

[236] A. Bartels and S. Zeki, "The Neural Correlates of Maternal and Romantic Love," *Neuroimage*, 21, no. 3, 1155-1166. 10.1016/j.neuroimage.2003.11.003. http://eprints.ucl.ac.uk/167379/.

[237] Anderson Teicher, "Sex Differences in Dopamine Receptors and Their Relevance to ADHD," *Neurosci Biobehav Rev.* 24, no. 1 (January 2000):137-41.

[238] Daniel P. Eisenberg, Philip D. Kohn, Erica B. Baller, Joel A. Bronstein, Joseph C. Masdeu, and Karen F. Berman, : "Seasonal Effects on Human Striatal Presynaptic Dopamine Synthesis."

[239] when one member of a group takes action against another, with a possible resulting benefit to the society, but at a possible cost to himself.

[240] Morishima et al., "Linking Brain Structure and Activation in Temporoparietal Junction to Explain the Neurobiology of Human Altruism."

[241] "Dopamine Type 2/3 Receptor Availability in the Striatum and Social Status in Human Volunteers" by Diana Martinez, Daria Orlowska, Rajesh Narendran, Mark Slifstein, Fei Liu, Dileep Kumar, Allegra Broft, Ronald Van Heertum, and Herbert D. Kleber. Martinez, Orlowska, Slifstein, Liu, Kumar, Broft, and Kleber are affiliated with the Department of Psychiatry, while Van Heertum is with the Department of Radiology, all at Columbia University, College of Physicians and Surgeons, New York, New York. Narendran is from the Department of Radiology, University of Pittsburgh, Pittsburgh, Pennsylvania. The article appears in *Biological Psychiatry*, 67, no. 3 (February 1, 2010).

[242] Karen Wynn, professor of psychology at Yale, was senior author of the study. J. Kiley Hamlin was lead author, and Paul Bloom, professor of psychology, was a third author. *Nature* (November 22, 2007). The animation can be viewed at www.yale.edu/infantlab/socialevaluation.

[243] Researchers Scott A. Huettel and Dharol Tankersley at Duke University, *Nature Neuroscience* (February 2007).

[244] Yosuke Morishima, Daniel Schunk, Adrian Bruhin, Christian C. Ruff, and Ernst Fehr, "Linking Brain Structure and Activation in the Temporoparietal Junction to Explain the Neurobiology of Human Altruism, *Neuron*, (July 12, 2012).

[245] Ernst Fehr, Dr. Thomas Baumgartner, and Daria Knoch reveal the neuronal networks behind self-control in an article recently published in *Nature Neuroscience*.

[246] Valtteri Kaasinen, M.D., Ph.D.; Kjell Någren, Ph.D.; Jarmo Hietala, M.D., Ph.D.; Lars Farde, M.D., Ph.D.; Juha O. Rinne, M.D., Ph.D. "Sex Differences in Extrastriatal Dopamine D2-Like Receptors in the Human Brain" Am J Psychiatry 2001;158:308-311. 10.1176/appi.ajp.158.2.308

[247] AGRICOH is a consortium of agricultural cohort studies initiated by the US National Cancer Institute (NCI) and coordinated by the International Agency for Research on Cancer (IARC) since October 2010. As of October 2012, 27 cohorts from 5 continents comprise AGRICOH. The joining studies are from South Africa (2), Canada (3), Costa Rica (2), USA (7), Australia (2), Korea (1), New Zealand (2), Denmark (1), France (3), Norway (3) and the UK (1).www. http://agricoh.iarc.fr/about/index.php

[248] "Early-Onset Puberty Puts Girls at Risk of Medical Problems," accessed December 1, 2009, http://www.redorbit.com/news/health/1067387/earlyonset_puberty_puts_girls_at_risk_of_medical_problems/index.html.

[249] Louis J. Guillette Jr., D. Andrew Crain, Mark P. Gunderson, Stefan A. E. Kools, Matthew R. Milnes, Edward F. Orlando, Andrew A. Rooney, and Allan R. Woodward, "Alligators and Endocrine Disrupting Contaminants: A Current Perspective."

[250] "The Brain: The Color of Stress," *The World of Science Discover Magazine*, (May 1997).

[251] Lim Li Ching, "Is Ecological Agriculture Productive?" *Third World Network* (November 2008).

[252] James Prescott, PhD, "The Origins of Love & Violence And The Developing Human Brain, A Conversation with Michael Mendizza," *Touch the Future* (1995) http://ttfuture.org/files/2/public/esa_jwp_origins_mm.pdf.

[253] He Johnson, r.a. (New York, 1974) 1.

[254] This can be found in the story of Isis. Horus has a time assigned to him to be born between Osiris and Seth, but is only born after Isis was born.

[255] Bohm, 1988 p.26.

[256] Stephen Hawking, *A Brief History of Time: From the Big Bang to Black Holes*, (New York: Bantam Books, 1990) 174.

[257] Freeman Dyson is a physicist and mathematician and professor emeritus at the Institute for Advanced Study, Princeton, New Jersey. His contributions to science include the unification of the

Notes, References, and Selected Bibliography

three versions of quantum electrodynamics invented by Feynman, Schwinger, and Tomonaga. Dyson's writings on the meaning of science and its relation to other disciplines, especially religion and ethics, challenge humankind to reconcile technology and social justice.

[258] We know that Harpocrates was identified as the "God of silence," but Horus has an index finger touching his chin, which for ancient Egyptians was associated to will. The association with silence is correct in the sense that the phase is of one associated to initiation.

[259] We can find this in numerous statues and depictions. For example, there is a statue of the solar god Mithras being born from the rock, naked but for the Phrygian cap on his head. (Marble, 180-192 AD, from the area of S. Stefano Rotondo, Rome.) It is ubiquitous: in statues of ancient Egypt, and seen in Tibetan and Hawaiian tradition as well. Not surprisingly the sign of Aquarius, the era we are entering upon, is symbolized by Ganymede, who wears a Phrygian hat. The solar association is obvious in all of these.

[260] Jung *Correspondance 1950-1954*, p. 85.

[261] Medical biocybernetics explains this by a lack or blockage of energy in the liver, thus an inability to control the pancreas phase. This, result in an inflammation of the pancreas. Despite fifty years of research, conventional medicine declares this link as being of unknown origin.

[262] Ibid. p. 61.

[263] G. Tsagalis, S. Zerefos, N. Zerefos, "Cardiorenal Syndrome at Different Stages of Chronic Kidney Disease, *Int J Artif Organs*, 30, no. 7 (July 2007): 654-76.

[264] "And I will kill her children with death; and all the churches shall know that I am he which searcheth the reins and hearts: and I will give unto every one of you according to your works." Rev. 2:23 (KJV).

[265] "Oh let the wickedness of the wicked come to an end; but establish the just: for the righteous God trieth the hearts and reins." Ps. 7:9 (KJV).

[266] "But, O Lord of hosts, that judgest righteously, that triest the reins and the heart, let me see thy vengeance on them: for unto thee have I revealed my cause." Jer. 11:20 (KJV). Note that in all of these, the word 'reins' is used instead of kidney. It is the archaic form not in use anymore.

[267] R. McCraty, *"Heart-Brain Neurodynamics: The making of Emotions*, Hardwood Academic Publishers.

[268] R. McCraty, "The Energetic Heart: Bioelectromagnetic Interactions within and between People," HeartMath Research Center, Institute of HeartMath, Publication No. 02-035, (2002).

[269] R. McCraty, citing Mike Atkinson and Dana Tomasino, "Modulation of DNA Conformation by Heart-Focused Intention." Also, significant changes in the conformation of the DNA (mean change 10.27%) were observed. In some cases, changes in DNA conformation of up to 25% were observed, indicating a robust effect.

[270] http://www.heartmath.org/research/science-of-the-heart/head-heart-interactions.html.

[271] See the Global Meditation Project. Dr. Buryl Payne, "The Power of Thought to Influence the Sun," the Academy for Peace Research: "Millions of people were invited to meditate at six dates each year (the Solstices, Equinoxes, etc.). A 3-1/2 year study culminated in June 1988. On the average the effects of several million people meditating appears to have resulted in a decrease in solar activity of up to 30% for a period of 7 to 10 days following the meditations." Also see a review of 130 studies on the subject: Daniel J. Benor, *Healing Research, vol.1-2* (Munich: Helix Verlag, 1993) and Larry Dossey, *Healing Words: The Power of Prayer and the Practice of Medicine* (San Francisco: HarperSanFrancisco, 1993).

[272] L. H. Powell, L. Shahabi, CE Thoresen, "Religion and Spirituality: Linkages to Physical Health," *The American Psychologist* 58, no. 1(January 2003): 36-52. doi:10.1037/0003-066X.58.1.36. PMID 12674817.

[273] Martha McClintock, researcher of University of Chicago (1971).

[274] H. Coetzee, PhD, "Biomagnetism and Bio-Electromagnetism: The Foundation of Life," Academy for the Future Science. Originally published in *Future History*, volume 8. http://www.affs.org/html/biomagnetism.html.

[275] Papineau Hazen et al., "Mineral Evolution," *American Mineralogist* 93: 1693-1720.

[276] Dr. Reams and Cliff Dudley, "Choose Life or Death."

[277] www.dailymotion.com/video/xnmr42_ericsson_demonstrates_technology_using_human_body-conductivity.

[278] A. P. Dubrov, *The Geomagnetic Field and Life*, 139.

[279] "Canadians are at risk of vitamin D deficiency from October to April because winter sunlight in northern latitudes does not allow for adequate vitamin D production," says Julie Foley, president & CEO of Osteoporosis Canada. http://medicine.ucalgary.ca/about/vitaminD/Hanley. also, recent

Isis Code

studies have also shown that higher vitamin D levels are associated with lower insulin levels which are linked with reduced risks of both diabetes and cancer (Cohen, Endocr. Relat. Cancer, 2012).

[280] New Crystal in the Pineal Gland:Characterization and Potential Role in Electromechano-Transduction Baconnier Simon(1), Lang Sidney B. (2), De Seze Rene(3) (1) DRC, Toxicologie Expérimentale, INERIS, 60550 Verneuil-en-Halatte, France. E-mail: simon.baconnieretudiant@ineris.fr (2)Department of Chemical Engineering, Ben-Gurion University of the Negev, 84105 Beer Sheva, Israel. E-mail: lang@bgumail.bgu.ac.il (3) As (1) above, but E-mail: Rene.De-Seze@ineris.fr

[281] "Fluoride in Drinking Water: A Scientific Review of EPA's Standards," *National Academies Press*, (2006): 221-222.

[282] J. Luke, "The Effect of Fluoride on the Physiology of the Pineal Gland," (University of Surrey, Guildford, 1997), 177.

[283] Miriam Jang, "Advances in Autism: Anti-Inflammatory Effects of Melatonin": "Good news, for those of you who use melatonin to help your Autistic child sleep at night, there seems to be an extra beneficial side effect: the anti-inflammatory effects, especially in areas like the gut! So often we worry about bad side effects in sleep aides, so it is such a pleasure to find good side effects!" (March 26, 2008).

[284] J. C. Mayo, R. M. Sainz, D. X. Tan, R. Hardeland, J. Leon, C. Rodriguez, R. J. Reiter, "Anti-Inflammatory Actions of Melatonin and Its Metabolites, N1-acetyl-N2-formyl-5-methoxykynuramine (AFMK) and N1-acetyl-5-methoxykynuramine (AMK), in Macrophages," *J Neuroimmunol* 165, no. 1-2 (August 2005):139-49.

[285] R. Sandy, G. I Awerbuch, "The Pineal Gland in Multiple Sclerosis," *Int J Neurosci*. 61, no. 1-2 (November 1991): 61-7.

[286] R. Sandy and G. I. Awerbuch, "The Relationship of Pineal Calcification to Cerebral Atrophy on CT Scan in Multiple Sclerosis," *Int J Neurosci*. 76, no. 1-2 (May 1994):71-9.

[287] C. M. Poser and J. C. Vernant, "Multiple Sclerosis in the Black Population," US National Library of Medicine National Institutes of Health.

[288] S. Lee, L. A. Donehower, A. J. Herron, D. D. Moore, L. Fu, Frank Beier, ed., "Disrupting Circadian Homeostasis of Sympathetic Signaling Promotes Tumor Development in Mice. PLoS ONE, 5, no. 6 (June 7, 2010): e10995. doi:10.1371/journal.pone.0010995.

[289] Edward Frederick, "The Nodal Centers and Their Neuroendocrine and Hormonal Correlates., Block IV, Ph.D. July 2005. Updated January 2010.

[290] John Harden, Nicholas Diorio, Alexander G. Petrov, and Antal Jakli, "Chirality of Lipids Makes Fluid Lamellar Phases Piezoelectric," *Journals Phys. Rev. E* 79, no. 1 (2009).

[291] H. Coetzee, 2003.

[292] Louis Slesin reporting on IARC (International Agency for Research on Cancer) meeting of 2001.

[293] Gudrun Bornhöft and Peter Matthiessen, Peter, eds., "Homeopathy in Healthcare Homeopathy in Healthcare Effectiveness, Appropriateness, Safety, Costs," (2012): 300.

[294] Joseph L. Kirschvink, Atusko Kobayashi-Kirschvink, Barbara J. Woodford, "Magnetite Crystals in the Brain as the Primary Facilitators of Consciousness," *Memory Proc. Natl. Acad. Sci. USA*. 89 (August 1992):7683-7687. "Biophysics Magnetite Biomineralization in the Human Brain (iron/extremely low frequency magnetic fields)," Division of Geological and Planetary Sciences, The California Institute of Technology, communicated by Leon T. Silver, May 7, 1992.

[295] R. R. Baker, J. G. Mather, J. H. Kennaugh, "Single Magnetite Crystal in the Human Brain" and "Magnetic Bones in Human Sinuses," *Nature* 301, no. 5895 (January 6, 1983):79-80.

[296] Kirschvink, Kobayashi-Kirschvink, Woodford, "Biophysics Magnetite Biomineralization in the Human Brain (iron/extremely low frequency magnetic fields), communicated by Leon T. Silver, May 7, 1992

[297] Ibid.

[298] Mariana Sincal, Diana Ganga, Diana Argherie, Doina Bica, "Antitumor Effect of Magnetite Nanoparticles in Cat Mammary Adenocarcinoma," *Journal of Magnetism and Magnetic Materials* 293, no. 1 (March 2005): 438-441.

[299] I. R. Schwab, G. R. O'Connor, "The Lonely Eye," *British Journal of Ophthalmology* 89, no.3 March 2005): 256.

[300] Second harmonic generation (SHG; also called frequency doubling) is a nonlinear optical process, in which photons interacting with a nonlinear material are effectively "combined" to form new

photons with twice the energy, and therefore twice the frequency and half the wavelength of the initial photons. Wikipedia

[301] Simon Baconnier, Sidney B. Lang, Rene De Seze, "New Crystal in the Pineal Gland: Characterization and Potential Role in Electromechano-Transduction."

[302] Kiiki Matsumoto and Stephen Birch, *Hara Diagnosis: Reflections on the Sea*, (Paradigm Publications, 1988),121.

[303] The Nan-ching, ancient Chinese medical classic, compiled during the first century AD by an unknown author.

[304] Manisha Mukewar, V. V. Baile, "Clinical Studies in Bioelectromagnetic Medicine Biological Effects of Electric Fields on Spleen and Liver of Rat," *Journal of Bioelectromagnetic Medicine* 10 (July 2004).

[305] He Ruo Yu, ZiWu Liu Zhu Zhen Jing, p.122.

[306] Matsumoto, Birch, *Hara Diagnosis: Reflections on the Sea*.

[307] This reminds me of a dream I had, in which my eldest son and I were silver hooks hanging at the border of God's orange-gold garment, two of countless others.

[308] Etymological Dictionary of Kadokawa Kanji (1985).

[309] Claude Larre and Elizabeth Rochat de la Vallée, *Heart Master Triple Heater*, (Monkey Press).

[310] Sinerik N. Ayrapetyan and M. S. Markov, "Bioelectromagnetics," *Springer*, 31-63 (2006): 33.

[311] Don Martin and Neil Gunton, "Anatomy of an MRI," (February 24, 2008), accessed October 6, 2009.

[312] Sinerik N. Ayrapetyan, and M. S. Markov, "Cell Aqua Medium as a Primary Target for the Effect of Electromagnetic Fields," *Bioelectromagnetics* 31, no. 63 (2006).

[313] V. M. Bakaĭkin, "Morphologic and Histochemical Organization of the Pericardial Nervous Apparatus of Vertebrates and Humans, *Arkh Anat Gistol Embriol*, 74, no. 1 (January 1978): 82-9.

[314] Roustem Miftahof, Hong Gil Nam Pohang, and David Lionel Wingate, "Mathematical Modeling and Simulation in Enteric Neurobiology," *World Scientific Pub.* (2009):12.

[315] Kate Karelina and Greg J. Norman "Oxytocin Influence on the Nucleus of the Solitary Tract: Beyond Homeostatic Regulation" The Journal of Neuroscience, 15 April 2009, 29(15): 4687-4689; doi: 10.1523/JNEUROSCI.0342-09.2009

[316] Oxytocin enhances cranial visceral afferent synaptic transmission to the solitary tract nucleus James H. Peters,1 Stuart J. McDougall,1 Daniel O. Kellett,2 David Jordan,2,# Ida J. Llewellyn-Smith,3 and Michael C. Andresen1

[317] Kate Karelina and Greg J. Norman "Oxytocin Influence on the Nucleus of the Solitary Tract: Beyond Homeostatic Regulation" The Journal of Neuroscience, 15 April 2009, 29(15): 4687-4689; doi: 10.1523/JNEUROSCI.0342-09.2009

[318] Traditionally, it is said that prostitutes tend to have this type of voice — does this mean their vagus nerve is not functional?

[319] "The Relationship between Vagus Nerve and the Heterogeneous Distribution of Cx40, Cx43 and Atrial Fibrosis in Rapid Atrial Pacing Dogs with or without SVC-AO Fat PAD," *Cardiovascular Diseases*.

[320] Julian F. Thayer and Richard D. Lane, "Claude Bernard and the Heart-Brain Connection: Further Elaboration of a Model of Neurovisceral Integration, *Neuroscience and Biobehavioral Reviews* 33 (2009): 81-88

[321] B. R. **Dev, M. Nandakumaran, L. Philip, and S. J. John**, "Brain Natriuretic Peptide-Mediated Changes in the Extracellular Neurotransmitter Turnover in the Rostral Ventrolateral Medulla," (1998).

[322] In 2007, a new role was found for the photoreceptive ganglion cell of the retina. Farhan H. Zaidi and colleagues, including Russell Foster, George Brainard, Charles Czeisler, and Steven Lockley, showed that, at least in humans, the retinal ganglion cell photoreceptor contributes to conscious sight as well as to non-image-forming functions like circadian rhythms, behavior, and pupillary reactions. The opsin found in the photoreceptor is called melanopsin. Atypical in vertebrates, melanopsin functionally resembles invertebrate opsins. In structure, it is an opsin, a retinylidene protein variety of a G protein-coupled receptor. G proteins are the first downstream component in photochromic signaling (Romero et al., 1991; Romero et al. 1993).

[323] Ibid.

[324] Bridget Engel, Natalie Staats Reiss, and Mark Dombeck, "Causes of Eating Disorders— Biological Factors," (February 2, 2007).

[325] Joseph Rhawn, "Neocortex," in *Neuropsychiatry, Neuropsychology, Clinical Neuroscience*, 3rd ed. (Academic Press).

Isis Code

126 H. Coetzee, "Biomagnetism and Bio-Electromagnetism: The Foundation of Life," in *Future History*, vol. 8.

127 Line S. Loken, Johan Wessberg, India Morrison, Francis McGlone, and Håkan Olausson, "Coding of Pleasant Touch by Unmyelinated Afferents in Humans," *Nature Neuroscience* 12, (2009): 547–548, doi:10.1038/nn.2312.

128 In the case of normal touch and certain types of pain, the third-order neuron has its cell body in the ventral posterior nucleus (VPN) of the thalamus, and ends in the postcentral gyrus of the parietal lobe.

129 Y. Cheng, H. Chou, J. Decety, Y. Chen, D. Hung, L. Tzeng, and P. Lin, "Sex Differences in the Neuroanatomy of Human Mirror-Neuron System," *Neuroscience*, 158 (2009): 713–720. http://home.uchicago.edu/decety/publications/Cheng_N2009.pdf.

130 Researchers make first direct recording of mirror neurons in human brain, April 12, 2010, Phys Org.

131 Lyrics by Dale Wasserman.

132 Ibid.

133 Ibid.

134 Sri Ramana. He also said: "How will the mind become quiescent? By the inquiry 'who am I?' The thought 'who am I?' will destroy all other thoughts, and like the stick used for stirring the burning pyre, it will itself in the end get destroyed. Then, there will arise Self-realization. As each thought arises, one should inquire with diligence, 'To whom has this thought arisen?' The answer that would emerge would be 'To me.' Thereupon if one inquires 'Who am I?' the mind will go back to its source; and the thought that arose will become quiescent. With repeated practice in this manner, the mind will develop the skill to stay in its source. When the mind that is subtle goes out through the brain and the sense-organs, the gross names and forms appear; when it stays in the heart, the names and forms disappear. Not letting the mind go out, but retaining it in the Heart is what is called 'inwardness' (antar-mukha). Letting the mind go out of the Heart is known as 'externalization' (bahir-mukha). Thus, when the mind stays in the Heart, the 'I' which is the source of all thoughts will go, and the Self which ever exists will shine. Whatever one does, one should do without the egoity 'I.' If one acts in that way, all will appear as of the nature of Siva (God). The followers of the 'I am Brahman' and 'Neti-Neti' schools share a common belief that the Self can be discovered by the mind, either through affirmation or negation. This belief that the mind can, by its own activities, reach the Self is the root of most of the misconceptions about the practice of self-inquiry. A classic example of this is the belief that self-inquiry involves concentrating on a particular center in the body called the Heart-center. The Heart is not really located in the body and that from the highest standpoint it is equally untrue to say that the 'I'-thought arises and subsides into this center on the right of the chest. The Heart 'as it is' is not a location, it is the immanent Self and one can only be aware of its real nature by being it. It cannot be reached by concentration."

135 David Godman, "The Nature of The Self," quote from Sri Ramana

136 Benedict Carey, "Watching New Love as It Sears the Brain," *New York Times*, May 31, 2005.

137 http://www.nytimes.com/2005/05/31/health/psychology/31love.html?pagewanted=print

138 *The Wisdom Paradox* and *The New Executive Brain*.

139 "Scientists Clone First-Ever Bull," *ScienceDaily*, September 13, 1999.

140 Ibid.

141 Cynthia Mills, "Second Chance Conservation," University of Washington, 7, no. 4 (October-December 2006).

142 M. A. Persinger, "Religious and Mystical Experiences as Artifacts of Temporal Lobe Function: A General Hypothesis," *Perceptual and Motor Skills*, 57 no. 3 (1991): 1255–62. doi:10.2466/pms.1983.57.3f.1255. PMID 6664802.

143 Robyn L. Bluhma, Elizabeth A. Osucha, Ruth A. Laniusa, Kristine F. Boksman, Richard W. J. Neufeld, Jean Theberge, and Peter Williamson, "Default Mode Network Connectivity: Elects of Age, Sex, and Analytic Approach."

144 Devarajan Sridharan, Daniel J. Levitin, and Vinod Menon, and Marcus E. Raichle, ed., "A Critical Role for the Right Fronto-Insular Cortex in Witching between Central-Executive and Default-Mode Networks," (June 20, 2008).

145 July/August 2009 issue of the journal *Child Development* was conducted by researchers at Queen's University at Kingston in Ontario, Canada.

Notes, References, and Selected Bibliography

346 C. Di Dio, E. Macaluso, and G. Rizzolatti, "The Golden Beauty: Brain Response to Classical and Renaissance Sculptures," PLoS ONE 2, no. 11(2007): e1201, doi:10.1371/journal.pone.0001201.

347 Helmholtz Association of German Research Centres, "Golden Ratio Discovered in Quantum World: Hidden Symmetry Observed for the First Time in Solid State Matter," ScienceDaily (January 7, 2010).

348 Di Dio, Macaluso, and Rizzolatti, "The Golden Beauty: Brain Response to Classical and Renaissance Sculptures."

349 Jeff Carpenter, "Beautiful Faces Trigger Reward Center of Brain," http://abcnews.go.com/Health/story?id=117131&page=1#.UDPaUPW04I8.

350 Dina Temple-Raston, "Neuroscientist Uses Brain Scan to See Lies Form," (October 2007), http://www.npr.org/templates/story/story.php?storyId=15744871. "The key point is that you need to exercise a system that is in charge of regulating and controlling your behavior when you lie more than when you just say the truth. Three areas of the brain generally become more active during deception: the anterior cingulated cortex, the dorsal lateral prefrontal cortex and the parietal cortex".

351 D. D. Langleben, L. Schroeder, J. A. Maldjian, R. Gur, C. S. McDonald, J. D. Ragland, C. P. O'Brien, and A. R. Childress, "Rapid Communication Brain Activity during Simulated Deception: An Event-Related Functional Magnetic Resonance Study," NeuroImage 15 (January 4, 2002): 727–732, doi:10.1006/nimg.2001.1003.

352 I. Karton and T. Bachmann, "Effect of Prefrontal Transcranial Magnetic Stimulation on Spontaneous Truth-Telling," Behavioural Brain Research, (2011), doi: 10.1016/j.bbr.2011.07.028.

353 Jung, Correspondance 1950–1954, p. 219.

354 Ibid. 108.

355 Considerable data show that in general, women score higher on tasks involving mathematical calculations, verbal fluency, and perceptual speed (Hyde et al., 1990; Springer and Deutsch, 1993; Christiansen, 2001; Hyde and Linn, 1988), while men perform better in mathematical reasoning and visuospatial tasks.

356 R. Cohen Kadosh, Kathrin Cohen Kadosh, A. Kaas, A. Henik, and R. Goebel, "Notation-Dependent and -Independent Representations of Numbers in the Parietal Lobes," Neuron, 53, no. 2 (2007): 307–14.

357 "Instead of consciousness collapsing a quantum superposition in a succession of quantum jumps, we have consciousness offering a quantum plenum of superposed possibilities to the match with the more restricted possibilities of sensory input. Instead of a saltatory world line in the Heisenberg succession of objective tendencies and actual events, there is a continuous unfolding of worlds from a holoworld." Gordon Globus.

358 Gordon Globus, "Quantum Consciousness Is Cybernetic."

359 Asaf Gilboa, Claude Alain, Yu He, Donald T. Stuss, and Morris Moscovitch, "Ventromedial Prefrontal Cortex Lesions Produce Early Functional Alterations during Remote Memory Retrieval," Journal of Neuroscience 29, no. 15 (April 15, 2009): 4871–4881; doi: 10.1523/JNEUROSCI.5210-08.2009.

360 Online edition of the Proceedings of the National Academy of Sciences, November 12, 2007.

361 "Brain Matures a Few Years Late in ADHD, But Follows Normal Pattern," NIMH press release November 12, 2007.

362 E. Leibenluft et al., 1014 Biol Psychiatry 53, (2003):1009–1020

363 Judy Ann Prasad, Emily Marilyn MacGregor, Yogita Chudasama, "Selective Lesions of the Thalamic Reuniens in Rats Increase Impulsive Responses in the 5-Choice Reaction Time Task," (June 13, 2010).

364 Max. J. Hilz, Orrin Devinsky, Hanna Szczepanska, Joan C. Borod, Harald Marthol, and Marcin Tutaj, "Right Ventromedial Prefrontal Lesions Result in Paradoxical Cardiovascular Activation with Emotional Stimuli," accessed January 19, 2006.

365 It can disturb the biological rhythms of humans and animals and intensify current ailments, due to the overall interconnection between solar and geomagnetic activity, Schumann resonances, ionospheric waveguide, and the human brain and heart. In fact, increased solar activity and geomagnetic activity have been correlated to a significant surge in heart attacks and incidence of death (Villoresi et al., 1998), as well as a 30% to 80% increase in hospital admissions for cardiovascular disease (Oraevskii et al., 1998). Those who already present sympathetic overstimulation and a fragile feminine polarity are more at risk than others.

366 Mark A. Smith, "Brain Plasticity- What Is It and How Extraordinary Can It Be."

367 Norman Doidge, The Brain That Changes Itself, (New York: Penguin Books) 125–126.

368 "John Forbes Nash, Jr. (born June 13, 1928) is an American mathematician whose works in game theory, differential geometry, and partial differential equations have provided insight into the

Isis Code

forces that govern chance and events inside complex systems in daily life. His theories are used in market economics, computing, evolutionary biology, artificial intelligence, accounting, politics, and military theory. Serving as a senior research mathematician at Princeton University during the latter part of his life, he shared the 1994 Nobel Memorial Prize in Economic Sciences with game theorists Reinhard Selten and John Harsanyi." Wikipedia.

[369] John Forbes Nash Jr., interview *Schizophrenia Daily News Blog*, April 10, 2005.
[370] H. Hafner, "Gender Differences in Schizophrenia," *Psychoneuroendocrinology* 28 (2003): 17-54, www.sciencedirect.com.

Chapter 6

[371] Vita Sackville-West, Saint Joan of Arc (London Folio Society).
[372] Sandra **Escher**, **Marius Romme**, Alex **Buiks**, **Philippe Delespaul, and Jim Van Os**, "Independent Course of Childhood Auditory Hallucinations: A Sequential 3-Year Follow-up Study."
[373] This study of polarities is quite interesting, as in the esoteric theory of religions, two healthy polarities working together have been known to express deity, perfection, and sacredness. As such, the bell that represents the feminine vibrates under the mallet, which is masculine. The vibrations spreading in the atmosphere are balanced in their polarities, purifying in some sort the imbalance of the environment and creating the perfect conditions for the "divine" world to manifest.
[374] Antonio R. Damasio, *Descartes Error: Emotion, Reason, and the Human Brain*, (New York: Avon Books 1995).
[375] Sam Harris, Sameer A. Sheth, and Mark S. Cohen, "Functional Neuroimaging of Belief, Disbelief, and Uncertainty."
[376] T. C. McLuhan, *Pieds Nus sur la Terre Sacrée*, 48.
[377] In their twentieth year, men who had a dedication used this ritual to express their link with Wakan Tanka (consciousness observed in the perfection of nature). Since they considered their bodies the only things that were really theirs, they gave their bodies to express their gratitude. They would cut holes in their flesh and push wooden pegs into them. The pegs were fixed to tethers that were linked to the central post of the sun dance.
[378] Red Jacket Sa-go-ye-wat-ha, Seneca chief (ca. 1750-1830): "Our religion teaches us to be thankful for all we receive, to love each other and to be united. We never fight over religion because it is a subject which concerns every man in front of the Great Spirit."
[379] "Hunting the Hairy Mammoth by Driving It into Marshland Using Men and Fire," Display at Venusium, the museum at Willendorf.
[380] A Venus figurine from the Swabian Jura rewrites prehistory, 13 May 2009 Universitaet Tübingen. http://www.alphagalileo.org/ViewItem.aspx?ItemId=57684&CultureCode=en.
[381] This is why those fertility statues could also be used to avert evil.
[382] This hadith (Mohammed saying) is reported by Imām At-Tirmidhi in his Sunan.
[383] To exemplify what a human devoid of his feminine polarity does and how he sees what beauty is, consider the following: Prior to committing multiple murders, Luke Woodham, age sixteen— thus in the heart phase— wrote in his journal that he and an accomplice had beat, burned, and tortured his female dog, Sparkle, to death. Woodham said it was "true beauty." He poured liquid fuel down his dog's throat and set fire to her neck, both inside and outside. The neck is linked to information-regulation, to the heart phase of our system. On 10/1/1977, Woodham stabbed his mother to death and then went to his high school, where he shot and killed two classmates— two girls aged sixteen and seventeen— and injured seven others. In June 1998, Woodham was found guilty of three murders and seven counts of aggravated assault. He was sentenced to three life sentences and an additional twenty years for each assault.
[384] "Are Casinos Like Cocaine for the Brain? (A Conversation with Harvard Neuroscientist Hans Breiter and Institute for American Values Fellow and Journalist Paul Davies on the Science of Gambling Addiction and the Public Health Costs and Consequences of Expanded Casino Gambling in New York)," http://www.centerforpublicconversation.org/pr/20120412.pdf.
[385] J. Barrois, *Dactylologie et langage primitive restitués d'après les monuments*, (Paris:1850) 282.
[386] c.1300, "sum, aggregate of a collection," from Anglo-Fr. noumbre, O.Fr. nombre and directly from L. numerus "a number, quantity," from PIE root "nem-" to divide, distribute, allot," http://www.etymonline.com. In French, chiffre originates from Zero, a notion of the infinite. "calque du skr. śūnya = id. -, par l'intermédiaire du lat. médiév. cifra = zéro - (xiies., Anon., Algor. Salem. ds Mittellat. W. s.v., 574, 12). Le zéro étant l'innovation la plus importante et la plus caractéristique du système numérique ar.,

Notes, References, and Selected Bibliography

le mot chiffre a fini par désigner toutes les figures de ce système, d'où 2. Le sens 3 est dû au fait que le zéro semblait doué d'un pouvoir magique." http://www.cnrtl.fr/etymologie/chiffre

[387] Wikipedia, Pythagorianism.

[388] This is actually a lost book whose contents are preserved in Damascius, de principiis, quoted in Kirk and Raven, *The Pre-Socratic Philosophers*, (Cambridge University Press, 1956) 55.

[389] The Pythagoreans held the pentacle sacred to Hygeia and the five points of the pentagram to each represent one of the five elements that make up the universe: fire, water, air, earth, and psyche or energy, fluid, breath, matter, and mind or liquid, gas, solid, plasma and aethyr.. Wikipedia: In Greek and Roman mythology, Hygieia (also Hygiea or Hygeia, Greek 'Υγιεία' or 'Υγεία, Latin Hygēa or Hygīa), was a daughter of the god of medicine, Asclepius. She was the goddess/personification of health (Greek: ὑγίεια - hugieia[1]), cleanliness and sanitation.

[390] Michael Chase, "Studies in pre-Platonic Demiurgy: The case of Pherecydes of Syros," http://cnrs.academia.edu/MichaelChase/Papers/1180399/Studies_in_prePlatonic_demiurgy_The_case_of_Pherecydes_of_Syros

[391] "Tout est psychique avant d'être quoi que ce soit d'autre." Everything is psychic before becoming anything else. Jung, *Correspondance 1950–1954*, p.127.

[392] http://galusaustralis.com/2009/09/1608/horny-jew-whats-the-deal-with-michelangelos-moses/

[393] www.jungtao.edu/index.php/classical-chinese-medicine/resources?.

[394] http://www.goldenmuseum.com.

[395] National Institute of Neurological Disorders and Stroke

[396] A. Lilja, S. Hagstadius, J. Risberg, L. G. Salford, and G. J. W. Smith, "Frontal Lobe Dynamics in Brain Tumor Patients: A Study of Regional Cerebral Flow and Affective Changes before and after Surgery," *J. Neuropsychiatry Neuropsychol Behavioral Neurology* 5, no. 4 (1992): 294–300.

[397] D. Walsh, Why Do They Act That Way? A Survival Guide to the Adolescent Brain for You and Your Teen (New York: Free Press, 2004).

[398] J. N. Giedd, "Structural Magnetic Resonance Imaging of the Adolescent Brain," *NY Acad Sci* 1021 (2004):77–85.

[399] Ibid.

[400] Christian Latin borrowed from the classic Greek meaning, which was, among other things, "which disunite" or "throws aside." Le Robert, diable.

[401] lit. "that which is thrown or cast together," from syn- "together" + bole "a throwing, a casting, the stroke of a missile, bolt, beam," from bol-, nom. stem of ballein "to throw" (see ballistics). The sense evolution in Greek is from "throwing things together," http://www.etymonline.com.

[402] "Look down from thy holy habitation, from heaven, and bless thy people Israel, and the land which thou hast given us, as thou swarest unto our fathers, a land that floweth with milk and honey" Deut. 26:15 (KJV).

[403] W. K. Simmons, A. Martin, and L. W. Barsalou, "Pictures of Appetizing Foods Activate Gustatory Cortices for Taste and Reward," *Cereb. Cortex.* 15 (2005): 1602–1608.

[404] Angela Wagner, Howard Aizenstein, Laura Mazurkewicz, Julie Fudge, Guido K. Frank, Karen Putnam, Ursula F. Bailer, Lorie Fischer, and Walter H. Kaye, "Altered Insula Response to Taste Stimuli in Individuals Recovered from Restricting Type Anorexia Nervosa," *Neuropsychopharmacology* 33 (2008):513–523, doi:10.1038/sj.npp.1301443.

[405] Walter H. Kaye, Julie L. Fudge, Martin Paulus, "New Insights into Symptoms and Neuro-Circuit Function of Anorexia Nervosa," *Nature reviews/ neuroscience* 10 (2009):575–583.

[406] E. Gould, P. Tanapat, B. S. McEwen, G. Flugge, and E. Fuchs, "Proliferation of Granule Cell Precursors in the Dentate Gyrus of Adult Monkeys Is Diminished by Stress," *PNAS* 95, no.6 (1998): 3168–3171. doi:10.1073/pnas.95.6.3168. PMC 19713. PMID 9501234.

[407] "Kitagiri Sutta-Majjhima Nikaya," Urbandharma.org.

[408] R. Michael Anson et al., "Intermittent Fasting Dissociates Beneficial Effects of Dietary Restriction on Glucose Metabolism and Neuronal Resistance to Injury from Calorie Intake," (2002).
Mark Mattson from the National Institute of Aging
C. Zauner, B. Scheneeweiss et al., "Resting Energy Expenditure in Short-Term Starvation Is Increased as a Result of an Increase in Serum Norepinephrine," *Am J Clin Nutr*. (2000).
Webber J. Macdonald, "The Cardiovascular, Metabolic and Hormonal Changes Accompanying Acute Starvation in Men and Women," *British Journal of Nutrition*, (1994).
C. T. Kimber, "Short-Term Fasting Induces Profound Neuronaynthesis in the Brain of Adult Rats," (1991).

Roberts R. Brooks and Schwe Fang Pong, "Effects of Fasting, Body Weight, Methylcellulose, and Carboxymethylcellulose on Hepatic Glutathione Levels in Mice and Hamsters," (1980).

[409] R. Michael Anson, Zhihong Guo, Rafael de Cabo, Titilola Iyun, Michelle Rios, Adrienne Hagepanos, Donald K. Ingram, Mark A. Lane, and Mark P. Mattson, "Intermittent Fasting Dissociates Beneficial Effects of Dietary Restriction on Glucose Metabolism and Neuronal Resistance to Injury from Calorie Intake," *Proc Natl Acad Sci*, 100, no. 10 (May 13, 2003): 6216–6220. Published online April 30, 2003. doi: 10.1073/pnas.1035720100 PMCID: PMC156352 Neuroscience

[410] "There exists a second psychic system of a collective, universal, and impersonal nature which is identical in all individuals. This collective unconscious does not develop individually but is inherited. It consists of pre-existent forms, the archetypes, which can only become conscious secondarily and which give definite form to certain psychic contents." Carl G. Jung, *The Archetypes and the Collective Unconscious*, (London: 1996) 43.

[411] Asaf Gilboa, Claude Alain, Yu He, Donald T. Stuss, and Morris Moscovitch, "Ventromedial Prefrontal Cortex Lesions Produce Early Functional Alterations during Remote Memory Retrieval," *Journal of Neuroscience* 29, no. 15 (April 15, 2009): 4871–4881. doi: 10.1523/JNEUROSCI.5210-08.2009

[412] Vincent Paquette, Johanne Levesque, Boualem Mensour, Jean-Maxime Leroux, Gilles Beaudoin, Pierre Bourgouin, and Mario Beauregard. "Change the Mind and You Change the Brain: Effects of Cognitivebehavioral Therapy on the Neural Correlates of Spider Phobia," *NeuroImage* 18 (2003): 401–409. http://neurodezign.com/Documents/Scientifiques/Paquette2003.pdf

[413] Jonathan Ma, "Lowering Prescription Drug Prices in the United States: Are Reimportation and Internet Pharmacies the Answer?" *Southern California Interdisciplinary Law Journal*, 15:345. http://www-bcf.usc.edu/~idjlaw/PDF/15-2/15-2%20Ma.pdf.

[414] "California Targets 39 Firms in Drug Fraud Lawsuit," *Las Vegas Sun*, August 26, 2005. "Among those named were drug giants Amgen Inc., Bristol-Myers Squibb Co., GlaxoSmithKline P.L.C., Novartis AG, Sandoz Inc., Mylan Laboratories Inc. and Schering-Plough Corp."

[415] Ibid.

[416] "Changing Patterns of Pharmaceutical Innovation," http://www.nihcm.org/innovations.pdf.

[417] "Profiting from Pain: Where Prescription Drug Dollars Go," July 2002, http://www.familiesusa.org/assets/pdfs/PPreport89a5.pdf. Quoted from Jonathan Ma. "Lowering Prescription Drug Prices in the United States: are Reimportation and Internet Pharmacies the Answer?"

[418] Marcia Angell, MD, *The Truth About the Drug Companies: How They Deceive Us and What to Do About It*, New York: Random House, 2004). And quoted at: http://www.nybooks.com/articles/archives/2004/jul/15/the-truth-about-the-drug companies/?pagination-false.

[419] Jones, *Textbook of Functional Medicine*.

[420] One night, the Philippine Coast Guard arrested nine Chinese fishermen for illegal poaching in Balabac Island. The coast guard vessel had approached the fishing vessel at night, and the personnel couldn't see what the fishermen were doing. As the coast guard vessel chased the fishing boat, crew members of the fishing boat threw some items overboard. When they board the fishing boat, the Coast Guard found containers of formaldehyde and stuffing materials used for embalming sea turtles, and noticed fresh blood and sea turtle scales on the vessel's deck. On another occasion, the Protected Areas and Wildlife Bureau believes that a ship had more than two hundred adult turtles and over ten thousand eggs.

[421] http://neuropolitics.org/defaultfeb08.asp.

[422] Lisa Kuly, "Locating Transcendence in Japanese Minzoku Geinō: Yamabushi and Miko Kagura," *Ethnologies* 25.1:191–208 (2003): 199.

[423] Lesley K. Fellows, "The Cognitive Neuroscience of Human Decision Making: A Review and Conceptual Framework," *Behav Cogn Neurosci Rev*, 3, no. 3 (September 2004). http://bcn.sagepub.com/content/3/3/159.abstract doi: 10.1177/1534582304273251

[424] Michel Guermonprez, *Matière médicale homéopathique*, editions Boiron, France, 1989.

[425] Here is an example. I hired someone who presented himself as a polite, competent, and eager-to-serve-the-beauty-of-my-garden man. After he had delivered 50% of the work, while I had paid 90% of the contract we had agreed on, he left the premises, saying that in his view he had worked for the amount I had paid. His evaluation of the work involved had been wrong. For him, time prevailed. For me, the work that was supposed to be done counted. His men had done work I did not ask for and could have done by myself. After many months of discussion, we eventually came to an understanding. Normally, the law would have taken my side as the written contract was clear. But in our society,

Notes, References, and Selected Bibliography

it is next to impossible to defend ourselves when we are wronged, even with a contract. Evaluators are nearly impossible to find. Lawyers are so expensive, some astute wrongdoers calculate that their punitive fees will prevent you from suing—and they are right.

Chapter 7

[426] Those who created Protected Designation of Origin (PDO) for such foods as Parmesan cheese and Parma ham unconsciously knew this. Cows had to be given the highest quality grass, had to be spoken to softly, treated with great respect. The cows we use presently in the industry, so enormous that their legs can hardly support them, would not have been accepted for this cheese. As for their famous ham, the prosciutto di Parma, has a slightly nutty flavor from the Parmigianino-Reggiano whey that is sometimes added to the pigs' diet. The quality of the surrounding air also is crucial to its final quality.

[427] http://www.youtube.com/watch?v=AQmQMCi5mNc

[428] Quote taken from J. Miles, *Wilderness as Healing Place: The Theory of Experiential Education*, (Dubuque, Iowa: Kendall Hunt Publishing, 1995).

[429] *Voyage de Teigué, fils de Cian*, Littératures Celtiques, Dottin

[430] *Nibelungenlied* is the work of an anonymous poet from the area of the Danube between Passau and Vienna, dating from about 1180 to 1210, possibly at the court of Wolfger von Erla, the bishop of Passau from 1191-1204.Wikipedia.

[431] Claude Mettra, Michel Albin, *La Chanson des Nibelungen*.

[432] Gerald Eddie Gerbrandt, *Kingship According to the Deuteronomistic History* (Atlanta: Scholars Press, 1986).

[433] Norman K. Gottwald, *The Hebrew Bible: A Brief Socio-Literary Introduction*, (Fortress Press) 223.

[434] Elsewhere in the Bible: "Do not bring a load out of houses or do any work on the Sabbath, but keep the Sabbath day holy, as I commanded your forefathers. Yet they did not listen or pay attention; they were stiff-necked and would not listen or respond to discipline" Jer. 17:22-23 (NIV).
"Go up to the land flowing with milk and honey. But I will not go with you, because you are a stiff-necked people and I might destroy you on the way" Exod. 33:3 (NIV).
"For the LORD has said to Moses, 'Tell the Israelites, "You are a stiff-necked people. If I were to go with you even for a moment, I might destroy you. Now take off your ornaments and I will decide what to do with you"' Exod. 33:5 (NIV).
"'O Lord, if I have found favor in your eyes,' he said, 'then let the Lord go with us. Although this is a stiff-necked people, forgive our wickedness and our sin, and take us as your inheritance'" Exod. 34:9 (NIV).
"Understand, then, that it is not because of your righteousness that the LORD your God is giving you this good land to possess, for you are a stiff-necked people" Deut. 9:6 (NIV).
"And the LORD said to me, 'I have seen this people, and they are a stiff-necked people indeed!'" Deut. 9:13 (NIV).

[435] Or the daily bread Jesus referred to.

[436] Eduardo E. Benarroch: "Basic neurosciences with clinical applications" and "Heteromodal association areas in the frontal, temporal, and parietal lobes integrate sensory data, motor feedback, and other information with instinctual and acquired memories. This integration facilitates learning and creates thought, expression, and behavior."

[437] The Internet-sold device counts breaths by sensing chest or abdominal movement, and sounds gradually slowing chimes that signal when to inhale and exhale. Users follow the tone until their breathing slows from the usual 16 to 19 breaths a minute to 10 or fewer.

[438] funded by the maker of a breathing device (InterCure Inc.),

[439] "Practice Slow Breathing to Lower Blood Pressure," NewsMax.com, August 1, 2006.

[440] A. Stancák Jr., M. Kuna, "Changes During Forced Alternate Nostril Breathing." *J Psychophysiol*. 1994 Oct;18, no. 1 (October 1994):75-9.

[441] D. A. Werntz, R. G. Bickford, D. Shannahoff-Khalsa, "Selective Hemispheric Stimulation by Unilateral Forced Nostril Breathing," *Hum Neurobiol* 6 (1987):165-171.

[442] are recurrent periods or cycles repeated throughout a 24-hour circadian day.(Wikipedia)

[443] David S. B. Shannahoff-Khalsa, "Selective Unilateral Autonomic Activation: Implications for Psychiatry," *CNS Spectr* 12, no. 8 (2007): 625-634.

[444] Victoria Gill, "Singing 'Rewires' Damaged Brain, BBC News, February 21, 2010.

[445] Of note: Daniel J. Levitin, *This is your brain on music*, (New York: Plume Publishing). *New York Times* bestseller.
[446] C. Grape, M. Sandgren, L. O. Hansson, M. Ericson, and T. Theorell, "Does Singing Promote Well-Being?: An Empirical Study of Professional and Amateur Singers During a Singing Lesson," *Integr Physiol Behav Sci*, 38, no. 1 (January-March 2003): 65-74.
[447] Gunter Kreutz, Stephen Bongard, Sonja Rohrmann, Volker Hodapp, Dorothee Grebe, "Effects of Choir Singing or Listening on Secretory Immunoglobulin A, Cortisol, and Emotional State," Department of Music Education, Johann Wolfgang Goethe-University, Frankfurt, Germany.
[448] "How Singing Improves Your Health (Even if Other People Shouldn't Hear You Singing)," www.SixWise.com.
[449] Gene D. Cohen, MD, PhD, "The Impact of Professionally Conducted Cultural Programs on Older Adults," The Creativity and Aging Study, Final Report (April 2006).
[450] Gould, Tanapat, McEwen, Flugge, and Fuchs, "Proliferation of granule cell precursors in the dentate gyrus of adult monkeys."
[451] Chase, "Studies in Pre-Platonic Demiurgy: The case of Pherecydes of Syros."
[452] Beth Riungu, "An Evolving Exploration of Eco-psychology," August 19, 2011. http://www.meadowbrookschool.com
[453] S. Harper, "The Way of Wilderness," in *Ecopsychology: Restoring the Earth, Healing the Mind*, eds. T. Roszak, M. E. Gomes, and A. D. Kanner (San Francisco: Sierra Club Books, 1995).
[454] Joseph Rhawn, "The Limbic System, Hypothalamus, Septal Nuclei, Amygdala, Hippocampus Emotion and the Unconscious Mind," in *Neuropsychiatry, Neuropsychology, Clinical Neuroscience*, (New York: Academic Press, 2000).
[455] A. D. Craig, "How Do You Feel—Now? The Anterior Insula and Human Awareness," *Nature Neuroscience Reviews* 10 (2009), www.nature.com/reviews/neuro.
[456] Ibid.
[457] Britta K. Hölzel, Ulrich Ott, Tim Gard, Hannes Hempel, Martin Weygandt, Katrin Morgen, and Dieter Vait, "Investigation of Mindfulness Meditation Practitioners with Voxel-Based Morphometry," *Oxford Journals Medicine Social Cognitive & Affective Neurosci*, 3, no. 1 (November 13, 2007): 55-61.
[458] Sara W. Lazar, Catherine E. Kerr, Rachel H. Wasserman, Jeremy R. Gray, Douglas N. Greve, Michael T. Treadway, Metta McGarvey, Brian T. Quinn, Jerry A. Dusek, Herbert Benson, Scott L. Rauch, Christopher I. Moore, and Bruce J. Fisch, "Meditation Experience Is Associated with Increased Cortical Thickness."
[459] David L. Clark, Nash N. Boutros, and Mario F. Mendez, "The Brain and Behavior: An Introduction to Behavioral Neuroanatomy," *Cambridge Medicine*, 68.
[460] T. Binstock, "Anterior Insular Cortex: Linking Intestinal Pathology and Brain Function in Autism-Spectrum Subgroups," *Medical Hypotheses* 57, no. 6 (December 2001): 714-717.
[461] Jung, *Correspondance 1950-1954*, p. 30.
[462] L. S. Allen and R. A. Gorski, "Sexual Dimorphism of the Anterior Commissure and Massa Intermedia of the Human Brain," *Journal of Comparative Neurology* 312 (1991): 97-104.
[463] F. C. Crick and C. Koch, "What Is the Function of the Claustrum?" *Philos Trans R Soc Lond B Biol Sci*, 360, no. 1458 (2005):1271-9.
[464] Gottfried Schlaug, Lutz Jancke, Yanxiong Huang, Jochen F. Staiger, and Helmuth Steinmetz, "Increased Corpus Callosum Size in Musicians," (1995).
[465] Eran Zaidel and Marco Iacoboni, eds., *The Parallel Brain, The Cognitive Neuroscience of the Corpus Callosum*, (Bradford Book).
[466] Marcel Kinsbourne, "The Corpus Callosum Equilibrates the Cerebral Hemispheres" in *The Parallel Brain* (Bradford Book),271-281.
[467] Result of a study by Ruben C. Gur, MD, professor of psychology in psychiatry, and Raquel E. Gur, MD, professor of psychiatry and neurology. Appeared in *Journal of Neuroscience*, May 15 2009.
[468] Doidge, *The Brain That Changes Itself*.
[469] J. H. Duan, H. O. Wang, J. Xu, X. Lin, S. Q. Chen, Z. Kang, Z. B. Yao, "White Matter Damage of Patients with Alzheimer's Disease Correlated with Decrease in Cognitive Function."
[470] Miles S. Guthman, Nancy C. Harvey, and Arthur Kling, "Social Development in the Rhesus Monkey Following Olfactory Bulbectomy," *Primates* 20, no. 2 (April 12, 2006): 211-219, doi:10.1007/BF02373374
[471] E. Gould, "How Widespread Is Adult Neurogenesis in Mammals?" *Neuroscience* 8, (2007): 481-8.
[472] James B. Aimone et al., "Adult neurogenesis," *Scholarpedia*, 2 no. 2 (2007): 2100.

Notes, References, and Selected Bibliography

[473] Alvaro Fernandez and Elkhonon Goldberg, "The Sharp Brain's Guide to Brain Fitness," (May 2009).

[474] Alvaro Fernandez, "Art Kramer on Why We Need Walking Book Clubs," June 25, 2008, http://www.sharpbrains.com/blog/2008/06/25/art-kramer-on-why-we-need-walking-book-clubs/.

[475] Wen Zhou and Denise Chen, "Encoding Human Sexual Chemosensory Cues in the Orbitofrontal and Fusiform Cortices," Journal of Neuroscience 28, no. 53 (2008): 14416-14421, doi: 10.1523/JNEUROSCI.3148-08.2008.

[476] Kyle T. Beggs, Kelly A. Glendining, Nicola M. Marechal, Vanina Vergoz, Ikumi Nakamura, Keith N. Slessor, Alison R. Mercer, "Queen Pheromone Modulates Brain Dopamine Function in Worker Honey Bees," ed. Gene E. Robinson. December 18, 2006.

[477] Tang, C. L. Eaves, J. C. Ng, D. M. Carpenter, X. Mai, D. H. Schroeder, C. A. Condon, R. J. Haier, "Brain Networks for Working Memory and Factors of Intelligence Assessed in Males and Females with fMRI and DTI Intelligence," 38 (2010): 293-303.

[478] Lynn Grodzki, "Approaching a Theory of Emotion: An Interview With Candace Pert, PhD."

[479] Itai Carmeli, Viera Skakalova, Ron Naaman, Zeev Vager, "Magnetization of Chiral Monolayers of Polypeptide: A Possible Source of Magnetism in Some Biological Membranes," Angewandte Chemie 114, no. 5 (March 1, 2002): 787-790.

[480] Grodzki, "Approaching A Theory of Emotion: An Interview With Candace Pert, PhD."

[481] J. Harasty, K. L. Double, G. M. Halliday, J. J. Kril, "Language-Associated Cortical Regions Are Proportionally Larger in the Female Brain."

[482] N. I. Eisenberger, M. D. Lieberman, and K. D. Williams, "Does Rejection Hurt? An fMRI Study of Social Exclusion," Science 302 (2003): 290-292.

[483] Deanne K. Thompson, Simon K. Warfield, John Carlin, Masa Pavlovic, Hong X. Wang, Merilyn Bear, Michael J. Kean, Lex W. Doyle, Gary F. Egan, and Terrie E. Inder, "Perinatal Risk Factors Altering Regional Brain Structure in the Preterm Infant."

[484] F. Baller, Handbook of Neuropsychology, vol. 5.

[485] James B Adams, Tapan Audhy, Sharon McDonough-Means, Robert A Rubin, David Quig, Elizabeth Geis, Eva Gehn, Melissa Loresto, Jessica Mitchell, Sharon Atwood, Suzanne Barnhouse, and Wondra Lee, "Nutritional and Metabolic Status of Children with Autism Vs. Neurotypical Children, and the Association with Autism Severity," Nutrition & Metabolism, 8, no. 34 (2011). doi:10.1186/1743-7075 http://www.nutritionandmetabolism.com/content/8/1/34 8 June 2011

"The autism group had many statistically significant differences in their nutritional and metabolic status, including biomarkers indicative of vitamin insufficiency, increased oxidative stress, reduced capacity for energy transport, sulfation and detoxification. Several of the biomarker groups were significantly associated with variations in the severity of autism. These nutritional and metabolic differences are generally in agreement with other published results and are likely amenable to nutritional supplementation. Research investigating treatment and its relationship to the co-morbidities and etiology of autism is warranted."

[486] Mas Ichise and Bob Innis at National Institutes of Mental Health Neuroimaging on the effects of maternal deprivation on the brain.

[487] Ana Emiliano and Julie Fudge, "From Galactorrhea to Osteopenia: Rethinking Serotonin-Prolactin Interactions," Neuropsychopharmacology 29, (2004): 833-846, doi:10.1038/sj.npp.1300412.

[488] R. Sandy, N. Tsagas, and P. A. Anninos, "Melatonin as a Proconvulsive Hormone in Humans," Int J Neurosci. 63, no. 1-2 (March 1992):125-35.

[489] March 6, 2000, issue of the journal Brain Research.

[490] T.R. Insel, G. Battaglia, J.N. Johannessen, S. Marra, E.B. DeSouza, "Methylenedioxymethamphetamine ('Ecstasy') Selectively Destroys Brain Serotonin Terminals in Rhesus Monkeys, J Pharmacol Exp Ther. 249, no. 3 (June 1989):713-20.

[491] G. J. Maestroni, "The Immunoneuroendocrine Role of Melatonin," Journal of Pineal Research, 14 (1993): 1-10, doi: 10.1111/j.1600-079X.1993.tb00478.x.

[492] B. Geiger, A. Bershadsky, R. Pankov, K. M. Yamada, "Transmembrane Crosstalk between the Extracellular Matrix—Cytoskeleton Crosstalk," Nat Rev Mol Cell Biol, 2, no. 11 (November 2001): 793-805.

[493] Yuki Okatani, Akihiko Wakatsuki, and Russell J. Reiter, Melatonin: Biological Basis of its Function in Health and Disease, "Melatonin and Mitochondrial Respiration."

[494] Andrzej Lewinski, "Melatonin and the Thyroid Gland."

Isis Code

[495] Ewa Sewerynek, "Cardiovascular Effects of Melatonin."
[496] Benjamin Geiger, Alexander Bershadsky, Roumen Pankov, and Kenneth M. Yamada, "Transmembrane Extracellular Matrix–Cytoskeleton Crosstalk," *Nature Reviews, Molecular Cell Biology*, vol. 2 (Macmillan Magazines Ltd, 2001).
[497] G. Benitez-King, G. Ramirez-Rodriguez, D. Garcia, and F. Antón-Tay, "Melatonin Synchronizes Cell Physiology through Cytoskeletal Rearrangements."
[498] Fabrizio Benedetti, Helen S. Mayberg, Tor D. Wager, Christian S. Stohler, and Jon-Kar Zubieta, "Neurobiological Mechanisms of the Placebo Effect," *Journal of Neuroscience* 25, no. 45 (November 9, 2005):10390-10402, doi:10.1523/JNEUROSCI.3458-05.2005 SYMPOSIA AND MINI-SYMPOSIA.
[499] Sylvie Riou-Millet, *Médecine, l'énigme du placebo, quand l'esprit guérit le corps*, Sciences et Avenir, (November 2005).
[500] J. D. Levine, N. C. Gordon, R. Smith, H. L. Fields, "Analgesic Responses to Morphine and Placebo in Individuals with Postoperative Pain," *Pain* 10, no. 3 (1981): 379-89, doi:10.1016/0304-3959(81)90099-3. PMID 7279424.
[501] I. Kirsch, G. Sapirstein, "Listening to Prozac but Hearing Placebo: A Meta-Analysis of Antidepressant Medication." Abstract. *Prevention & Treatment* 1, (1998).
[502] "Against Depression, a Sugar Pill Is Hard to Beat," *Washington Post*, May 7, 2002.
[503] Goldberg, *The New Executive Brain*, 49.
[504] Mae-Wan Ho, "Natural Being and a Coherent Society," in *Evolution, Order and Complexity*, eds. E. L. Khalil and K. E. Boulding (London: Routledge, 1996). Schlang, Lutz Jancke, Yanxiong Huang, Jochen F. Staiger, Helmuth Steinmetz. Department of Neurology, Heinrich-Heine University, Düsseldorf. 1995
[505] W. C. Allee, *The Social Life of Animals* (London: The Book Club, 1951).
[506] See the news at http://arabnews.com/variety/islam/article77010.ece?service=print
[507] Senn TE Espy KA Kaufmann, "Using Path Analysis to Understand Executive Function Organization in Preschool Children," *Developmental Neuropsychology* 26, (2004): 445-464.
[508] E. A. Wallis Budge in the early part of the twentieth century.
[509] Father Joseph MacVeigh, "Renewing the Irish Church," (1993).
[510] *Encyclopaedia Iranica*, Jacques Duchesne-Guillemin, (London: Routledge & Kegan Paul, 1982), s.v. "Ahriman."
[511] Bruce MacLennan, "Summary of the Plotinian Ascent," (2001). http://www.cs.utk.edu/~mclennan/Classes/US310/Plotinus-Ascent.html.
[512] Ozanam, "Dante ou la philosophie cathare. au XIIIe siècle." p. 424.

Chapter 8

[513] "Thou shalt beat him with the rod, and shalt deliver his soul from hell" Prov. 23:14 (KJV).
[514] http://answering-islam.org/Silas/wife-beating.htm#_Toc160373830
[515] The Koran, translated with notes by N.J. Dawood. Penguin Books.
[516] Evola, *Revolt Against the Modern World*, xxxiv.
[517] Gary Greenberg, "The Generations of the Heavens and of the Earth: Egyptian Deities in the Garden of Eden," presented at the annual meeting of the American Research Center in Egypt, St. Louis, 1996.
[518] Another gospel text is also clear on this. Acts 8:17: "Then Peter and John placed their hands on them, and they received the Holy Spirit."
[519] tying twisted white and red woolen threads, are a result of centuries-old tradition and suggest Thracian (paleo-Balkan) Hellenic or even Roman origin.
[520] To me, Ambrosius Aurelianus, Riothamus and King Arthur refer to the same individual. The personality who lived was probably Ambrosius Aurelianus, the legends tell of his individuality – had it been manifested- therefore of a legendary hero, but this could be the subject of a totally different book.
[521] If Vortigern is accepted to have lived in the fifth century, then these people are the British whom the Saxons failed to subdue and who became the Welsh.
[522] This could be seen in the fact that when the white dragon inflicts wounds to the red one, it is said that the screams of pain of the red dragon cause women to miscarry, animals to perish and plants to become barren.

Notes, References, and Selected Bibliography

[523] Between 1485 and 1603, the dragon formed part of the arms of the Tudor dynasty, but it was replaced on the royal coat of arms with a unicorn by order of James I. The motto of the flag is "Y Ddraig Goch Ddyry Cychwyn," meaning "The red dragon gives the lead". http://dragonsinn.net/today_nat.htm

[524] "The Apocalypse Master" (fl. in Paris, ca. 1480–1510) provided the designs for the famous Apocalypse Rose of the Sainte-Chapelle in Paris, ordered by the French king Charles VIII. This artist is alternatively named the Master of the Très Petites Heures d'Anne de Bretagne after one of his manuscripts (Paris, Bibliothèque nationale de France, nouv. acq. lat. 3120) or the Master of the Hunt of the Unicorn, after designs for tapestries (New York, Cloisters). The Apocalypse Master played an important role in the production of printed books of hours, which flourished in Paris from the 1480s. He provided many series of designs to illustrate numerous editions, designs that he also used in his small œuvre of manuscripts. Typically, his figures are stocky and spaces are carefully constructed. He is the direct successor of the Coëtivy Master, who might be identified with Colin d'Amiens. The oldest son of this renowned master, Jean d'Ypres (d. 1508), he was registered in Paris as a so-called maitre-juré and could be the Apocalypse Master. The British Library Catalogue of Illuminated Manuscripts.

[525] Barbara Hanawalt, *The Middle Ages: An Illustrated History*, (New York: Oxford University Press, 1998), 36. (See: Muhammad leads Abraham, Moses and Jesus in prayer, from medieval Persian manuscript.) http://en.wikipedia.org/wiki/File:Medieval_Persian_manuscript_Muhammad_leads_Abraham_Moses_Jesus.jpg

[526] Debora Hammond, *The Science of Synthesis*, (Colorado: University Press of Colorado, 2003),126–127.

[527] *Correspondance 1950–1954*, p.145.

[528] http://www.brainyquote.com/quotes/authors/d/david_bohm

[529] http://www.ncbi.nlm.nih.gov/books/NBK26911/

[530] "Organic produce and meat typically aren't any better for you than conventional food when it comes to vitamin and nutrient content, although they do generally reduce exposure to pesticides and antibiotic-resistant bacteria, according to a US study. There is slightly more phosphorous in the organic products. Organic milk and chicken may also contain more omega-3 fatty acids, but that was based on only a few studies." "Right now I think it's all based on anecdotal evidence," said Chensheng Lu, who studies environmental health and exposure at the Harvard School of Public Health. The study was led by Crystal Smith-Spangler and a team of researchers from Stanford University and the Veterans Affairs Palo Alto Health Care. *Toronto Sun*. "Organic food no more nutritious than non-organic," Reuters, September 4, 2012.

[531] http://www.fluoridation.com/.

[532] http://208.109.172.241/pesticides/sodium.f.pineal.htm.

[533] http://www.aglabs.com/newletters/energy_concepts.html.

[534] http://www.food.gov.uk/multimedia/pdfs/gmnewcastlereport.PDF.

[535] http://www.progress.org/gene84.htm.

[536] http://www.voanews.com/content/foreign-agro-firms-scoop-up-ethiopian-farmland-84973402/159791.html.

[537] Al Huebner, "How Agri-Food Corporations Make the World Hungry," February 9, 2010, http://towardfreedom.com/home/content/view/1851/1/.

[538] Indonesia, as well as Brazil, is seeking foreign and local investors to lease massive parcels of fertile countryside, thus making their country a major food producer. Between now and 2030 Indonesia expects to become a main producer of rice, maize, sugar, coffee, shrimp, meats, and palm oil, senior agriculture ministry official Hilman Manan said. "In order to avoid any forms of monopolies or land grabbing, we're limiting each company to a maximum of 10,000 hectares of land," Manan said, stressing that the government was selling land-use rights, not the land itself. Could individual companies regroup as a hidden consortium and then "lease" many of these parcels? Such worries are well known in other countries, such as Brazil and Madagascar, where there is deep suspicion about food and biofuel companies monopolizing agricultural land. This is not for the good of the host country's land or for the peasants who used to live on it, but to become major market manipulators in items that are essential for the survival of billions of people. These are not charitable endeavors; they are exclusively for-profit ones. Considering the general imbalance of the world's LIFE biosystem, presuming otherwise would be foolish.

[539] Jerome Rivet, "Indonesia aims to be world's breadbasket," AFP, February 20, 2010.

[540] Jonathan Watts, "China's soil deterioration may become growing food crisis, adviser claims. China faces struggle to feed population as pollution and urbanisation threaten supply, says government

Isis Code

expert." guardian.co.uk, February 23, 2010. 13.35 GMT http://www.guardian.co.uk/environment/2010/feb/23/china-soil-deterioration-food-supply.

[541] Shaheen Nazar, "Kingdom must have agriculture base, says Al-Rasheed," Arab News, February 18, 2010.

[542] "Farming furor: World Bank launches new agriculture fund," Bretton Woods Project, 15 February 2010. http://www.brettonwoodsproject.org/art-565915.

[543] Emphasis is given to property rights and the need to "strengthen land rental and sales markets." This sounds like the World Bank's involvement in foreign land distribution. The environmental focus is more on allowing the Bank to "tap into carbon markets" than integrating environmental considerations across their agriculture portfolio

[544] "Farming Furor: World Bank Launches New Agriculture Fund," Bretton Woods Project.

[545] It holds around fifty stocks selected from across the globe, and the fund managers did not restrict themselves in terms of subsectors, currencies, countries, or market cap. The companies they target are those involved with agriculture-related *chemical products*, equipment, and infrastructure, as well as soft commodities and food, biofuels, forestry, agricultural sciences, and arable land. Here is a partial division of their goods:

Top Fund Holdings for WLDA2SG (BlackRock Global Funds— World Agriculture Fund)
Filing Date: 12/30/2011

Name	Position	Value	% of Total
Monsanto Co	605,700	42,441,399	9.388%
Deere & Co	503,100	38,914,785	8.608%
Potash Corp of Saskat.	851,000	35,129,279	7.770%
Wilmar International Ltd	6,956,000	26,822,967	5.933%
Syngenta AG	89,000	26,173,671	5.789%
BRF—Brazil Foods SA	1,071,800	20,953,689	4.635%
Mosaic Co.	414,009	20,878,474	4.618%
CF Industries Holdings Inc.	130,280	18,887,994	4.178%
Archer-Daniels-Midland Co.	640,000	18,304,000	4.049%
Yara International ASA	430,000	17,292,370	3.825%

[546] http://www.wpwealthadvisors.com/.

[547] www.globalissues.org/article/431/bush-the-media-cover-up-the-jihad-schoolbook-scandal.

[548] Ties with the ruling family of Great Britain can be severed if the chosen governor general can be recognized as the bearer of spirituality. By this I mean that people from different faiths can feel this person is above all faiths, but certainly not faithless! This is possible for someone who puts unity and harmony above all in his or her life. It doesn't mean that this person has to accept everything, but it means that this person first sees unifying factors.

[549] Allan DeBavelaere and Nicole DeBavelaere, *The Biocybernetical Revolution Applying an Elementary BiocyberneticSystem to Medicine and Society: Complexity, Democracy, Sustainability*.

[550] Sigmund Freud, "Lines of Advance in Psycho-Analytic Therapy," from *Narcissism: A New Theory* by Neville Symington (London, 2003), 110.

[551] or about 9.5 percent of the U.S. population

[552] http://www.sehn.org/wing.html.

[553] Fluoride action network, pesticide project http://208.109.172.241/pesticides/sodium.f.pineal.htm

[554] Laetitia Mailhes, "Dragging Monsanto to Justice over GM Alfalfa," Care2.com, January 30, 2011.

[555] http://www.care2.com/greenliving/dragging-monsanto-to-justice.html.

[556] http://blog.thegreenplate.org/2011/01/dragging-monsanto-to-justice/.

[557] Margaret Munro, "Genetically Modified Canola Goes Wild; 'Feral' plants found growing along roads from North Dakota to Manitoba," *Vancouver Sun*, October 11, 2011.

[558] Ibid.

[559] Ibid.

[560] http://growbetterfood.com/Advancing_Eco_Agriculture_Learn/Energy%20And%20How%20It%20Affects%20Crop%20Growth.pdf

[561] "Animal Protein Allergies Explained," *ScienceDaily* (October 16, 2007). Reference: "Evolutionary Distance from Human Homologues Reflects Allergenicity of Animal Food Proteins," by John A. Jenkins, Heimo Breiteneder, E. N. Clare Mills. Published online by the *Journal of Allergy and Clinical Immunology*, October 16, 2007. http://www.sciencedaily.com/releases/2007/10/071015081742.htm

Index

A

addiction 65, 84, 138, 147, 153, 191, 207, 208, 210, 211, 262, 293, 382, 622, 630
ADHD 65, 140, 157, 194, 214, 249, 258, 259, 362, 624, 629
adrenal glands 98, 156, 213, 246, 253, 324
aging 86, 149, 216, 217, 293, 300, 320, 357, 420, 423, 466, 473, 479, 490, 618, 622, 631, 634
Akasha 45, 83, 390
Albert Einstein 20, 22, 35, 36, 48, 54, 63, 170, 227, 281, 294, 296, 307, 316, 357, 369, 510, 533, 620
Allan N. Schore 139, 151, 619
alpha wave 202, 203, 204
Alzheimer's disease 84, 86, 149, 199, 216, 217, 347, 420, 474, 477, 608, 616, 634
amygdala 128, 131, 136, 143, 144, 157-159, 162, 163, 174, 215, 218, 225, 229, 230, 245, 250, 255, 350, 351, 361, 362, 368, 422, 471, 472, 483, 620, 634
Analytical brain 9, 80, 102, 128, 137, 152, 157, 173, 179, 195, 196, 199, 216, 226, 230, 234, 243, 248, 253, 257, 260, 266, 267, 271, 272, 284, 287, 316, 333, 336, 338, 339, 354, 355, 359, 366, 370, 381, 385, 392, 399, 401, 403, 404, 409-413, 422, 427, 429, 431, 434-437, 439, 444, 453, 458, 474, 480, 484, 492, 502, 504, 505, 510, 512, 514, 522, 537, 539, 547, 561, 562, 568, 622
anorexia 14, 18, 76, 77, 143, 149, 413, 415-418, 631
anterior insula 254, 471-473, 492, 634
Anthony Walsh 161, 619, 631
Antonio R. Damasio 102, 230, 231, 362, 369, 370, 630
archetypal human 281
archetype 105, 107, 109, 177, 179, 193, 205, 226, 266, 315, 334, 354, 355, 377, 383, 384, 447, 451, 517, 525, 536, 557, 613, 614, 632
asthma 65, 179, 191, 461, 610, 620
atrial natriuretic peptide 98, 329, 478
Autism 155, 253, 254, 259, 347, 348, 474, 477, 616, 626, 634, 635
awareness xiii, xix, 4, 8, 9, 24, 27, 28, 30, 33, 40, 42, 66, 77, 79, 100, 102, 103, 132, 134, 135, 137, 175, 200, 203, 205, 206, 209, 226, 227, 230, 254, 265, 272, 282, 287, 291, 333, 339, 343, 356, 359, 364, 371, 373, 379, 385, 386, 402, 413, 416, 421, 430, 471-473, 481, 483, 492, 512, 528, 540, 543, 563-565, 634

B

Bacchus 526, 527
backward-forward syndrome 238, 293, 411
basal ganglia 90, 136, 137, 158, 225, 232, 248-250, 257, 475, 623
beauty, objective 80, 282, 349, 350, 374
beauty, subjective 350, 351
Big Bang 19, 20, 21, 26, 37, 48, 295, 624
biopsychosocial 67
biosynarchy xiii, 99, 152, 317, 449, 520, 532, 556, 557, 560, 617
biosystem xii, xix, 22, 31, 49, 50, 52, 53, 56, 59, 60, 67, 77, 79, 80, 82, 89, 90, 92, 95, 96, 99, 128, 132, 136, 137, 141, 146, 151, 164, 165, 169, 173, 175, 178-180, 194-196, 198, 203, 219, 222, 224, 227, 251, 254-256, 262, 285, 289, 293, 306, 314, 326, 331, 345, 346, 348-350, 356, 361, 362, 369, 373, 376, 382, 392, 394, 397, 399, 406, 416-418, 435-437, 450, 455, 456, 468, 471, 472, 484, 496, 500, 502, 504, 508, 510, 513, 522, 523, 528, 529, 532, 534, 536, 538, 539, 541, 542, 547, 553, 556, 557, 561, 565, 610, 617, 637
bonobos 12, 186, 615
brain xii, xvi, xix, xx, 3, 4, 6, 7, 9, 12, 19, 23, 25, 27, 28, 31, 37, 40-42, 44-46, 50, 52-54, 65-67, 69, 72, 77-80, 82-99, 102, 123, 124, 127, 128, 131-143, 145, 147-159, 161-165, 167-175, 179, 180, 184-186, 187, 191-197, 199-202, 204, 207-209, 214-220, 221, 225-237, 240, 241, 243, 244, 246-252, 253-258, 260-262, 264-273, 281-285, 287-289, 291, 292, 299, 301, 302, 304, 305, 309, 310, 315-317, 319-325, 327-334, 336, 338, 339, 341, 345-351, 354-359, 361-364, 366, 369, 370, 372, 373, 381, 385, 392, 394, 395, 399-412, 416-418, 420-422, 427-429, 431, 432, 434-437, 439-441, 444, 447, 448, 451, 453-455, 457, 458, 462-465, 470, 473-484,

639

Isis Code

486-489, 492-494, 497, 500, 502, 503, 504, 505, 509-513, 519, 522, 529, 532, 537, 539, 540, 542, 543, 547, 553-555, 557, 559-561, 568, 614-636
brain stem 86, 94, 131, 133, 134, 138, 192, 194, 199, 204, 324, 329, 346, 347, 470
breathing 14, 54-56, 70, 84, 95, 124, 128, 134, 146, 157, 198, 202, 208, 223, 246, 310-312, 319, 325, 346, 457, 459, 460, 460-463, 467, 473, 479, 497, 498, 501, 633
Buddhism 535, 537, 538

C

cancer 37, 64, 84, 118, 138, 149, 189, 191, 194, 197-199, 212, 247, 262, 264, 298, 375, 384, 412, 420, 425, 428, 487, 541, 548, 601-603, 607, 610, 614, 615, 624, 626
Candace Pert 482, 635
Carl G. Jung xvii, 8, 9, 17, 18, 23, 33, 36, 47, 49, 78, 79, 135, 170, 184, 204, 286, 290, 345, 355, 377, 390, 408, 423, 441, 470, 474, 510, 513, 538, 539, 613-615, 618, 620, 625, 629, 631, 632, 634
Cathar xviii, 289, 336, 506-509, 509-511, 513, 517, 526
celiac plexus 296, 315, 322, 327, 328, 473
choir singing 464, 467, 634
Christ 170, 334, 379, 408, 506, 534, 538
Christian 23, 42, 49, 106, 311, 313, 377, 398, 424, 492, 505, 505-507, 509, 520, 536, 537, 624, 631, 636
Chtonie 45, 467
cingulate cortex 225, 254-256, 347, 354, 417, 434, 470, 473, 484
circumventricular organs 301
coherence 19-21, 53, 71, 231, 238, 291, 494, 613, 614, 617, 623
collective unconscious 27, 33, 47, 104, 132, 169, 175, 195, 204, 219, 226, 377, 382, 385, 421, 424, 448, 468, 470, 632
conscious 13, 14, 16, 25, 36, 45, 67, 74, 76, 79, 80, 85, 137, 150, 181, 205, 220, 232, 246, 282, 311, 313, 317, 338, 341, 358, 359, 367, 382, 398, 408, 409, 412, 422, 430, 460, 472, 473, 511, 512, 519, 523, 531, 558, 565, 620, 621, 627, 632
consciousness 8, 9, 21, 27, 32, 33, 40, 42, 45, 57, 67, 77, 79, 80, 99, 100, 102, 103, 135, 137, 168, 172, 174, 185, 195, 200, 208, 209, 220, 230, 233, 265, 272, 284, 308, 309, 313, 339, 360, 370, 377, 389, 402, 413, 421, 448, 469, 471, 517, 569, 629, 630
Consciousness 8, 9, 40, 41, 42, 52, 79, 81, 92, 99-104, 111, 171, 200, 285, 287, 290, 313, 315, 317, 318, 332, 333, 337, 339, 343, 358, 373, 377, 398, 406, 409, 445, 451, 460, 470, 497, 521, 528, 614, 626, 629

corpus callosum 67, 143, 229, 230, 250, 254, 321, 366, 407, 448, 458, 470, 476-478, 634
cortisol 56, 141, 149, 156, 157, 253, 361, 487, 488, 618, 619, 634
courtly love xviii, 226, 283, 343, 513, 514, 516, 517
cytokine 95, 196, 197, 198, 489, 621
cytoplasm 57, 327, 402

D

Dante Alighieri 509
Dantian 99, 310
dark energy 21, 38, 284, 285, 613
David Bohm 22, 30, 37, 41, 49, 52, 53, 83, 106, 283, 519, 539, 613, 614, 624, 637
Debora Hammond 534, 637
democracy xv, 4, 533, 555, 614, 638
depression 67-70, 84, 94, 142, 144, 147-150, 154, 192, 195, 196, 199, 202, 203, 211, 215, 218, 221, 249, 250, 261, 262, 311, 326, 330, 347, 416, 428, 439, 466, 479, 487, 567, 569, 615, 616, 621, 636
Deuteronomy 181, 437, 448, 454-457
DNA 45, 50, 57-60, 77, 132, 195, 291, 300, 302, 312, 450, 496, 536, 537, 545, 608, 625
domney 514
Don Quixote xix, 335, 336, 337, 340
dopamine 98, 138, 159, 164, 202, 207, 209, 210, 211, 214, 248, 250, 253, 255, 258, 260, 330, 362, 385, 404, 432, 433, 481, 487, 616, 618, 622, 624, 635
dorsolateral prefrontal cortex 137, 248, 260, 261, 354, 359, 392, 422, 492
Dr. A. DeBavelaere 18, 49, 52, 61-64, 66, 71, 72, 169, 205, 306, 318, 375, 393-395, 552, 614, 620, 638
dream 5, 6, 7, 43, 62, 74, 76, 139, 167, 182, 199, 202-205, 222, 281, 282, 360, 367, 368, 388, 424, 444, 453, 458, 468, 474, 498, 534, 547, 552, 570, 613, 627
Dr. Guermonprez 439, 632
Dr. Hakan Olausson 332, 628
Dr. Hamer 194, 197
Dr. Hans Breiter 349, 350, 630
Dr. Hans Eysenck 247
Dr. J.C. Darras 61, 63, 394
Dr. Joseph Diener 197
Dr. Persinger 346, 396, 483, 628
Dulcinea 334, 336, 337, 340

E

ecopsychology 467, 467-469, 634
Eleanor of Aquitaine 516
electromagnetic spectrum 26, 36, 54, 293, 621
Elkhonon Goldberg 86, 89, 194, 250, 341, 405, 458, 614-616, 618, 621, 623, 635, 636
emotional human 225, 409

Index

emotional self 92, 163, 204, 209, 221, 224, 225, 227, 231, 250, 257, 287, 288, 338, 350, 362, 409, 417, 450, 482, 508, 535
endocrine glands 246, 482
endorphins 461, 491
energy 3, 12, 16, 21, 22, 24, 26, 29, 30, 33, 36-40, 42, 46-49, 51-55, 57-62, 64, 66-69, 71, 87, 89, 90, 101, 102, 107, 110, 112, 113, 118, 125, 128, 132, 134, 136, 141, 143, 148, 151, 153, 154, 163, 166, 170, 174, 177, 178, 180, 182, 193, 196, 199, 207, 208, 213, 216, 220, 221, 223, 225, 226, 231, 240, 243, 246, 256-258, 272, 283-288, 291, 293-297, 299, 300, 306-313, 321, 322, 324, 327, 328, 331, 332, 350, 362, 374-376, 379, 383, 385, 392, 393, 396, 402, 407, 416, 418-420, 422, 434, 438, 441, 444, 448, 454, 460, 467, 470, 473, 479, 488, 492, 509, 510, 513, 536, 548, 557, 563, 607, 613, 614, 617, 625, 627, 631, 635, 637, 638
entorhinal cortex 131, 215, 252, 622
enviromics 189
environment 3, 6, 14, 25, 26, 38, 41, 42, 49, 51, 53, 55, 56, 58-60, 67, 68, 70, 85, 103-105, 124, 128, 132, 135, 137, 144, 150, 155, 157, 167, 178, 181, 185, 189, 193, 197, 200, 205, 209, 211, 215, 218, 219, 221, 226, 236, 239, 240, 250, 263, 267, 272, 300, 301, 302, 304, 308, 316, 319, 332, 336, 338, 340-342, 344, 347, 353, 360, 366, 370, 372, 380, 382, 387, 395, 396, 410, 412, 418, 419, 425, 433, 441, 442, 444, 455, 459, 467, 468, 478, 483, 525, 529, 532, 534, 536, 542, 548, 551-553, 559-563, 564, 568, 569, 601, 603, 604, 607, 610, 620, 621, 630, 638
Eric Nestler 207, 622
esoteric 23, 53, 282, 285, 313, 389, 391, 398, 456, 495, 499, 523, 527, 630
Eve 101, 102, 179, 180, 182, 183, 308, 311, 314, 318, 411
Exodus 181, 226, 266-270, 290, 437-439, 456, 457
expressiveness 257
extracellular matrix 95, 125, 304, 309, 310, 319, 320, 322, 329, 331, 332, 490, 491, 635, 636

F

facial recognition 84, 228, 480, 483, 485
faith xviii, 74, 76, 169, 178, 255, 289, 294, 336, 393, 423, 428, 434, 491, 501, 505, 507-511, 536, 537, 545
fasting 208, 418-421, 501, 631, 632
feminine polarity xix, 9, 12, 18, 32, 51, 55, 60-64, 68, 70, 78, 82-84, 91, 93, 94, 96, 97, 100, 101, 105, 106, 111, 125, 138, 139, 141, 146, 148-151, 154, 155, 157-159, 161, 163, 164, 166, 169, 172-174, 176-180, 184, 186, 191, 192, 193, 195, 198-201, 204, 207, 216-222, 224, 228-235, 237-239, 243, 245-249, 254-257, 258-262, 266, 269, 271, 272, 282, 284-290, 292-295, 299, 300, 304, 306, 308, 312-314, 316-319, 326, 328, 330, 333, 334, 338, 340, 341, 343, 345-351, 356, 357, 360, 363, 365, 373-377, 385, 389, 392-395, 404-407, 409, 411, 413, 416, 419, 425, 427, 428, 430-433, 435, 441, 444, 448, 453, 454, 462, 464, 465, 468, 477, 478-480, 483, 485, 487, 491, 494-497, 500-503, 505, 509, 510, 512, 515, 517, 521-526, 528-534, 537, 540-542, 548, 554-563, 567, 569, 570, 607, 615, 619, 630
fight or flight 98, 156, 462, 619
fin'amor xviii, 242, 244, 343, 397, 449, 514, 516-518, 530
fractal 22, 23, 25, 46, 67, 100, 104, 128, 141, 170, 179, 312, 378, 535, 613
Franklin D. Roosevelt xv, 544, 613
Freud 204, 219, 271, 470, 565, 638
fusiform gyrus 448, 483, 485
fusional 206, 449

G

Gene Ontology 216
Genesis 22, 101, 132, 172, 179, 180, 182, 266, 268, 290, 313, 317, 376, 520, 521, 610
genital spectrum 164, 166, 172, 173, 176, 244
genotype 69, 83
Geschwind Syndrome 345
Gilgamesh 169, 386, 456
God 22, 33, 38, 43, 49, 50, 53, 74, 91, 92, 101, 108-112, 115, 117, 119-121, 123, 132, 169-174, 177-179, 182, 187, 204, 222, 226, 267-270, 283, 286, 288, 290, 291, 313, 314, 315-318, 335, 337-339, 345, 371, 373-375, 397, 398, 408, 412, 435-438, 440-442, 449, 454-457, 496, 497, 499, 501, 503, 504, 507, 510, 511, 512, 515, 517, 519-521, 525, 527, 532, 536, 568, 617, 618, 625, 627, 628, 631, 633
government xv, xvi, 2, 71, 187, 190, 197, 384-386, 541, 545, 548-551, 556, 557-559, 561, 563, 568, 603, 607, 610, 621, 637
great mother 556, 560, 563
György Buzsáki 41, 614, 618

H

heart aspect 58, 68, 99, 101, 102, 112, 140, 141, 146, 147, 148, 149, 152, 169, 173, 192, 211, 218, 221, 224, 243, 252, 254, 267, 281, 283, 287-291, 293, 302, 305, 306, 307, 316, 317, 325, 333, 334, 337, 338, 343, 345, 347, 348, 393, 399, 404, 406, 408, 421, 422, 435, 440-442, 449, 450, 455-457, 467, 500, 502, 524, 526, 529-531, 537, 554-556
heart phase 153, 201, 202, 204, 208, 212, 213, 215, 219, 229, 243, 247, 268, 272, 282, 287, 290, 299, 315,

Isis Code

331, 340, 346, 350, 383, 387, 402, 422, 433, 436, 455, 499, 523, 530, 630
heteromodal 89, 422, 458, 494, 633
hippocampus 131, 184, 192, 202, 216, 225, 230, 246, 251, 252, 256, 351, 479, 481-483, 489, 623, 634
holistic 3, 22, 40, 50, 63, 71, 72, 157, 177, 340, 343, 346, 359, 487, 497, 532, 565
Holoworld 57, 629
homeopathy 60, 62, 65, 70-72, 75, 302, 303, 427, 429, 439, 440, 615, 626
homeostasis 31, 62, 97, 99, 146, 165, 173, 177, 183, 198, 272, 430, 614, 626
homosexuality 127, 164, 167, 168, 175, 540, 623
Horus xvii, 5, 105, 109, 111-114, 118, 120-123, 123, 283, 286, 287, 315, 408, 410, 449, 455, 511, 519, 525, 535, 537, 566, 624, 625
human brain 19, 31, 52, 170, 217, 304, 319, 331, 334, 358, 480, 628, 629
Human brain 23, 25, 78, 128, 134, 148, 151, 152, 154, 157, 169, 196, 218, 226, 230, 231, 250, 255, 258, 261, 271, 272, 285, 334, 338, 354, 355, 356, 361, 362, 363, 364, 366, 369, 370, 395, 422, 428, 434, 439, 453, 454, 458, 489, 492, 512, 522, 539, 555, 568
human brain aspect 361, 364
human self 132, 534
hypothalamus 98, 131, 155, 159, 162, 199, 225, 253, 262, 301, 324, 325, 330, 348, 471, 481, 634

I

idealistic self 175, 208, 209, 210, 218, 226, 229, 281, 284, 285, 286-291, 293, 312, 331, 338, 342, 355, 359, 366, 369, 382, 397, 408, 409, 416-418, 422, 435, 443, 450, 454, 456, 472, 491, 493, 525, 530, 532, 535, 567-569
identity 2, 67, 104, 106, 153, 174, 209, 210, 221, 234, 247, 261, 266, 271, 283, 286, 287, 333, 336-338, 345, 354, 359, 360, 362, 365, 366, 367, 370, 371, 374, 398, 410, 417, 421, 449, 469, 493, 500, 506, 512, 540, 551, 568, 620
implicate order 19, 22, 37, 45, 46, 52-54, 57, 58, 80, 83, 99-101, 123, 135, 169, 177, 181, 193, 205, 266, 285, 294, 306-308, 314, 315, 343, 350, 358, 389, 390, 397, 429, 437, 438, 441, 450, 451, 456, 457, 494, 501, 502, 505, 515, 523, 534-536, 539, 558, 569, 604, 613, 614, 617
individuality xvii, 2, 3, 40, 79, 170, 171, 182, 195, 200, 204, 221, 222, 235, 237, 238, 251, 266, 269, 270, 283, 288, 290, 291, 314, 333, 336, 337, 338, 340, 343, 355, 356, 359, 375, 378, 388, 392, 394, 397-399, 404, 406, 410, 411, 421, 441, 444, 450, 451, 453, 454, 469, 494, 512, 517, 519, 521, 522, 525, 526, 528, 531, 534, 636
information 3, 5, 10, 21, 24, 25, 37, 40, 41, 42, 46, 51-53, 55, 57-59, 64, 69, 70, 78-80, 83, 89, 94-98, 101-103, 108, 133, 134, 136, 142, 155, 158, 165, 166, 168, 170, 171, 174, 177, 181, 186, 192, 195, 200, 203, 208, 218, 223, 225, 226, 229-231, 236, 240, 247-249, 252, 258, 264, 270, 281, 282, 284, 285, 289, 291-293, 297, 301, 302, 303, 304, 306-309, 311, 316, 320, 323, 325-328, 342, 344-346, 354-359, 364, 370, 377, 378, 383, 392, 404, 405, 407, 417, 422-424, 428, 439, 441-443, 449, 456, 459, 470-474, 476, 480-486, 490, 492, 530, 532, 543, 553, 561, 562, 603, 615, 617, 619, 630, 633
inhibition 56, 69, 89, 132, 149, 166, 213, 264, 354, 363, 382, 470, 473, 487, 503, 619
instinctual human 131
Isis xii, xiii, xvi, xvii, xx, 7, 27, 41, 103-106, 109, 111-115, 117, 118-122, 123, 180, 182, 286, 314, 343, 391, 407-409, 419, 442, 448, 451, 508, 539, 566, 569, 570, 624

J

James W. Prescott 148, 270, 363, 487, 618, 619, 624
Jean d'Ypres 527, 528, 637
Jeffrey A. Gray 156, 196
Jesus 170, 172, 173, 178, 181, 266, 273, 290, 313, 317, 336, 337, 343, 389, 393, 397, 408, 427, 441, 491, 496, 498, 508, 511, 520-523, 526, 536, 537, 558, 633, 637
Joan of Arc 368, 528, 630
John D. Rockefeller 555
John Robbins 159

K

Kabbalah xviii, 23, 25, 45, 83, 107-109, 111, 115, 219, 267, 456, 500
Karl Pribram 41, 83, 615
kidney aspect 68, 131, 133, 136, 137, 141, 147, 162, 165, 192, 218, 221, 230, 259, 286, 290, 291, 293, 307, 312, 313, 316, 325, 373, 374, 390, 421, 434, 443, 456, 524, 531, 537
kidney phase 153, 158, 182, 201, 208, 213, 218, 219, 240, 268, 290, 313, 438, 455, 499, 568
kingship 449, 455, 499, 530, 568, 633
Koran 169, 316, 373, 465, 497, 502, 515, 532, 536, 636
Kuramoto model 56
Kwashiorkor 142, 143, 618

L

Lady and the Unicorn xix, 392, 527, 528
lateral prefrontal 257, 351, 423, 458, 629
left hemisphere 28, 31, 91-94, 97, 129, 137, 246, 254, 255, 346, 351, 354, 357, 392, 403, 405, 412, 431-434, 440, 464, 476, 494, 531, 533, 618
Leonardo da Vinci 141, 166, 287, 402, 454, 519, 522, 523-525, 528
lesion 82, 86, 89, 91, 143, 199, 359, 363, 370, 422, 432, 471, 483, 485, 629, 632

642

Index

Leviticus 175, 176, 181, 266, 283, 287-291, 315, 437
life xii, xv, xvi, xvii, xviii, xix, xx, 1-7, 10-16, 19, 22, 24, 26, 28, 30-32, 34, 40, 42, 43, 45, 47, 49-55, 58-60, 62, 66-70, 72-74, 76, 77, 79, 80, 82, 83, 87-90, 92, 95-97, 99, 102, 105, 113, 115, 118, 120, 123, 126-128, 132, 133, 135-141, 143, 146, 148-154, 157, 159, 164, 165, 169, 170, 173-184, 192-196, 197, 199, 203, 205, 210, 211, 219, 220-224, 227, 232, 233, 236, 238, 239, 241, 242, 246, 251, 254, 255, 259, 262-264, 270-272, 281, 283-285, 289, 290, 292, 293, 295, 296, 299, 301, 304-306, 308, 311-316, 318, 321, 326, 328, 331, 334-336, 340-343, 345, 346, 348-350, 352, 353, 355, 356, 358, 360-362, 367-371, 373, 376, 378-380, 382-385, 387, 392, 395-397, 399, 403-406, 409-411, 413-418, 420, 425, 430, 435-439, 442, 444, 447, 449-451, 453-458, 466, 467, 469, 470, 472, 476, 478, 484, 487, 494, 496-498, 500-503, 504, 506, 508-512, 513, 517, 520-525, 527-529, 532, 534-536, 538-543, 547, 551, 553, 556-558, 561-565, 567, 569, 570, 606, 609, 610, 615, 617, 618, 625, 628, 630, 636-638
LIFE biosystem xii, 22, 31, 49, 52, 53, 58, 60, 66, 67, 77, 79, 80, 82, 89, 90, 92, 95, 96, 99, 128, 136, 137, 141, 146, 151, 164, 165, 168, 169, 173, 175, 178-180, 194-196, 197, 203, 219, 222, 224, 227, 251, 254, 255, 262, 285, 289, 293, 306, 314, 326, 331, 345, 346, 348-350, 356, 361, 362, 369, 373, 376, 382, 392, 397, 399, 406, 416-418, 435-437, 450, 455, 456, 467, 470, 472, 484, 496, 500, 502, 504, 508, 510, 513, 522, 523, 528, 529, 532, 536, 538, 539, 541, 542, 547, 553, 556, 557, 561, 565, 610, 617, 637
liver aspect 136, 141, 154, 157, 163, 223, 225, 230, 241, 243, 246, 250, 256, 259, 261, 263, 272, 316, 325, 356, 421, 434, 440, 531
liver phase 153, 162, 201-204, 212-214, 243, 267, 302, 314, 438, 455, 499
locus coeruleus 217, 218
love xvii, xviii, xix, 1-4, 10, 10-12, 38, 39, 99, 111, 112, 114, 121, 124, 131, 132, 136, 139, 140, 145, 148, 151-153, 161, 168, 171, 185, 186, 200, 205, 207, 220, 226, 233-238, 243, 258, 259, 271, 282, 283, 291, 293, 311, 328, 333-338, 340, 341, 343, 353, 364, 365, 367, 377, 380, 381, 386, 396-399, 403, 405, 408, 414, 424, 427, 437, 449, 452, 453, 465, 471, 495, 498, 508, 511, 512-516, 520, 524, 525, 529, 534, 551, 558, 566, 569, 570, 616, 618, 619, 624, 628, 630
lung aspect 68, 101, 141, 148, 152, 159, 173, 199, 204, 211, 216, 222, 231, 246, 250, 254, 256, 259, 261, 264, 293, 305, 311, 314, 317, 322, 325, 328, 332, 345, 346, 348, 364, 366, 370, 374, 378, 386, 393, 404, 416, 419, 421, 440-443, 448, 450, 455, 456, 460, 467-470, 472-474, 477, 483, 492, 494, 500, 502, 512, 514, 529, 531, 557, 568
lung phase 135, 153, 181, 201-203, 212, 213, 218, 243, 272, 286, 290, 298, 299, 302, 330, 348, 350, 432, 433, 455, 457, 502, 528

M

Ma'at 45, 48, 114, 123, 177, 181, 224, 266, 390, 501, 503, 505
mammalian brain 12, 25, 68, 92, 128, 134-136, 140, 143, 149, 151, 154, 163, 164, 186, 196, 202, 203, 204, 206, 215, 217, 218, 222, 225, 227-233, 235, 237, 240, 241, 248, 250-253, 255-258, 260, 261, 266-269, 271, 272, 282, 283, 288, 299, 301, 316, 338, 356, 362, 364, 366, 406, 416, 418, 434, 439, 440, 453, 457, 475, 477, 481, 483, 489, 492, 494, 504, 509, 539, 547, 554, 555, 568
marasmus 142
masculine polarity xix, 3, 9, 31, 32, 62, 69, 78, 91, 96, 100, 101, 106, 113, 132, 135, 140, 142, 145, 148, 151, 155, 159, 162, 164, 167, 173, 176, 179, 180, 184, 186, 192, 198, 200, 204, 206, 215-217, 218-221, 225, 226, 231-237, 244-248, 254, 255, 256, 258, 260, 261, 266, 285-290, 299, 302, 306, 312-315, 316, 325, 333-335, 338, 340, 342, 343, 345, 347, 348, 350, 354, 357, 359, 373-375, 378, 384, 387, 389, 392, 403, 404, 407-410, 412, 413, 418, 419, 423, 425, 427, 428, 430, 433-435, 437, 443, 444, 456, 458, 464-467, 472, 480, 491, 494-497, 499-502, 504, 505, 509, 512, 515, 521-526, 528, 531-533, 537, 540-544, 548, 552, 556, 557, 559, 561, 562, 568
massa intermedia 448, 475, 634
Master of the Heart 31, 37, 45, 51, 58, 60, 62, 70, 79, 83, 87, 93, 97, 98, 100, 101, 112, 125, 138, 141, 142, 155, 200, 252, 282, 292-294, 296, 298, 303, 304, 306-308, 310-315, 318, 319, 320, 322-328, 330, 332, 359, 364, 365, 386, 390, 417, 418, 441, 442, 456, 459, 467-469, 471-473, 481, 482, 490, 492, 497, 510, 524, 561, 620
medical biocybernetics 18, 19, 31, 49, 52, 67, 68, 71, 72, 75, 129, 196, 299, 303, 393, 425, 429, 430, 460, 625
melatonin 56, 250, 259, 297-301, 329-331, 485-487, 489-491, 626, 636
memory xv, 41, 66, 74-76, 78, 84, 122, 134, 135, 137, 138, 140, 149, 157, 163, 193, 200, 202, 208, 211, 213, 214, 216, 222, 223, 229, 246-248, 250, 251, 256, 265, 300, 302, 306, 320, 326, 348, 351, 355, 357-361, 367, 378, 379, 396, 422, 481-483, 485, 486, 492, 503, 521, 616, 624, 626, 629, 632, 633, 635
meridian 61, 65, 71, 87, 98, 200, 304, 307-309, 310-312, 318, 323, 328, 331, 374-376, 394, 530, 614
Michael M. Merzenich 155

Isis Code

Michelangelo 392
middle prefrontal 362, 413, 472, 474, 492
mirror neurons 254, 282, 334, 471, 474, 628
modular 3, 24, 89, 135, 232, 272, 342, 345, 403, 427, 494, 524, 542
monarchy 176, 529, 530, 557, 559
monkey 8, 18, 125, 144, 146, 147, 184, 186, 251, 272, 475, 478, 483, 487, 521, 619, 627, 631, 634, 635
Moses 23, 45, 170, 181, 265-269, 287, 289, 290, 373, 389, 392, 393, 435-439, 441, 454-456, 504, 535-537, 631, 633, 637
multiple sclerosis 194, 196, 298, 300, 477, 626
myth xvii, 349, 407, 409, 508, 519, 561, 618, 620

N

nature xiii, xv, xvii, xx, 1, 5-7, 13, 14, 17, 18, 27-29, 31, 33, 36-40, 45, 49-52, 50, 55, 56, 58, 63, 70, 72, 81, 96, 98, 100, 101, 105-107, 112, 124, 126, 134, 140, 141, 150, 152, 157, 166, 173, 176, 177, 180, 182, 184, 187, 196, 200, 212, 219, 222, 225, 230, 231, 233, 238-240, 266, 270, 272, 285, 288, 294, 295, 297, 303, 311, 314-317, 325, 333, 335, 339, 343, 350, 366, 380, 383, 389, 395, 409, 411, 419, 424, 425, 429, 430, 433, 438, 444, 448, 450, 454, 456-458, 460, 467-469, 480, 487, 494, 501, 519, 523, 525, 529, 531, 532, 540, 541, 546, 551, 561, 569-572, 607, 610, 614, 617, 618, 620, 622, 624, 626, 628, 630-632, 634, 636
neocortex 84, 90, 137, 195, 201, 216, 225, 232, 272, 283, 321, 475, 477, 615, 619, 627
Nephthys 105, 109, 113, 114, 118, 121, 132, 315, 409
neurotransmitter 85, 94, 98, 134, 140, 202, 208, 210-212, 248, 258, 291, 465, 486, 488, 616, 627
Nibelungen 633
Niels Bohr 29, 33, 35, 36
Numbers 181, 219, 266, 291, 403, 435, 436, 437, 439, 440, 629

O

occipital lobe 225, 355, 364
OCD 191, 192, 202, 249
olfactory bulb 98, 251, 478, 479
Omraam Mikhaël Aïvanhov 16, 564
oscillation 41, 53, 54, 56, 58, 69, 141, 199-201, 203, 231, 292, 302, 341, 395, 396, 614
Osiris xvi, xvii, 41, 104-106, 109-117, 118-124, 226, 267, 268, 286, 287, 314, 407, 409, 413, 451, 499, 508, 526, 527, 566, 624
oxytocin 99, 150, 260, 324, 359, 465, 627

P

pancreas aspect 69, 99, 102, 112, 137, 141, 153, 173, 198, 206, 224, 243, 267, 272, 287-290, 293, 315, 317, 385, 393, 402, 403, 407, 410, 421, 434, 440, 441, 473, 514, 531, 556, 568
pancreas phase 64, 100, 173, 201, 204, 212, 213, 243, 268, 299, 316, 318, 347, 351, 374, 406, 417, 419, 430, 435, 436, 443, 455, 563, 625
parasympathetic 53, 97, 162, 183, 245, 246, 292, 297, 325, 326, 329, 485
parietal lobe 82, 128, 132, 248, 250, 253, 255, 272, 286, 305, 325, 334, 347, 355-358, 628, 629, 633
Paris 11, 13-16, 18, 36, 60, 169, 198, 205, 527, 528, 552, 553, 554, 564, 630, 637
pentagram 108, 388, 389-393, 513, 631
pentemychos 307, 390
peptide 98, 207, 300, 301, 329, 478, 482, 627
perfect human 266, 282, 289, 520, 532
pericardium 93, 124, 307, 318-320, 322
personality 2, 3, 10, 12, 30, 40, 53, 77, 79, 90, 100, 141, 149, 152, 170, 195, 196, 200, 204, 222, 223, 235-238, 247, 265-267, 269, 270, 283, 288, 291, 316, 332, 335-337, 338-340, 343, 347, 350, 355, 359-361, 378, 379, 382, 383, 386, 388, 392, 394, 397-399, 404, 406, 410, 411, 419, 421, 441-443, 450, 451, 453, 454, 456, 469, 493, 501, 512, 517, 521, 522, 528, 531, 534, 562, 564, 636
personal self 316, 407, 409, 413, 493, 562
pH 125, 126, 132, 162, 164-167, 174, 175, 178, 185, 198, 227, 620, 624, 626
phase xix, 23, 28, 31, 32, 40, 53, 55, 56, 58, 64, 66, 78, 89, 90, 92, 95, 100, 109, 124, 127, 129, 131-135, 139, 141, 144, 149, 151-154, 158, 159, 162, 164, 173, 175, 179, 181, 192, 193, 195, 196, 197, 199, 201-204, 206, 208, 210, 212-214, 216, 218, 219, 221, 223, 225-227, 229, 232, 237, 240, 242-244, 247, 252-257, 259, 261, 264-271, 272, 281-283, 286, 287, 289-291, 293, 298, 299, 302, 307, 308, 313-316, 318, 330, 331, 333, 338-340, 342, 346-348, 350, 361-363, 365, 369, 371, 374, 382, 383, 387, 394, 397, 399, 402-404, 406, 407, 410, 411, 417, 419, 422, 430-433, 435-438, 441-443, 448, 449, 455-458, 469, 475, 499-501, 515, 520, 521, 523, 528, 530, 535, 537-540, 563, 567-569, 604, 610, 614, 617, 625, 626, 630
phenotype 59, 69, 83, 166
physical self 8, 103, 131, 162, 163, 172, 209, 220, 225, 290, 293, 319, 329, 409, 509, 529, 535
Pierre C. Renard 16, 60
pineal gland 60, 79, 98, 258, 298-301, 305, 306, 310, 321, 330, 432, 448, 478, 485-487, 489, 490, 626, 627
placebo 62, 65, 302, 303, 491-494, 636
planum parietale 356
Plato 26, 80, 281, 307, 308, 316, 350, 354, 390, 525, 558

Index

prefrontal cortex 84, 99, 128, 137, 157, 179, 202, 207, 211, 215, 218, 248, 252, 255, 258, 260, 261, 282, 292, 326, 331, 341, 346–348, 354, 357, 359, 361, 362, 369, 370, 392, 403, 405, 407, 413, 418, 422, 423, 432, 458, 473, 474, 489, 492, 542, 617, 622, 623, 629, 632
prenatal awareness xiii, 356, 386, 540, 562–564
psychocybernetics 67, 565
psychopath 157
pyramidal neurons 252, 331, 486

Q

quantas 21, 29, 45, 54, 103, 135, 141, 223, 224, 294, 302, 311, 379, 534
quantum neurophysics 57
quantum physics 29, 30, 36, 40, 57, 103

R

receptivity 183, 234, 492, 531
receptors 3, 99, 141, 163, 164, 208, 211–214, 251, 258–261, 264, 297, 302, 309, 311, 322, 324, 325, 328, 330, 332, 418, 481–483, 487, 602, 624, 627
reductionism 167, 424, 429
religion 17, 23, 26, 29–33, 43–45, 50–53, 170, 176, 184, 187, 191, 194, 209, 222, 282, 285, 287, 289, 297, 369, 375–379, 398, 412, 413, 418, 419, 423–425, 427–429, 431–434, 438, 440, 444, 454, 457, 465, 495–498, 499–502, 507, 508, 510, 517, 521, 522, 530, 531, 535–537, 541, 542, 555, 557, 559, 568, 625, 630
REM sleep 99, 199, 202–204, 486
Reptilian brain 79, 82, 88, 94, 127, 128, 131–137, 139, 148, 149, 151, 154, 156, 157, 159, 161–164, 164, 167, 171, 172, 174, 175, 180, 187, 191–195, 195, 196, 199, 215–217, 221, 230, 231, 235–237, 243, 244, 250, 252, 260, 265, 269, 271, 283, 284, 316, 333, 338, 345–347, 356, 361–364, 366, 369, 372, 373, 392, 399, 402, 404, 405, 407, 408, 411, 413, 439, 453, 454, 470, 477, 504, 529, 539, 553–555, 568
research xii, 3, 7, 10, 13, 15, 16, 18, 34, 37, 50, 53, 54, 59, 64, 68, 72, 82, 84, 85, 88–90, 92, 94, 98, 102, 125, 136, 140, 144, 149, 155, 161, 164, 167, 197, 203, 207, 216, 251, 253, 254, 257–260, 263, 271, 289, 293, 296, 297, 302–304, 305, 319, 322, 330, 332, 345, 347, 351, 354, 357, 361, 366, 369, 386, 391, 407, 416, 419, 422, 431, 434, 460, 480, 487–489, 492, 539, 540, 543, 545, 550, 564, 602, 603, 606, 615, 617, 619, 620, 622–626, 629, 630, 635, 636
resonance circuitry 448, 474
right hemisphere 28, 87, 91–94, 97, 98, 129, 139, 140, 154, 156, 217, 245, 254, 256, 260, 271, 282, 284, 292, 326, 341, 346, 347, 351, 356, 363, 366, 392, 403, 405, 431–434, 439, 448, 454, 464, 470, 473, 478, 485, 494, 497, 503, 531, 533, 541
rituals 10, 42, 77, 120, 174, 175, 182, 192, 193, 210, 242, 288, 297, 315, 318, 371–373, 376, 377, 396, 398, 411, 419, 436, 438, 439, 442, 449, 496, 501, 521, 530, 531, 537, 539, 557, 622, 630
RNA 50, 57–60, 282, 300, 302, 421, 608

S

sacredness 1, 15, 19, 28, 105, 178, 322, 335, 340, 442, 528, 570, 630
schizophrenia 34, 35, 138, 194, 250, 253, 261, 326, 347, 356, 365, 367–369, 433, 475, 478, 484, 614, 630
science 7, 13, 22, 26–37, 39, 40, 43–46, 48–52, 54, 57, 61–63, 67, 72, 78–80, 82–84, 89, 92, 96, 99, 145, 151, 161, 165, 170, 171, 183, 193, 200, 223, 232, 260, 270, 272, 281, 282, 284, 285, 294, 311, 334, 344, 349, 358, 365, 368, 370, 378, 379, 401, 403, 412, 413, 417, 423–425, 428–430, 433, 479, 496, 519, 530, 534–537, 545, 608, 613, 615, 617, 619, 620, 622–626, 629, 630, 635–638
self xiii, xviii, 2, 11, 13, 16, 22, 28, 42, 49, 52, 79, 92, 96, 99, 101, 103, 114, 120, 121, 129, 132, 146, 147, 151, 162, 163, 172, 175, 181, 204, 205, 206, 208, 209, 214, 218, 221, 225, 226, 227, 229, 231, 234, 236, 237, 244, 250, 254, 257, 260, 266, 268, 269, 272, 282, 284, 285, 286, 287, 288, 289, 290, 291, 293, 298, 312, 316, 319, 329, 331, 332, 333, 334, 337, 338, 339, 340, 341, 342, 343, 345, 347, 355, 357, 358, 359, 360, 362, 364, 366, 369, 370, 377, 382, 390, 395, 397, 401, 404, 406, 407, 408, 409, 411, 413, 416, 417, 418, 421, 422, 435, 436, 437, 438, 439, 441, 443, 444, 450, 452, 454, 456, 457, 458, 471, 472, 480, 482, 491, 492, 493, 502, 506, 509, 512, 515, 517, 529, 530, 532, 534, 535, 540, 552, 567, 569, 608, 609, 624, 628
Self xvii, 49, 79, 99, 105, 131, 195, 196, 204, 225, 266, 281, 282, 283, 311, 313, 333, 334, 335, 336, 337, 338, 339, 340, 341, 342, 345, 358, 359, 386, 396, 398, 399, 407, 408, 416, 447, 491, 517, 519, 538, 539, 557, 569, 622, 623, 628
septal nuclei 143, 225, 634
serosal membrane 320
serotonin 138, 142, 202, 211, 299, 300, 321, 329–331, 418, 432, 465, 485, 487, 488, 490, 616, 619, 622, 635
Seth xvi, xvii, xix, 105, 106, 109, 111–118, 120, 121–124, 287, 315–317, 402, 407–411, 510, 522, 566, 624
Shen 308, 311–313, 317, 318, 327
social human 447
soul xiii, xvii, xx, 2, 5, 6, 10, 12, 28, 77–79, 81, 99–101, 117, 118, 121, 123, 127, 141, 152, 177, 204, 205, 220,

645

Isis Code

222, 227, 235-237, 243, 245, 281, 285, 311, 314, 316, 336, 355, 359, 360, 377, 383, 390, 397, 398, 408, 454, 457, 459, 505, 510, 512, 515, 517, 519, 562, 613, 636

spindle neurons 254, 473

spirituality xviii, xix, 171, 193, 239, 285, 345, 346, 378, 428, 433, 434, 491, 495-497, 499-501, 508, 530, 531, 557, 560, 625, 638

Sri Ramana Maharshi 338, 339

stress 24, 95, 139, 141, 148-150, 157, 163, 195-197, 199, 203, 213, 214, 217, 231, 247, 251, 251-253, 262-264, 273, 295, 300, 324, 326, 331, 346, 379, 382, 416, 420, 421, 461, 466, 479, 484, 488, 614, 618, 619, 623, 624, 631, 635

symbol 12, 22, 33, 60, 78, 100, 107, 170, 180, 182, 205, 219, 240, 266, 269, 286, 288, 308, 314, 317, 358, 383-386, 387-390, 392, 407-409, 430, 438, 443, 444, 451, 496, 499, 511, 519, 524, 525, 527, 530, 533, 534, 555, 557

sympathetic 97, 98, 149, 155, 162, 183, 246, 292, 297, 299, 322, 329, 363, 461, 463, 467, 485, 626, 629

synarchic brain 363, 519

system science 7, 22, 33, 36, 54, 63, 72, 282, 358, 430

T

Taijitu 22, 107, 182, 185, 218, 219, 314, 315, 317, 407

Tao 23, 45, 53, 61, 101, 106-108, 219, 312, 313, 317, 337, 394

temporal lobe 84, 129, 158, 215, 250, 252, 255, 261, 320, 345, 347, 355, 368, 433, 448, 458, 470, 481, 483, 485, 628

thalamus 98, 136, 154, 201, 215, 225, 232, 272, 348, 363, 366, 417, 470, 472, 475, 619, 628

Theodore Roszak 467, 468, 634

theta wave 201

Thoth 110, 120

Tore A. Nielsen 204

Traditional Chinese Medicine (TCM) 18, 22, 49, 54, 58, 61-66, 72, 90, 94, 95, 96, 125, 128, 129, 136, 137, 267, 307, 310, 312, 393-395, 420, 428-430, 540, 615, 616

triple warmer 62, 304, 306-308, 310-312, 315, 318, 321, 322, 326, 387

tryptophan 142, 330, 486, 487, 616

U

ultradian 199, 202, 463

uniqueness 36, 48, 51, 70, 303, 425

universal brain 79, 80, 129, 148, 151, 154, 158, 195, 196, 199, 218, 226, 229, 231, 249, 251, 254-256, 257, 261, 262, 283, 285, 302, 317, 327, 342, 345, 361, 362, 370, 403, 439, 440, 447, 451, 454, 455, 458, 470, 475, 478, 480, 492, 532, 539, 568

universe xvii, 5, 14, 19-22, 26, 28, 30, 37-39, 40, 43-46, 51, 71, 80, 83, 101, 102, 104, 106, 106-108, 110, 137, 175, 179, 181, 183, 193, 220, 224, 266, 284-286, 291, 307, 311-313, 315, 317, 371, 424, 428, 447, 496, 510, 519, 528, 534, 631

V

vagus nerve 97, 134, 192, 296, 304, 318, 322, 323-325, 326-329, 417, 473, 627

ventricle 85, 298, 320, 327, 475, 478

ventromedial prefrontal cortex 157, 261, 292, 346, 361, 362, 369, 422, 492, 629, 632

vitalistic 28

W

Walter Cannon 156

water 2, 21, 66, 72, 79, 98, 101, 103, 109, 111, 117, 119, 122, 128, 132, 164, 184, 188, 221, 223, 268, 269, 297-299, 302, 313, 320, 321, 329, 338, 378, 397, 414, 430, 436, 439-441, 452-454, 459, 464, 490, 499, 504, 512, 536, 541, 544, 548-552, 560, 602, 607, 617, 621, 626, 631

William Heisenberg 21, 30, 33, 35, 57, 613, 629

Wolfgang Ernst Pauli 29, 33, 36, 45, 49, 613, 614

Y

Yodh 314